Heterogeneous Photocatalysis: A Solution for a Greener Earth

Heterogeneous Photocatalysis: A Solution for a Greener Earth

Editors

Stéphanie Lambert
Julien Mahy

MDPI • Basel • Beijing • Wuhan • Barcelona • Belgrade • Manchester • Tokyo • Cluj • Tianjin

Editors
Stéphanie Lambert
Department of Chemical
Engineering
University of Liège
Liège
Belgium

Julien Mahy
Department of Chemical
Engineering
University of Liège
Liège
Belgium

Editorial Office
MDPI
St. Alban-Anlage 66
4052 Basel, Switzerland

This is a reprint of articles from the Special Issue published online in the open access journal *Catalysts* (ISSN 2073-4344) (available at: www.mdpi.com/journal/catalysts/special_issues/photocatal_green).

For citation purposes, cite each article independently as indicated on the article page online and as indicated below:

LastName, A.A.; LastName, B.B.; LastName, C.C. Article Title. *Journal Name* **Year**, *Volume Number*, Page Range.

ISBN 978-3-0365-6296-4 (Hbk)
ISBN 978-3-0365-6295-7 (PDF)

© 2023 by the authors. Articles in this book are Open Access and distributed under the Creative Commons Attribution (CC BY) license, which allows users to download, copy and build upon published articles, as long as the author and publisher are properly credited, which ensures maximum dissemination and a wider impact of our publications.

The book as a whole is distributed by MDPI under the terms and conditions of the Creative Commons license CC BY-NC-ND.

Contents

About the Editors vii

Julien G. Mahy and Stéphanie D. Lambert
Heterogeneous Photocatalysis: A Solution for a Greener Earth
Reprinted from: *Catalysts* **2022**, *12*, 686, doi:10.3390/catal12070686 1

Lucas A. Almeida, Anja Dosen, Juliana Viol and Bojan A. Marinkovic
TiO_2-Acetylacetone as an Efficient Source of Superoxide Radicals under Reduced Power Visible Light: Photocatalytic Degradation of Chlorophenol and Tetracycline
Reprinted from: *Catalysts* **2022**, *12*, 116, doi:10.3390/catal12020116 7

Petronela Pascariu, Corneliu Cojocaru, Anton Airinei, Niculae Olaru, Irina Rosca and Emmanouel Koudoumas et al.
Innovative Ag–TiO_2 Nanofibers with Excellent Photocatalytic and Antibacterial Actions
Reprinted from: *Catalysts* **2021**, *11*, 1234, doi:10.3390/catal11101234 21

Antonietta Mancuso, Wanda Navarra, Olga Sacco, Stefania Pragliola, Vincenzo Vaiano and Vincenzo Venditto
Photocatalytic Degradation of Thiacloprid Using Tri-Doped TiO_2 Photocatalysts: A Preliminary Comparative Study
Reprinted from: *Catalysts* **2021**, *11*, 927, doi:10.3390/catal11080927 41

Hayette Benkhennouche-Bouchene, Julien G. Mahy, Cédric Wolfs, Bénédicte Vertruyen, Dirk Poelman and Pierre Eloy et al.
Green Synthesis of N/Zr Co-Doped TiO_2 for Photocatalytic Degradation of *p*-Nitrophenol in Wastewater
Reprinted from: *Catalysts* **2021**, *11*, 235, doi:10.3390/catal11020235 55

Seong-Rak Eun, Shielah Mavengere, Bumrae Cho and Jung-Sik Kim
Photocatalytic Reactivity of Carbon–Nitrogen– Sulfur-Doped TiO_2 Upconversion Phosphor Composites
Reprinted from: *Catalysts* **2020**, *10*, 1188, doi:10.3390/catal10101188 77

Julien G. Mahy, Valériane Sotrez, Ludivine Tasseroul, Sophie Hermans and Stéphanie D. Lambert
Activation Treatments and SiO_2/Pd Modification of Sol-gel TiO_2 Photocatalysts for Enhanced Photoactivity under UV Radiation
Reprinted from: *Catalysts* **2020**, *10*, 1184, doi:10.3390/catal10101184 93

Julien G. Mahy, Louise Lejeune, Tommy Haynes, Stéphanie D. Lambert, Raphael Henrique Marques Marcilli and Charles-André Fustin et al.
Eco-Friendly Colloidal Aqueous Sol-Gel Process for TiO_2 Synthesis: The Peptization Method to Obtain Crystalline and Photoactive Materials at Low Temperature
Reprinted from: *Catalysts* **2021**, *11*, 768, doi:10.3390/catal11070768 109

Cristina De Ceglie, Sudipto Pal, Sapia Murgolo, Antonio Licciulli and Giuseppe Mascolo
Investigation of Photocatalysis by Mesoporous Titanium Dioxide Supported on Glass Fibers as an Integrated Technology for Water Remediation
Reprinted from: *Catalysts* **2021**, *12*, 41, doi:10.3390/catal12010041 139

Asma Ghazzy, Lina Yousef and Afnan Al-Hunaiti
Visible Light Induced Nano-Photocatalysis Trimetallic $Cu_{0.5}Zn_{0.5}$-Fe: Synthesis, Characterization and Application as Alcohols Oxidation Catalyst
Reprinted from: *Catalysts* **2022**, *12*, 611, doi:10.3390/catal12060611 153

Julien G. Mahy, Marlène Huguette Tsaffo Mbognou, Clara Léonard, Nathalie Fagel, Emmanuel Djoufac Woumfo and Stéphanie D. Lambert
Natural Clay Modified with ZnO/TiO_2 to Enhance Pollutant Removal from Water
Reprinted from: *Catalysts* **2022**, *12*, 148, doi:10.3390/catal12020148 165

Duc Quang Dao, Thi Kim Anh Nguyen, Thanh-Truc Pham and Eun Woo Shin
Synergistic Effect on Photocatalytic Activity of Co-Doped $NiTiO_3/g$-C_3N_4 Composites under Visible Light Irradiation
Reprinted from: *Catalysts* **2020**, *10*, 1332, doi:10.3390/catal10111332 179

Otman Bazta, Ana Urbieta, Susana Trasobares, Javier Piqueras, Paloma Fernández and Mohammed Addou et al.
In-Depth Structural and Optical Analysis of Ce-modified ZnO Nanopowders with Enhanced Photocatalytic Activity Prepared by Microwave-Assisted Hydrothermal Method
Reprinted from: *Catalysts* **2020**, *10*, 551, doi:10.3390/catal10050551 191

Lara Faccani, Simona Ortelli, Magda Blosi and Anna Luisa Costa
Ceramized Fabrics and Their Integration in a Semi-Pilot Plant for the Photodegradation of Water Pollutants
Reprinted from: *Catalysts* **2021**, *11*, 1418, doi:10.3390/catal11111418 211

Luis A. González-Burciaga, Juan C. García-Prieto, Manuel García-Roig, Ismael Lares-Asef, Cynthia M. Núñez-Núñez and José B. Proal-Nájera
Cytostatic Drug 6-Mercaptopurine Degradation on Pilot Scale Reactors by Advanced Oxidation Processes: UV-C/H_2O_2 and UV-C/TiO_2/H_2O_2 Kinetics
Reprinted from: *Catalysts* **2021**, *11*, 567, doi:10.3390/catal11050567 223

Nuno P. F Gonçalves, Zsuzsanna Varga, Edith Nicol, Paola Calza and Stéphane Bouchonnet
Comparison of Advanced Oxidation Processes for the Degradation of Maprotiline in Water—Kinetics, Degradation Products and Potential Ecotoxicity
Reprinted from: *Catalysts* **2021**, *11*, 240, doi:10.3390/catal11020240 237

Javier Tejera, Daphne Hermosilla, Ruben Miranda, Antonio Gascó, Víctor Alonso and Carlos Negro et al.
Assessing an Integral Treatment for Landfill Leachate Reverse Osmosis Concentrate
Reprinted from: *Catalysts* **2020**, *10*, 1389, doi:10.3390/catal10121389 253

Vidhya Selvanathan, M. Shahinuzzaman, Shankary Selvanathan, Dilip Kumar Sarkar, Norah Algethami and Hend I. Alkhammash et al.
Phytochemical-Assisted Green Synthesis of Nickel Oxide Nanoparticles for Application as Electrocatalysts in Oxygen Evolution Reaction
Reprinted from: *Catalysts* **2021**, *11*, 1523, doi:10.3390/catal11121523 271

About the Editors

Stéphanie Lambert

Prof. Stéphanie D. Lambert, chemical engineer, is a Professor and FRS-FNRS Research Director in the Department of Chemical Engineering (DCE) at the University of Liege (Belgium). She achieved her Ph.D. in Applied Sciences at the University of Liège in 2003, with her study entitled "Development of Pd, Ag and Cu based mono- and bimetallic cogelled catalysts and their use in hydrodechlorination and oxidation reactions". After working in the position of Engineer in a Belgian chemical company (2004–2005), and two postdoctoral stays at the DCE of the University of Illinois at Chicago in 2006 and at the Institute Charles Gerhardt in Montpellier in 2007, she joined DCE as FRS-FNRS Associate Researcher, in which she develops heterogeneous catalysts for sustainable chemistry. Since October 2022, she has been Director of the Research Unit of Chemical Engineering. Prof. Stéphanie Lambert has published over 115 publications, 12 book chapters, holds 1 patent, and has an h-index of 27. She has also delivered 25 invited lectures and she is the Guest Editor of two Special Issues in the journal *Catalysts* and one Special Issue in the journal *Processes*.

Key Topics: materials engineering; heterogeneous (photo)catalysis; sol–gel process; environmental engineering; inorganic biomaterials

ORCID : 0000-0001-9564-1270

Institutional website: https://www.chemeng.uliege.be/cms/c_3482790/fr/chemeng-repertoire?uid=u182186

Research Gate: https://www.researchgate.net/profile/Stephanie-Lambert-8

URL Google Scholar: https://scholar.google.be/citations?user=MhgEhjMAAAAJ&hl=fr

Julien Mahy

Julien G. Mahy is a chemical engineer at the University of Liège (ULiège, Belgium). His Ph.D. thesis, conducted under the supervision of Prof. Stéphanie Lambert and Prof. Benoît Heinrichs, focused on the development of a TiO_2 aqueous sol–gel process in order to produce, at large scale, photocatalysts with hydrophilic property and high activity, both under visible and UV/visible light, for water and air remediation. In 2018, after a short period working in the industry in the CRM group as a project leader, he accepted a postdoctoral position in collaboration between the NCE (ULiège) and the "Institut für Energie- und Umwelttechnik e.V. (IUTA)", Duisburg (Germany) to work on the development of a quaternary treatment for water in order to degrade residual organic micropollutants using oxidation processes. From July 2019 to June 2021, he worked as a postdoctoral researcher at the Institute of Condensed Matter and Nanosciences (IMCN) at the Université catholique de Louvain under the supervision of Pr. Sophie Hermans. Their work focused on the development of low-cost sensors for the detection of air pollution by producing onion-like composite nanoparticles.

Since October 2021, Julien Mahy has been an FNRS postdoctoral researcher at the Université de Liège under the supervision of Prof. Stéphanie Lambert and Prof. Nathalie Job and at the INRS, Québec, under the supervision of Prof. Patrick Drogui. His work focuses on the development of inorganic materials and processes in environmental applications such as water and air decontamination process by photocatalysis, adsorption and electrochemistry.

Key Topics: TiO_2; photocatalysts; water treatment; sol–gel process; adsorption

ORCID : 0000-0003-2281-9626

Institutional website: https://www.chemeng.uliege.be/cms/c_3482790/fr/chemeng-repertoire?uid=u217056

Research Gate: https://www.researchgate.net/profile/Julien-Mahy?ev=hdr_xprf

URL Google Scholar: https://scholar.google.com/citations?hl=fr&user=ohH-zUgAAAAJ

Editorial

Heterogeneous Photocatalysis: A Solution for a Greener Earth

Julien G. Mahy [1,2,*] and Stéphanie D. Lambert [1]

[1] Department of Chemical Engineering—Nanomaterials, Catalysis, Electrochemistry, B6a, University of Liège, B-4000 Liège, Belgium; stephanie.lambert@uliege.be
[2] Institute of Condensed Matter and Nanosciences (IMCN), Université Catholique de Louvain, Place Louis Pasteur 1, B-1348 Louvain-la-Neuve, Belgium
* Correspondence: julien.mahy@uliege.be; Tel.: +32-4-366-4771

Since the beginning of the industrial era, various human activities have steadily increased, leading to rapid technological developments and high population growth. In consequence, the expanding industry has heavily polluted the atmosphere, soil, and water, with negative consequences for humans and the environment.

To decrease pollution emissions, various chemical, physical, and biological treatment methods have been developed. The major technics for treating wastewater are based on wastewater treatment plants using dry cleaning, decantation, and biological treatments. Occasionally, pollutant molecules are not eliminated by these processes; therefore, other technics can be used as secondary treatments to remove these small residual fractions of pollution. Among these methods, photocatalysis is a well-developed technic in the past few years. Through a photocatalyst and using light's energy, photocatalysis allows the production of highly reactive species that can react and decompose organic molecules, yielding, in the best case, the final decomposition products CO_2 and H_2O. The most commonly used photocatalysts are titanium dioxide (TiO_2), zinc oxide (ZnO), and tin oxide (SnO_2).

The papers contributing to this Special Issue address innovative photocatalytic processes for environmental applications. In what follows, we provide a synopsis of the obtained results in the 17 papers published in this Special Issue.

1. TiO_2-based photocatalysts

In this collection of articles, TiO_2 was used as the photocatalyst for depollution applications, and different doping and shaping were used to produce photoefficient materials.

In a contribution by Almeida et al. [1], the efficiency of photodegradation, the volatilization profile of bidentates, and the role of reactive oxidizing species (ROS) were explored for nanocrystalline TiO_2 modified with bidentate ligands (acetylacetone). In this study, TiO_2-ACAC CTC calcined at 300 °C (TiO_2-A300) was applied for the photocatalytic degradation of chlorophenol (4-CP) and tetracycline (TC) under low power visible light (26 W). Furthermore, the ROS scavengers isopropanol and benzoquinone were added for studying the photocatalytic role of $^\bullet OH$ and $^\bullet O_2^-$ radicals. The photocatalytic abatement of tetracycline (68.6%), performed via TiO_2-A300, was two times higher than that observed for chlorophenol (31.3%) after 6 h, indicating the distinct participation of ROS in the degradation of these pollutants. The addition of the ROS scavenger revealed $^\bullet O_2^-$ radicals as primarily responsible for the high efficiency of TiO_2-ACAC CTC under reduced visible light. On the other hand, the $^\bullet OH$ radicals were not efficiently generated in the CTC. Therefore, the development of heterostructures based on TiO_2-ACAC CTC can increase the generation of ROS through coupling with semiconductors capable of generating $^\bullet OH$ under visible light.

In the study of Pascariu et al. [2], Ag–TiO_2 nanostructures were prepared via electrospinning, followed by calcination at 400 °C. Morphological characterization revealed the presence of one-dimensional uniform Ag–TiO_2 nanostructured nanofibers, with a diameter from 65 to 100 nm, depending on the Ag loading, composed of small crystals interconnected with each other. Structural characterization indicated that Ag was successfully integrated as small nanocrystals without affecting much of the TiO$_2$ crystal lattice. Moreover, the presence of nano-Ag was found to contribute to reducing the band gap energy, which enables the activation by the absorption of visible light while, at the same time, delaying the electron–hole recombination. Tests of their photocatalytic activity in methylene blue, amaranth, Congo red, and orange II degradation revealed an increase by more than 20% in color removal efficiency at an almost double rate for the case of 0.1% Ag–TiO$_2$ nanofibers compared with pure TiO$_2$.

In another study [3], Mancuso et al. successfully prepared different tri-doped TiO$_2$ photocatalysts (Fe-N-P/TiO$_2$, Fe-N-S/TiO$_2$, Fe-Pr-N/TiO$_2$, Pr-N-S/TiO$_2$, and P-N-S/TiO$_2$) and tested them in the photocatalytic removal of thiacloprid (THI) under UV-A, visible, and direct solar light irradiation. The physical–chemical properties of the prepared catalysts revealed that dopants were effectively incorporated into the anatase TiO$_2$ lattice, resulting in a decrease in the energy band gap. The reduction in photoluminescence intensity indicated a lower combination rate and longer lifespan of photogenerated carriers of all doped samples in comparison with the undoped TiO$_2$. The doped photocatalysts promoted photodegradation under UV-A light irradiation and also extended the optical response of TiO$_2$ to the visible light region. Fe-N-P tri-doped TiO$_2$ sample exhibited the highest THI photodegradation degree (64% under UV-A light, 29% under visible light, and 73% under solar light).

In the contribution of Benkhennouche-Bouchene et al. [4], TiO$_2$ that was prepared using a green aqueous sol–gel peptization process was co-doped with nitrogen and zirconium to improve and extend its photoactivity to the visible region. For all doped and co-doped samples, TiO$_2$ nanoparticles with sizes ranging from 4 to 8 nm were formed of anatase–brookite phases. X-ray photoelectron (XPS) measurements showed that nitrogen was incorporated into the TiO$_2$ materials through Ti–O–N bonds, allowing light absorption in the visible region. The XPS spectra of the Zr-(co)doped powders showed the presence of TiO$_2$–ZrO$_2$ mixed oxide materials. Under visible light, the best co-doped sample yielded a degradation of p-nitrophenol (PNP) equal to 70%, instead of 25% with pure TiO$_2$ and 10% with P25 under the same conditions. Similarly, the photocatalytic activity improved under UV–Vis, reaching 95% with the best sample, compared with 50% with pure TiO$_2$.

In the study of Eun et al. [5], sol–gel-synthesized N-doped and carbon–nitrogen–sulfur (CNS)-doped TiO$_2$ solutions were deposited on upconversion phosphor using a dip-coating method. Scanning electron microscopy (SEM) imaging showed that there was a change in the morphology of TiO$_2$ coated on NaYF$_4$:Yb,Er from spherical to nanorods caused by additional urea and thiourea doping reagents. Fourier transform infrared (FTIR) spectroscopy further verified the existence of nitrate–hyponitrite, carboxylate, and SO$_4^{2-}$ because of the doping effect. NaYF$_4$:Yb,Er composites coated with N- and CNS-doped TiO$_2$ exhibited a slight shift in UV–Vis spectra toward the visible light region. The photocatalytic reactivity with CNS-doped TiO$_2$/NaYF$_4$:Yb,Er surpassed that of the undoped TiO$_2$/NaYF$_4$:Yb,Er for the MB solution and toluene. The photocatalytic activity was increased by CNS doping of TiO$_2$, which improved light sensitization as a result of band gap narrowing due to impurity sites.

In the presented study by Mahy et al. [6], the objective was to improve the efficiency of TiO$_2$ photocatalysts via activation treatments and through modification with palladium nanoparticles and doping with SiO$_2$. X-ray diffraction provided evidence that the crystallographic structure of TiO$_2$ was anatase and that Pd was present, either in its oxidized form after calcination or in its reduced form after reduction. The results on methylene blue degradation showed that the photocatalytic activity of the catalysts was inversely proportional to the content of silica present in the matrix. A small amount of silica improved the photocatalytic activity, compared with the pure TiO$_2$ sample. By contrast, a high amount of silica delayed the crystallization of TiO$_2$ in its anatase form. The introduction of Pd species increased the photocatalytic activity of the samples because it allowed for a decrease in the rate of electron–hole recombination in TiO$_2$. The reduction treatment improved the activity of photocatalysts, regardless of the palladium content, owing to the reduction of Ti^{4+} to Ti^{3+}, and the formation of defects in the crystallographic structure of anatase.

A review of the green synthesis of TiO_2 is also presented in this issue [7]. Indeed, in this study, the authors reported an eco-friendly process for producing TiO_2 via colloidal aqueous sol–gel synthesis, resulting in crystalline materials without a calcination step. Three types of colloidal aqueous TiO_2 were reviewed: the as-synthesized type obtained directly after synthesis, without any specific treatment; the calcined, obtained after a subsequent calcination step; and the hydrothermal, obtained after a specific autoclave treatment. This eco-friendly process is based on the hydrolysis of a Ti precursor in excess of water, followed by the peptization of the precipitated TiO_2. Depending on the synthesis parameters, the three crystalline phases of TiO_2 (anatase, brookite, and rutile) can be obtained. The morphology of the nanoparticles can also be tailored by the synthesis parameters. The most important parameter is the peptizing agent. Indeed, depending on its acidic or basic character and also on its amount, it can modulate the crystallinity and morphology of TiO_2. Colloidal aqueous TiO_2 photocatalysts are mainly being used in various photocatalytic reactions for organic pollutant degradation. The as-synthesized materials seem to have equivalent photocatalytic efficiency to the photocatalysts post-treated with thermal treatments and the commercial Evonik Aeroxide P25, which is produced via a high-temperature process.

TiO_2 in form of film is also presented in the contribution of De Ceglie et al. [8]. In this study, the photocatalytic efficiency of an innovative UV-light catalyst consisting of a mesoporous TiO_2 coating on glass fibers was investigated. Photocatalytic activity of the synthesized material was tested, for the first time, on a secondary wastewater effluent spiked with nine pharmaceuticals (PhACs), and the results were compared with the photolysis used as a benchmark treatment. Interestingly, the novel photocatalyst led to an increase in the degradation of carbamazepine and trimethoprim (about 2.2 times faster than the photolysis). Several transformation products (TPs) resulting from both the spiked PhACs and the compounds naturally occurring in the secondary wastewater effluent were identified through UPLC–QTOF/MS/MS. Some of them, produced mainly from carbamazepine and trimethoprim, were still present at the end of the photolytic treatment, while they were completely or partially removed by the photocatalytic treatment.

2. Composite materials as photocatalysts

In the papers under this theme, different composite materials were produced to be used as photocatalysts.

In one study [9], Ghazzy et al. reported a visible-light-induced, trimetallic catalyst ($Cu0.5Zn0.5Fe_2O_4$) prepared through green synthesis using Tilia plant extract. The spinel crystalline material was ~34 nm. In benign reaction conditions, the prepared photocatalyst oxidized various benzylic alcohols with excellent yield and selectivity toward aldehyde with 99% and 98%, respectively. Aromatic and aliphatic alcohols (such as furfuryl alcohol and 1-octanol) were photocatalytically oxidized using $Cu0.5Zn0.5Fe_2O_4$, LED light, and H_2O_2 as oxidant, 2 h reaction time, and ambient temperature. The advantages of the catalyst were found in terms of reduced catalyst loading, activating catalyst using visible light in mild conditions, high conversion of the starting material, and recyclability of up to 5 times without loss of selectivity.

In the contribution of Mahy et al. [10], raw clays, from Cameroon, were modified with semiconductors (TiO_2 and ZnO) to improve their depollution properties with the addition of photocatalytic properties. Cu^{2+} ions were also added to the clay via an ionic exchange to increase the specific surface area. The presence of TiO_2 and ZnO was confirmed by the detection of anatase and wurtzite, respectively. The composite clays showed increased specific surface areas. The adsorption property of the raw clays was evaluated on two pollutants—namely, fluorescein (FL) and p-nitrophenol (PNP). The experiments showed that the raw clays can adsorb FL but were not efficient for PNP. To demonstrate the photocatalytic property resulting from the added semiconductors, photocatalytic experiments were performed under UV-A light on PNP. These experiments showed degradation up to

90% after 8 h of exposure with the best ZnO-modified clay. The proposed treatment of raw clays seems promising to treat pollutants, especially in developing countries.

In another contribution [11], Dao et al. prepared co-doped $NiTiO_3/g\text{-}C_3N_4$ composite photocatalysts using a modified Pechini method to improve their photocatalytic activity toward methylene blue photodegradation under visible light irradiation. The combination of Co-doped $NiTiO_3$ and $g\text{-}C_3N_4$ and Co-doping into the $NiTiO_3$ lattice synergistically enhanced the photocatalytic performance of the composite photocatalysts. X-ray photoelectron spectroscopy results for the Co-doped $NiTiO_3/g\text{-}C_3N_4$ composite confirmed Ti–N linkages between the Co-doped $NiTiO_3$ and $g\text{-}C_3N_4$. In addition, characteristic X-ray diffraction peaks for the $NiTiO_3$ lattice structure clearly indicated the substitution of Co into the $N\text{-}TiO_3$ lattice structure. The composite structure and Co-doping of the C-x composite photocatalysts (x wt % Co-doped $NiTiO_3/g\text{-}C_3N_4$) decreased the emission intensity of the photoluminescence spectra but also the semicircle radius of the Nyquist plot in electrochemical impedance spectroscopy, giving the highest K_{app} value (7.15×10^{-3} min^{-1}) for the Co-1 composite photocatalyst.

In the study of Bazta et al. [12], pure and Ce-modified ZnO nanosheet-like polycrystalline samples were successfully synthesized via a microwave-based process. The XRD results showed that the obtained photocatalysts were composed of hexagonal, wurtzite-type crystallites in the 34–44 nm size range. The microscopy showed nanosheet-shaped crystallites, with a composition close to $Ce_{0.68}Zn_{0.32}O_x$. Importantly, the STEM–XEDS characterization of the photocatalyst samples revealed that Ce did not incorporate into the ZnO crystal lattice as a dopant but that a heterojunction formed between the ZnO nanosheets and the Ce–Zn mixed oxide phase nanoparticles. The optical analysis revealed that in the ZnO:Ce samples optical band gap was found to decrease to 3.17 eV in the samples with the highest Ce content. It was also found that the ZnO:Ce (2 at.%) sample exhibited the highest photocatalytic activity for the degradation of methylene blue (MB) when compared with both the pure ZnO and commercial TiO_2–P25 under simulated sunlight irradiation.

3. Pilot and comparative process studies, and other photocatalytic processes

In this last collection of papers, pilot reactors were tested for larger-scale photocatalytic processes, and other advanced oxidative processes were explored.

In one study [13], Faccani et al. offered easily scalable solutions for adapting TiO_2-based photocatalysts, which were deposited on different kinds of fabrics and implemented in a 6 L semipilot plant, using the photodegradation of rhodamine B (RhB) as a model of water pollution. They took advantage of a multivariable optimization approach to identify the best design options in terms of photodegradation efficiency and turnover frequency (TOF). Surprisingly, in the condition of use, the irradiation with a light-emitting diode (LED) visible lamp appeared as a valid alternative to the use of UV LED. The identification of the best design options in the semipilot plant allowed scaling up the technology in a 100 L pilot plant suitable for the treatment of industrial wastewater.

In the presented paper by Gonzalez-Burciaga et al. [14], several advanced oxidative processes were used to degrade 6-mercaptopurine (6-MP), a commonly used cytostatic agent. To degrade 6-MP, three processes were applied: photolysis (UV-C), photocatalysis (UV-C/TiO_2), and their combination with H_2O_2. Each process was performed with variable initial pH (3.5, 7.0, and 9.5). Pilot-scale reactors were used, using UV-C lamps as a radiation source. Kinetic calculations for the first 20 min of reaction proved the significance of the addition of H_2O_2: In UV-C experiments, the highest k was reached under pH 3.5, k = 0.0094 min^{-1}, while under UV-C/H_2O_2, k = 0.1071 min^{-1} was reached under the same initial pH; similar behavior was observed for photocatalysis since k values of 0.0335 and 0.1387 min^{-1} were calculated for UV-C/TiO_2 and UV-C/TiO_2/H_2O_2 processes, respectively, also under acidic conditions. Degradation percentages here reported for UV-C/H_2O_2 and UV-C/TiO_2/H_2O_2 processes were above 90% for all tested pH values.

In another contribution [15], Gonçalves et al. investigated the impact of different oxidation processes on the maprotiline degradation pathways via liquid chromatography–high-resolution mass spectrometry (LC/HRMS) experiments. Semiconductors photocatalysts—

namely, Fe–ZnO, Ce–ZnO, and TiO$_2$—proved to be more efficient than heterogeneous photo-Fenton processes in the presence of hydrogen peroxide and persulfate. No significant differences were observed in the degradation pathways in the presence of photocatalysis, while the SO$_4{}^-$-mediated process promoted the formation of different transformation products (TPs). Species resulting from ring openings were observed with higher persistence in the presence of SO$_4{}^-$. In silico tests on mutagenicity, developmental/reproductive toxicity, Fathead minnow LC50, D. Magna LC50, and fish acute LC50 were carried out to estimate the toxicity of the identified transformation products. Low toxicant properties were estimated for TPs resulting from hydroxylation onto the bridge rather than aromatic rings, as well as those resulting from ring openings.

In the contribution of Tejera et al. [16], an integral treatment process for landfill leachate reverse osmosis concentrate (LLROC) was designed and assessed, aiming to reduce organic matter content and conductivity, as well as to increase its biodegradability. The process consisted of three steps. First, a coagulation/flocculation treatment (removal of the 76% chemical oxygen demand (COD), 57% specific ultraviolet absorption (SUVA), and 92% color) was utilized. In the second step, a photo-Fenton process resulted in enhanced biodegradability (i.e., the ratio between the biochemical oxygen demand (BOD5) and the COD increased from 0.06 to 0.4), and an extra 43% of the COD was removed at the best-trialed reaction conditions. A UV-A light-emitting diode lamp was tested and compared with conventional high-pressure mercury vapor lamps, achieving a 16% power consumption reduction. Finally, an optimized 30 g L^{-1} lime treatment was implemented, which reduced conductivity by 43%, and the contents of sulfate, total nitrogen, chloride, and metals by 90%. Overall, the integral treatment of LLROC achieved the removal of 99.9% color, 90% COD, 90% sulfate, 90% nitrogen, 86% Al, 77% Zn, 84% Mn, 99% Mg, and 98% Si, in addition to significantly increasing biodegradability up to BOD5/COD = 0.4.

In the presented paper by Selvanathan et al. [17], the green synthesis of nickel oxide nanoparticles using phytochemicals from three different sources was employed to synthesize nickel oxide nanoparticles (NiO$_x$ NPs) as an efficient nanomaterial-based electrocatalyst for water splitting. Nickel (II) acetate tetrahydrate was reacted in presence of aloe vera leaves extract, papaya peel extract, and dragon fruit peel extract, respectively, and the physicochemical properties of the biosynthesized NPs were compared with sodium hydroxide (NaOH)-mediated NiO$_x$. Based on the average particle size calculation from Scherrer's equation, using X-ray diffractograms and field-emission scanning electron microscope analysis revealed that all three biosynthesized NiO$_x$ NPs had smaller particle sizes than that synthesized using the base. Aloe-vera-mediated NiO$_x$ NPs exhibited the best electrocatalytic performance, with an overpotential of 413 mV at 10 mA cm^{-2} and a Tafel slope of 95 mV dec^{-1}. Electrochemical surface area (ECSA) measurement and electrochemical impedance spectroscopic analysis verified that the high surface area, efficient charge-transfer kinetics, and higher conductivity of aloe-vera-mediated NiO$_x$ NPs contribute to its low overpotential values.

In conclusion, this Special Issue entitled "Heterogeneous Photocatalysis: A Solution for a Greener Earth" gives an overview of the latest advances in the development of innovative photocatalytic processes. These studies pave the path for the development of innovative processes for a greener Earth.

Finally, we are grateful to all authors for their valuable contributions and to the editorial team of *Catalysts* for their kind support, especially to Pamela Li for her constant help and availability during this Special Issue submission and publication process.

Funding: This research received no external funding.

Acknowledgments: J.G.M. and S.D.L. are grateful to F.R.S.-F.N.R.S. for, respectively, his postdoctoral position and her Senior Research Associate position.

Conflicts of Interest: The authors declare no conflict of interest.

References

1. Almeida, L.A.; Dosen, A.; Viol, J.; Marinkovic, B.A. TiO_2-Acetylacetone as an Efficient Source of Superoxide Radicals under Reduced Power Visible Light: Photocatalytic Degradation of Chlorophenol and Tetracycline. *Catalysts* **2022**, *12*, 116. [CrossRef]
2. Pascariu, P.; Cojocaru, C.; Airinei, A.; Olaru, N.; Rosca, I.; Koudoumas, E.; Suchea, M.P. Innovative Ag–TiO_2 Nanofibers with Excellent Photocatalytic and Antibacterial Actions. *Catalysts* **2021**, *11*, 1234. [CrossRef]
3. Mancuso, A.; Navarra, W.; Sacco, O.; Pragliola, S.; Vaiano, V.; Venditto, V. Photocatalytic Degradation of Thiacloprid Using Tri-Doped TiO_2 Photocatalysts: A Preliminary Comparative Study. *Catalysts* **2021**, *11*, 927. [CrossRef]
4. Benkhennouche-Bouchene, H.; Mahy, J.G.; Wolfs, C.; Vertruyen, B.; Poelman, D.; Eloy, P.; Hermans, S.; Bouhali, M.; Souici, A.; Bourouina-Bacha, S.; et al. Green Synthesis of N/Zr Co-Doped TiO_2 for Photocatalytic Degradation of p-Nitrophenol in Wastewater. *Catalysts* **2021**, *11*, 235. [CrossRef]
5. Eun, S.R.; Mavengere, S.; Cho, B.; Kim, J.S. Photocatalytic Reactivity of Carbon–Nitrogen– Sulfur-Doped TiO_2 Upconversion Phosphor Composites. *Catalysts* **2020**, *10*, 1188. [CrossRef]
6. Mahy, J.G.; Sotrez, V.; Tasseroul, L.; Hermans, S.; Lambert, S.D. Activation Treatments and SiO_2/Pd Modification of Sol-Gel TiO_2 Photocatalysts for Enhanced Photoactivity under UV Radiation. *Catalysts* **2020**, *10*, 1184. [CrossRef]
7. Mahy, J.G.; Lejeune, L.; Haynes, T.; Lambert, S.D.; Marcilli, R.H.M.; Fustin, C.A.; Hermans, S. Eco-Friendly Colloidal Aqueous Sol-Gel Process for TiO_2 Synthesis: The Peptization Method to Obtain Crystalline and Photoactive Materials at Low Temperature. *Catalysts* **2021**, *11*, 768. [CrossRef]
8. de Ceglie, C.; Pal, S.; Murgolo, S.; Licciulli, A.; Mascolo, G. Investigation of Photocatalysis by Mesoporous Titanium Dioxide Supported on Glass Fibers as an Integrated Technology for Water Remediation. *Catalysts* **2022**, *12*, 41. [CrossRef]
9. Ghazzy, A.; Yousef, L.; Al-Hunaiti, A. Visible Light Induced Nano-Photocatalysis Trimetallic $Cu_{0.5}Zn_{0.5}$-Fe: Synthesis, Characterization and Application as Alcohols Oxidation Catalyst. *Catalysts* **2022**, *12*, 611. [CrossRef]
10. Mahy, J.G.; Mbognou, M.H.T.; Léonard, C.; Fagel, N.; Woumfo, E.D.; Lambert, S.D. Natural Clay Modified with ZnO/TiO_2 to Enhance Pollutant Removal from Water. *Catalysts* **2022**, *12*, 148. [CrossRef]
11. Dao, D.Q.; Nguyen, T.K.A.; Pham, T.T.; Shin, E.W. Synergistic Effect on Photocatalytic Activity of Co-Doped $NiTiO_3$/$g-C_3N_4$ Composites under Visible Light Irradiation. *Catalysts* **2020**, *10*, 1332. [CrossRef]
12. Bazta, O.; Urbieta, A.; Trasobares, S.; Piqueras, J.; Fernández, P.; Addou, M.; Calvino, J.J.; Hungría, A.B. In-Depth Structural and Optical Analysis of Ce-Modified Zno Nanopowders with Enhanced Photocatalytic Activity Prepared by Microwave-Assisted Hydrothermal Method. *Catalysts* **2020**, *10*, 551. [CrossRef]
13. Faccani, L.; Ortelli, S.; Blosi, M.; Costa, A.L. Ceramized Fabrics and Their Integration in a Semi-Pilot Plant for the Photodegradation of Water Pollutants. *Catalysts* **2021**, *11*, 1418. [CrossRef]
14. González-Burciaga, L.A.; García-Prieto, J.C.; García-Roig, M.; Lares-Asef, I.; Núñez-Núñez, C.M.; Proal-Nájera, J.B. Cytostatic Drug 6-Mercaptopurine Degradation on Pilot Scale Reactors by Advanced Oxidation Processes: Uv-c/H_2O_2 and Uv-c/TiO_2/H_2O_2 Kinetics. *Catalysts* **2021**, *11*, 567. [CrossRef]
15. Gonçalves, N.P.F.; Varga, Z.; Nicol, E.; Calza, P.; Bouchonnet, S. Comparison of Advanced Oxidation Processes for the Degradation of Maprotiline in Water—Kinetics, Degradation Products and Potential Ecotoxicity. *Catalysts* **2021**, *11*, 240. [CrossRef]
16. Tejera, J.; Hermosilla, D.; Miranda, R.; Gascó, A.; Alonso, V.; Negro, C.; Blanco, Á. Assessing an Integral Treatment for Landfill Leachate Reverse Osmosis Concentrate. *Catalysts* **2020**, *10*, 1389. [CrossRef]
17. Selvanathan, V.; Shahinuzzaman, M.; Selvanathan, S.; Sarkar, D.K.; Algethami, N.; Alkhammash, H.I.; Anuar, F.H.; Zainuddin, Z.; Aminuzzaman, M.; Abdullah, H.; et al. Phytochemical-Assisted Green Synthesis of Nickel Oxide Nanoparticles for Application as Electrocatalysts in Oxygen Evolution Reaction. *Catalysts* **2021**, *11*, 1523. [CrossRef]

Article

TiO$_2$-Acetylacetone as an Efficient Source of Superoxide Radicals under Reduced Power Visible Light: Photocatalytic Degradation of Chlorophenol and Tetracycline

Lucas A. Almeida, Anja Dosen, Juliana Viol and Bojan A. Marinkovic *

Department of Chemical and Materials Engineering, Pontifical Catholic University of Rio de Janeiro (PUC-Rio), Rio de Janeiro 22453-900, Brazil; lucasalmeida@aluno.puc-rio.br (L.A.A.); adosen@puc-rio.br (A.D.); juliana-viol@puc-rio.br (J.V.)
* Correspondence: bojan@puc-rio.br

Abstract: Visible light-sensitive TiO$_2$-based nanomaterials are widely investigated for photocatalytic applications under high power ($\geq$300 W) UV and visible light. The formation of charge transfer complexes (CTCs) between bidentate ligands and nanocrystalline TiO$_2$ promotes visible light absorption and constitutes a promising alternative for environmental remediation under reduced visible light power. However, the efficiency of photodegradation, the volatilization profile of bidentates, and the role of reactive oxidizing species (ROS) are not fully understood. In this study, thermogravimetric analyses coupled with mass spectroscopy (TGA-MS) were performed on TiO$_2$-Acetylacetone (ACAC) CTC. TiO$_2$-ACAC CTC calcined at 300 °C (TiO$_2$-A300) was applied for the photocatalytic degradation of chlorophenol (4-CP) and tetracycline (TC) under low power visible light (26 W). Furthermore, the ROS scavengers isopropanol and benzoquinone were added for studying the photocatalytic role of •OH and •O$_2^-$ radicals. The TGA-MS showed the release of ACAC fragments, such as ethyl ions and acetone, in the range between 150 °C and 265 °C, while between 300 °C and 450 °C only CO$_2$ and H$_2$O were released during oxidation of ACAC. The photocatalytic abatement of tetracycline (68.6%), performed by TiO$_2$-A300, was ~two times higher than that observed for chlorophenol (31.3%) after 6 h, indicating a distinct participation of ROS in the degradation of these pollutants. The addition of the ROS scavenger revealed •O$_2^-$ radicals as primarily responsible for the high efficiency of TiO$_2$-ACAC CTC under reduced visible light. On the other hand, the •OH radicals are not efficiently generated in the CTC. Therefore, the development of heterostructures based on TiO$_2$-ACAC CTC can increase the generation of ROS through coupling with semiconductors capable of generating •OH under visible light.

Keywords: ligand-to-metal charge transfer; anatase; oxygen-based bidentate diketone; sol-gel; TGA-MS; remediation of aqueous pollutants

1. Introduction

The intensive use of drugs for the treatment of human health and industrial wastewater disposal are potential sources of contamination of aqueous effluents [1–4]. Currently, water treatment plants are not designed for the remediation of pharmaceutical and personal care products (PPCPs) [1,2]. An example of PPCPs is the antibiotic tetracycline, used to prevent bacterial infections, found in sewage from the pharmaceutical industry, hospitals, and livestock [5,6]. The remediation of organic compounds, such as chlorophenol, is a priority due to its high toxicity, low biodegradability, and carcinogenic and endocrine disruptive properties [4,7].

In this context, photocatalysis (an advanced oxidative process) is an alternative and sustainable technology for the remediation of these pollutants. The development of new photocatalysts mainly seeks to increase the capacity of the absorption of sunlight, i.e., to expand absorption from the ultra-violet (UV) region to the visible spectrum, since the visible region comprises ~45% of the solar spectrum [6,8–12]. In addition, the maximization of the photogeneration of reactive oxygen species (ROS), such as $^{\bullet}O_2^-$ and $\bullet OH$, plays a key role in photocatalytic activity [13,14]. This maximization occurs through the increase in the formation of e^-/h^+ pairs and, consequently, their interaction with the O_2 and H_2O molecules to generate $^{\bullet}O_2^-$ and $\bullet OH$, respectively.

Thus, the visible light sensitization mechanism denominated as the ligand-to-metal charge transfer (LMCT) has gained recognition in the development of TiO_2-based nanomaterials [15], although the effect of point defects should also be considered [16–18]. The LMCT mechanism promotes, in theory, the efficient photogeneration of superoxide radicals under visible light [16] and consists of the direct injection of e^- from the highest occupied molecular orbital (HOMO) of a chelating ligand into the conduction band (CB) of a semiconductor, such as TiO_2 [19]. The LMCT mechanism occurs via bonds formed between the chelating ligands (small organic molecules) and the TiO_2 surface [15,16,19].

The LMCT complex based on TiO_2 and acetic acid (monodentate ligand) showed photodegradation of phenol higher than 90% after 1.5 h exposure to a blue light-emitting diode (LED) lamp (20 W) [20]. In addition, another LMCT complex, made of TiO_2 and salicylic acid (bidentate ligand), revealed the potential for the selective aerobic oxidation of amines to imines under blue LED irradiation (3 W), having a selective yield of 92% of the desired product, N-benzylidenebenzylamine [21]. Other LMCT complexes of TiO_2 and bidentate ligands, such as glucose [22] and alizarin [23], showed ~100% photocatalytic reduction of Cr (VI) during 1 h under visible light irradiation (Xenon lamp 300 W). Furthermore, the nanocrystalline charge transfer complex (CTC) between TiO_2 and acetylacetone (bidentate) calcinated at 300 °C in air revealed ~100% NOx gas photodegradation for 2 h under visible light irradiation (24 W) [16]. Another study on the amorphous TiO_2-Acetylacetone complex revealed ~90% degradation of 2,4-dichlorophenol after 24 h in the absence of irradiation in the presence of high catalyst concentration (1 $g \cdot L^{-1}$) [24]. However, as the authors are aware, a CTC such as TiO_2-bidentate has not been investigated thus far for photocatalytic applications in drug degradation.

Moreover, the understanding of the role of reactive radicals on the photocatalytic activity of these complexes is not fully understood. Experimental analyses performed by electron paramagnetic resonance (EPR) [16,25] and theoretical studies based on density functional theory (DFT) reported for the TiO_2-Acetylacetone [25] and TiO_2-Thiosalicylic acid [26] complexes demonstrate the capacity of the LMCT mechanism to produce superoxide ($^{\bullet}O_2^-$). On the other hand, Fourier transform infrared (FTIR) analysis reveals the presence of adsorbed hydroxyls on the surface of these LMCT complexes [16,20–22,25] that may participate in the formation of $\bullet OH$ if electronic holes exist in the valence band (VB) of TiO_2. However, to the best of our knowledge, the individual potential of ROS in the LMCT systems has not been evaluated for photocatalytic applications. The ROS scavengers have been, however, already employed to other photocatalysts in photocatalytic measurements in the liquid medium to assess the individual potential of each ROS [4,6,27–30].

The measurements of photocatalytic degradation of chlorophenol and tetracycline by CTC, combined with the use of $^{\bullet}O_2^-$ and $\bullet OH$ scavengers, would clarify the actual potential of each ROS in the photocatalytic abatement.

Photocatalytic degradation in a liquid medium under visible light irradiation is mostly carried out with a light power superior to 300 W [30–33]. The high energy consumption of the lamps follows the opposite path to the sustainable nature of photocatalysis. Therefore, a lamp with reduced power such as residential lamps of 26 W will be applied in this research for the degradation of aqueous pollutants.

This study has as the main goals (1) the evaluation of the potential of TiO_2-Acetylacetone CTC calcinated at 300 °C in air for the photodegradation of chlorophenol and tetracycline under reduced power visible light (26 W) and (2) an understanding of ROS ($^{\bullet}O_2^-$ and $\bullet OH$) roles in photocatalysis performed by TiO_2-Acetylacetone using scavengers, such as benzoquinone and isopropanol. Furthermore, the volatilization profile of TiO_2-ACAC CTC was studied up to 550 °C by thermogravimetric analysis coupled with mass spectroscopy (TGA-MS) to confirm the presence of acetylacetone at temperatures above 150 °C.

2. Results

2.1. Evaluation of TiO$_2$-ACAC CTC Volatilization Profile by TGA-MS

TiO$_2$-A-RT, TiO$_2$-ACAC, and TiO$_2$-A300 were analyzed by TGA-MS to confirm the presence of ACAC above 150 °C through the release of ACAC fragments or the release of the entire ACAC molecule on heating.

The TGA curve of the TiO$_2$-A-RT sample (Figure 1a) revealed an expressive mass loss of 29.5 wt.% at temperatures below 150 °C due to dehydration of the material. Furthermore, the mass loss between 150 °C and 450 °C is associated with the release of acetylacetone, as documented by TGA-MS results (Figure 1b). The mass loss of organic species between 150 °C and 450 °C was ~15 wt.% and approximately equal to the mass loss of the TiO$_2$-ACAC xerogel for the same temperature region [16]. The TGA curves of TiO$_2$-ACAC and TiO$_2$-A300 and their first derivatives (DTG) (Figure S1, see Supplementary Materials) were in accordance with the results previously reported by Almeida et al. [16].

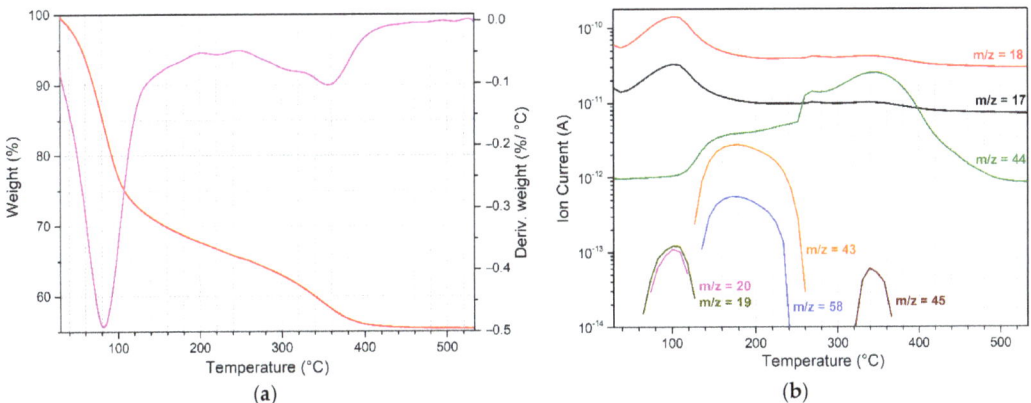

Figure 1. (a) TGA and DTG curves of TiO$_2$-A-RT sample; (b) Evolution profiles of gaseous species as monitored by TGA-MS for TiO$_2$-A-RT sample.

Figure 1b illustrates the evolution profiles of released gases as a function of temperature for the TiO$_2$-A-RT sample. In the first temperature stage of volatilization between 30 °C and 150 °C, species with m/z of 17, 18, 19, and 20, corresponding to water were detected [34]. In the second temperature stage between 150 °C and ~265 °C, the m/z of 43, 44, and 58 were identified, owing to the release of acetyl ions (CH$_3$C≡O$^+$), CO$_2$, and acetone, respectively [35]. Bowie et al. [35] showed that m/z of 43 and 58 are associated with the fragmentation of the ACAC molecule, where the most intense peak of the ACAC volatilization in the mass spectrum is m/z 43, belonging to acetyl ions. Acetone and CO$_2$ release in this temperature range are in accordance with Acik et al. [36] who observed the release of both species from the amorphous TiO$_2$-ACAC xerogel. In the third temperature region of mass loss between 300 °C and 450 °C, an increase of CO$_2$ release was identified (m/z = 44 and 45) together with water release, which is in accordance with the observation of Acik et al. [36] and Madarász et al. [37] for amorphous TiO$_2$-ACAC xerogel and crystalline titanium oxobis(acetylacetonate), respectively.

The profiles of gaseous species released from nanocrystalline TiO$_2$-ACAC and TiO$_2$-A300 samples are shown in Figure 2. The volatilization stages of TiO$_2$-ACAC (Figure 2a) follow a similar path as TiO$_2$-A-RT (Figure 1). TiO$_2$-ACAC water release occurs in the temperature range between 30 °C and 150 °C, followed by a two-stage release of molecules due to ACAC oxidation above 150 °C. In the interval from 150 °C to ~265 °C, only acetyl ions

(m/z = 43) and CO_2 (m/z = 44) were released. In addition, in the third temperature stage (300–450 °C), CO_2 (m/z = 44 and 45) and water release were reported. The absence of acetone release in the TiO_2-ACAC xerogel probably occurred due to overnight drying at 100 °C, reducing the excess ACAC content in the sample.

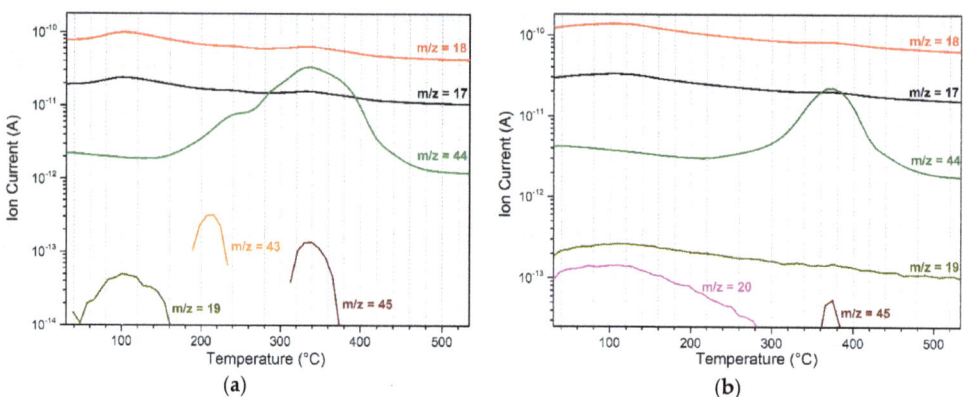

Figure 2. Evolution profiles of gaseous species as monitored by TGA-MS for (**a**) TiO_2-ACAC and (**b**) TiO_2-A300 samples.

On the other hand, the TGA-MS data of the TiO_2-A300 sample (Figure 2b) revealed only two distinct temperature stages of gases volatilization. The first stage between 30 °C and 150 °C was related to a release of water (m/z = 17, 18 e 19, and 20), while the second stage between 300 °C and 450 °C was due to CO_2 and water release. The volatilization step of the ACAC oxidation products above 300 °C, for all studied TiO_2-ACAC CTC samples, indirectly contributes to proving that the ACAC is responsible for the sensitization of TiO_2 in visible light in the samples calcined at 300 °C, as previously reported [16].

2.2. Chlorophenol Photodegradation and ROS Efficiency under Reduced Power Visible Light

Figure 3 shows the photocatalytic degradation of chlorophenol using TiO_2-A300 CTC during the period of 6 h. Photolysis did not promote pollutant degradation (green curve).

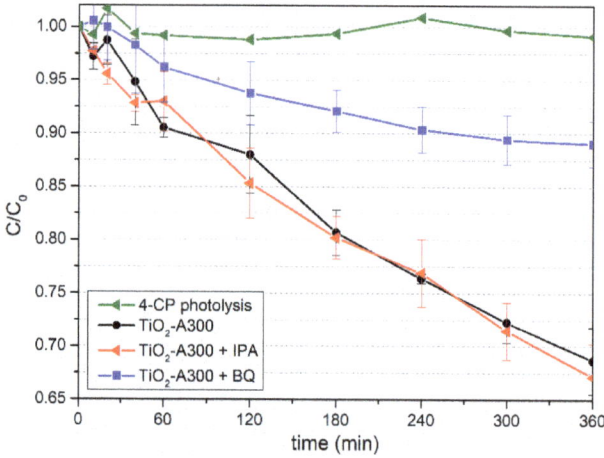

Figure 3. Chlorophenol photodegradation by TiO_2-A300 photocatalyst with and without the addition of scavengers. The green curve stands for photolysis.

The TiO$_2$-A300 CTC revealed photocatalytic abatement of 31.3% ± 3.1% for 4-CP after 6 h (Figure 3). Li et al. [32] quantified visible light abatement of 4-CP as 29.9% after 2 h using graphene grafted titania/titanate nanosheets under UV-Vis light of 500 W for the same concentrations of pollutants and photocatalyst used in our study. Furthermore, 4-CP degradation using TiO$_2$-A300 was nearly half of the degradation reported by Li et al. [32] after 2 h, however, with a light source with 20 times lower power.

The addition of the •OH radical scavenger (IPA) did not significantly impact the degradation efficiency of TiO$_2$-A300 (Figure 3). The observed photodegradation of 4-CP was 32.8% ± 3.1% after 6 h and was within the error of the degradation measured for TiO$_2$-A300 without the addition of the IPA scavenger. On the other hand, Moraes et al. [4] revealed that with the addition of the IPA, the degradation of 4-CP (10 mg·L^{-1}) using nanometric anatase (0.2 g·L^{-1}) was about two times lower in comparison to the degradation without the addition of the IPA, however, under UV light and after 3 h. Therefore, these data indicate that hydroxyl radicals played an important role in the degradation of 4-CP for TiO$_2$ under UV light, since it promoted the formation of e$^-$/h$^+$ pairs and, consequently, •OH radials. The results reported by Moraes et al. [4] are in accordance with studies that show that •OH radicals have a dominant role in photodegradation of 4-CP [32,38]. However, for TiO$_2$-A300 CTC under visible light, it was not possible to detect the role of •OH radicals in the photodegradation of 4-CP. Therefore, our results strongly indicate that TiO$_2$-A300 CTC does not efficiently generate •OH radicals under visible light.

On the other hand, the addition of the •O$_2^-$ scavenger (BQ) caused a strong reduction of the photodegradation potential of TiO$_2$-A300 (Figure 3). The 4-CP abatement after the addition of BQ was only 10.9% ± 2.1% after 6 h interval. Therefore, the •O$_2^-$ plays an important role in the photocatalytic activity of TiO$_2$-A300 during the degradation of 4-CP. According to Su et al. [39] and Fónagy et al. [40], benzoquinone is predominantly reduced by •O$_2^-$ radicals and e$^-$ and, consequently, degraded. Therefore, •O$_2^-$ generated by TiO$_2$-A300 CTC preferentially degrades BQ molecules instead of degrading 4-CP, as illustrated in Figure 4. The expressive 76.7% ± 2.1% BQ degradation and low 4-CP degradation after 6 h demonstrate the efficient generation of •O$_2^-$ radicals under visible light from TiO$_2$-A300.

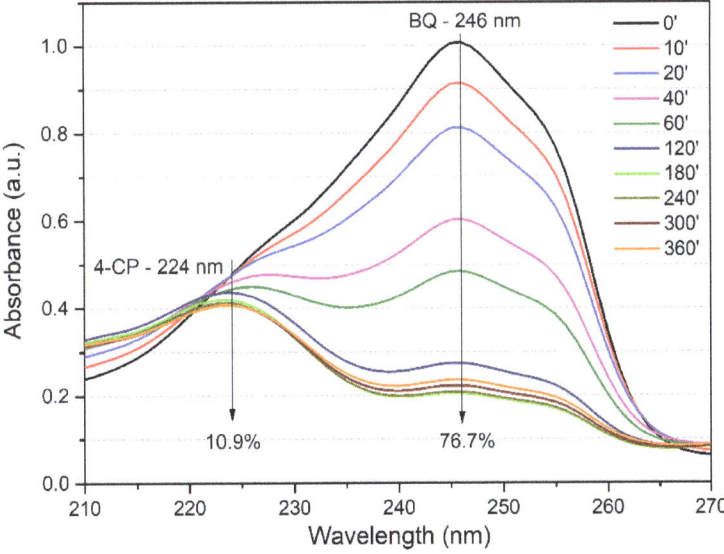

Figure 4. The absorbance of the BQ band situated at 246 nm and the 4-CP band at 224 nm, over time.

2.3. Tetracycline Photodegradation and ROS Efficiency under Reduced Power Visible Light

Figure 5 shows the photocatalytic degradation of tetracycline using TiO$_2$-A300. The photolysis (green curve) did not show pollutant degradation during the test time.

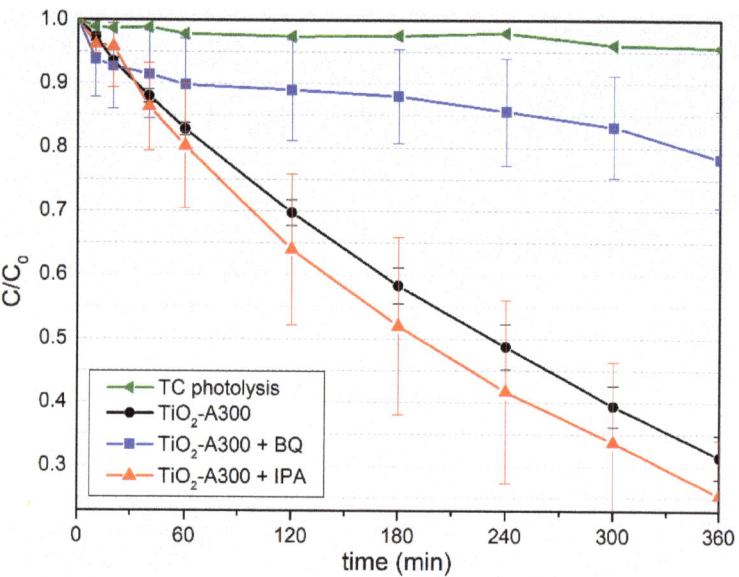

Figure 5. Tetracycline photodegradation by TiO$_2$-A300 with and without the addition of scavengers. The green curve stands for photolysis.

The TiO$_2$-A300 CTC showed a high photocatalytic degradation of 68.6% ± 3.4% of TC after 6 h (Figure 5). In addition, the maximum photodegradation of TC up to the abatement stabilization was 83% after 11 h (Figure S2). This maximum abatement of TC from TiO$_2$-A300 was similar to the 86.7% degradation reported by Wang et al. [6] for the WO$_3$-TiO$_2$ core-shell heterostructure decorated by Pt nanoparticles (0.08 g·L^{-1}) under UV-Vis light at 300 W after 2 h. However, the performance of TiO$_2$-A300 was achieved with 11.5 times lower light power with the use of visible instead of UV light and without the use of precious metals.

The use of isopropanol as the •OH scavenger revealed an abatement of 74.6% ± 8.7% after 6 h, within the error bars of the degradation without the addition of the IPA scavenger. In contrast, TC degradation under UV-Vis light by heterostructures, such as ZnO/GO/Ag$_3$PO$_4$ [41], Ag$_3$PO$_4$/AgBr/g-C$_3$N$_4$ [42], and Bi$_2$W$_2$O$_9$/g-C$_3$N$_4$ [43] revealed a decrease in TC abatement due to the addition of IPA. These results reported for the heterostructures indicate that •OH radicals contribute to the degradation of TC when generated [5]. Therefore, the result obtained for TiO$_2$-A300 suggests that TiO$_2$-A300 is not capable of producing •OH under visible light.

The TC photodegradation with the addition of BQ was significantly reduced to 21.8% ± 7.6% using TiO$_2$-A300 after 6 h (Figure 5). The reduction observed in our study agrees with the literature, where different ceramic heterojunctions showed low TC degradation efficiency with the addition of BQ under high power UV or UV-Vis light sources (>300 W) [6,42,44]. He et al. [5] reported that •O$_2^-$ and H$_2$O$_2$ radicals are capable of attacking the benzene ring of tetracycline, oxidizing the N-dimethyl group and the -C(O)NH$_2$ group, suggesting that •O$_2^-$ plays an important role in the photodegradation pathway of TC. Therefore, the TC degradation behavior observed in Figure 5 is mainly due to the efficient generation of •O$_2^-$ radicals by TiO$_2$-A300 CTC.

3. Discussion

The TGA-MS results of TiO_2-A-RT, TiO_2-ACAC, and TiO_2-A300 samples evidenced the fragments of ACAC oxidation release, such as acetyl ions, acetone, CO_2, and water between 150 °C and ~265 °C, while only CO_2 and water release were detected at higher temperatures between 300 °C and 450 °C (Figures 1b and 2). Two distinct steps of ACAC fragments release for TiO_2-A-RT and TiO_2-ACAC, between 150 °C and 450 °C, are in accordance with the TGA-MS and FTIR-MS analysis reported by Acik et al. [36] for amorphous TiO_2-ACAC xerogel. The volatilization profile of ACAC from TiO_2-A300 (Figure 2b) converges well with the mass loss reported by Almeida et al. [16] for the same material since CO_2 and water were released as ACAC oxidation products above 300 °C. Therefore, it is worth noting for the understanding of the CTC mechanism that nanocrystalline TiO_2-ACAC samples preserved a part of ACAC when calcined at temperatures higher than 150 °C, such as 300 °C, as applied for TiO_2-A300.

The results on the photocatalytic activity of TiO_2-A300 support the hypothesis that a TiO_2-ACAC CTC would be able to degrade aqueous pollutants under reduced power visible light. The TiO_2-A300 showed photocatalytic potential for degradation of both aqueous pollutants 4-CP and TC. In addition, TiO_2-A300 did not show adsorption of pollutants in the dark 1 h before turning on visible light (Figure S3). Therefore, the abatement of all the pollutants was the result of the photocatalytic activity of TiO_2-ACAC CTC.

In our previous study [16], TiO_2-A300 revealed a high efficiency (~100%) for photodegradation of NOx gas (100 ppm) during 2 h under reduced power visible light (24 W). Therefore, the results presented in the current research additionally increase the photocatalytic applicability of TiO_2-ACAC CTC.

Zhu et al. [41] reported 96.3% degradation of TC (30 mg·L^{-1}) after 75 min, with ~50% of the reported degradation values associated with the adsorption phenomenon in the dark. The authors used for this purpose $ZnO/GO/Ag_3PO_4$ photocatalyst (1 g·L^{-1}) under low power visible light (65 W). Therefore, photodegradation of TC reported in our study after 6 h (68.6%) was higher than documented by Zhu et al. (46.3%) [41] with 2.5 lower visible light power and using four times less photocatalyst. Additionally, the capacity of degradation of TC is ~two times higher than that observed for 4-CP using TiO_2-A300, after 6 h, indicating a distinct participation of ROS in the degradation of these two pollutants.

ROS Generation Efficiency in TiO_2-A300 CTC

One of the main goals of our study was understanding the roles of ROS ($^{\bullet}O_2^{-}$ and •OH) in photocatalysis performed by TiO_2-A300 CTC, using BQ and IPA scavengers. The photodegradation of 4-CP and TC under visible light with the addition of $^{\bullet}O_2^{-}$ scavenger confirms the efficient generation of $^{\bullet}O_2^{-}$ radicals by TiO_2-ACAC CTC (Figures 3 and 5), in accordance with the EPR characterization and DFT studies carried out on LMCT complexes [16,25,26]. On the other hand, the •OH radicals were not efficiently generated by TiO_2-A300 CTC under visible light, resulting in the lower photodegradation of 4-CP than the photodegradation of TC, since •OH radicals are the dominant ROS in the 4-CP photodegradation [4,7,32], while $^{\bullet}O_2^{-}$ radicals play a key role in TC degradation [5].

Figure 6 represents an update of the photocatalytic reactions conducted by TiO_2-A300 CTC for the ROS generation and consequent degradation of 4-CP and TC. Additionally, the photocatalytic reactions for hydrogen production are also presented in Figure 6.

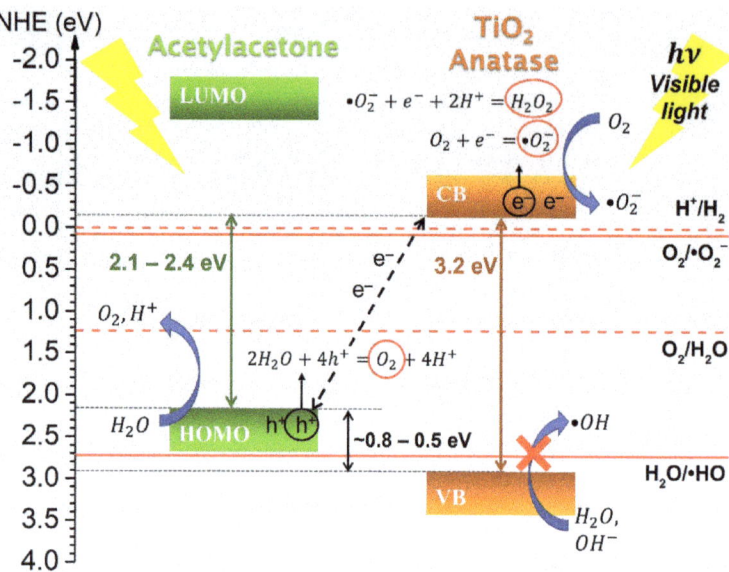

Figure 6. Scheme of electronic bands in TiO$_2$-A300 CTC and photocatalytic reactions for ROS generation. The position of BV and BC of anatase at redox potential scale, in accordance with normal hydrogen electrode (NHE), was adopted from literature [8,9], as well as the positions of the ROS species. The position of HOMO of acetylacetone was previously established in [16]. The position of LUMO of acetylacetone was arbitrarily added, since it does not have a role in photocatalytic and water splitting events and is higher than 3.2 eV.

Figure 6 shows the electron transfer from ACAC HOMO to TiO$_2$ (anatase) CB and, consequently, the formation of $^\bullet$O$_2^-$ radicals due to the reaction between O$_2$ adsorbed on the surface of anatase and free e$^-$ from CB (Equation (1)).

$$O_2 + e^- \rightarrow {}^\bullet O_2^-, \qquad (1)$$

The efficient $^\bullet$O$_2^-$ generation is also a consequence of a high O$_2$ presence in the reaction medium [9]. A possible additional explanation for the O$_2$ presence in an aqueous solution is the water splitting mechanism (Equation (2)) [8,9]. The h$^+$ generated in the HOMO of the ACAC may promote water splitting, i.e., decomposing H$_2$O into O$_2$ and H$^+$ through an oxidation half reaction as shown in Figure 6.

$$2\,H_2O + 4\,h^+ \rightarrow O_2 + 4\,H^+, \qquad (2)$$

Additionally, the efficient production of $^\bullet$O$_2^-$ radicals can also form H$_2$O$_2$ [9,45], which is another ROS capable of degradation of aqueous pollutants due to its ability to attack benzene rings and the N-dimethyl group [5]. The energy barrier of the formation of H$_2$O$_2$ from $^\bullet$O$_2^-$ radicals (Equation (3)) is 0.3 eV below the normal hydrogen electrode potential (NHE) [13]. Therefore, a part of $^\bullet$O$_2^-$ generated by TiO$_2$-A300 can be transformed into H$_2$O$_2$ and can continue to degrade 4-CP and TC.

$$^\bullet O_2^- + e^- + 2\,H^+ \rightarrow O_2 + H_2O_2, \qquad (3)$$

As seen in Figure 6, the arrangement of CB of TiO$_2$ and HOMO of ACAC in the TiO$_2$-ACAC CTC may provide photocatalytic production of H$_2$ from water splitting. The electron holes from HOMO of ACAC are capable of oxidizing water, forming H^+ (Equation (2)), and the e^- in the CB of TiO_2 may reduce H^+ to H_2 (Equation (4)).

$$2\,H^+ + 2\,e^- \rightarrow H_2, \qquad (4)$$

Finally, some previous studies demonstrated the photocatalytic potential of Z-scheme heterostructures based on WO_3 [6,46], SnO_2 [47], and $BiVO_4$ [48] for TC and NOx degradation and hydrogen production under UV-Vis light. These low-bandgap semiconductors can absorb visible light and generate •OH radicals via VB and CB with positive redox potential. The Z-scheme heterojunction consists of the recombination of the CB electrons of these low-bandgap semiconductors with the VB of another semiconductor with a lower positive redox potential, providing a reduction in the recombination rate of e^-/h^+ pairs in the low-bandgap semiconductors, such as WO_3, SnO_2, and $BiVO_4$ and, consequently, an increase in the formation of •OH radicals. Therefore, coupling TiO_2-ACAC CTC with semiconductors capable of generating •OH radicals under visible light may maximize the generation of ROS and increase of photocatalytic efficiency of these new heterojunctions.

4. Materials and Methods

4.1. Materials

All reagents were purchased from Sigma-Aldrich and used as obtained. Titanium isopropoxide ($Ti(OiPr)_4$, 97%), acetylacetone (ACAC, ≥99%), ethanol (≥99.8%), and nitric acid (65%) were used for the synthesis of TiO_2-ACAC CTC nanoparticles. The aqueous pollutants were chlorophenol (4-CP, ≥99%) and tetracycline (TC, ≥98%). Benzoquinone (BQ, ≥98%) and isopropanol (IPA, ≥99.5%) were used as ROS scavengers.

4.2. Synthesis of TiO_2-Acetylacetone Charge Transfer Complex

The synthesis of TiO_2 anatase nanoparticles coupled with acetylacetone (CTC) was carried out by the sol-gel route as reported by Scolan and Sanchez [49]. In addition, as previously reported [12,16], the molar ratios for hydrolysis (H = $[H_2O]/[Ti]$), acidity ($H^+ = [H^+]/[Ti]$), and complexing (A = $[ACAC]/[Ti]$) were kept at 100, 0.027 and 2, respectively, to produce TiO_2-ACAC xerogel with mean crystal size ~2.5 nm.

In the adopted procedure, 30 mL of $Ti(OiPr)_4$ was added dropwise into a solution of 20 mL of ACAC with 100 mL of ethanol (1:5 v/v). The obtained yellow solution remained under magnetic stirring for 40 min at room temperature. Afterward, 180 mL of HNO_3 solution (0.015 M) was dropped slowly into the yellowish solution under continuous stirring. The dark orange solution obtained was heated to 60 °C and kept under magnetic stirring for 8 h. Next, the sol of TiO_2 was dried overnight in Petri dishes at room temperature, and a gel was formed (TiO_2-A-RT). Finally, the gel was dried at 100 °C overnight to obtain a red-yellowish xerogel of TiO_2-Acetylacetone (TiO_2-ACAC). The TiO_2-ACAC xerogel, grounded in an agate mortar, was calcined in air at 300 °C for 2 h in a Tubular Maitec-INTI FET 1600/H furnace. The as-prepared powder was denoted TiO_2-A300.

4.3. Thermogravimetric Analysis Coupled with Mass Spectroscopy (TGA-MS)

Thermogravimetric analysis (TGA) and differential scanning calorimetry (DSC) were performed on a Perkin-Elmer Simultaneous Thermal Analyzer STA-6000 under synthetic air flow (130 mL·min^{-1}) with a heating rate of 10 °C·min^{-1} and within the temperature range between 30 °C and 550 °C. A mass of ~25 mg was used.

To identify the volatilized species from TiO_2-ACAC and confirm the presence of ACAC in the TiO_2-A300 sample, a mass spectrometer OmniStar/ThermoStar-GSD 320 O3 (Pfeiffer Vacuum) was coupled to the STA-6000. The m/z range until 105 was scanned with the measuring time of 0.5 s·amu^{-1}.

4.4. Measurement of the Photodegradation of Chlorophenol and Tetracycline and the Efficiency of ROS

The photocatalytic potential of TiO_2-A300 for degradation of 5 mg·L^{-1} of 4-CP and TC under visible light irradiation was evaluated. In accordance with [16], the inorganic part of TiO_2-A300 CTC is composed of nanocrystalline anatase. Two other materials, TiO_2-A-RT and TiO_2-ACAC (where the inorganic part also consists of nanocrystalline anatase), were not evaluated since they previously showed inferior photocatalytic efficiency [16]. In addition, a measurement of the TC photodegradation until stabilization of TC abatement was performed.

All measurements were performed with 0.2 g·L^{-1} of the photocatalyst. The light source used was a DULUX D/E 26 W residential fluorescent lamp with an irradiance of 0.23 W·cm^{-2} and an emission of light in the wavelength range from 400 nm to 700 nm. The photocatalytic system is shown in Figure S4. The lamp is housed in a cylindrical quartz bulb (length 11 cm, internal diameter 3.2 cm, and thickness 0.15 cm) unable to absorb visible light. The photoreactor was immersed in a bath with cold water circulation to maintain the system at room temperature. The photocatalytic tests were carried out under vigorous magnetic stirring in two stages. In the first stage, the pollutant and the photocatalyst were kept in the dark for 1 h to attain the absorption–desorption equilibrium. After this period, the lamp was turned on, and 5 mL aliquots of supernatant were acquired at intervals of 1 h for 6 h. The supernatant was filtered through a Merck Millipore filter (0.45 μm) and analyzed by an Agilent UV-Vis spectrophotometer (model 8453). The absorption at the wavelengths of 224 nm and 358 nm was accompanied for photodegradation of 4-CP and TC during the test time, respectively. The observed reduction in these wavelengths for each pollutant was used as input to determine the percentage of photodegradation performed by TiO_2-A300 CTC, as seen in Figure S5 and Table S1. The degradation data obtained were presented through the mean value together with the respective standard deviations (Table S1) since all photocatalytic tests were performed in triplicate to ensure repeatability of results.

The ROS efficiency in TiO_2-ACAC CTC was evaluated by adding the scavenger molecule BQ (5.4 mg·L^{-1}; 0.05 mmol·L^{-1}) to inhibit the participation of the $^{\bullet}O_2^-$ radical in the degradation of 4-CP and TC. On the other hand, the addition of the IPA scavenger (6.0 mg·L^{-1}; 0.1 mmol·L^{-1}) was performed to inhibit the participation of the •OH radical. The scavengers were added to the 5.0 mg·L^{-1} solutions of 4-CP or TC before starting the photocatalytic tests to ensure a mass ratio close to 1:1 between the pollutant and the scavenger.

5. Conclusions

The TGA-MS indirectly proved the presence of ACAC on TiO_2-A-RT, TiO_2-ACAC xerogel, and TiO_2-A300 at temperatures higher than 150 °C. The release of acetyl ions and acetone between 150 °C and ~265 °C was detected, while at temperatures above 300 °C, and not higher than 450 °C, the release of CO_2 and water was detected as the products of ACAC oxidation.

The TiO_2-A300 photodegrades 4-CP and TC under reduced power visible light. The photocatalytic abatement of tetracycline (68.6%) was ~two times higher than that observed for chlorophenol (31.3%) after 6 h, indicating a distinct potential of ROS in the degradation of these pollutants.

The addition of BQ scavenger on the photocatalytic degradation of 4-CP and TC proved TiO_2-A300 CTC to be a powerful source of superoxide ($^{\bullet}O_2^-$) radicals under visible light, promoting a high photodegradation of TC (68.6%) after 6 h. However, the •OH scavenger does not promote the reduction in photodegradation of 4-CP and TC, proving TiO_2-ACAC CTC is not an efficient generator of •OH radicals under visible light.

The photocatalytic potential reported for TiO_2-A300 may stimulate new studies on hydrogen production process from water splitting under visible light due to the redox potential arrangement of CB of TiO_2 and HOMO of ACAC. Furthermore, the development of heterostructures based on TiO$_2$-A300 CTC, and similar CTC, can increase the generation of ROS through coupling with semiconductors capable of generating •OH under visible light and, therefore, capable of formation of more powerful photocatalysts.

Supplementary Materials: The following are available online at https://www.mdpi.com/article/10.3390/catal12020116/s1, Figure S1: TGA and DTG curves of (a) TiO2-ACAC xerogel and (b) TiO2-A300 sample, Figure S2: Tetracycline photodegradation by TiO2-A300, Figure S3: Curves of adsorption (1 h in dark) and photocatalytic activity of 4-CP and TC using TiO2-A300 of (a) 4-CP and (b) TC by TiO2-A300 without scavengers addition during 6 h, Figure S4: Photocatalytic system for aqueous pollutants degradation, Figure S5: Absorbance of (a) 4-CP band, situated at 224 nm and (b) TC band at 358 nm, over the time, Table S1: 4-CP and TC degradation data obtained in the first repetition of the photocatalytic test using TiO$_2$-A300.

Author Contributions: Conceptualization, L.A.A. and B.A.M.; synthesis and overall characterizations, L.A.A.; funding acquisition, B.A.M.; data interpretation, L.A.A. and B.A.M.; TGA-MS analysis, A.D. and J.V.; TGA-MS interpretation, A.D. and J.V.; writing—original draft preparation, L.A.A.; writing—review and editing, L.A.A., A.D. and B.A.M. All authors have read and agreed to the published version of the manuscript.

Funding: The authors are grateful to FAPERJ for financial support through the projects E-26/210.585/2019 and E-26/210.046/2021.

Institutional Review Board Statement: Not applicable.

Informed Consent Statement: Not applicable.

Data Availability Statement: The data underlying this article will be shared on reasonable request from the corresponding author.

Acknowledgments: This study was financed in part by the Coordenação de Aperfeiçoamento de Pessoal de Nível Superior-Brasil (CAPES)–Finance Code 001. B.A.M. is grateful to CNPq (National Council for Scientific and Technological Development) for a Research Productivity Grant. L.A. is grateful to ANP (Agência Nacional do Petróleo, Gás Natural e Biocombustíveis) and Finep (Funding Authority for Studies and Projects) for a doctoral degree scholarship.

Conflicts of Interest: The authors declare no conflict of interest.

References

1. Liu, N.; Wang, J.; Wu, J.; Li, Z.; Huang, W.; Zheng, Y.; Lei, J.; Zhang, X.; Tang, L. Magnetic Fe$_3$O$_4$@MIL-53(Fe) nanocomposites derived from MIL-53(Fe) for the photocatalytic degradation of ibuprofen under visible light irradiation. *Mater. Res. Bull.* **2020**, *132*, 111000. [CrossRef]
2. Zhou, Z.; Li, Y.; Li, M.; Li, Y.; Zhan, S. Efficient removal for multiple pollutants via Ag$_2$O/BiOBr heterojunction: A promoted photocatalytic process by valid electron transfer pathway. *Chin. Chem. Lett.* **2020**, *31*, 2698–2704. [CrossRef]
3. Mohammad, A.; Khan, M.E.; Cho, M.H.; Yoon, T. Adsorption promoted visible-light-induced photocatalytic degradation of antibiotic tetracycline by tin oxide/cerium oxide nanocomposite. *Appl. Surf. Sci.* **2021**, *565*, 150337. [CrossRef]
4. de Moraes, N.P.; Torezin, F.A.; Dantas, G.V.J.; de Sousa, J.G.M.; Valim, R.B.; da Silva Rocha, R.; Landers, R.; da Silva, M.L.C.P.; Rodrigues, L.A. TiO$_2$/Nb$_2$O$_5$/carbon xerogel ternary photocatalyst for efficient degradation of 4-Chlorophenol under solar light irradiation. *Ceram. Int.* **2020**, *46*, 14505–14515. [CrossRef]
5. He, X.; Kai, T.; Ding, P. Heterojunction photocatalysts for degradation of the tetracycline antibiotic: A review. *Environ. Chem. Lett.* **2021**, *19*, 4563–4601. [CrossRef]
6. Wang, Y.; Fu, H.; Yang, X.; An, X.; Zou, Q.; Xiong, S.; Han, D. Pt nanoparticles-modified WO$_3$@TiO$_2$ core–shell ternary nanocomposites as stable and efficient photocatalysts in tetracycline degradation. *J. Mater. Sci.* **2020**, *55*, 14415–14430. [CrossRef]
7. de Moraes, N.P.; Valim, R.B.; da Silva Rocha, R.; da Silva, M.L.C.P.; Campos, T.M.B.; Thim, G.P.; Rodrigues, L.A. Effect of synthesis medium on structural and photocatalytic properties of ZnO/Carbon xerogel composites for solar and visible light degradation of 4-Chlorophenol and bisphenol A. *Colloids Surf. A Physicochem. Eng. Asp.* **2020**, *584*, 124034. [CrossRef]
8. Christoforidis, K.C.; Fornasiero, P. Photocatalytic Hydrogen Production: A Rift into the Future Energy Supply. *ChemCatChem* **2017**, *9*, 1523–1544. [CrossRef]
9. Ren, H.; Koshy, P.; Chen, W.-F.; Qi, S.; Sorrell, C.C. Photocatalytic materials and technologies for air purification. *J. Hazard. Mater.* **2017**, *325*, 340–366. [CrossRef]
10. Kapilashrami, M.; Zhang, Y.; Liu, Y.-S.; Hagfeldt, A.; Guo, J. Probing the Optical Property and Electronic Structure of TiO$_2$ Nanomaterials for Renewable Energy Applications. *Chem. Rev.* **2014**, *114*, 9662–9707. [CrossRef] [PubMed]

11. Zhou, H.; Qu, Y.; Zeid, T.; Duan, X. Towards highly efficient photocatalysts using semiconductor nanoarchitectures. *Energy Environ. Sci.* **2012**, *5*, 6732–6743. [CrossRef]
12. Habran, M.; Pontón, P.I.; Mancic, L.; Pandoli, O.; Krambrock, K.; da Costa, M.E.H.M.; Letichevsky, S.; Costa, A.M.L.M.; Morgado, E.; Marinkovic, B.A. Visible light sensitive mesoporous nanohybrids of lepidocrocite-like ferrititanate coupled to a charge transfer complex: Synthesis, characterization and photocatalytic degradation of NO. *J. Photochem. Photobiol. A Chem.* **2018**, *365*, 133–144. [CrossRef]
13. Nosaka, Y.; Nosaka, A.Y. Generation and Detection of Reactive Oxygen Species in Photocatalysis. *Chem. Rev.* **2017**, *117*, 11302–11336. [CrossRef]
14. Khavar, A.H.C.; Moussavi, G.; Mahjoub, A.R.; Luque, R.; Rodríguez-Padrón, D.; Sattari, M. Enhanced visible light photocatalytic degradation of acetaminophen with Ag_2S-ZnO@rGO core-shell microsphere as a novel catalyst: Catalyst preparation and characterization and mechanistic catalytic experiments. *Sep. Purif. Technol.* **2019**, *229*, 115803. [CrossRef]
15. Luciani, G.; Imparato, C.; Vitiello, G. Photosensitive Hybrid Nanostructured Materials: The Big Challenges for Sunlight Capture. *Catalysts* **2020**, *10*, 103. [CrossRef]
16. Almeida, L.A.; Habran, M.; dos Santos Carvalho, R.; da Costa, M.E.H.M.; Cremona, M.; Silva, B.C.; Krambrock, K.; Pandoli, O.G.; Morgado, E., Jr.; Marinkovic, B.A. The Influence of Calcination Temperature on Photocatalytic Activity of TiO_2-Acetylacetone Charge Transfer Complex towards Degradation of NO_x under Visible Light. *Catalysts* **2020**, *10*, 1463. [CrossRef]
17. Piskunov, S.; Lisovski, O.; Begens, J.; Bocharov, D.; Zhukovskii, Y.F.; Wessel, M.; Spohr, E. C-, N-, S-, and Fe-Doped $TiO2$ and $SrTiO3$ Nanotubes for Visible-Light-Driven Photocatalytic Water Splitting: Prediction from First Principles. *J. Phys. Chem. C* **2015**, *119*, 18686–18696. [CrossRef]
18. Serga, V.; Burve, R.; Krumina, A.; Pankratova, V.; Popov, A.I.; Pankratov, V. Study of phase composition, photocatalytic activity, and photoluminescence of TiO_2 with Eu additive produced by the extraction-pyrolytic method. *J. Mater. Res. Technol.* **2021**, *13*, 2350–2360. [CrossRef]
19. Zhang, G.; Kim, G.; Choi, W. Visible light driven photocatalysis mediated via ligand-to-metal charge transfer (LMCT): An alternative approach to solar activation of titania. *Energy Environ. Sci.* **2014**, *7*, 954–966. [CrossRef]
20. Liu, J.; Han, L.; An, N.; Xing, L.; Ma, H.; Cheng, L.; Yang, J.; Zhang, Q. Enhanced visible-light photocatalytic activity of carbonate-doped anatase TiO_2 based on the electron-withdrawing bidentate carboxylate linkage. *Appl. Catal. B Environ.* **2017**, *202*, 642–652. [CrossRef]
21. Li, X.; Xu, H.; Shi, J.-L.; Hao, H.; Yuan, H.; Lang, X. Salicylic acid complexed with TiO_2 for visible light-driven selective oxidation of amines into imines with air. *Appl. Catal. B Environ.* **2019**, *244*, 758–766. [CrossRef]
22. Kim, G.; Lee, S.-H.; Choi, W. Glucose–TiO_2 charge transfer complex-mediated photocatalysis under visible light. *Appl. Catal. B Environ.* **2015**, *162*, 463–469. [CrossRef]
23. Li, M.; Li, Y.; Zhao, J.; Li, M.; Wu, Y.; Na, P. Alizarin-TiO_2 LMCT Complex with Oxygen Vacancies: An Efficient Visible Light Photocatalyst for Cr(VI) Reduction. *Chin. J. Chem.* **2020**, *38*, 1332–1338. [CrossRef]
24. Aronne, A.; Fantauzzi, M.; Imparato, C.; Atzei, D.; De Stefano, L.; D'Errico, G.; Sannino, F.; Rea, I.; Pirozzi, D.; Elsener, B.; et al. Electronic properties of TiO_2-based materials characterized by high Ti_3^+ self-doping and low recombination rate of electron–hole pairs. *RSC Adv.* **2017**, *7*, 2373–2381. [CrossRef]
25. Ritacco, I.; Imparato, C.; Falivene, L.; Cavallo, L.; Magistrato, A.; Caporaso, L.; Camellone, M.F.; Aronne, A. Spontaneous Production of Ultrastable Reactive Oxygen Species on Titanium Oxide Surfaces Modified with Organic Ligands. *Adv. Mater. Interfaces* **2021**, *8*, 2100629. [CrossRef]
26. Milićević, B.; Đorđević, V.; Lončarević, D.; Dostanić, J.M.; Ahrenkiel, S.P.; Dramićanin, M.D.; Sredojević, D.; Švrakić, N.M.; Nedeljković, J.M. Charge-transfer complex formation between TiO_2 nanoparticles and thiosalicylic acid: A comprehensive experimental and DFT study. *Opt. Mater.* **2017**, *73*, 163–171. [CrossRef]
27. Lam, S.M.; Sin, J.-C.; Satoshi, I.; Abdullah, A.Z.; Mohamed, A.R. Enhanced sunlight photocatalytic performance over Nb_2O_5/ZnO nanorod composites and the mechanism study. *Appl. Catal. A Gen.* **2014**, *471*, 126–135. [CrossRef]
28. Abd Elkodous, M.; El-Sayyad, G.S.; Abdel Maksoud, M.I.A.; Kumar, R.; Maegawa, K.; Kawamura, G.; Tan, W.K.; Matsuda, A. Nanocomposite matrix conjugated with carbon nanomaterials for photocatalytic wastewater treatment. *J. Hazard. Mater.* **2021**, *410*, 124657. [CrossRef]
29. Zhang, Y.; Li, J.; Liu, H. Synergistic Removal of Bromate and Ibuprofen by Graphene Oxide and TiO2 Heterostructure Doped with F: Performance and Mechanism. *J. Nanomater.* **2020**, *2020*, 6094984. [CrossRef]
30. de Sousa, J.G.M.; da Silva, T.V.C.; de Moraes, N.P.; da Silva, M.L.C.P.; da Silva Rocha, R.; Landers, R.; Rodrigues, L.A. Visible light-driven ZnO/g-C_3N_4/carbon xerogel ternary photocatalyst with enhanced activity for 4-Chlorophenol degradation. *Mater. Chem. Phys.* **2020**, *256*, 123651. [CrossRef]
31. Myilsamy, M.; Mahalakshmi, M.; Subha, N.; Rajabhuvaneswari, A.; Murugesan, V. Visible light responsive mesoporous graphene–Eu_2O_3/TiO_2 nanocomposites for the efficient photocatalytic degradation of 4-Chlorophenol. *RSC Adv.* **2016**, *6*, 35024–35035. [CrossRef]
32. Li, F.; Du, P.; Liu, W.; Li, X.; Ji, H.; Duan, J.; Zhao, D. Hydrothermal synthesis of graphene grafted titania/titanate nanosheets for photocatalytic degradation of 4-Chlorophenol: Solar-light-driven photocatalytic activity and computational chemistry analysis. *Chem. Eng. J.* **2018**, *331*, 685–694. [CrossRef]

33. Alam, U.; Shah, T.A.; Khan, A.; Muneer, M. One-pot ultrasonic assisted sol-gel synthesis of spindle-like Nd and V codoped ZnO for efficient photocatalytic degradation of organic pollutants. *Sep. Purif. Technol.* **2019**, *212*, 427–437. [CrossRef]
34. Acik, I.O.; Madarász, J.; Krunks, M.; Tõnsuaadu, K.; Pokol, G.; Niinistö, L. Titanium(IV) acetylacetonate xerogels for processing titania films: A Thermoanalytical Study. *J. Therm. Anal. Calorim.* **2009**, *97*, 39–45. [CrossRef]
35. Bowie, J.H.; Williams, D.H.; Lawesson, S.-O.; Schroll, G. Studies in Mass Spectrometry. IX.1 Mass Spectra of β-Diketones. *J. Org. Chem.* **1966**, *31*, 1384–1390. [CrossRef]
36. Acik, I.O.; Madarász, J.; Krunks, M.; Tõnsuaadu, K.; Janke, D.; Pokol, G.; Niinistö, L. Thermoanalytical studies of titanium(IV) acetylacetonate xerogels with emphasis on evolved gas analysis. *J. Therm. Anal. Calorim.* **2007**, *88*, 557–563. [CrossRef]
37. Madarász, J.; Kaneko, S.; Okuya, M.; Pokol, G. Comparative evolved gas analyses of crystalline and amorphous titanium(IV)oxo-hydroxo-acetylacetonates by TG-FTIR and TG/DTA-MS. *Thermochim. Acta* **2009**, *489*, 37–44. [CrossRef]
38. Schneider, J.; Matsuoka, M.; Takeuchi, M.; Zhang, J.; Horiuchi, Y.; Anpo, M.; Bahnemann, D.W. Understanding TiO_2 Photocatalysis: Mechanisms and Materials. *Chem. Rev.* **2014**, *114*, 9919–9986. [CrossRef]
39. Su, R.; Tiruvalam, R.; He, Q.; Dimitratos, N.; Kesavan, L.; Hammond, C.; Lopez-Sanchez, J.A.; Bechstein, R.; Kiely, C.J.; Hutchings, G.J.; et al. Promotion of Phenol Photodecomposition over TiO2 Using Au, Pd, and Au–Pd Nanoparticles. *ACS Nano* **2012**, *6*, 6284–6292. [CrossRef] [PubMed]
40. Fónagy, O.; Szabó-Bárdos, E.; Horváth, O. 1,4-Benzoquinone and 1,4-Hydroquinone based determination of electron and superoxide radical formed in heterogeneous photocatalytic systems. *J. Photochem. Photobiol. A Chem.* **2021**, *407*, 113057. [CrossRef]
41. Zhu, P.; Duan, M.; Wang, R.; Xu, J.; Zou, P.; Jia, H. Facile synthesis of $ZnO/GO/Ag_3PO_4$ heterojunction photocatalyst with excellent photodegradation activity for tetracycline hydrochloride under visible light. *Colloids Surf. A Physicochem. Eng. Asp.* **2020**, *602*, 125118. [CrossRef]
42. Yu, H.; Wang, D.; Zhao, B.; Lu, Y.; Wang, X.; Zhu, S.; Qin, W.; Huo, M. Enhanced photocatalytic degradation of tetracycline under visible light by using a ternary photocatalyst of $Ag_3PO_4/AgBr/g-C_3N_4$ with dual Z-scheme heterojunction. *Sep. Purif. Technol.* **2020**, *237*, 116365. [CrossRef]
43. Obregón, S.; Ruíz-Gómez, M.A.; Rodríguez-González, V.; Vázquez, A.; Hernández-Uresti, D.B. A novel type-II $Bi_2W_2O_9/g-C_3N_4$ heterojunction with enhanced photocatalytic performance under simulated solar irradiation. *Mater. Sci. Semicond. Process.* **2020**, *113*, 105056. [CrossRef]
44. Lu, F.; Chen, K.; Feng, Q.; Cai, H.; Ma, D.; Wang, D.; Li, X.; Zuo, C.; Wang, S. Insight into the enhanced magnetic separation and photocatalytic activity of Sn-doped TiO_2 core-shell photocatalyst. *J. Environ. Chem. Eng.* **2021**, *9*, 105840. [CrossRef]
45. Mamaghani, A.H.; Haghighat, F.; Lee, C.-S. Photocatalytic oxidation technology for indoor environment air purification: The state-of-the-art. *Appl. Catal. B Environ.* **2017**, *203*, 247–269. [CrossRef]
46. Zhi, L.; Zhang, S.; Xu, Y.; Tu, J.; Li, M.; Hu, D.; Liu, J. Controlled growth of AgI nanoparticles on hollow WO_3 hierarchical structures to act as Z-scheme photocatalyst for visible-light photocatalysis. *J. Colloid Interface Sci.* **2020**, *579*, 754–765. [CrossRef] [PubMed]
47. Li, Y.; Wu, X.; Ho, W.; Lv, K.; Li, Q.; Li, M.; Lee, S.C. Graphene-induced formation of visible-light-responsive $SnO_2-Zn_2SnO_4$ Z-scheme photocatalyst with surface vacancy for the enhanced photoreactivity towards NO and acetone oxidation. *Chem. Eng. J.* **2018**, *336*, 200–210. [CrossRef]
48. Zhou, F.Q.; Fan, J.C.; Xu, Q.J.; Min, Y.L. $BiVO_4$ nanowires decorated with CdS nanoparticles as Z-scheme photocatalyst with enhanced H_2 generation. *Appl. Catal. B Environ.* **2017**, *201*, 77–83. [CrossRef]
49. Scolan, E.; Sanchez, C. Synthesis and Characterization of Surface-Protected Nanocrystalline Titania Particles. *Chem. Mater.* **1998**, *10*, 3217–3223. [CrossRef]

Article

Innovative Ag–TiO₂ Nanofibers with Excellent Photocatalytic and Antibacterial Actions

Petronela Pascariu [1,*], Corneliu Cojocaru [1], Anton Airinei [1], Niculae Olaru [1], Irina Rosca [1], Emmanouel Koudoumas [2] and Mirela Petruta Suchea [2,3,*]

1. "Petru Poni" Institute of Macromolecular Chemistry, 41A Grigore Ghica Voda Alley, 700487 Iasi, Romania; ccojoc@gmail.com (C.C.); airineia@icmpp.ro (A.A.); nolaru@icmpp.ro (N.O.); rosca.irina@icmpp.ro (I.R.)
2. Center of Materials Technology and Photonics, School of Engineering, Hellenic Mediterranean University, 71410 Heraklion, Greece; koudoumas@hmu.gr
3. National Institute for Research and Development in Microtechnologies (IMT-Bucharest), 023573 Bucharest, Romania
* Correspondence: dorneanu.petronela@icmpp.ro or pascariu_petronela@yahoo.com (P.P.); mira.suchea@imt.ro or mirasuchea@hmu.gr (M.P.S.)

Abstract: Ag–TiO₂ nanostructures were prepared by electrospinning, followed by calcination at 400 °C, and their photocatalytic and antibacterial actions were studied. Morphological characterization revealed the presence of one-dimensional uniform Ag–TiO₂ nanostructured nanofibers, with a diameter from 65 to 100 nm, depending on the Ag loading, composed of small crystals interconnected with each other. Structural characterization indicated that Ag was successfully integrated as small nanocrystals without affecting much of the TiO₂ crystal lattice. Moreover, the presence of nano Ag was found to contribute to reducing the band gap energy, which enables the activation by the absorption of visible light, while, at the same time, it delays the electron–hole recombination. Tests of their photocatalytic activity in methylene blue, amaranth, Congo red and orange II degradation revealed an increase by more than 20% in color removal efficiency at an almost double rate for the case of 0.1% Ag–TiO₂ nanofibers with respect to pure TiO₂. Moreover, the minimum inhibitory concentration was found as low as 2.5 mg/mL for *E. coli* and 5 mg/mL against *S. aureus* for the 5% Ag–TiO₂ nanofibers. In general, the Ag–TiO₂ nanostructured nanofibers were found to exhibit excellent structure and physical properties and to be suitable for efficient photocatalytic and antibacterial uses. Therefore, these can be suitable for further integration in various important applications.

Keywords: nano-Ag–TiO₂ nanostructured nanofibers; electrospinning; UV-visible light assisted photocatalytic activity; photodegradation kinetics; antimicrobial activity

1. Introduction

Titanium dioxide (TiO₂) is considered a suitable compound for decomposing wastes and antimicrobial action due to its photocatalytic nature and because it is a chemically stable, non-toxic, inexpensive, and quite safe substance. In particular, various types of photocatalysts have been developed using TiO₂ for wastewater treatment employing various methods including immobilized catalyst systems, membrane separation, and gravitational separation systems. As the TiO₂ semiconductor band gap energy (3.2 eV) limits its activity when used in visible light, in many of these studies, an important issue to be considered is the activation of the photocatalysts with solar radiation, as this can improve the energy efficiency and consequently the economic viability of the process. In that respect, TiO₂ is doped with various elements to enhance visible light activity. As a result, numerous photocatalysts have been developed, such as composite, co-doped, and co-catalysts compounds, their photocatalytic performance depending strongly on their morphological, structural, and textural properties [1–8].

There are many studies regarding the antifungal and antibacterial properties of TiO₂ nanoparticles in various forms against a broad range of both Gram-positive and Gram-negative bacteria, properties needed in sectors such as food, textiles, medicine, water disinfection, and food packaging [9,10]. In general, the antimicrobial activity of nanostructured TiO_2 is greatly dependent on the photocatalytic performance of TiO_2, which depends strongly on its morphological, structural, and textural properties, as was mentioned before [1–4].

Studies have shown that the crystalline structure and morphology of TiO_2 nanoparticles, are influenced by growth process parameters such as temperatures, starting concentration of precursors, pH, etc. The potential health impact and toxicity to the environment of nanomaterials is also currently an important matter to be addressed. As an example, metal oxide nanoparticles (NPs) conventionally synthesized using chemical methods, have shown different levels of toxicity to test organisms [11–13]. Their toxicity seems to be mainly related to the small size that permits easy penetration through cellular membranes and its light dependent properties. One way to avoid nanoparticulate free circulation is to immobilize them onto substrates or larger structures [14].

Regarding the improvement of the electron–hole pair generation and the enlargement of the spectral absorption domain of TiO_2, many studies have been performed related to the inclusion of metal/non-metal ions in the structure of TiO_2, the dye functionalization on the TiO_2 surface in dye-sensitized solar cells (DSSC), and the growth of noble metals onto the TiO_2 surface [15]. Silver (Ag) is among the most interesting metals used as a dopant to modify the structure of TiO_2 because it has the particular property to prevent the recombination of electron–hole pairs. Moreover, Ag can generate surface plasmon resonance with TiO_2 under visible light. These changes provided by Ag doping lead to a significant improvement of the photocatalytic activity, a fact confirmed by other authors [16–18]. On the other hand, Ag nanoparticles possess a broad spectrum of antibacterial, antifungal, and antiviral properties. Ag nanoparticles have the ability to penetrate bacterial cell walls, changing the structure of cell membranes and even resulting in cell death. Their efficacy is due not only to their nanoscale size but also to their large ratio of surface area to volume. They can increase the permeability of cell membranes, produce reactive oxygen species, and interrupt the replication of deoxyribonucleic acid by releasing silver ions.

There are studies showing that the inclusion of Ag in the structure of TiO_2 leads to enhance photocatalytic efficiency, as well as antimicrobial properties. Over the years, a large volume of reported research was focused on obtaining 0D systems (nanoparticles) based on Ag doped TiO_2, which were tested in terms of photocatalytic and antimicrobial performances [13,19–22]. As an example, studies showed that TiO_2-NPs had efficient antimicrobial activity against *E. coli*, *S. aureus*, methicillin-resistant *S. aureus*, *K. pneumoniae* [23,24]. However, little attention was paid to the development of Ag doped TiO_2 nanofibers by the electrospinning-calcination technique, as well as studies of their performances in photocatalytic dye degradation and antimicrobial action. As an example, Zhang et al. [15] prepared hierarchical structures composed of TiO_2 fibers on which Ag nanoparticles were grown to improve the photocatalytic efficiency for Rhodamine B (RhB) dye degradation. Moreover, nano-Ag-decorated TiO_2-nanofibres proved that the inclusion of Ag exhibited an increased antimicrobial effect on *S. aureus* and *E. coli* [25]. Recently, Roongraung et al. [18] reported the photocatalytic performance of Ag doped TiO_2 nanofibers for photocatalytic glucose conversion.

Although the research on TiO_2 has a very long history and its applications are almost countless as the respective publications are too, this semiconductor has the potential to offer even today very interesting results worth being further investigated. This paper reports the development and optimization of pure TiO_2 and Ag–TiO_2 photocatalytic nanostructured nanofibers, fabricated by electrospinning followed by calcination at high temperature. Nanofibers can be well stacked together in larger 3D structures, thus, they can better be immobilized in membranes or other kinds of support, offering at the same time a much larger active surface than that for the case of nano-powders, while they can easily be recovered from the active medium used in applications. At the same time, the presence of Ag can lead to both an effective photocatalytic activity and antimicrobial action and a sunlight-driven activation, as compared to pure TiO_2. Therefore, the integration of such materials into food packaging, medical textiles, and other healthcare related items can be a quite innovative approach for real-life applications. In the present study, we report the fabrication of materials with remarkable photocatalytic efficiency of 99% at a constant rate of k = 1.29 × 10^{-2} min^{-1} that was found for the 0.1% Ag–TiO_2 sample for the case of Methylene Blue dye degradation. Additionally, these materials proved able to degrade up 99% of other organic dyes with constant rates in the range of 4.57 × 10^{-3} to 7.28 × 10^{-2} min^{-1} depending on the dye nature, all degradation tests being performed under UV-visible light irradiation from a halogen lamp, without intervening to acidify the solution or add H_2O_2. These materials proved to be also reusable.

2. Results

2.1. X-rays Diffraction (XRD) Analysis

Typical XRD diffractograms of the reference pure TiO_2 and the Ag–TiO_2 nanostructured nanofibers with different Ag content, analyzed in the range of 2θ = 5–80° are shown in Figure 1.

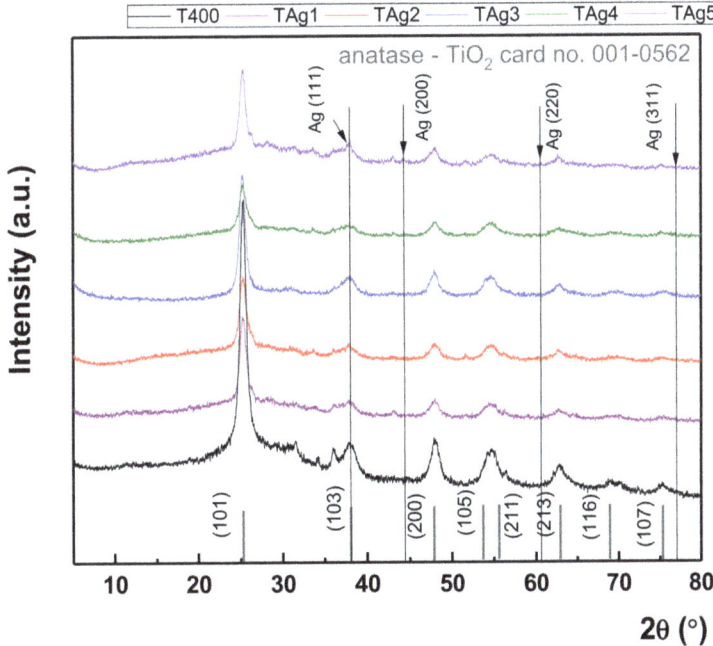

Figure 1. X-ray diffraction patterns of pure TiO_2 and Ag–TiO_2 nanostructured nanofibers materials with different Ag content.

The grazing incidence X-ray diffraction patterns present multiple diffraction peaks, assigned unambiguously as anatase phase with lattice parameters: a = b = 0.38 nm; c = 0.94 nm according to ICDD (International Center for Diffraction Data) database card No. 001-0562. The diffraction peaks located at 25.30°, 37.96°, 47.94°, 54.41°, 54.93°, 62.77°, 69.24°, and 75.17°, correspond to the (101), (103), (200), (105), (211), (213), (116) and (107) Miller indices. An investigation of the position of the diffraction peaks shows that the lattice constant of anatase is preserved for different Ag content (the respective values for the (101) diffraction peak are shown in Table 1). In order to identify the diffraction peaks, in Figure 1 the positions of Ag diffraction peaks were added according to card no. 001-1164. It is clear that numerous very small diffraction peaks remain unassigned. It exists a possibility to be an unknown organic compound contaminant, but usually, the organic compounds give diffraction peaks below 30°. Additionally, it exists the possibility that some traces of unknown titanates compounds to appear (brookite was unsuccessfully checked). The second possibility is more plausible, as the Energy Dispersive X-ray Analysis (EDX) investigations did not show the presence of other elements. However, the very small peaks cannot be assigned to any expected material.

Table 1. Structural parameters of pure TiO_2, and Ag–TiO_2 nanostructured nanofibers.

Sample	2θ (°)	d_{hkl} (Å)	a (Å)	c (Å)	D (nm) Scherrer	D (nm) Rietveld	ε (%)	E_g (eV)
T400	25.28	3.52	3.790	9.459	7.41	7.4	0.13	3.01
TAg1	25.13	3.54	3.803	9.484	7.32	7.3	0.52	2.70
TAg2	25.25	3.52	3.805	9.486	6.41	6.4	0.57	2.69
TAg3	25.22	3.53	3.799	9.481	7.54	7.5	0.37	2.69
TAg4	25.19	3.53	3.796	9.464	6.33	7.2	0.65	2.70
TAg5	25.24	3.52	3.800	9.477	8.90	6.3	0.53	2.68

Spacing distance between crystallographic planes (d_{hkl}), the lattice parameters a and c, the crystallite size (D), microstrain (ε) and band gap energy (E_g).

As the lattice constant remains unmodified, these findings suggest that interstitial doping of Ag did not take place. This observation is strengthened by the observation that the size of the crystalline domains remains practically unchanged at different Ag content. The mean crystallite size was evaluated with two independent methods, Scherrer's equation and Rietveld refinement [26,27]. The respective values are presented in Table 1. In addition, Rietveld refinement allowed the evaluation of the lattice strain, which was increasing from 0.13% (pure TiO_2 sample) to 0.52% (TAg1), reaching 0.53% at the highest Ag content (Table 1). Even though Ag was detected by EDX (as will be shown later), there is no diffraction peak of Ag observed in the XRD patterns, a fact that can be attributed to its existence as small crystals having a size below the resolution limit of the technique, observation that agrees with previous literature data [28,29]. In any case, the XRD findings indicate that the Ag content variation did not affect the unit cell of TiO_2 or the crystal quality.

2.2. Morphological Characterization

The morphological characterization of the nanostructured nanofibers was carried out by Scanning Electron Microscopy (SEM) analysis. Representative samples of Ag–TiO_2 nanostructured nanofibers, after calcination at 400 °C, were investigated and their images are presented in Figure 2.

As can observe in Figure 2, one-dimensional uniform Ag–TiO_2 nanostructured nanofibers were obtained. The average value of the fiber diameters was evaluated from the measurement of 100 diameters for each sample using Image-J software. Histogram of fiber diameter distribution plotted from SEM images analysis of pure TiO_2 (T400) and Ag doped TiO_2 (TAg1 and TAg5) are given in Figure S1 of the Supplementary material. The average width of the fibers was found to vary between 65 and 100 nm, depending on the Ag loading, the mean fiber diameter decreasing as the Ag concentration increases. It was also observed that Ag doping leads to a slight variation of the dimensional distribution of nanofiber widths. As the Ag concentration increases, less thinner fibers are present. Moreover, it is interesting to indicate the particular nanostructuring of the fibers, that is, each fiber is composed of small crystals interconnected with each other. This observation was confirmed by Transmission Electron Microscopy (TEM). The elemental composition of the obtained nanofibers was studied using the energy dispersive X-ray (EDX) technique. Typical EDX spectra of Ag–TiO_2 nanostructured nanofibers (TAg1, TAg3 and TAg5) are presented in Figure 3.

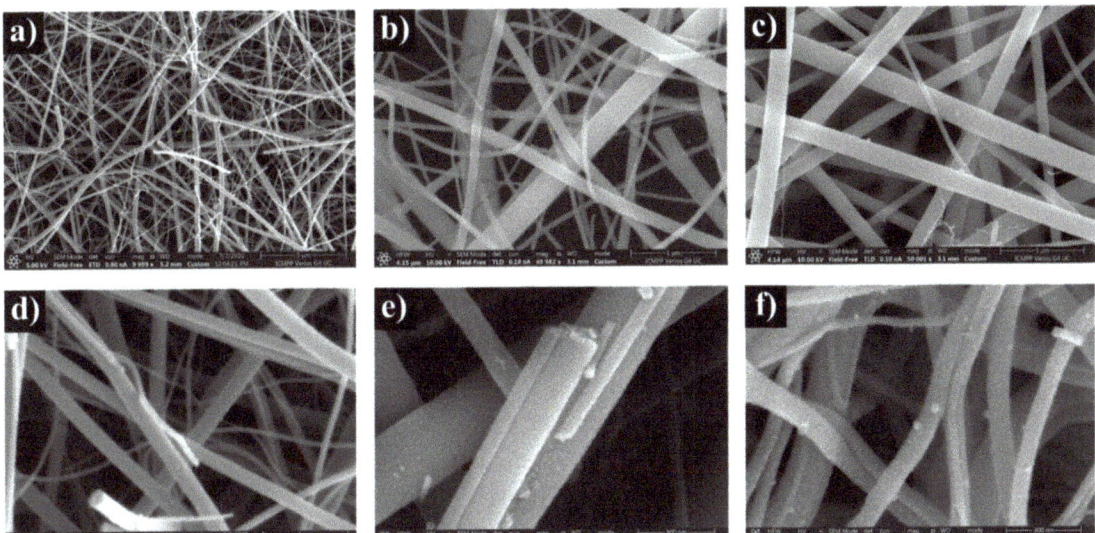

Figure 2. Small and larger magnification SEM images of Ag–TiO$_2$ nanostructured nanofibers: pure TiO$_2$ × 10,000 (**a**), 0.1% Ag–TiO$_2$ × 50,000 (**b**), 5.0% Ag–TiO$_2$ × 50,000 (**c**), TiO$_2$ × 50,000 (**d**), 0.1% Ag–TiO$_2$ × 150,000 (**e**), and 5.0% Ag–TiO$_2$ × 150,000 (**f**).

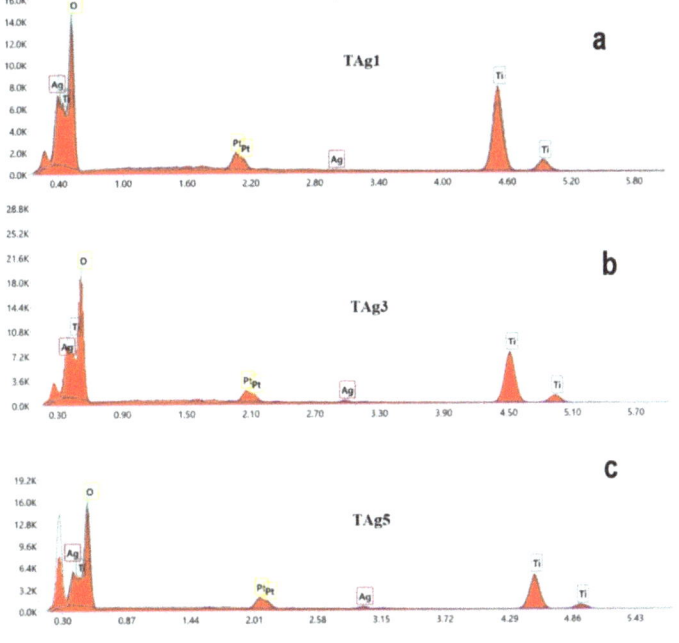

Figure 3. EDX spectra of 0.1% Ag–TiO$_2$ (**a**), 1.0% Ag–TiO$_2$ (**b**) and 5.0% Ag–TiO$_2$ (**c**).

The EDX characterization confirmed the presence of Ti, O, Ag, and Pt elements. No other impurity elements were identified. The appearance of Pt in the EDS spectra is due to the sample's preparation by Pt metallization for SEM/EDS measurements. The atomic percentages of Ag in the analyzed samples were found to be 0.05%, 0.21%, 0.40%, 0.51%

and 0.6% respectively, values depending on the amount of precursor solutions. Some examples of EDX spectra are presented in Figure 3.

TEM characterization of pure and Ag doped TiO_2 nanofibers confirmed their polycrystalline nature and nanostructuring. It can be noticed that the crystallites forming the doped materials are slightly smaller than in the case of pure TiO_2 and that their order of magnitude is <10 nm as observed by XRD characterization too. Some examples are presented in the TEM images shown in Figure 4.

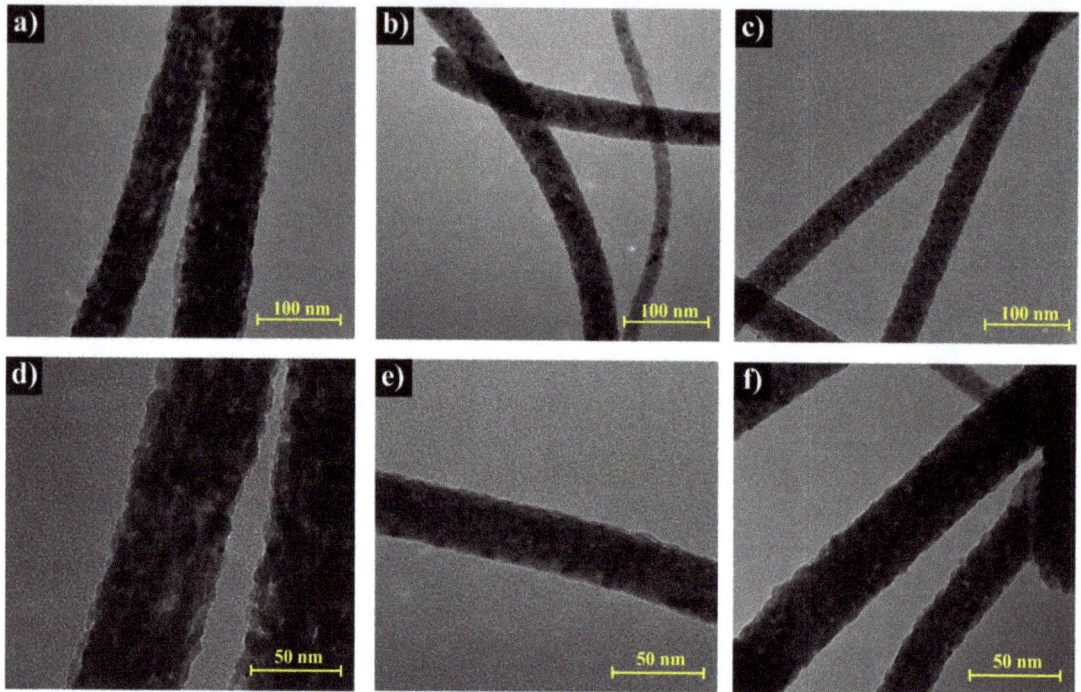

Figure 4. Smaller (scale 100 nm) and larger magnification (scale 50 nm) TEM images of Ag–TiO_2 nanostructured nanofibers: pure TiO_2 (**a**), 0.1% Ag–TiO_2 (**b**), 5.0% Ag–TiO_2 (**c**), TiO_2 (**d**), 0.1% Ag–TiO_2 (**e**), and 5.0% Ag–TiO_2 (**f**).

The specific surface area characterized by BET as a function of Ag concentration is illustrated in Table 2, while the weight change/relative humidity graphs are included in Supplementary material (Figure S2). It was found that as the Ag concentration increased, the specific surface area of the TiO_2 nanofibers slightly increased for low Ag concentration but it dropped for the 5% Ag TiO_2. This might be due to the lower diameters size of the fibers and the higher degree of crystallites agglomeration.

Table 2. BET data of pure TiO_2 and Ag doped TiO_2 nanostructured materials.

Sample	Sorption Capacity (%)	Average Pore Size(nm)	BET Data	
			Area (m^2g^{-1})	Monolayer (g/g^{-1})
T400	6.43	2.38	54.25	0.0154
TAg1	7.99	2.86	56.07	0.0159
TAg3	13.17	3.88	68.01	0.1937
TAg5	7.05	2.94	48.09	0.0136

2.3. Raman Spectroscopy Measurements

Several authors [22,30–38] have shown in their research that the anatase phase has a tetragonal structure consisting of six Raman active modes, including A_{1g} located at 516 cm^{-1}, $2B_{1g}$ corresponding to 397 and 516 cm^{-1} and $3E_g$ from 144, 198 and 638 cm^{-1}, where the A_{1g} mode overlapped with the B_{1g} mode at 516 cm^{-1} [22,32,39].

In Figure 5, which displays the Raman spectra of Ag–TiO$_2$ nanostructured nanofibers with different Ag content, one can be seen that the Raman bands observed correspond to the anatase phase of pure TiO$_2$, even for those samples containing Ag. In particular, the E_g modes at 144 and 638 cm^{-1} correspond to the symmetric stretching vibration of O–Ti–O bond, whereas the B_{1g} and A_{1g} Raman modes at 397 and 516 cm^{-1} can be assigned to the symmetric and anti-symmetric bending vibration of O–Ti–O [40,41]. In any case, the peaks for the Ag containing samples were found to present small shifts (about 3 nm) to longer wavelengths. Additional peaks corresponding to Ag were not detected in any of the Ag containing samples.

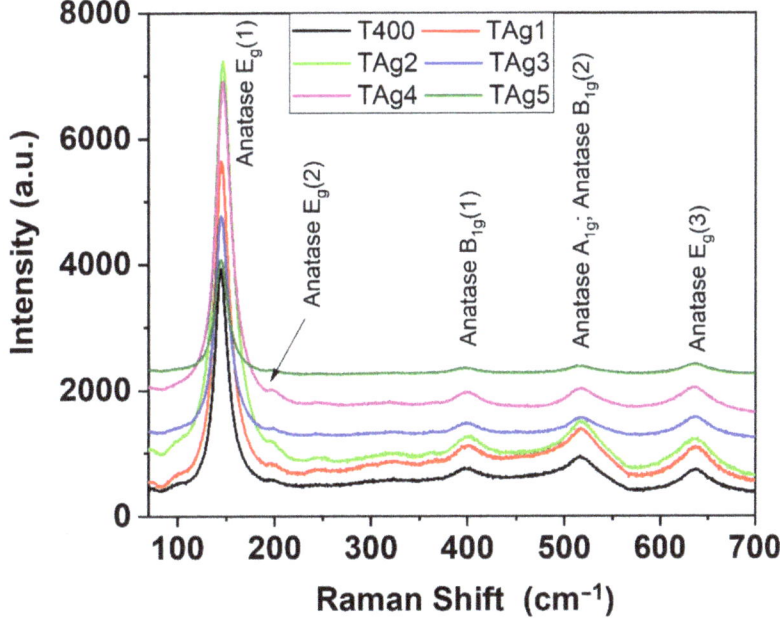

Figure 5. Raman spectra of pure TiO$_2$ and Ag–TiO$_2$ nanostructured nanofibers materials.

It can be noted that the results of Raman characterization are in good agreement with the XRD observations. All Ag-containing materials have a similar structure of the TiO$_2$ anatase phase, suggesting that the Ag was successfully integrated as small nanocrystals without affecting the TiO$_2$ crystal lattice as intended. The slight red-shift of the peaks in the Raman spectra may be due to the slight distortion caused by the incorporation of metal, particles of various sizes, nature of defects, and so on [17–19,32].

2.4. Optical Properties and Band Gap Energy

It is known that the band gap energy is the most important optical parameter that significantly influences photocatalytic activity. Diffuse reflectance spectra recorded for pure TiO$_2$ and Ag–TiO$_2$ nanostructured nanofibers are illustrated in Figure 6, where a slight shift of about 10 nm to blue can be noticed for the Ag containing samples. The reflectance data were processed according to the method indicated in reference [39] for indirect bandgap semiconductors and the corresponding values are given in Table 1. The Eg values for the

Ag/TiO$_2$ nanostructures are much lower than those corresponding to pure TiO$_2$ due to the Ag doping process. As can be observed, the presence of nano-Ag leads to decreased values of around 2.70 eV for the optical band gap, as compared to the 3.01 eV gap of pure TiO$_2$. This means that photons with lower energy can generate electron–hole pairs and the photocatalytic activity of such materials can be activated even under visible light irradiation. Many studies [13,40] have shown that this decrease of the band gap may be due to the occurrence of new energy levels in the band gap range of the composite materials.

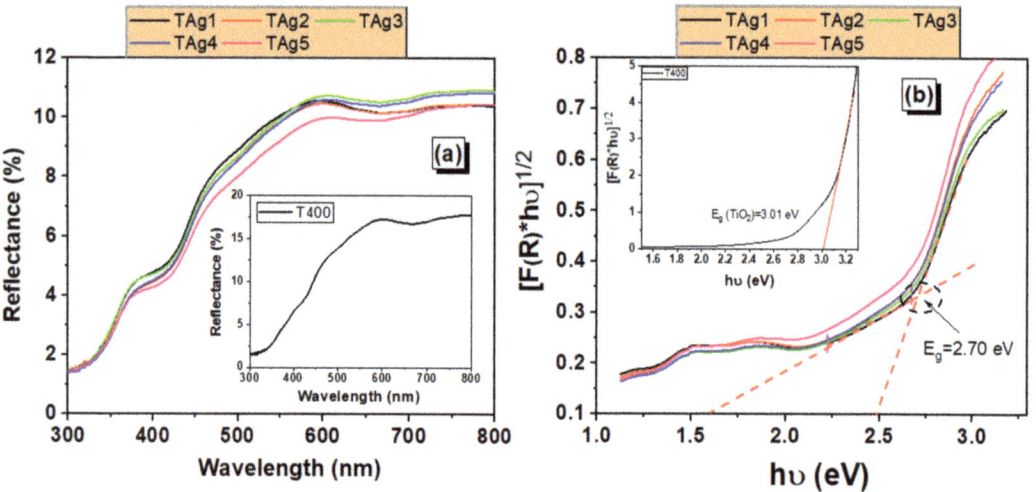

Figure 6. Optical properties: (**a**) reflectance spectra and (**b**) Tauc plots of Ag–TiO$_2$ nanostructured nanofibers materials.

2.5. Photoluminescence Analysis

In the context of studies of a photocatalytic material, it is of great importance to gather information on the active surface sites of the catalyst and on how they affect the dynamics of adsorption and photoactivated transformations of the targeted species. In this regard, studies of photoluminescence (PL) properties of the material are very well suited and useful. PL phenomena in semiconductors are driven by diffusion and recombination of photogenerated charges, which typically occurs in a thin region beneath the semiconductor surface (typical widths of few tenths of nm if the excitation is provided at photon energy larger than the bandgap), making it very sensitive to small local variations. To observe how the Ag doping affects the carrier recombination and diffusion phenomena in TiO$_2$, PL characterization using different excitation wavelengths was performed to see the excitation states involved in the emission and to observe the occurrence of sub-bandgaps. Figure 7 shows the PL spectra for the studied materials, excited at different wavelengths (λ_{ex} = 280, 300, 320 and 340 nm).

TiO$_2$ has an indirect band-edge configuration and hence its PL emission occurs at wavelengths longer than the bandgap wavelength: that is, the PL of TiO$_2$ is not caused by band-to-band transitions but involves localized states. [42] The fluorescence spectra of TiO$_2$ nanostructures normally display three bands, assigned to self-trapped excitons, oxygen vacancies and surface defects [18,24,33,35–37]. In particular, these emission bands are located in the violet, the blue (460 nm) and the blue-green (485 nm) regions respectively, which can be attributed to self-trapped excitons localized on TiO$_6$ octahedral (422 nm) [36,37], and to oxygen related defect sites or surface defects (460 and 485 nm) [38]. Moreover, the band edge emission around 364 nm corresponds to free exciton recombination in TiO$_2$ materials [35,36]. As can be seen, all materials present the same emission bands, but with slightly different intensities. In particular, the PL intensity of the Ag–TiO$_2$ nanostructured nanofibers was found lower as compared to that of pure TiO$_2$. As is known, the emission intensity is related to the recombination of electron–hole pairs in the structure of TiO_2 [13]. In addition, the low intensity in the fluorescence spectra suggests that the photoexcited electron–hole pairs can be achieved at a longer time, which is beneficial in the processes of photocatalytic degradation. Thus, these findings suggest that the presence of nano Ag has a distinct effect on limiting the electron–hole recombination, as the photoexcited electron may be captured by the Ag nanoparticles that behave as an electron storage source on the TiO_2 surface [13]. Nano Ag presence also contributed significantly to reducing the band gap energy and facilitating the activation by the absorption of light in the visible region, along with delaying the electron–hole recombination. Therefore, the presence of nano-Ag offers several advantages in the functionality of the Ag–TiO_2 nanostructured nanofibers. Additionally, it is expected that the best photocatalytic activity under the visible irradiation would be performed for an optimal nano Ag concentration level in TiO_2.

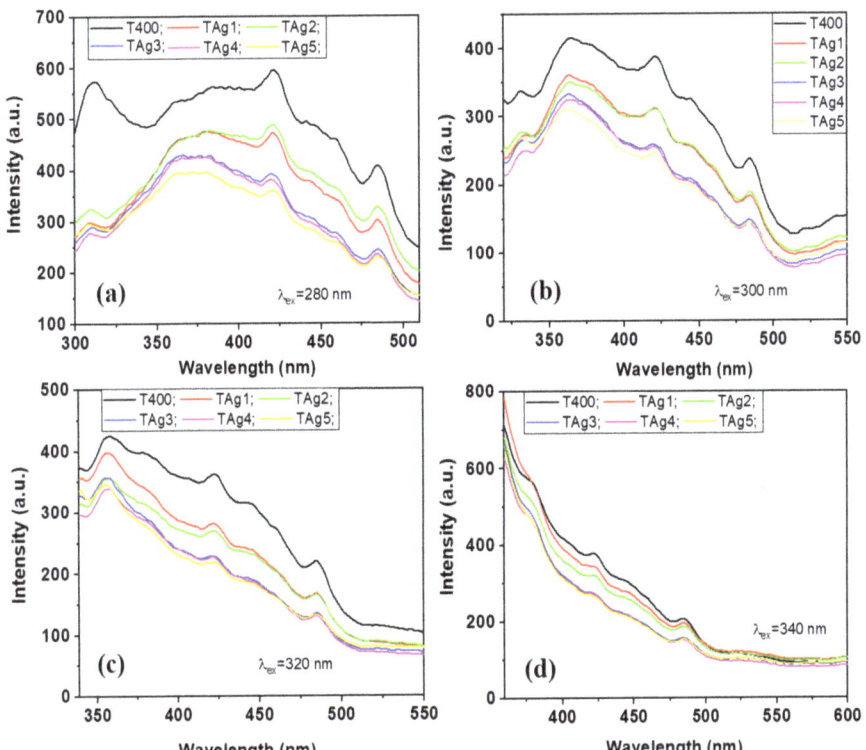

Figure 7. Emission spectra of pure TiO_2 and Ag–TiO_2 nanostructured nanofibers at different excitation wavelengths λ_{ex} = 280 nm (**a**), 300 nm (**b**), 320 nm (**c**) and 340 nm (**d**).

2.6. Photocatalytic Properties

2.6.1. Methylene Blue Dye Degradation

Methylene blue (MB) (C_0 = 10 mg/L) was used to evaluate the photocatalytic activity of the grown materials. The dye degradation was performed under a halogen lamp light irradiation (400 W) and the amount of photocatalyst was maintained at 0.4 g/L for all samples. Typical UV-VIS absorption spectra recorded for MB dye solution degradation up 300 min under halogen lamp light irradiation in presence of pristine TiO_2 and 0.1% Ag–TiO_2 nanostructured nanofibers are shown in Figure 8. It can be observed that the intensity of the absorption band corresponding to a wavelength at 665 nm decreases with the increase of the irradiation time. In addition, all Ag–TiO_2 nanostructured nanofibers show a faster decreasing tendency of colorant concentration as compared to pure TiO_2. Regarding the color removal efficiency, this is shown in Figure 8c. The maximum degradation efficiency was found for the TAg1 sample, having a value of 97.05%. The kinetics of the photodegradation process under visible light irradiation was also evaluated.

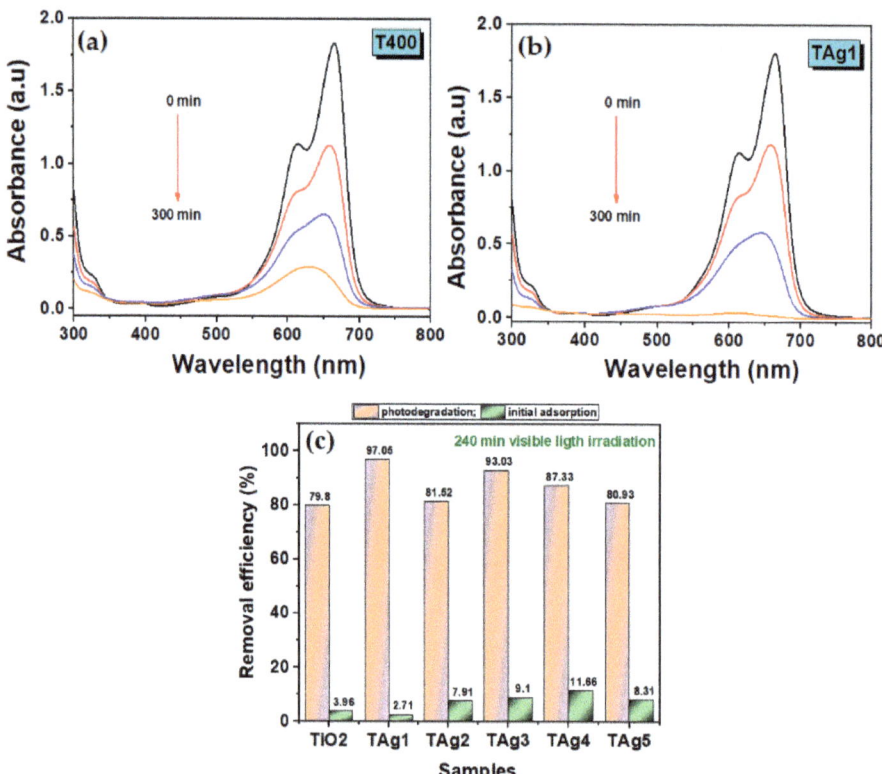

Figure 8. UV-VIS absorption spectra for the degradation of MB dye (10 mg/L) at various irradiation times in the presence of pure TiO_2 (**a**), 0.1% Ag–TiO_2 nanostructured nanofibers (**b**), and (**c**) color removal efficiency obtained for all materials after the end of the photodegradation.

2.6.2. Kinetics of the Photodegradation Process

Kinetics plots of the photodegradation of MB in aqueous solutions under the halogen lamp irradiation in the presence of Ag–TiO_2 nanostructured nanofibers are presented in Figure 9. The data were interpolated to the pseudo-first-order (PFO) kinetic model by using the nonlinear regression technique. The goodness-of-fit was estimated by chi-square statistic test (χ^2-value). Thus, the decay of MB dye concentration versus time was fitted to PFO equation, which can be expressed as:

$$C_t = C_0 \cdot e^{-kt} \quad (1)$$

where C_0 is the initial MB dye concentration (~10 mg/L), k is the pseudo-first-order reaction rate constant (min^{-1}), and t is the irradiation time (min). The calculated parameters of the PFO model are listed in Table 3.

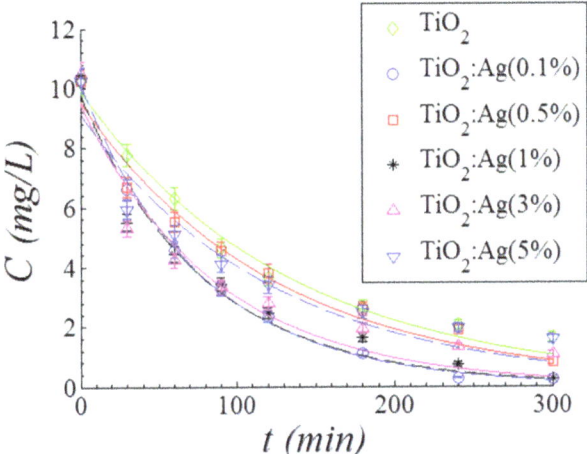

Figure 9. Kinetics plots of MB dye decay against irradiation time during the photodegradation process under halogen lamp in the presence of Ag–TiO$_2$ nanostructured nanofibers catalysts. Solid and dash lines represent predictions given by PFO kinetic model. Experimental conditions: catalyst dosage = 0.4 g/L; $T = 23 \pm 2$ °C; pH 7.0 ± 0.2.

Table 3. Kinetic parameters for MB dye photodegradation under visible light in the presence of Ag–TiO$_2$ nanostructured nanofibers catalysts.

Code	k (min^{-1})	χ^2-Test Value
T400	7.47×10^{-3}	0.52
TAg1	1.90×10^{-2}	0.10
TAg2	7.99×10^{-3}	0.39
TAg3	1.26×10^{-2}	0.81
TAg4	1.14×10^{-2}	3.65
TAg5	8.13×10^{-3}	1.59

As observed from Table 3, the presence of nano Ag in TiO$_2$ nanofibers increases the rate constant (k). This effect was found more important for a 0.1% content of Ag in the electrospun solution. Thus, the optimal formulation of the developed materials when acting as photocatalyst seems to be the 0.1% Ag–TiO$_2$ (TAg1 sample), as this presents the maximum value of both the constant rate (k = 1.29×10^{-2} min^{-1}) and the degradation efficiency of MB dye (97.05%). Furthermore, the kinetics for the degradation of methylene blue (MB), Congo red (CR), amaranth and orange II dyes under TAg1 photocatalyst are presented below, in which the dye concentration was maintained at 10 mg/L (Figure 10). The photolysis test (without catalysts) for all four dyes was performed under fluorescent bulb light irradiation for 300 min of irradiation, and the corresponding spectra are presented in Figure S3, Supplementary material.

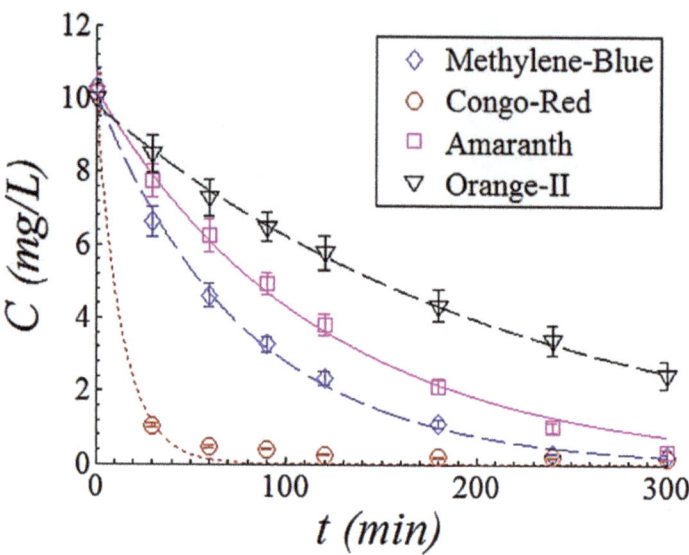

Figure 10. Comparative study showing the kinetics curves for degradation of various dyes under visible light in the presence of sample TAg1. Solid and dash lines represent predictions given by PFO kinetic model. The experimental conditions are: catalyst dosage = 0.4 g/L, T = 23 ± 2 °C, pH 7.0 ± 0.2.

The UV-visible absorption spectra (Figure S4, Supplementary material) have been recorded for the degradation of MB, CR, amaranth, and orange II dye solutions in the following experimental conditions: initial dye concentration=10 mg/L for all dyes, catalyst dosage = 0.4 g/L, irradiation time = 300 min, T = 23 ± 2 °C and pH 7.0 ± 0.2. From Figure S3, it can be noted that color removal efficiency varies between 75% and 98%, depending on the type of dye. The highest degradation efficiency (99%) was found for Congo red dye, the efficiency being achieved in a shorter time (30 min) as compared to other dyes.

A comparative study is reported in Figure 10 showing the photodegradation kinetics of different dyes under visible light using TAg1 as a catalyst. Experimental data were interpolated to PFO-kinetic model and the calculated parameters are summarized in Table 4.

Table 4. Kinetic parameters for photodegradation of different dyes under the fluorescent bulb light in the presence of 0.1% Ag–TiO$_2$ nanostructured nanofibers photocatalyst (TAg1).

Dye Subjected to Degradation.	k (min^{-1})	χ^2-Test Value
Methylene Blue	1.29×10^{-2}	1.04×10^{-1}
Congo Red	7.28×10^{-2}	8.80×10^{6}
Amaranth	8.63×10^{-3}	3.01×10^{-1}
Orange II	4.57×10^{-3}	1.58×10^{-2}

According to Table 4, the highest rate constant (7.28×10^{-2} min^{-1}) was observed for Congo red dye photodegradation, and the lowest one (4.57×10^{-3} min^{-1}) for the orange-II dye, respectively. Comparing the present results with others reported on appropriate photocatalysts (Ag doped TiO$_2$ nanostructures) [16–22,43], one can observe the excellent performance of our samples, as these are capable to degrade up 99% of dyes depending on the dye nature, with constant rates between 4.57×10^{-3} and 7.28×10^{-2} min^{-1}. In addition, all the degradation tests on the fabricated samples were done using a moderate amount of catalyst 0.4 g/L, fluorescent bulb light irradiation (400 W), and temperature (23 ± 2 °C), pH (7.0 ± 0.2) and without intervening to acidify the solution or add H_2O_2. For comparative purposes, Harikishore et al. reported a rate constant value of 8.16×10^{-3} min^{-1} for methylene blue (C_0 = 5 ppm) dye degradation in the presence of 1.5g/L amount of catalyst under UV lamp irradiation [43]. Recently, Ali et al. [22] obtained Ag-doped TiO_2 nanoparticles with improved photocatalytic and antibacterial properties. They found a degradation efficiency of 96% for MB degradation (C_0 = 10 mg/L) in the presence of a 1g/L catalyst under visible light irradiation (500 W).

2.6.3. Photocatalyst Reuse

The degradation efficiency after several photodegradation cycles is an important parameter to evaluate the stability of the photocatalyst. Figure 11c) shows a degradation efficiency registered after 5 cycles of reuse under visible light irradiation for one hour. It was noticed that the photocatalytic efficiency was maintained at about 98% even after five cycles for Congo red (C_0 = 10 mg/L) dye degradation. In addition, the stability of the reused photocatalyst was confirmed by XRD and SEM characterization. XRD pattern and an SEM image for recycled photocatalyst are presented in Figure 11a,b. It can be observed that both, crystallinity and morphology are preserved after its use.

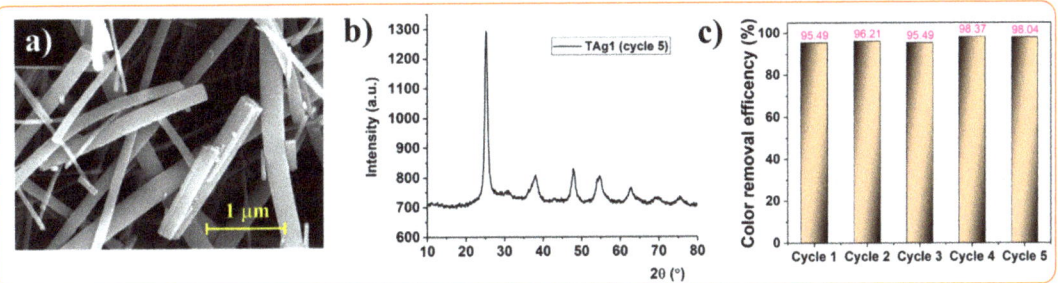

Figure 11. SEM image and XRD pattern of TAg1 reused material after five cycles (**a**,**b**); color removal efficiency after 60 min in the presence of both materials (TAg1 reused and pristine) (**c**).

2.7. Antimicrobial Activity

All samples were also tested for their antibacterial activity against *S. aureus* and *E. coli*. Samples T400 and TAg1 did not show any antibacterial action. TAg3 and TAg5 proved to be quite effective against the Gram-positive bacterial strain represented by *S. aureus*, while TAg2 and TAg5 against the Gram-negative bacterial strain represented by *E. coli*. The highest efficiency observed, regardless of the bacterial strain, was found in sample TAg5 that is at the highest Ag concentration (5%). The nanostructures were in general more efficient against *E. coli* (19 mm of inhibition zone) than against *S. aureus* (16 mm of inhibition zone). An image of a culture plate is presented in Figure 12.

From this image, the diameters of the inhibition zones of each tested sample can be calculated and are presented in Table 5.

Table 5. Antimicrobial activity of the tested compounds against the reference strains.

Strain	Inhibition Zwone (mm)					
	T400	TAg1	TAg2	TAg3	TAg4	TAg5
S. aureus	-	-	-	13.20 ± 0.99	12.19 ± 0.27	16.58 ± 0.18
E. coli	-	-	14.74 ± 1.71	14.05 ± 0.04	15.01 ± 0.08	19.02 ± 0.69

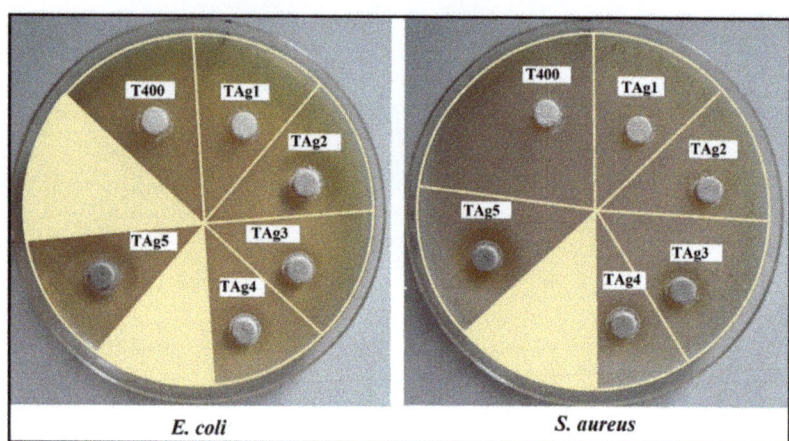

Figure 12. Image of a culture plate for evaluation of antimicrobial activity against *S. aureus* and *E. coli*.

The MIC of the samples against *E. coli* was as low as 2.5 mg/mL for sample TAg5, and 5 mg/mL for the same sample against *S. aureus* (see Table 6).

Table 6. Antimicrobial activities of the samples against *S. aureus* and *E. coli*. MIC: minimum inhibitory concentration.

Samples	Minimum Inhibitory Concentration (MIC) (mg/mL)	
	S. aureus	E. coli
T400	Not tested	Not tested
TAg1	Not tested	10
TAg2	Not tested	10
TAg3	10	10
TAg4	10	5
TAg5	5	2.5

The final trial was focused on the determination of the minimum bactericidal concentration (MBC), which is recognized as the lowest concentration that destroyed all bacterial cells. This was investigated by spreading 10 µL samples from the wells on MH agar plates. However, MBC against both bacterial strains could not be estimated, as a higher concentration of Ag within the samples was probably required. The samples had better antibacterial activity against the Gram-negative strain represented by *E. coli* probably due to its cell wall composition, distinctive rod shape and extracellular matrix [44]. Literature states that when *E. coli* is exposed to light activated Ag–TiO$_2$ the oxidative damage to the bacterial cell envelope may occur which plays a significant role in biocidal activity [45]. In this particular case, the antibacterial activity is directly correlated with the silver concentration within the samples [46].

3. Materials and Methods

3.1. Materials

Titanium (IV) isopropoxide (TTIP), silver nitrate (AgNO$_3$), glacial acetic acid (CH$_3$COOH), ethanol, polyvinylpyrrolidone (PVP) (Mw = 1.300.000) of analytical grade purchased from Sigma-Aldrich (Sigma-Aldrich/Merck KGaA, Darmstadt, Germany) were used for the preparation of the Ag-doped TiO$_2$ nanostructures. Methylene blue, Congo red, orange II, and amaranth dyes were procured from Sigma-Aldrich (Sigma-Aldrich/Merck KGaA, Darmstadt, Germany) and utilized without further purification.

3.2. Preparation of Ag–TiO$_2$ Nanostructured Nanofibers

Pure TiO$_2$ and Ag–TiO$_2$ nanostructured nanofibers were developed using electrospun solutions, which were obtained by mixing two initial solutions, solution 1 containing 0.75 mL of TTIP in 1.5 mL acetic acid and solution 2 consisting of 0.25 g PVP in 2.5 mL ethanol. This solution was used to prepare the pure TiO$_2$ blank sample (named T400), while in the case of Ag–TiO$_2$ nanostructures, these were prepared by adding to solution 2 different amounts of AgNO$_3$ as follows: 0.5 mg corresponding to a 0.1% mass percentage (sample TAg1), 2.5 mg for 0.5% (sample TAg2), 5 mg for 1.0% (sample TAg3), 15 mg for 3.0% (sample TAg4) and 25 mg for 5.0% (sample TAg5). The electrospinning was performed using the set-up already described in our previous publications [47,48]. The electrospinning parameters used to obtain the fibrous materials were: 25 kV high voltage, 15 cm tip to collector distance, and 0.75 mL/h feed flow rate. The Ag–TiO$_2$ nanostructured nanofibers were obtained after removal of PVP matrix using calcination in air at 400 °C with a heating rate of 15 °C/min for 4 h.

3.3. Characterization

The crystallographic structure of all materials was studied using X-ray diffraction (XRD) technique performed with a Rigaku SmartLab-9kW diffractometer (Rigaku Corporation, Tokyo, Japan). The experimental spectra were analyzed using PDXL software developed by Rigaku Corp., Tokyo, Japan. The respective morphological characteristics were examined using a Verios G4 UC Scanning Electron Microscope (Thermo Fisher Scientific, Waltham, MA, USA) equipped with an energy dispersive spectrometer (EDX) (AMETEK, Tokyo, Japan), EDAX Octane Elite. The morphology of the fibers based on pure and doped TiO$_2$ was studied using a Hitachi High-Tech HT7700 Transmission Electron Microscope (TEM) (Hitachi, Tokyo, Japan), operated in high contrast mode at 120 kV accelerating voltage. BET analysis was performed using a fully automated gravimetric analyzer IGASorp supplied by Hidden Analytical, Warrington (UK), with an ultrasensitive microbalance, which was used to measure dynamic water vapor sorption capacity of the samples by the weight change with variation of humidity at a constant temperature. Each sample was dried in flowing nitrogen (250 mL/min) until the weight of the sample was in equilibrium at RH < 1%. Experiments were carried out at 25 °C in the relative humidity (RH) range 0–90%, with 10% humidity steps, each having a pre-established equilibrium time between 10 and 20 min (minimum time and time out, respectively).

A Horiba Scientific LabRam HR Evolution spectrometer (HORIBA Scientific, Palaiseau, France) was used for Raman spectroscopy measurements, which were done in the 25–1650 cm^{-1} frequency range, employing excitation at 532 nm, acquisition time 100 s, 3 accumulations and a grating of 600 gr/mm. The optical properties were evaluated by UV-vis diffuse reflectance spectra, recorded on an Analytik Jena (Jena, Germany) SPECORD UV/Vis 210$^+$ spectrophotometer with an integrating sphere. Emission spectra were recorded employing a luminescence spectrometer Perkin Elmer LS55 (PerkinElmer, Inc., Waltham, MA, USA).

3.4. Photocatalytic Activity

The photocatalytic activity of Ag–TiO$_2$ nanostructured nanofibers was evaluated by studying the photodegradation of methylene blue (MB), amaranth, Congo red (CR) and orange II dye solutions under UV-visible light irradiation supplied by a 400 W halogen lamp (Model TG-2503.0l, Elbi Electric and Lighting, Bragadiru, Romania) with the emission spectrum presented in Figure S5 in supporting material file. The distance between the light source (lamp) and the reactor was 30 cm. Deionized water was used to prepare the dye solution. In all photocatalytic measurements, the following experimental conditions were employed: initial dye concentration=10 mg/L for all dyes, catalyst dosage = 0.4 g/L (the catalyst loading that proved previously optimal for reference sample T400), maximum irradiation time = 300 min, T = 23 $\pm$ 2 °C and pH 7.0 $\pm$ 0.2. The pH was recorded for the initial and final solutions and was found to be 7 in both cases. Initially, the dye solution and the catalyst were continuously magnetically stirred (500 rotation/min) in the absence of light for 30 min (time needed to establish an adsorption equilibrium). After this, the solution was irradiated with visible light for 5 hours, and during photodegradation, a quantity of 3 ml of the solution was extracted at the given time intervals to assess the change in the concentration. All absorption spectra for collected samples were measured after 12 h of settling when the catalyst was completely found as sediment. Absorption spectra were recorded using UV–Vis spectrophotometer SPECORD 210⁺ (Analytik Jena gmbh, Jena, Germany).

The photocatalytic efficiency (%) for degradation of the dyes was evaluated using the following relationship [13]:

$$\text{Photocatalytic efficiency (\%)} = \frac{C_0 - C_e}{C_0} \times 100 \qquad (2)$$

where C_0 = initial dye concentration (mg/L); C_e = the dye concentration after irradiation time (mg/L).

3.5. Antimicrobial Activity

The antimicrobial activity of the materials was determined by disk diffusion assay against Staphylococcus aureus ATCC25923 and Escherichia coli ATCC25922. Both microorganisms were stored at −80 °C in 20–40% glycerol. The bacterial strains were refreshed in tryptic soy agar (TSA) at 36 ± 1 °C for 24 h. Microbial suspensions were prepared with these cultures in sterile solutions to obtain turbidity optically comparable to that of 0.5 McFarland standards.

Volumes of 0.3 ml from each inoculum were spread on the Petri dishes. The sterilized paper disks were placed on plates and an aliquot (20 μL) of the samples with a concentration of 10 mg/ml was added. To evaluate the antimicrobial properties, the growth inhibition was measured under standard conditions after 24 h of incubation at 36 ± 1 °C under solar light illumination. All tests were carried out in triplicate to verify the results. After incubation, the diameters of inhibition zones were measured by using Image J version 1.53e software (Madison, WI, USA), and were expressed as the mean ± standard deviation (SD) performed with XLSTAT software Ecology version 2019.4.1 software (New York, NY, USA) [49,50].

3.6. Determination of Minimum Inhibitory Concentration (MIC)

The concentration of a sample that prevented the growth of bacteria is in general recognized as minimum inhibitory concentration (MIC). This was investigated by a broth dilution method, which was performed in 96-well microtiter plates using the resazurin reduction assay concept [49]. Bacterial culture grown to log phase was adjusted to 1×10^8 cells/mL in Muller–Hinton (MH) broth. Inoculants of 50 μL were mixed with 50 μL of serial dilutions of samples and were subsequently incubated at 36 ± 1 °C for 24 h. Resazurin was prepared at 0.015% by dissolving 0.015 g, vortexed and filter sterilized (0.22 μm filter) and stored at 4 °C for a maximum of 2 weeks after preparation. After incubation for 24 h at 36 ± 1 °C, resazurin (0.015%) was added to all wells (20 μL per well), and these were further incubated for 2–4 h more. MIC was determined by reading the fluorescence at λ_{ex} = 575 nm and λ_{em} = 590 nm with FLUOstar Omega Microplate Reader (BMG LABTECH, Ortenberg, Germany). To determine accurately the MIC values, the experiments were performed in triplicate.

4. Conclusions

Ag–TiO$_2$ nanostructured nanofibers with different Ag content were produced using the electrospinning technique followed by calcination at 400 °C for 4 h and were tested regarding their photocatalytic and antimicrobial action. XRD characterization confirmed the formation of anatase TiO$_2$ crystalline structure, without identifying additional peaks corresponding to other phases (brookite, rutile) or other secondary peaks corresponding to the Ag presence, which indicates that Ag was successfully integrated as small nanocrystals without affecting much of the TiO_2 crystal lattice. Regarding the lattice strain, this was found to increase from 0.13% (pure TiO_2 sample) to 0.52% (TAg1), reaching 0.53% at the highest Ag content. These observations were also confirmed by the Raman spectroscopy analysis. SEM characterization proved the formation of one-dimensional uniform Ag–TiO_2 nanostructured nanofibers of excellent quality, with a diameter in the range of 65 to 100 nm. The optical band gap energy decreased to 2.4 eV for the Ag–TiO_2 nanostructured nanofibers as compared to 3.01 eV of pure TiO_2, which leads to an improvement of the photocatalytic activity under visible light irradiation. A remarkable photocatalytic efficiency of 99% at a constant rate of $k = 1.29 \times 10^{-2}$ min^{-1} was found for the 0.1% Ag–TiO_2 sample for the case of methylene blue dye degradation. Additionally, these materials proved able to degrade up 99% of other organic dyes with constant rates in the range of 4.57×10^{-3} to 7.28×10^{-2} min^{-1} depending on the dye nature, all degradation tests being performed under visible light irradiation, without intervening to acidify the solution or add H_2O_2. The present study focused on materials optimization for photocatalytic applications. Further studies regarding the structure and nano-properties of these materials as well as solving the photocatalytic mechanism and by-products are ongoing. The Ag–TiO_2 nanostructured nanofibers materials showed promising antimicrobial responses against Gram-positive (*S. aureus*) and Gram-negative (*E. coli*) strains. The sample with a higher concentration of Ag (5%) was found to present a MIC value of 2.5 mg/mL against *E. coli* and 5 mg/mL against *S. aureus*. Based on these results, one can conclude that Ag–TiO_2 nanostructured nanofibers were developed with excellent structure and suitable physical properties for photocatalytic and antibacterial action. Further studies regarding the possibility of integration of these materials into food packaging, medical textiles, and other healthcare related items will be performed.

Supplementary Materials: The following are available online at https://www.mdpi.com/article/10.3390/catal11101234/s1, Figure S1. Histogram of fiber diameter distribution plotted from SEM images analysis of pure TiO_2 (T400) and Ag doped TiO_2 (TAg1 and TAg5). Figure S2. Dynamic vapor sorption isotherms for pure TiO_2 and Ag doped TiO_2 materials. Figure S3. UV-VIS absorption spectra of the photolysis test for all 4 dyes performed under visible light irradiation for 300 minutes. Figure S4. UV-VIS absorption spectra recorded for the degradation of (a) MB, (b) CR, (c) Amaranth and (d) Orange II dyes. Figure S5. Lamp spectrum.

Author Contributions: Data Curation, P.P., C.C., N.O., E.K. and M.P.S.; Formal Analysis, P.P., A.A., I.R., E.K. and M.P.S.; Funding Acquisition, P.P.; Investigation, P.P. and I.R.; Methodology, P.P. and I.R.; Project Administration, P.P.; Resources, P.P., C.C., A.A. and N.O.; Supervision, P.P.; Validation, P.P., E.K. and M.P.S.; Visualization, P.P. and M.P.S.; Writing—Original Draft, P.P., C.C., E.K. and M.P.S.; Writing—Review and Editing, P.P., E.K. and M.P.S. All authors have read and agreed to the published version of the manuscript.

Funding: This work was partially supported by a grant of the Romanian Ministry ofResearch, Innovation and Digitization, CNCS/CCCDIUEFISCDI, project number PN-III-P1-1.1-TE-2019-0594, within PNCDI III.

Data Availability Statement: The raw and processed data required to reproduce these findings cannot be shared at this time due to technical or time limitations. The raw and processed data will be provided upon reasonable request to anyone interested anytime until technical problems are being solved.

Acknowledgments: Special thanks to Cosmin Romanitan for helping with the XRD characterization. M. P. Suchea contribution was partially financed by the Romanian Ministry of Research, Innovation and Digitization through "MICRO-NANO-SIS PLUS" core Programme.

Conflicts of Interest: The authors declare that they have no known competing financial interests or personal relationships that could have appeared to influence the work reported in this paper.

References

1. Estrada-Flores, S.; Martinez-Luevanosa, A.; Perez-Berumen, C.M.; Garcia-Cerda, L.A.; Flores-Guia, T.E. Relationship between morphology, porosity, and the photocatalytic activity of TiO_2 obtained by sol–gel method assisted with ionic and nonionic surfactants. *Bol. Soc. Esp. Ceram. Vidr.* **2020**, *59*, 209–218. [CrossRef]
2. Morales-Garcia, A.; Escatllar, A.M.; Illas, F.; Bromley, S.T. Understanding the interplay between size, morphology and energy gap in photoactive TiO_2 nanoparticles. *Nanoscale* **2019**, *11*, 9032–9041. [CrossRef]
3. Adan, C.; Marugan, J.; Sanchez, E.; Pablos, C.; Van Grieken, R. Understanding the effect of morphology on the photocatalytic activity of TiO_2 nanotube array electrodes. *Electrochim. Acta* **2016**, *191*, 521–529. [CrossRef]
4. Li, Y.F.; Liu, Z.P. Particle size, shape and activity for photocatalysis on titania anatase nanoparticles in aqueous surroundings. *J. Am. Chem. Soc.* **2011**, *133*, 15743–15752. [CrossRef] [PubMed]
5. Wang, Y.; Liu, L.; Xu, L.; Meng, C.; Zhu, W. Ag/TiO_2 nanofiber heterostructures: Highly enhanced photocatalysts under visible light. *J. Appl. Phys.* **2013**, *113*, 174311. [CrossRef]
6. Kudhier, M.A.; Sabry, R.S.; Al-Haidarie, Y.K.; AL-Marjani, M.F. Significantly enhanced antibacterial activity of Ag-doped TiO_2 nanofibers synthesized by electrospinning. *Mater. Technol.* **2018**, *33*, 220–226. [CrossRef]
7. Wang, S.; Bai, J.; Liang, H.; Xu, T.; Li, C.; Sun, W.; Liu, H. Synthesis, characterization, and photocatalytic properties of Ag/TiO_2 composite nanofibers prepared by electrospinning. *J. Dispers. Sci. Technol.* **2014**, *35*, 777–782. [CrossRef]
8. Norouzi, M.; Fazeli, A.; Tavakoli, O. Phenol contaminated water treatment by photocatalytic degradation on electrospun Ag/TiO_2 nanofibers: Optimization by the response surface method. *J. Water Process. Eng.* **2020**, *37*, 101489. [CrossRef]
9. Ge, M.; Cao, C.; Huang, J.; Li, S.; Chen, Z.; Zhang, K.Q.; Al-Deyab, S.S.; Lai, Y. A review of one-dimensional TiO_2 nanostructured materials for environmental and energy applications. *J. Mater. Chem. A* **2016**, *4*, 6772–6801. [CrossRef]
10. Verma, R.; Gangwar, J.; Srivastava, A.K. Multiphase TiO_2 nanostructures: A review of efficient synthesis, growth mechanism, probing capabilities, and applications in bio-safety and health. *RSC Adv.* **2017**, *7*, 44199–44224. [CrossRef]
11. Baranowska-Wojcik, E.; Szwajgier, D.; Oleszczuk, P.; Winiarska-Mieczan, A. Effects of titanium dioxide nanoparticles exposure on human health—A Review. *Biol. Trace. Elem. Res.* **2020**, *193*, 118–129. [CrossRef]
12. Jafari, S.; Mahyad, B.; Hashemzadeh, H.; Janfaza, S.; Gholikhani, T.; Tayebi, L. Biomedical applications of TiO_2 nanostructures: Recent advances. *Int. J. Nanomed.* **2020**, *15*, 3447–3470. [CrossRef]
13. Boxi, S.S.; Paria, S. Visible light induced enhanced photocatalytic degradation of organic pollutants in aqueous media using Ag doped hollow TiO_2 nanospheres. *RSC Adv.* **2015**, *5*, 37657–37668.
14. Pascariu Dorneanu, P.; Cojocaru, C.; Samoila, P.; Olaru, N.; Airinei, A.; Rotaru, A. Novel fibrous composites based on electrospun PSF and PVDF ultrathin fibers reinforced with inorganic nanoparticles: Evaluation as oil spill sorbents. *Polym. Adv. Technol.* **2018**, *29*, 1435–1446. [CrossRef]
15. Zhang, F.; Wang, X.; Liu, H.; Liu, C.; Wan, Y.; Long, Y.; Cai, Z. Recent advances and applications of semiconductor photocatalytic technology. *Appl. Sci.* **2019**, *9*, 2489. [CrossRef]
16. Nalbandian, M.J.; Zhang, M.; Sanchez, J.; Kim, S.; Choa, Y.H.; Cwiertny, D.M.; Myung, N.V. Synthesis and optimization of Ag–TiO_2 composite nanofibers for photocatalytic treatment of impaired water sources. *J. Hazard. Mater.* **2015**, *299*, 141–148. [CrossRef]
17. Zhang, F.; Cheng, Z.; Kang, L.; Cui, L.; Liu, W.; Xu, X.; Hou, G.; Yang, H. A novel preparation of Ag-doped TiO_2 nanofibers with enhanced stability of photocatalytic activity. *RSC Adv.* **2015**, *5*, 32088–32091. [CrossRef]
18. Roongraung, K.; Chuangchote, S.; Laosiripojana, N.; Sagawa, T. Electrospun Ag-TiO_2 nanofibers for photocatalytic glucose conversion to high-value chemicals. *ACS Omega* **2020**, *5*, 5862–5872. [CrossRef]
19. Suwanchawalit, C.; Wongnawa, S.; Sriprang, P.; Meanha, P. Enhancement of the photocatalytic performance of Ag-modified TiO_2 photocatalyst under visible light. *Ceram. Int.* **2012**, *38*, 5201–5207. [CrossRef]
20. Suwarnkar, M.B.; Dhabbe, R.S.; Kadam, A.N.; Garadkarn, K.M. Enhanced photocatalytic activity of Ag doped TiO_2 nanoparticles synthesized by a microwave assisted method. *Ceram. Int.* **2014**, *40*, 5489–5496. [CrossRef]
21. Ahamed, M.; Majeed Khan, M.A.; Akhtar, M.J.; Alhadlaq, H.A.; Alshamsan, A. Ag-doping regulates the cytotoxicity of TiO_2 nanoparticles via oxidative stress in human cancer cells. *Sci. Rep.* **2017**, *7*, 17662. [CrossRef] [PubMed]
22. Ali, T.; Ahmed, A.; Alam, U.; Uddin, I.; Tripathi, P.; Muneer, M. Enhanced photocatalytic and antibacterial activities of Ag-doped TiO_2 nanoparticles under visible light. *Mater. Chem. Phys.* **2018**, *212*, 325–335. [CrossRef]
23. Jesline, A.; John, N.P.; Narayanan, P.M.; Vani, C.; Murugan, S. Antimicrobial activity of zinc and titanium dioxide nanoparticles against biofilm-producing methicillin-resistant Staphylococcus aureus. *Appl. Nanosci.* **2014**, *5*, 157–162. [CrossRef]
24. Verdier, T.; Countand, M.; Bertron, A.; Roques, C. Antibacterial activity of TiO_2 photocatalyst alone or in coatings on E. coli: The influence of methodological aspects. *Coatings* **2014**, *4*, 670–686. [CrossRef]
25. Srisitthiratkul, C.; Pongsorrarith, V.; Intasanta, N. The potential use of nanosilver decorated titanium dioxide nanofibers for toxin decomposition with antimicrobial and self-cleaning properties. *Appl. Surf. Sci.* **2011**, *257*, 8850–8856. [CrossRef]
26. Md Saad, S.K.; Umar, A.A.; Umar, M.I.A.; Tomitori, M.; Rahman, M.Y.A.; Salleh, M.M.; Oyama, M. Two-dimensional, hierarchical Ag-doped TiO_2 nanocatalysts: Effect of the metal oxidation state on the photocatalytic properties. *ACS Omega* **2018**, *3*, 2579–2587. [CrossRef]
27. Aguilar, T.; Navas, J.; Alcantara, R.; Fernández-Lorenzo, C.; Gallardo, J.J.; Blanco, G.; Martín-Calleja, J. A route for the synthesis of Cu-doped TiO_2 nanoparticles with a very low band gap. *Chem. Phys. Lett.* **2013**, *571*, 49–53. [CrossRef]

28. Liaqat, M.A.; Hussain, Z.; Khan, Z.; Akram, M.A.; Shuj, A. Efects of Ag doping on compact TiO_2 thin flms synthesized via one-step sol–gel route and deposited by spin coating technique. *J. Mater. Sci. Mater. Electron.* **2020**, *31*, 7172–7181. [CrossRef]
29. Pham, T.D.; Lee, B.K. Effects of Ag doping on the photocatalytic disinfection of E. coli in bioaerosol by $Ag–TiO_2$/GF under visible light. *J. Colloid Interface Sci.* **2014**, *428*, 24–31. [CrossRef] [PubMed]
30. Pedroza-Herrera, G.; Medina-Ramírez, I.E.; Lozano-Álvarez, J.A.; Rodil, S.E. Evaluation of the photocatalytic activity of copper doped TiO_2 nanoparticles for the purification and/or disinfection of industrial effluents. *Catal. Today* **2020**, *341*, 37–48. [CrossRef]
31. Shi, Q.; Ping, G.; Wang, X.; Xu, H.; Li, J.; Cui, J.; Abroshan, H.; Ding, H.; Li, G. CuO/TiO_2 heterojunction composites: An efficient photocatalyst for selective oxidation of methanol to methyl formate. *J. Mater. Chem. A* **2019**, *7*, 2253–2260. [CrossRef]
32. Popović, Z.V.; Dohčević-Mitrović, Z.; Šćepanović, M.; Grujić-Brojčin, M.; Aškrabić, S. Raman scattering on nanomaterials and nanostructures. *Ann. Phys.* **2011**, *523*, 62–74. [CrossRef]
33. Lettieri, S.; Pavone, M.; Fioravanti, A.; Amato, L.S.; Maddalena, P. Charge Carrier Processes and Optical Properties in TiO_2 and TiO_2-Based Heterojunction Photocatalysts: A Review. *Materials* **2021**, *14*, 1645. [CrossRef]
34. Pascariu, P.; Cojocaru, C.; Samoila, P.; Airinei, A.; Olaru, N.; Rusu, D.; Rosca, I.; Suchea, M. Photocatalytic and antimicrobial activity of electrospun ZnO:Ag nanostructures. *J. Alloys Compd.* **2020**, *834*, 155144. [CrossRef]
35. Mathpal, M.C.; Tripathi, A.K.; Singh, M.K.; Gairola, S.P.; Pandey, S.N.; Agarwal, A. Effect of annealing temperature on Raman spectra of TiO_2 nanoparticles. *Chem. Phys. Lett.* **2013**, *555*, 182–186. [CrossRef]
36. Munguti, B.L.; Dejene, F. Influence of annealing temperature on structural, optical and photocatalytic properties of $ZnO–TiO_2$ composites for application in dye removal in water. *Nano-Struct. Nano-Objects* **2020**, *24*, 100594. [CrossRef]
37. Chen, R.; Han, J.; Yan, X.; Zou, C.; Bian, J.; Alyamani, A.; Gao, W. Photocatalytic activities of wet oxidation synthesized ZnO and $ZnO–TiO_2$ thick porous films. *Appl. Nanosci.* **2011**, *1*, 37–44. [CrossRef]
38. Ahmed, S.A. Structural, optical and magnetic properties of Cu-doped TiO_2 samples. *Crys. Res. Technol.* **2017**, *52*, 1600335. [CrossRef]
39. Makula, P.; Pacia, M.; Macyk, W. How to correctly determine the band gap energy of modified semiconductor photocatalysts based on UV–Vis spectra. *J. Phys. Chem. Lett.* **2018**, *9*, 6814–6817. [CrossRef]
40. Reda, S.M.; Khairy, M.A.; Mousa, M.A. Photocatalytic activity of nitrogen and copper doped TiO_2 nanoparticles prepared by microwave-assisted sol-gel process. *Arabian J. Chem.* **2020**, *13*, 86–95. [CrossRef]
41. Jaiswal, R.; Bharambe, J.; Patel, N.; Dashora, A.; Kothari, D.C.; Miotello, A. Copper and nitrogen co-doped TiO_2 photocatalyst with enhanced optical absorption and catalytic activity. *Appl. Catal. B Environ.* **2015**, *128/129*, 333–341. [CrossRef]
42. Yu, P.Y.; Cardona, M. *Fundamentals of Semiconductors*; Graduate Texts in Physics; Springer: Berlin/Heidelberg, Germany, 2010; ISBN 978-3-642-00709-5.
43. Harikishore, M.; Sandhyarani, M.; Venkateswarlu, K.; Nellaippan, T.A.; Rameshbabu, N. Effect of Ag doping on antibacterial and photocatalytic activity of nanocrystalline TiO_2. *Procedia Mater. Sci.* **2014**, *6*, 557–566. [CrossRef]
44. Hufnagel, D.A.; DePas, W.H.; Chapman, M.R. The biology of the escherichia coli extracellular matrix. *Microbiol. Spectr.* **2015**, *3*, 1–24. [CrossRef]
45. Hu, C.; Guo, J.; Qu, J.; Hu, X. Photocatalytic degradation of pathogenic bacteria with AgI/TiO_2 under visible light irradiation. *Langmuir* **2007**, *23*, 4982–4987. [CrossRef]
46. Gomathi Devi, L.; Nagaraj, B. Disinfection of Escherichia coli Gram negative bacteria using surface modified TiO_2: Optimization of Ag metallization and depiction of charge transfer mechanism. *Photochem. Photobiol.* **2014**, *90*, 1089–1098. [PubMed]
47. Pascariu, P.; Olaru, L.; Matricala, A.L.; Olaru, N. Photocatalytic activity of ZnO nanostructures grown on electrospun CAB ultrafine fibers. *Appl. Surf. Sci.* **2018**, *455*, 61–69. [CrossRef]
48. Cojocaru, C.; Pascariu Dorneanu, P.; Airinei, A.; Olaru, N.; Samoila, P.; Rotaru, A. Design and evaluation of electrospun polysulfone fibers and polysulfone/NiFe2O4 nanostructured composite as sorbents for oil spill cleanup. *J. Taiwan Inst. Chem. Eng.* **2017**, *70*, 267–281. [CrossRef]
49. XLSTAT Statistical and Data Analysis Solution. New York: Addinsoft. 2020. Available online: https://www.xlstat.com (accessed on 15 September 2020).
50. Riss, T.L.; Moravec, R.A.; Niles, A.L.; Duellmanet, S.; Benink, H.A.; Worzella, T.J.; Minor, L. Cell Viability Assays. In *Assay Guidance Manual*; Markossian, S., Sittampalam, G.S., Grossman, A., Eds.; Eli Lilly & Company and the National Center for Advancing Translational Sciences, Bethesda: Rockville, MD, USA, 2004.

Article

Photocatalytic Degradation of Thiacloprid Using Tri-Doped TiO$_2$ Photocatalysts: A Preliminary Comparative Study

Antonietta Mancuso [1,*], Wanda Navarra [2], Olga Sacco [2,*], Stefania Pragliola [2], Vincenzo Vaiano [1] and Vincenzo Venditto [2]

[1] Department of Industrial Engineering, University of Salerno, Via Giovanni Paolo II, 132, 84084 Fisciano, Italy; vvaiano@unisa.it

[2] Department of Chemistry and Biology "A. Zambelli" & INSTM Research Unit, University of Salerno, Via Giovanni Paolo II, 132, 84084 Fisciano, Italy; wnavarra@unisa.it (W.N.); spragliola@unisa.it (S.P.); vvenditto@unisa.it (V.V.)

* Correspondence: anmancuso@unisa.it (A.M.); osacco@unisa.it (O.S.)

Abstract: Different tri-doped TiO$_2$ photocatalysts (Fe-N-P/TiO$_2$, Fe-N-S/TiO$_2$, Fe-Pr-N/TiO$_2$, Pr-N-S/TiO$_2$, and P-N-S/TiO$_2$) were successfully prepared and tested in the photocatalytic removal of thiacloprid (THI) under UV-A, visible, and direct solar light irradiation. The physical-chemical properties of the prepared catalysts were analyzed by different characterization techniques, revealing that dopants are effectively incorporated into the anatase TiO$_2$ lattice, resulting in a decrease of the energy band gap. The reduction of photoluminescence intensity indicates a lower combination rate and longer lifespan of photogenerated carriers of all doped samples in comparison with the un-doped TiO$_2$. The doped photocatalysts not only significantly promote the photodegradation under UV-A light irradiation but also extend the optical response of TiO$_2$ to visible light region, and consequently improve the visible light degradation of THI. Fe-N-P tri-doped TiO$_2$ sample exhibits the highest THI photodegradation degree (64% under UV-A light, 29% under visible light and 73% under solar light).

Keywords: tri-doped TiO$_2$; thiacloprid; photocatalysis; water treatment

1. Introduction

Today, the most important commercial insecticides available on the market are neonicotinoids, thanks to their high insecticidal activity, adequate water solubility, and high stability [1]. Usage has spread beyond agriculture to home garden [2]. Unsurprisingly, widespread use of neonicotinoids has led to an almost ubiquitous environmental presence of these pollutants, including in surface water and groundwater [3]. The neonicotinoid insecticides, both at high and low concentrations (e.g., LC50 of 5 ng/bee) [4], bind to nervous systems causing receptor blockage, paralysis, and death. Thiacloprid (THI) is an insecticide, belonging to the second-generation neonicotinoid pesticides, introduced by Bayer Crop Science with the name Calipsol [5]. Like all neonicotinoid insecticides, THI selectively acts on the insect nervous system as an agonist of the nicotinic acetylcholine receptors. This unique mode of action makes THI highly applicable for controlling the biological effect on insects which developed resistance to conventional chlorinated hydrocarbons, organophosphate, carbamate, and pyrethroid insecticides [6]. Due to its high water solubility and recalcitrance to biodegradation, THI results still persistent in water after the conventional wastewater treatment [7]. Thus, various technologies have been tested to degrade THI from water. Unfortunately, all the proposed techniques suffer to relative low removal efficiency and require continued use of electricity and chemicals [8,9]. Therefore, it is of great importance to develop a reliable and effective method for removing THI from water.

Heterogeneous photocatalysis is an advanced oxidation technology, which offers a promising alternative to remove a great variety of organic pollutants with high efficient

degradation rate. TiO$_2$ is the most widely studied photocatalyst because of its excellent properties, such as non-toxic nature, water insolubility, low cost, favorable band edges position, photochemical stability, and high light conversion efficiency [10,11].

However, the large band gap energy of TiO$_2$ (3.2 eV for anatase) makes it active only under UV-A light irradiation (~5% of solar spectrum) and scarcely efficient with visible light, which comprises large portion of solar spectrum (about 45%). Different strategies have been developed to enhance the photocatalytic efficiency of TiO$_2$ under solar irradiation. Previous studies revealed that doping with various metals and non-metals elements (Fe-N and Fe-Pr) or coupling different semiconductors [12] lessens the band gap of TiO$_2$ for the photo-excitation (red shift) and simultaneously reduces the recombination rate of photogenerated electron–hole pairs. One possible doping route is the incorporation into TiO$_2$ crystal structure of non-metal elements, which could extend the light absorption to the visible region. Among non-metal elements, the TiO$_2$ doping with nitrogen was considered to be the most effective one, due to its high stability and efficiency. In fact, nitrogen (atomic radius 0.56 Å) can be easily introduced into the TiO$_2$ structure, due to the atomic size similar to oxygen (atomic radius 0.48 Å) [13]. Further, it was proved that doping with sulfur increases photocatalytic activity of TiO$_2$ under visible light irradiation [14]. In the last few years, phosphorous doped TiO$_2$ photocatalysts have also gained considerable research interest because of their high photocatalytic activity under visible light irradiation. The visible light response of P-doped TiO$_2$ has been attributed to the formation of Ti–O–P bonds [15]. Although N-doped, S-doped, and P-doped TiO$_2$ exhibited improved visible-light-induced photocatalytic activity, electron–hole recombination is still an obstacle that limits the effective use of non-metals doped TiO$_2$ photocatalysts.

Doping lanthanides (such as Pr, Nd, Sm, and Yb), having a 4f electron configuration, can act as electron sink, thereby reducing the recombination of photogenerated electron-hole pairs and efficiently improving the photochemical or electrochemical properties of semiconductors [16].

On other hand, the doping with transition metals (such as Cr, Mn, Fe, Co, and Cu), having unfilled d-electron structures, can accommodate more electrons, introducing impurity levels in the band gap of TiO$_2$ [17]. For example, Fe^{3+} having ionic radius (0.64 Å) close to Ti^{4+} (0.68 Å) could replace Ti in the crystalline structure, allowing the formation of an inter band, within the conduction band and the valence band edge, capable of absorbing visible light [18]. Doping TiO$_2$ with two or more dopants has a significant synergistic effect on photocatalytic activity compared to un-doped samples [19]. The insertion of dopants in the TiO$_2$ matrix changes the recombination dynamics of light charge carriers (electron–hole pairs) and causes the energy gap shift, resulting in nanoparticles active under visible light. Recently, tri-doped TiO$_2$ has been the focus for further improving the visible light photocatalytic activity. Umare et al. [20] reported an enhanced visible light activity of Ga, N, and S co-doped TiO$_2$ towards the decomposition of azo dyes. The highest dye removal efficiency, compared with the single doped and co-doped TiO$_2$ was described by Maki et al., who synthetized Fe-Ce-N tri-doped TiO$_2$ [21]. To our knowledge, a systematic study on effect of different elements (e.g., Fe, Pr and N, P, S) in tri-doped TiO$_2$ for the removal of THI has never been reported.

In this work, different tri-doped TiO$_2$ photocatalysts were successfully synthesized by sol-gel process. The effect of doping elements on the photoactivity of the tri-doped samples towards the degradation of THI in aqueous solution under UV-A, visible, and solar light irradiation was investigated. In particular, an initial concentration of THI equal to 0.5 mg/L was chosen since this pesticide belongs to the so-called "emerging contaminants", and it is present at very low concentration (0.02–4.5 µg/L) in surface water [22,23].

2. Results

2.1. Characterization of Photocatalysts

Specific surface area (S$_{BET}$) of synthesized tri-doped TiO$_2$ samples, calculated by BET method, are reported in Table 1 and also compared with S$_{BET}$ of an un-doped TiO$_2$ sample.

The un-doped TiO$_2$ sample shows S$_{BET}$ value of 107 m^2 g^{-1}, which is comparable with S$_{BET}$ value of P-N-S/TiO$_2$ and Fe-N-P/TiO$_2$ samples, while all the other tri-doped TiO$_2$ samples have lower S$_{BET}$ values. Lower values of Fe-N-S/TiO$_2$, Fe-N-P/TiO$_2$, Fe-Pr-N/TiO$_2$, and Pr-N-S/TiO$_2$ S$_{BET}$ compared to un-doped TiO$_2$ may be associated with the formation of metal oxide clusters which could obstruct (clog) structural pores [24]. Figure 1a reports UV-Vis DRS spectra of TiO$_2$, Fe-N-P/TiO$_2$, Fe-N-S/TiO$_2$, Fe-Pr-N/TiO$_2$, Pr-N-S/TiO$_2$, and P-N-S/TiO$_2$ samples in the range 300–900 nm.

Table 1. Crystallite size, lattice parameters, specific surface area, and band gap data of all prepared photocatalysts.

Sample	Crystallite Size (nm)	Lattice Parameter (Å)		S$_{BET}$ (m^2 g^{-1})	E$_{bg}$ (eV)
		$a = b$	c		
TiO$_2$	10	3.74	8.68	107	3.2
Fe-N-P/TiO$_2$	8	3.80	10.00	105	2.8
Fe-N-S/TiO$_2$	9	3.79	9.40	81	2.65
Fe-N-Pr/TiO$_2$	10	3.81	9.92	91	2.73
P-N-S/TiO$_2$	8	3.80	10.09	109	2.55
Pr-N-S/TiO$_2$	9	3.81	9.86	85	2.5

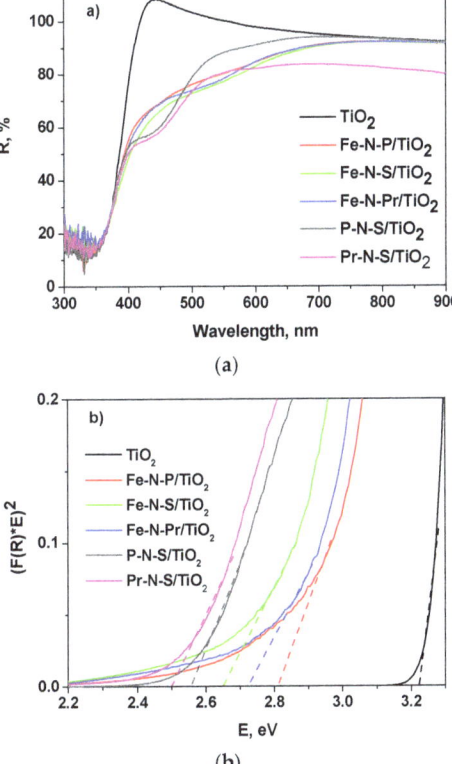

Figure 1. (a) UV-vis DRS spectra and (b) band-gap evaluation of all prepared samples.

The main absorption of the un-doped TiO$_2$ is located in the UV-A region, and no absorption is detected in the visible region. The Fe-N-P/TiO$_2$, Fe-N-S-TiO$_2$, and Fe-Pr-N-TiO$_2$ samples exhibit similar absorption performances in the range 450–600 nm, probably due to the presence of dopant elements into TiO$_2$ lattice [25,26]. Both Pr-N-S/TiO$_2$ and P-N-S/TiO$_2$ sample absorption spectra present a shoulder in the visible region. Band gap energies (E_{bg}) are calculated by Kubelka–Munk function (Figure 2) [27], and the values are shown in Table 1. In detail, the band gap value decreases from 3.2 eV for un-doped TiO_2 to 2.5–2.8 eV for the tri-doped TiO_2 samples. These results demonstrate that the introduction of metallic and non-metallic elements in the TiO_2 crystal structure can induce the absorption of visible light, due to the decrease of band gap value. XRD patterns of the un-doped TiO_2 and tri-doped samples in the 2θ range 20–80° are reported in Figure 2.

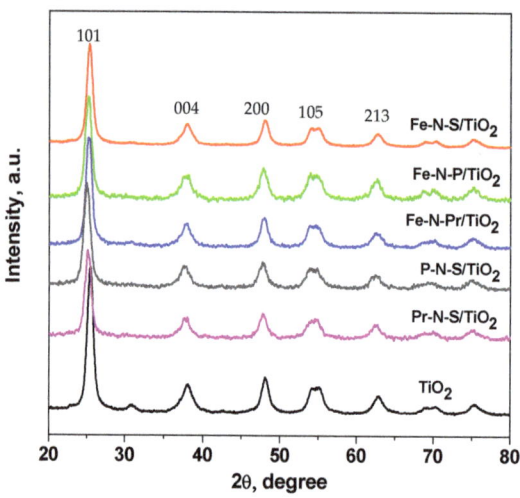

Figure 2. XRD patterns of all prepared samples.

Pure TiO_2 and all tri-doped samples present only the characteristic XRD peaks of TiO_2 anatase phase [28]. None of characteristic peaks of dopant element oxides have been detected in XRD patterns of all tri-doped samples; despite this, small quantity of oxides, lower than that detectable by the XRD, could be also present. Moreover, a remarkable shift of (101) reflection to lower 2θ angles, with respect to the TiO_2 XRD peak position, was observed in the XRD pattern of Pr-N-S/TiO_2 and P-N-S/TiO_2 samples. This shift can be explained by the insertion of dopant elements into semiconductor TiO_2 crystalline structure resulting in an increasing of TiO_2 lattice parameters. It is, in fact, well known that the doping with non-metallic elements (such as nitrogen, phosphorus, and sulfur) implies lattice deformations, due to the substitution of oxygen atoms by non-metallic dopants, all of them having atomic/ionic radius larger than oxygen [29]. The average crystallite size and lattice parameters data (determined by using Scherrer's equation and the (101) reflections of XRD patterns) of un-doped and tri-doped TiO_2 samples are reported in Table 1. The crystallite sizes of most of tri-doped samples are lower than that of TiO_2. This result is in agreement with literature studies [26,29,30], reporting that the tri-doping process commonly hinders the crystallite growth. In detail, P-N-S/TiO_2 and Fe-N-P/TiO_2 samples exhibit crystallite size values lower to that of un-doped TiO_2 (10 nm). The lattice parameters (Table 1) of all tri-doped samples undergo an increase with respect to the value of un-doped titania (a = b = 3.74 Å, c = 8.68 Å), which could be attributed to the difference between the radius of dopant elements and those of the host semiconductor atoms [31].

Figure 3 reports the Raman spectra of all prepared samples in the range 100–900 cm^{-1}.

The Raman bands at 144, 197, 399, 516, and 639 cm^{-1} were observed for un-doped TiO_2 and tri-doped TiO_2 samples, confirming the presence of anatase crystalline phase in all the prepared samples. It is worth noting that the most intense Raman band at 144 cm^{-1} of un-doped TiO_2 was slightly shifted (blue shift) after the doping process. This experimental evidence could be related to the disorder in the TiO_2 lattice due to the incorporation of dopant elements that generate defects (such as oxygen vacancies in anatase structure) [31].

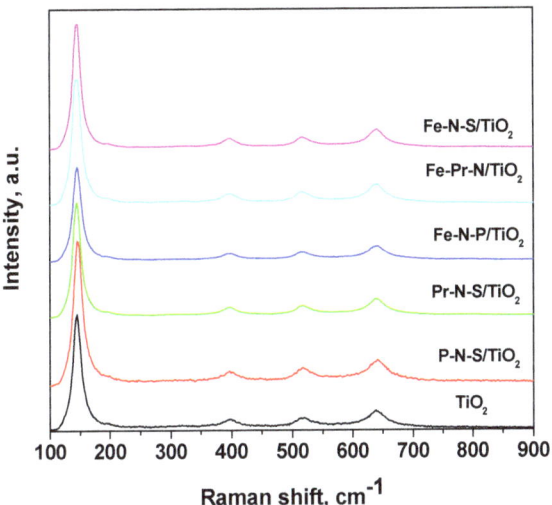

Figure 3. Raman spectra of all the prepared samples.

Figure 4 shows the FTIR absorbance spectra of all tri-doped TiO$_2$ samples. The IR band at about 3400 cm^{-1} is due to the O−H stretching vibration, and the band at 1630 cm^{-1} corresponds to the H−O−H bending vibration of adsorbed water on the photocatalysts surface [32]. The low-frequency broad band in the range 400–900 cm^{-1} corresponds to the Ti-O-Ti vibrational mode [33]. It is worth it to observe that all tri-doped TiO$_2$ samples show different shape of low frequency band together with a little shift of the position to lower wavenumbers that could be an indication of structure defects [34]. Consistent with XRD results and according to studies already reported in the literature [32–34], the observed band shift should be related to the introduction of dopant species into the TiO$_2$ framework. Figure 5 reports photoluminescence spectra (PL) of all the samples, acquired at room temperature in the emission range 325–500 nm, by using an excitation wavelength of 280 nm. All samples evidence a strong emission peak at ca 380 nm, due to the electron–hole pairs recombination. The intensity of this band, together with intensity of the entire emission spectrum, is instead reduced in all tri-doped TiO$_2$ samples. In detail, the intensity of the 380 nm PL peak decreases in the following order: TiO$_2$ > Fe-N-S/TiO$_2$ > Fe-N-Pr/TiO$_2$ > P-N-Pr/TiO$_2$ > Pr-N-S/TiO$_2$ > Fe-N-P/TiO$_2$. As it is generally accepted that a low PL intensity indicates a lower electrons-holes recombination rate and longer duration of photogenerated carries [35], the result of Figure 5 suggests that, for tri-doped TiO$_2$, a photogenerated gap separation higher than in un-doped TiO$_2$ is expected. Keeping in mind that, generally, PL intensity is directly related with the recombination of electrons and holes, and lower PL intensity indicates a lower recombination rate, as well as higher lifespan of photogenerated carriers, this result is expected. The observed lower PL band intensity in tri-doped TiO$_2$ with respect to un-doped TiO$_2$ should be related to a better separation of photogenerated electron-hole in the doped samples.

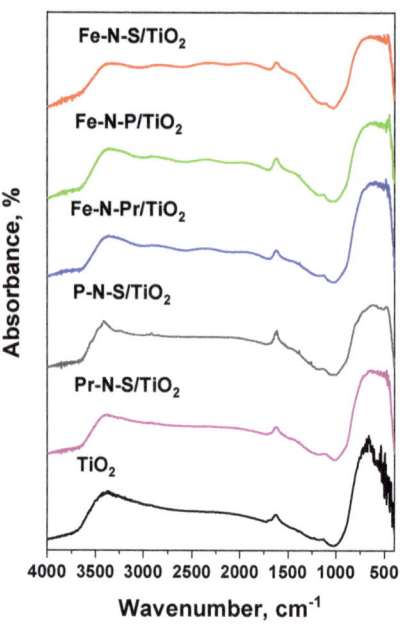

Figure 4. Fourier-transform infrared (FTIR) spectra of all prepared samples.

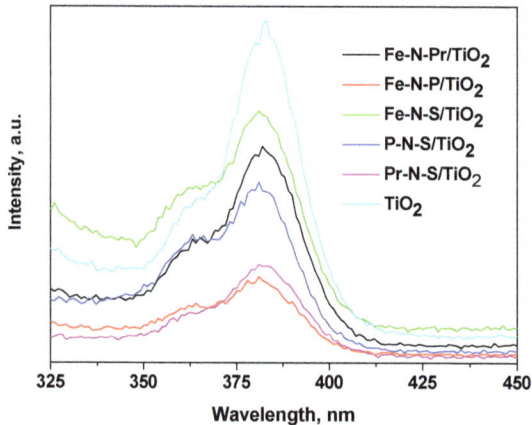

Figure 5. PL spectra of all the prepared samples.

2.2. Photocatalytic Activity Results

The photocatalytic degradation of THI in aqueous solution under UV-A and visible light irradiation was evaluated for all the doped photocatalysts and compared with bare TiO_2. Figure 6 reports the THI relative concentration as a function of irradiation time for all the tested photocatalysts in presence of UV-A, visible, and direct solar light.

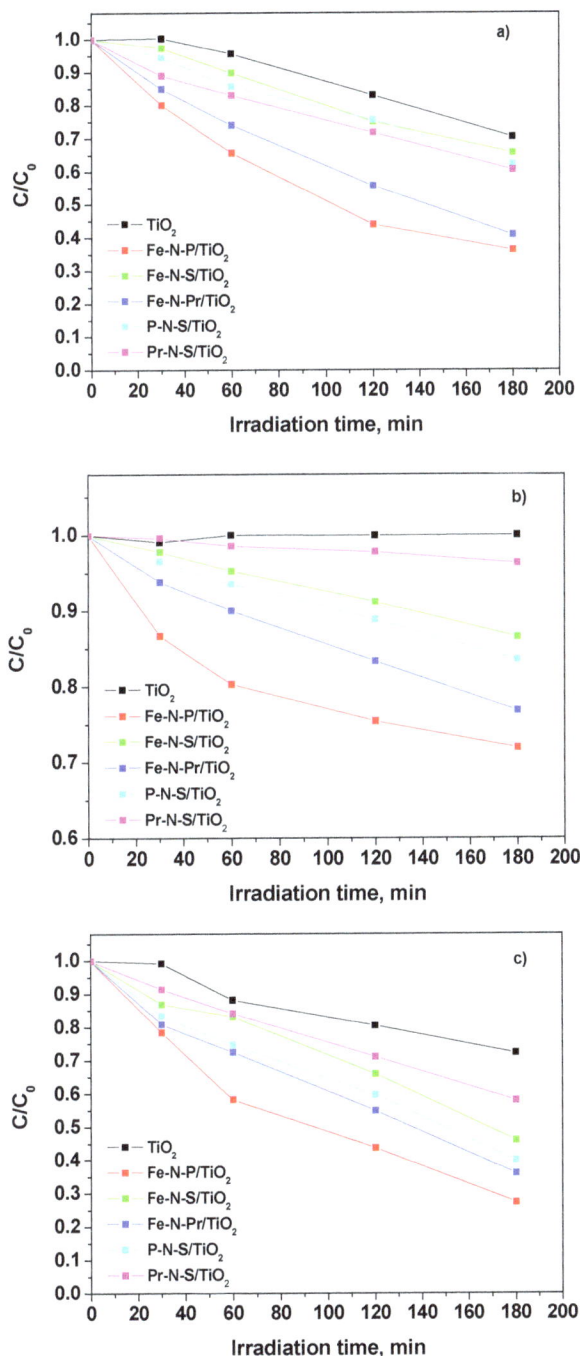

Figure 6. Photocatalytic degradation of THI under (**a**) UV-A, (**b**) visible, and (**c**) direct solar light irradiation.

A progressive decrease of the THI concentration was observed in the presence of all the doped photocatalysts under UV-A, visible light, and direct solar light irradiation, evidencing that the doping of TiO$_2$ is able to increase photocatalytic activity. The Fe-N-P tri-doped TiO$_2$ sample exhibits the highest photodegradation degree (64% under UV-A light, 29% under visible light, and 73% under solar light) among all the samples, indicating that Fe, N, and P tri-doping comes possibly into a synergistic effect, resulting in a catalyst with superior performances than other doped one.

Considering that P-N-S/TiO$_2$ and Fe-N-P/TiO$_2$ samples showed a comparable S$_{BET}$ value, and despite the fact that P-N-S/TiO$_2$ photocatalysts have an E$_{bg}$ lower than that of Fe-N-P/TiO$_2$ (Table 1), the highest photoactivity of such sample is possibly linked to a recombination rate of photogenerated electron-hole lower than the other tested samples (as shown by PL results).

Keeping in mind that, to our knowledge, no articles have been reported in the literature, so far, dealing with the use of the TiO$_2$ tri-doped sample for THI removal under direct sunlight, these results are quite exciting. It is worth underlining that the optimized Fe-N-P/TiO$_2$ catalyst showed a superior photocatalytic activity, especially under UV-A irradiation and sunlight using a pollutant concentration and catalyst dosage very low if compared with data reported in the literature (Table 2). Moreover, Fe-N-P/TiO$_2$ sample evidenced a visible light activity very similar to a catalyst based on a noble metal (e.g., Ag$_3$PO$_4$) [1]. The degradation mechanism of THI is mainly driven by •OH radicals and holes. Indeed, it is reported in literature [36,37] that THI photodegradation mechanism proceeds via three different decomposition pathways because of non-selective attach by •OH radicals. During the photocatalytic reaction, the main by-products are formed by hydroxylation/oxidation reactions of the THI molecules and of the portions of the molecule resulting from the detachment of the (thiazolidin-2-ylidene) cyanamide group and/or the 2-chloro-5-methylpyridine group.

Table 2. Comparison with the available literature dealing with the photocatalytic degradation of THI.

Catalyst	Catalyst Dosage	THI Concentration	Light Source	Photocatalytic Degradation Efficiency (%)	Reference
Fe-TiO$_2$	1.67 g L^{-1}	~80 mg L^{-1}	UV light	~45%	[38]
ZnO	2.0 g L^{-1}	~100 mg L^{-1}	UV-A light	~100%	[39]
ZnO	2.0 g L^{-1}	~100 mg L^{-1}	visible light	~59%	[39]
TiO$_2$	1.0 g L^{-1}	~25 mg L^{-1}	UV-A light	~100%	[40]
ZnO	2.0 g L^{-1}	~100 mg L^{-1}	UV light	~86%	[41]
Ag$_3$PO$_4$	0.8 g L^{-1}	~5 mg L^{-1}	visible light	~30%	[1]
Fe-N-P-TiO$_2$	0.5 g L^{-1}	~0.5 mg L^{-1}	UV-A light	~64%	[this paper]
Fe-N-P-TiO$_2$	0.5 g L^{-1}	~0.5 mg L^{-1}	visible light	~29%	[this paper]
Fe-N-P-TiO$_2$	0.5 g L^{-1}	~0.5 mg L^{-1}	solar light	~73%	[this paper]

2.3. Kinetics Evaluation of THI Degradation

The apparent kinetic constant for THI degradation was calculated to underline the influence of doping elements on photocatalytic activity. It was assumed that the photocatalytic degradation rate depends on THI concentration in aqueous solution according to the pseudo first order kinetics [40]. Therefore, the following relationship (Equation (1)) is used for the estimation of k values:

$$-\ln\left(\frac{C}{C_0}\right) = k \cdot t, \tag{1}$$

where:

C = concentration of THI (mg L^{-1}) at the generic irradiation time;
C_0 = concentration of THI (mg L^{-1}) after the dark period;
T = irradiation time (min);

k = apparent kinetic constant (min^{-1}).

The values of the apparent kinetic constant k are calculated by plotting $-\ln\left(\frac{C}{C_0}\right)$ as a function of the irradiation time (t). The obtained values (k_{UV}, k_{Vis}, and k_{solar} for THI degradation) are reported in Table 3 for the TiO_2, Fe-N-P/TiO_2, Fe-N-S/TiO_2, Fe-Pr-N/TiO_2, P-N-S/TiO_2, and Pr-N-S/TiO_2 samples.

Table 3. Apparent kinetic constant for TiO_2, Fe-N-P/TiO_2, Fe-N-S/TiO_2, Fe-N-Pr/TiO_2, P-N-S/TiO_2, and Pr-N-S/TiO_2 samples.

Sample	k_{UV} (min^{-1})	k_{Vis} (min^{-1})	k_{solar} (min^{-1})
TiO_2	1.7×10^{-3}	7.58×10^{-6}	1.8×10^{-3}
Fe-N-P/TiO_2	6.0×10^{-3}	2.1×10^{-3}	7.2×10^{-3}
Fe-N-S/TiO_2	2.2×10^{-3}	7.88×10^{-4}	3.9×10^{-3}
Fe-N-Pr/TiO_2	4.9×10^{-3}	1.5×10^{-3}	5.5×10^{-3}
P-N-S/TiO_2	2.5×10^{-3}	1.0×10^{-3}	4.8×10^{-3}
Pr-N-S/TiO_2	2.8×10^{-3}	2.02×10^{-4}	2.9×10^{-3}

As it is possible to note, the Fe-N-P/TiO_2 photocatalyst showed the highest apparent kinetic constant values (k_{UV} = 6.0×10^{-3} min^{-1}, k_{Vis} = 2.1×10^{-3} min^{-1}, k_{Solar} = 7.2×10^{-3} min^{-1}) among all the analyzed samples.

3. Materials and Methods

3.1. Materials

Titanium tetraisopropoxide ($C_{12}H_{28}O_4Ti$ > 97% Sigma Aldrich, Milan, Italy), iron acetylacetonate (99.95% Sigma Aldrich), praseodymium nitrate hexahydrate (99.9% Sigma Aldrich), urea (CH_4N_2O Sigma Aldrich), phosphoric acid (H_3PO_4 > 99% Sigma-Aldrich), sodium sulfate (Na_2SO_4) thiacloprid ($C_{10}H_9ClN_4S$ > 99% Sigma-Aldrich), and distilled water were employed without additional purification treatment.

3.2. Preparation of Photocatalysts

Fe-N-x and y-N-S/TiO_2 (where x is S, Pr, or P, and y is P or Pr) tri-doped photocatalysts were synthetized through sol-gel method. Fe-N-x photocatalysts were prepared starting from 50 mL of distilled water containing 1.2 g of urea [20] and 17 mL of phosphoric acid [42] or 0.025 g of sodium sulfate [43] or 0.0085 g of praseodymium nitrate hexahydrate [44]. Then, the obtained solution was mixed with a solution obtained dissolving 0.025 g of iron(II) acetylacetonate in 12.5 mL of titanium tetraisopropoxide [20,44]. The obtained photocatalysts were called Fe-N-P/TiO_2 or Fe-N-S/TiO_2 or Fe-N-Pr/TiO_2. y-N-S/TiO_2 photocatalysts were prepared starting from a 50 mL of distilled water and 1.2 g of urea and 17 mL of phosphoric acid or with 0.025 g of sodium sulfate or 0.0085 g of praseodymium nitrate hexahydrate. The obtained aqueous solution was finally added into 12.5 mL of titanium tetraisopropoxide. The obtained photocatalysts were named P-N-S/TiO_2 or P-N-Pr/TiO_2. The preparation of all the photocatalysts were carried out at room temperature, maintaining the systems under continuous stirring for 10 min. The obtained suspensions were centrifuged and washed with distilled water three times. Finally, the precipitates were placed in a furnace at 450 °C for 30 min in static air. Table 4 summarizes the solution volume and the amount of metal or non-metal precursors used for the synthesis, together with molar ratio values for all the prepared photocatalysts.

Table 4. List of all prepared photocatalysts, amount of titanium tetraisopropoxide(T), urea (Ur), iron acetylacetonate (F), phosphoric acid (Pac), sodium sulfate (Ss), and praseodymium nitrate hexahydrate (PrN) used for the synthesis and N/Ti, Fe/Ti, P/Ti, S/Ti, and Pr/Ti molar ratio values.

Sample	T (mL)	Ur (g)	F (mg)	Pac (µL)	Ss (mg)	PrN (µL)	N/Ti	Fe/Ti	P/Ti	S/Ti	Pr/Ti
TiO_2	12.5	0	0	0	0	0	0	0	0	0	0
Fe-N-P/TiO_2	12.5	1.2	25	17	0	0	0.97	0.0017	0.01	0	0
Fe-N-S/TiO_2	12.5	1.2	25	0	25	0	0.97	0.0017	0	0.005	0
Fe-N-Pr/TiO_2	12.5	1.2	0	0	0	8.5	0.97	0.0017	0	0	0.0069
P-N-S/TiO_2	12.5	1.2	0	17	25	0	0.97	0	0.01	0.005	0
Pr-N-S/TiO_2	12.5	1.2	0	17	0	8.5	0.97	0	0.01	0	0.0069

3.3. Photocatalytic Tests

Photocatalytic tests were carried out using a volume of 75 mL of THI aqueous solution (initial concentration: 0.5 mg/L) and 300 mg of catalyst. The batch photoreactor used for the all the tests was a Pyrex cylinder. The suspension was continuously mixed using an external recirculation system assured by a peristaltic pump. The reactor was irradiated by an UV-A (emission: 365 nm; irradiance 13 W/m^2) and visible (emission range: 400–800 nm; irradiance: 16 W/m^2) LEDs strip wrapped around and in contact with external surface of the reactor body. Moreover, additional tests were carried out under the direct solar light (latitude 40° N, longitude 14° E), the average solar UV-A irradiance for all the tests was about 2.2 W/m^2. The total exposure time for each experiment was 180 min. Experiments under the direct solar light were performed on May and typically started at 10:00–11:00 a.m. till 01:00–02:00 p.m. Sunlight irradiance spectra were measured by radiometer BLACK-Comet Stellar Net UV-VIS (StellarNet, Tampa, FL, USA). During each test, the system was left in the dark for 60 min to reach the adsorption equilibrium of THI on the photocatalysts surface and then irradiated for 180 min. At different times, about 1.5 mL of the suspension was withdrawn from the photoreactor and filtered to remove the catalyst particles. The aqueous solution was then analyzed by HPLC UltiMate 3000 Thermo Scientific system (equipped with DAD detector, quaternary pump, column thermostat, and automatic sample injector with 100 µL loop) and using a reversed-phase Luna 5u C18 column (150 mm × 4.6 mm i.d., pore size 5 µm) (Phenomenex) at 25 °C. The mobile phase consisted of an acetonitrile/water mixture (70/30 v/v). The flow rate, injection volume, and detection wavelength were 1 mL/min, 40 µL, and 242 nm, respectively.

3.4. Chemical-Physical Characterization Methods

Laser Raman spectra were obtained at room temperature with a Dispersive MicroRaman (Invia, Renishaw), equipped with 514 nm laser, in the range 100–2000 cm^{-1} Raman shift. The ultraviolet-visible diffuse reflectance spectra (UV-Vis DRS) of the samples were recorded using a Perkin Elmer spectrometer Lambda 35 spectrophotometer using an RSA-PE-20 reflectance spectroscopy accessory (Labsphere, Inc., North Sutton, NH, USA). The band gap values of the samples were determined through the corresponding Kubelka–Munk function (KM) and by plotting (KM × hv^2) against hv. The average crystallite size of the synthetized powders was calculated using Scherrer's equation [45].

Wide-angle X-ray diffraction (WAXD) patterns were performed with an automatic Bruker D8 Advance diffractometer (VANTEC-1 detector) using reflection geometry and nickel filtered Cu-Kα radiation.

The lattice parameters were calculated using the following equation:

$$\frac{1}{d^2_{(h\,k\,l)}} = \frac{h^2 + k^2}{a^2} + \frac{l^2}{c^2}, \tag{2}$$

where the value of $d_{(h\,k\,l)}$ for an XRD peak was determined from Bragg's law:

$$2d_{(h\,k\,l)} \cdot \sin\theta = n \cdot \lambda. \qquad (3)$$

h, k, and l are the crystal planes, and $d_{(h\,k\,l)}$ is the distance between crystal plane of ($h\,k\,l$), while a and c are the lattice parameters (for tetragonal anatase phase of TiO_2: a = b ≠ c). To evaluate the lattice parameters, the planes (1 0 1) and (2 0 0) for anatase were considered [46].

The Brunauer Emmett and Teller (BET) surface area of the samples was measured from dynamic N_2 adsorption measurement at $-196\ °C$, performed by a Costech Sorptometer 1042 after a pre-treatment for 30 min in He flow at 150 °C.

Fourier-transform infrared spectra (FT-IR) of the samples were obtained using TENSOR 27 BRUKER INSTRUMENT in the frequency range of 400–4000 cm^{-1} using KBr as reference pilot. Photoluminescence spectra (PL) of the samples were acquired using a VARIAN CARY-ECLIPSE spectrophotometer. The excitation wavelength was 250 nm. Reflected beams were measured in the range 300–500 nm.

4. Conclusions

The effect of different doping elements was assessed towards the photocatalytic degradation of thiacloprid under UV-A, visible, and solar light irradiation. The physical-chemical properties of all the samples were analyzed by different characterization techniques. In particular, XRD results showed the characteristic peaks of TiO_2 in anatase phase, while no signals of dopant species oxide can be detected in all the tri-doped TiO_2 samples. The increase in a and c axis lattice parameters, found from XRD patterns, and the shifting of the most intense mode in the Raman spectra of tri-doped TiO_2 powder indicate the incorporation of doped elements in TiO_2 lattice. The increase of absorption in the visible region and the shifting of the absorption edge of tri-doped TiO_2 samples evidence a narrowing of the band gap due to dopants elements incorporation in TiO_2 lattice, which is useful for visible light photocatalysis in practical applications. In addition, the reduction of PL intensity indicates a lower recombination rate and higher life span of photogenerated carriers for all the doped samples in comparison with the un-doped TiO_2. Photocatalytic activity results showed that Fe-N-P tri-doped TiO_2 exhibited the highest THI degradation degree (64% under UV-A light, 29% under visible light, and 73% under the direct solar light).

Author Contributions: A.M., W.N. and O.S. performed the experiments and wrote the manuscript. V.V. (Vincenzo Vaiano), V.V. (Vincenzo Venditto) and S.P. provided the concept, experimental design of the study, and reviewed the paper prior to submission. All authors discussed the results, analyzed the data, commented on, and revised the manuscript. All authors have read and agreed to the published version of the manuscript.

Funding: This research received no external funding.

Conflicts of Interest: The authors declare no conflict of interest.

References

1. Lee, Y.-J.; Kang, J.-K.; Park, S.-J.; Lee, C.-G.; Moon, J.-K.; Alvarez, P.J.J. Photocatalytic degradation of neonicotinoid insecticides using sulfate-doped Ag_3PO_4 with enhanced visible light activity. *Chem. Eng. J.* **2020**, *402*, 126183. [CrossRef]
2. Hladik, M.L.; Kolpin, D.W.; Kuivila, K.M. Widespread occurrence of neonicotinoid insecticides in streams in a high corn and soybean producing region, USA. *Environ. Pollut.* **2014**, *193*, 189–196. [CrossRef]
3. Todey, S.A.; Fallon, A.M.; Arnold, W.A. Neonicotinoid insecticide hydrolysis and photolysis: Rates and residual toxicity. *Environ. Toxicol. Chem.* **2018**, *37*, 2797–2809. [CrossRef]
4. Goulson, D. REVIEW: An overview of the environmental risks posed by neonicotinoid insecticides. *J. Appl. Ecol.* **2013**, *50*, 977–987. [CrossRef]
5. Elbert, A.; Haas, M.; Springer, B.; Thielert, W.; Nauen, R. Applied aspects of neonicotinoid uses in crop protection. *Pest Manag. Sci.* **2008**, *64*, 1099–1105. [CrossRef] [PubMed]
6. Jeschke, P.; Moriya, K.; Lantzsch, R.; Seifert, H.; Lindner, W.; Jelich, K.; Göhrt, A.; Beck, M.E.; Etzel, W. Thiacloprid (Bay YRC 2894)-A new member of the Chloronicotinyl Insecticide (CNI) family. *Pflanzenschutz-Nachr. Bayer* **2001**, *54*, 147–160.

7. Yin, K.; Deng, Y.; Liu, C.; He, Q.; Wei, Y.; Chen, S.; Liu, T.; Luo, S. Kinetics, pathways and toxicity evaluation of neonicotinoid insecticides degradation via UV/Chlorine process. *Chem. Eng. J.* **2018**, *346*, 298–306. [CrossRef]
8. Cernigoj, U.; Stangar, U.L.; Jirkovský, J. Effect of dissolved ozone or ferric ions on photodegradation of thiacloprid in presence of different TiO_2 catalysts. *J. Hazard. Mater.* **2010**, *177*, 399–406. [CrossRef]
9. Abramović, B.F.; Banić, N.D.; Šojić, D.V. Degradation of thiacloprid in aqueous solution by UV and UV/H_2O_2 treatments. *Chemosphere* **2010**, *81*, 114–119. [CrossRef]
10. Guo, Q.; Zhou, C.; Ma, Z.; Yang, X. Fundamentals of TiO_2 photocatalysis: Concepts, mechanisms, and challenges. *Adv. Mater.* **2019**, *31*, 1901997. [CrossRef] [PubMed]
11. Pant, B.; Park, M.; Park, S.-J. Recent advances in TiO_2 films prepared by Sol-Gel methods for photocatalytic degradation of organic pollutants and antibacterial activities. *Coatings* **2019**, *9*, 613. [CrossRef]
12. Pant, B.; Ojha, G.P.; Kuk, Y.-S.; Kwon, O.H.; Park, Y.W.; Park, M. Synthesis and characterization of ZnO-TiO_2/carbon fiber composite with enhanced photocatalytic properties. *Nanomaterials* **2020**, *10*, 1960. [CrossRef] [PubMed]
13. Anas, M.; Han, D.; Mahmoud, K.; Park, H.; Abdel-Wahab, A. Photocatalytic degradation of organic dye using titanium dioxide modified with metal and non-metal deposition. *Mater. Sci. Semicond. Process.* **2015**, *41*, 209–218. [CrossRef]
14. Yu, J.C.; Ho, W.; Yu, J.; Yip, H.; Wong, P.K.; Zhao, J. Efficient visible-light-induced photocatalytic disinfection on sulfur-doped nanocrystalline titania. *Environ. Sci. Technol.* **2005**, *39*, 1175–1179. [CrossRef]
15. Kuo, C.-Y.; Wu, C.-H.; Wu, J.-T.; Chen, Y.-R. Synthesis and characterization of a phosphorus-doped TiO_2 immobilized bed for the photodegradation of bisphenol a under UV and sunlight irradiation. *React. Kinet. Mech. Catal.* **2015**, *114*, 753–766. [CrossRef]
16. Jiang, H.; Wang, Q.; Li, S.; Li, J.; Wang, Q. Pr, N, and P tri-doped anatase TiO_2 nanosheets with enhanced photocatalytic activity under sunlight. *Chin. J. Catal.* **2014**, *35*, 1068–1077. [CrossRef]
17. Zhang, D.; Chen, J.; Xiang, Q.; Li, Y.; Liu, M.; Liao, Y. Transition-Metal-Ion (Fe, Co, Cr, Mn, Etc.) doping of TiO_2 nanotubes: A general approach. *Inorg. Chem.* **2019**, *58*, 12511–12515. [CrossRef]
18. Shen, X.-Z.; Guo, J.; Liu, Z.-C.; Xie, S.-M. Visible-light-driven Titania photocatalyst Co-Doped with nitrogen and Ferrum. *Appl. Surf. Sci.* **2008**, *254*, 4726–4731. [CrossRef]
19. Xu, X.; Zhou, X.; Zhang, L.; Xu, L.; Ma, L.; Luo, J.; Li, M.; Zeng, L. Photoredox degradation of different water pollutants (MO, RhB, MB, and Cr(VI)) using Fe-N-S-Tri-Doped TiO_2 nanophotocatalyst prepared by novel chemical method. *Mater. Res. Bull.* **2015**, *70*, 106–113. [CrossRef]
20. Mancuso, A.; Sacco, O.; Sannino, D.; Pragliola, S.; Vaiano, V. Enhanced visible-light-driven photodegradation of acid orange 7 azo dye in aqueous solution using Fe-N Co-Doped TiO_2. *Arab. J. Chem.* **2020**, *13*, 8347–8360. [CrossRef]
21. Khaledi Maki, L.; Maleki, A.; Rezaee, R.; Daraei, H.; Yetilmezsoy, K. LED-activated immobilized Fe-Ce-N Tri-Doped TiO_2 nanocatalyst on glass bed for photocatalytic degradation organic dye from aqueous solutions. *Environ. Technol. Innov.* **2019**, *15*, 100411. [CrossRef]
22. Sánchez-Bayo, F.; Hyne, R.V. Detection and analysis of neonicotinoids in river waters—development of a passive sampler for three commonly used insecticides. *Chemosphere* **2014**, *99*, 143–151. [CrossRef]
23. Schmuck, R. Ecotoxicological profile of the insecticide thiacloprid. *Pflanzenschutz Nachr.-Bayer-Engl. Ed.* **2001**, *54*, 161–184.
24. Khlyustova, A.; Sirotkin, N.; Kusova, T.; Kraev, A.; Titov, V.; Agafonov, A. Doped TiO_2: Effect of doping element on the photocatalytic activity. *Mater. Adv.* **2020**, *1*. [CrossRef]
25. Cheng, X.; Yu, X.; Xing, Z. One-step synthesis of Fe-N-S-Tri-Doped TiO_2 catalyst and its enhanced visible light photocatalytic activity. *Mater. Res. Bull.* **2012**, *47*, 3804–3809. [CrossRef]
26. Xing, M.; Wu, Y.; Zhang, J.; Chen, F. Effect of synergy on the visible light activity of B, N and Fe Co-Doped TiO_2 for the degradation of MO. *Nanoscale* **2010**, *2*, 1233–1239. [CrossRef]
27. de la Osa, R.A.; Iparragirre, I.; Ortiz, D.; Saiz, J.M. The extended Kubelka–Munk theory and its application to spectroscopy. *ChemTexts* **2019**, *6*, 2. [CrossRef]
28. Ricci, P.C.; Carbonaro, C.M.; Stagi, L.; Salis, M.; Casu, A.; Enzo, S.; Delogu, F. Anatase-to-rutile phase transition in TiO_2 nanoparticles irradiated by visible light. *J. Phys. Chem. C* **2013**, *117*, 7850–7857. [CrossRef]
29. Giannakas, A.E.; Antonopoulou, M.; Daikopoulos, C.; Deligiannakis, Y.; Konstantinou, I. Characterization and catalytic performance of B-Doped, B–N Co-Doped and B–N–F Tri-Doped TiO_2 towards simultaneous Cr(VI) reduction and benzoic acid oxidation. *Appl. Catal. B Environ.* **2016**, *184*, 44–54. [CrossRef]
30. Adyani, S.M.; Ghorbani, M. A comparative study of physicochemical and photocatalytic Properties of visible light responsive Fe, Gd and P Single and Tri-Doped TiO_2 nanomaterials. *J. Rare Earths* **2018**, *36*, 72–85. [CrossRef]
31. Akshay, V.R.; Arun, B.; Mandal, G.; Mutta, G.R.; Chanda, A.; Vasundhara, M. Observation of optical Band-Gap narrowing and enhanced magnetic moment in Co-Doped Sol–Gel-Derived anatase TiO_2 nanocrystals. *J. Phys. Chem. C* **2018**, *122*, 26592–26604. [CrossRef]
32. Ramandi, S.; Entezari, M.; Ghows, N. Sono-synthesis of solar light responsive S–N–C–Tri doped TiO_2 Photo-Catalyst under optimized conditions for degradation and mineralization of diclofenac. *Ultrason. Sonochem.* **2017**, *38*, 234–245. [CrossRef]
33. Wang, J.A.; Limas-Ballesteros, R.; López, T.; Moreno, A.; Gómez, R.; Novaro, O.; Bokhimi, X. Quantitative determination of titanium lattice defects and solid-state reaction mechanism in iron-doped TiO_2 photocatalysts. *J. Phys. Chem. B* **2001**, *105*, 9692–9698. [CrossRef]

34. Gaur, L.K.; Kumar, P.; Kushavah, D.; Khiangte, K.R.; Mathpal, M.C.; Agrahari, V.; Gairola, S.P.; Soler, M.A.G.; Swart, H.C.; Agarwal, A. Laser induced phase transformation influenced by co doping in TiO_2 nanoparticles. *J. Alloys Compd.* **2019**, *780*, 25–34. [CrossRef]
35. Jiang, X.; Zhang, Y.; Jiang, J.; Rong, Y.; Wang, Y.; Wu, Y.; Pan, C. Characterization of oxygen vacancy associates within hydrogenated TiO_2: A positron annihilation study. *J. Phys. Chem. C* **2012**, *116*, 22619–22624. [CrossRef]
36. Decomposition and Detoxification of the Insecticide Thiacloprid by TiO_2-Mediated Photocatalysis: Kinetics, Intermediate Products and Transformation Pathways-Berberidou-2019-Journal of Chemical Technology & Biotechnology-Wiley Online Library. Available online: https://onlinelibrary.wiley.com/doi/full/10.1002/jctb.6034?casa_token=mlM5yFvrjRkAAAAA%3AqjvoJlIhUQ_zQmYIRRpXJG94RegEqrQWiH9QZaKrcgQ3wAgZVLhyxtgjZKsfd_vzqTH8s8Tw6XgHLg (accessed on 2 February 2021).
37. Zhong, Z.; Li, M.; Fu, J.; Wang, Y.; Muhammad, Y.; Li, S.; Wang, J.; Zhao, Z.; Zhao, Z. Construction of cu-bridged Cu_2O/MIL(Fe/Cu) catalyst with enhanced interfacial contact for the synergistic photo-Fenton degradation of thiacloprid. *Chem. Eng. J.* **2020**, *395*, 125184. [CrossRef]
38. Banić, N.; Abramović, B.; Krstić, J.; Šojić, D.; Lončarević, D.; Cherkezova-Zheleva, Z.; Guzsvány, V. Photodegradation of thiacloprid using Fe/TiO_2 as a heterogeneous photo-Fenton catalyst. *Appl. Catal. B Environ.* **2011**, *107*, 363–371. [CrossRef]
39. Banić, N.D.; Abramović, B.F.; Šojić, D.V.; Krstić, J.B.; Finčur, N.L.; Bočković, I.P. Efficiency of neonicotinoids photocatalytic degradation by using annular slurry reactor. *Chem. Eng. J.* **2016**, *286*, 184–190. [CrossRef]
40. Rózsa, G.; Kozmér, Z.; Alapi, T.; Schrantz, K.; Takács, E.; Wojnárovits, L. Transformation of Z-thiacloprid by three advanced oxidation processes: Kinetics, intermediates and the role of reactive species. *Catal. Today* **2017**, *284*, 187–194. [CrossRef]
41. Abramović, B.F.; Banić, N.D.; Krstić, J.B. Degradation of thiacloprid by ZnO in a laminar falling film slurry photocatalytic reactor. *Ind. Eng. Chem. Res.* **2013**, *52*, 5040–5047. [CrossRef]
42. Mendiola-Alvarez, S.; Hernandez-Ramírez, A.; Guzmán Mar, J.; Garza-Tovar, L.L.; Reyes, L. Phosphorous-doped TiO_2 nanoparticles: Synthesis, characterization, and visible photocatalytic evaluation on sulfamethazine degradation. *Environ. Sci. Pollut. Res.* **2019**, *26*. [CrossRef] [PubMed]
43. Mayoufi, A.; Nsib, M.F.; Ahmed, O.; Houas, A. Synthesis, characterization and photocatalytic performance of W, N, S-Tri-Doped TiO_2 under visible light irradiation. *Comptes Rendus Chim.* **2015**, *18*, 875–882. [CrossRef]
44. Mancuso, A.; Sacco, O.; Vaiano, V.; Sannino, D.; Pragliola, S.; Venditto, V.; Morante, N. Visible light active Fe-Pr Co-doped TiO_2 for water pollutants degradation. *Catal. Today* **2021**. [CrossRef]
45. Freeda, M.; Suresh, G. Structural and luminescent properties of Eu-doped $CaAl_2O_4$ nanophosphor by Sol-Gel method. *Mater. Today Proc.* **2017**, *4*, 4260–4265. [CrossRef]
46. Ganesh, I.; Gupta, A.K.; Kumar, P.P.; Sekhar, P.S.C.; Radha, K.; Padmanabham, G.; Sundararajan, G. Preparation and characterization of Ni-Doped TiO_2 materials for photocurrent and photocatalytic applications. *Sci. World J.* **2012**, *2012*, 127326. [CrossRef] [PubMed]

Article

Green Synthesis of N/Zr Co-Doped TiO$_2$ for Photocatalytic Degradation of *p*-Nitrophenol in Wastewater

Hayette Benkhennouche-Bouchene [1], Julien G. Mahy [2,*], Cédric Wolfs [3], Bénédicte Vertruyen [4], Dirk Poelman [5], Pierre Eloy [2], Sophie Hermans [2], Mekki Bouhali [6], Abdelhafid Souici [6], Saliha Bourouina-Bacha [1] and Stéphanie D. Lambert [3]

[1] Faculté de Technologie, Département de Génie des Procédés, Université de Bejaia, Bejaia 06000, Algeria; hayette.bouchene@gmail.com (H.B.-B.); reacteurschimiques@gmail.com (S.B.-B.)
[2] Institute of Condensed Matter and Nanosciences (IMCN), Université Catholique de Louvain, Place Louis Pasteur 1, 1348 Louvain-la-Neuve, Belgium; pierre.eloy@uclouvain.be (P.E.); sophie.hermans@uclouvain.be (S.H.)
[3] Department of Chemical Engineering-Nanomaterials, Catalysis & Electrochemistry, University of Liège, B6a, Quartier Agora, Allée du six Août 11, 4000 Liège, Belgium; cedric.wolfs@uliege.be (C.W.); stephanie.lambert@uliege.be (S.D.L.)
[4] GreenMAT, CESAM Research Unit, University of Liège, B6a, Quartier Agora, Allée du six Août 13, 4000 Liège, Belgium; b.vertruyen@uliege.be
[5] LumiLab, Department of Solid State Sciences, Ghent University, 9000 Gent, Belgium; dirk.poelman@ugent.be
[6] Laboratory of Physical Chemistry of Materials and Catalysis (LPCMC), Faculty of Exact Sciences, University of Bejaia, Bejaia 06000, Algeria; bouhali_net@yahoo.fr (M.B.); souici2015@gmail.com (A.S.)
* Correspondence: julien.mahy@uclouvain.be; Tel.: +32-4-3664771

Citation: Benkhennouche-Bouchene, H.; Mahy, J.G.; Wolfs, C.; Vertruyen, B.; Poelman, D.; Eloy, P.; Hermans, S.; Bouhali, M.; Souici, A.; Bourouina-Bacha, S.; et al. Green Synthesis of N/Zr Co-Doped TiO$_2$ for Photocatalytic Degradation of *p*-Nitrophenol in Wastewater. *Catalysts* **2021**, *11*, 235. https://doi.org/10.3390/catal11020235

Academic Editor: Suresh C. Pillai
Received: 30 November 2020
Accepted: 6 February 2021
Published: 10 February 2021

Publisher's Note: MDPI stays neutral with regard to jurisdictional claims in published maps and institutional affiliations.

Copyright: © 2021 by the authors. Licensee MDPI, Basel, Switzerland. This article is an open access article distributed under the terms and conditions of the Creative Commons Attribution (CC BY) license (https://creativecommons.org/licenses/by/4.0/).

Abstract: TiO$_2$ prepared by a green aqueous sol–gel peptization process is co-doped with nitrogen and zirconium to improve and extend its photoactivity to the visible region. Two nitrogen precursors are used: urea and triethylamine; zirconium (IV) tert-butoxide is added as a source of zirconia. The N/Ti molar ratio is fixed regardless of the chosen nitrogen precursor while the quantity of zirconia is set to 0.7, 1.4, 2, or 2.8 mol%. The performance and physico-chemical properties of these materials are compared with the commercial Evonik P25 photocatalyst. For all doped and co-doped samples, TiO$_2$ nanoparticles of 4 to 8 nm of size are formed of anatase-brookite phases, with a specific surface area between 125 and 280 m^2 g^{-1} vs. 50 m^2 g^{-1} for the commercial P25 photocatalyst. X-ray photoelectron (XPS) measurements show that nitrogen is incorporated into the TiO$_2$ materials through Ti-O-N bonds allowing light absorption in the visible region. The XPS spectra of the Zr-(co)doped powders show the presence of TiO$_2$-ZrO$_2$ mixed oxide materials. Under visible light, the best co-doped sample gives a degradation of *p*-nitrophenol (PNP) equal to 70% instead of 25% with pure TiO$_2$ and 10% with P25 under the same conditions. Similarly, the photocatalytic activity improved under UV/visible reaching 95% with the best sample compared to 50% with pure TiO$_2$. This study suggests that N/Zr co-doped TiO$_2$ nanoparticles can be produced in a safe and energy-efficient way while being markedly more active than state-of-the-art photocatalytic materials under visible light.

Keywords: ambient crystallization; photocatalysis; Zr/N doping; titania; aqueous sol-gel process; *p*-nitrophenol degradation

1. Introduction

Industrial development is the main cause of the increase of pollution in water [1]. Although these polluted effluents undergo various physical, chemical, and biological treatments [2], some pollutants are not degraded and unfortunately return to the ecosystem [3].

Advanced oxidation processes (AOPs) are among the most efficient processes for the total mineralization of organic compounds [4–6]. The interest of AOPs lies in their ability to degrade almost all organic molecules, e.g., by reacting with the double bonds (-C=C-) and attacking the aromatic rings, major constituents of refractory pollutants. Photocatalysis is a promising advanced oxidation process in view of its ability to degrade a great number of organic molecules, low cost, and versatile application in the field of pollution control [5,7].

Indeed, the principle of photocatalysis for pollution abatement is to oxidize target molecules, leading to CO_2 and H_2O only in case of total degradation. Semiconductors are typically used as photocatalysts due to their physico-chemical characteristics in the presence of light. Indeed, photons with sufficient photon energy can promote an electron from the semiconductor's valence band to its conduction band. These photogenerated species migrate at the surface of the photocatalysts and can do redox reactions with the surrounding medium. In water, this leads to the production of hydroxyl radicals ·OH, the most powerful oxidizing species, and the formation of the superoxide radical $O^{2-}\cdot$. This process is persistent as long as light is available [3,5,8].

Several successful semiconductors for this application are described in literature, including ZnO, WO_3, ZnS, CdS, Fe_2O_3, and TiO_2 [3]. Titanium dioxide is the ninth most abundant component on earth and is used in various fields of applications such as food packaging (UV protection of food), in sunscreens, in orthopedic implants, but also in photocatalysis for the production of various self-cleaning coatings, the removal of impurities, and the depollution of water and air [9,10].

Anatase TiO_2, its most photoactive phase, activation requires photons with an energy greater or equal to the 3.2 eV band gap of TiO_2. This energy is equivalent to the energy of a photon with a wavelength of 388 nm, corresponding to the ultraviolet range [11]. TiO_2 presents some advantages over other semiconductors. Besides its non-toxic and chemical stability aspect, it has good corrosion resistance, and is also relatively inexpensive. TiO_2 has a high photocatalytic activity and chemical stability under ultraviolet light (<388 nm). On the other hand, titania's large band gap only allows to harvest UV light, representing only about 5% of the solar spectrum [12]. Another disadvantage of TiO_2 is the strong recombination rate of the photo-generated electron–hole (e^-/h^+) species.

The development of photocatalysts with a high and stable activity under visible light (>400 nm) should allow to use a larger part of the solar spectrum, even under low-intensity indoor lighting [8,13]. To reach these goals, modifications of TiO_2 were envisaged with the incorporation of metallic or non-metallic atoms. The list of doping elements in literature is long and includes transition metal ions such as Zr [14], Cu [15], Co [16], Ni [17], Cr [18], Mn [19], Mo [20], Nb [21], V [22], Fe [23], Ru [24], Au [25], Ag [26], and Pt [11], or non-metallic ions such as N [27], S [28], C [29], B [30], P [31], I [32], and F [33]. The combination with other semiconductors, having a lower band energy, was also studied [34], as well as the sensitization of TiO_2 with organic or organometallic dyes [35].

The production of TiO_2 materials can be carried out by several methods such as mechanochemical techniques, precipitation, aerosol powder coating, hydrothermal methods, crystallization, and sol–gel methods [36]. The sol–gel process has several advantages, as it offers simplicity of implementation, cost effectiveness, high purity, and careful control of the chemical composition [36–38].

The first work on the doping of TiO_2 with a non-metallic element, and more precisely nitrogen, was carried out in 1986 by Sato et al. [39], who obtained N-doped powders with better oxidation of carbon monoxide and ethane compared to the commercial Evonik P25. Then, in 2001, Morikawa et al. [40] reported that nitrogen doping reduced the band gap of TiO_2 and improved the absorption of visible light, due to a modification of the band structure by substitution of the $2p$ states of N in the TiO_2 lattice mixed with the O $2p$ states. Indeed, nitrogen can be easily introduced into the TiO_2 structure due to its atomic size comparable to that of oxygen, its low ionization energy, and its high stability [27].

Transition metal modification can also extend the spectral response of TiO_2 in the visible light region by inducing electronic transitions between the d-electrons of the transition metal ions and the conduction band of TiO_2 [11]. This activity is also related to the formation of a new energy level produced in the TiO_2 band gap by dispersing metal nanoparticles in the TiO_2 lattice. This modification also improves the photoactivity due to electron trapping on delocalized metallic dopant, reducing e^-/h^+ recombination rate during irradiation [11]. TiO_2 N-doping proved to be a promising method to increase its photoactivity, but the photocatalytic efficiency of N-doped TiO_2 was limited due to the strongly localized N $2p$ states at the top of the valence band, which can act as traps, and rapid recombination for the excited electrons [12,40]. Therefore, the simultaneous use of a metal and non-metal as co-doping elements may be an effective alternative for improving photocatalytic activity [12].

Co-doping of TiO_2 with zirconium and nitrogen simultaneously has been reported in only a limited number of papers compared to other doping elements [12,41–44]. This work is a contribution to complement the development of co-doped TiO_2 powders, more active in the visible range but also more efficient in the UV range.

In this study, TiO_2 powders were doped with two nitrogen precursors (urea and triethylamine) and with a zirconium source. The photocatalysts were prepared by an aqueous sol–gel method, combining and optimizing two methods developed previously in Mahy et al. [14,27]. Emphasis was placed on the development of a green process by using water as solvent and no calcination step to crystallize the titania. The conditions were optimized by fixing the N/Ti ratio for both urea and triethylamine to the best ratio in terms of photoactivity, and by varying the zirconium source fractions (0.7, 1.4, 2, or 2.8 mol%). Subsequently, the samples were characterized by X-ray diffraction (XRD), nitrogen adsorption–desorption measurements, diffuse reflectance UV–Visible spectroscopy (DR-UV-Vis), X-ray photoelectron spectroscopy (XPS), transmission electron microscopy (TEM), Fourier transform infrared spectroscopy (FTIR), thermogravimetric analysis (TG), Photoluminescence (PL), and by inductively coupled plasma–atomic emission spectroscopy (ICP–AES).

The last part of this study is devoted to testing the photocatalytic activity of those doped samples for the degradation of a model solution of p-nitrophenol (PNP, $C_6H_5NO_3$) under UV–visible and visible light. The aims are to show the influence of N/Zr co-doping on the photocatalytic activity and to identify the best doping ratios. The commercial Evonik P25 photocatalyst will be used as reference for benchmarking the obtained performances.

2. Results

Concerning the notation of samples used in this study, samples doped with urea are designated TiO_2/Ux, where x is the molar ratio between titanium dioxide and urea. In this work, x is chosen equal to 4, following Mahy et al. [27]. Samples doped with triethylamine are designated TiO_2/Ny, where y is the molar ratio between titanium dioxide and triethylamine. In this work, y is chosen equal to 42, following Mahy et al. [27]. For samples doped with zirconium (IV) tertbutoxide, they are denoted by TiO_2/Zrz, where z is the molar percentage of Zr and is equal to 0.7, 1.4, 2, and 2.8 mol%. The samples co-doped with urea and zirconium are denoted $TiO_2/U4/Zrz$ and those co-doped with triethylamine and zirconium are denoted $TiO_2/N42/Zrz$.

2.1. Crystallographic Properties of Samples

Figure 1 shows the XRD patterns of pure TiO_2 and selected N/Zr-doped and co-doped TiO_2 samples; reference patterns for the anatase and brookite TiO_2 phases are also presented.

These powders, prepared by precipitation-peptization, are similar in crystallographic structure with the identified presence of the anatase, brookite, and amorphous phases. The other samples (not shown) have similar patterns. The crystallographic fractions shown in Table 1 were obtained by Rietveld refinement of the scale factors with the TOPAS software using a CaF_2 internal standard.

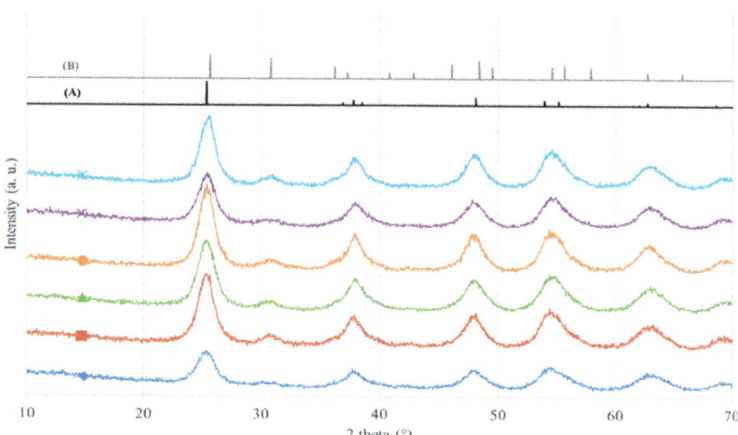

Figure 1. X-ray diffraction (XRD) patterns: (♦) pure TiO_2, (■) TiO_2/Zr2.8, (▲) TiO_2/U4, (●) TiO_2/U4/Zr2.8, (×) TiO_2/N42, (○) TiO_2/N42/Zr2.8. (**A**) Reference pattern of anatase and (**B**) reference pattern of brookite.

Table 1. Quantitative analysis of samples' crystallinity and dopant content by XRD.

Sample	Anatase Content (%) ±5	Brookite Content (%) ±5	Amorphous Content (%) ±5	Rutile Content (%) ±5	Theoretical Zr Content (mol.%)	Actual Zr Content (mol.%)
P25	80	-	-	20	-[1]	-[1]
Pure TiO_2	30	25	45	-	-[1]	-[1]
TiO_2/Zr0.7	30	20	50	-	0.70	0.70
TiO_2/Zr1.4	30	20	50	-	1.40	1.53
TiO_2/Zr2	25	20	55	-	2.00	1.64
TiO_2/Zr2.8	25	15	60	-	2.80	2.68
TiO_2/U4	35	10	55	-	-[1]	-[1]
TiO_2/U4/Zr0.7	40	30	30	-	0.70	0.89
TiO_2/U4/Zr1.4	35	30	35	-	1.40	1.50
TiO_2/U4/Zr2	40	25	35	-	2.00	2.32
TiO_2/U4/Zr2.8	40	25	35	-	2.80	2.97
TiO_2/N42	35	20	45	-	-[1]	-[1]
TiO_2/N42/Zr0.7	40	30	30	-	0.70	1.36
TiO_2/N42/Zr1.4	35	20	45	-	1.40	1.69
TiO_2/N42/Zr2	35	20	45	-	2.00	2.09
TiO_2/N42/Zr2.8	35	20	45	-	2.80	3.38

[1] Not measured.

The highest content of the anatase phase (around 40%, Table 1) is observed for the TiO_2 samples co-doped with urea and zirconium tert-butoxide and for the TiO_2/N42/Zr0.7 sample. The brookite phase is more present (around 30%, Table 1) in urea/Zr co-doped powders compared to other samples. The amorphous fraction of TiO_2 ranges between 30% and 60%. Zirconium doping seems to increase the amorphous fraction of TiO_2, both in doped Zr/TiO_2 samples and in co-doped N/Zr/TiO_2 samples. Thus, the crystallization of amorphous TiO_2 into anatase and brookite structures is less favored with a higher zirconium loading, as reported by other authors [41]. It is also observed in Table 2 that the TiO_2 crystallite sizes, d_{XRD}, slowly increase from 4 to 6 nm with increasing zirconium content, as already observed by Mahy et al. [14].

Table 2. Textural and optical properties of TiO_2-based photocatalysts.

Sample	S_{BET} (m²g⁻¹) ±5	V_P (cm³g⁻¹) ±0.01	V_{DR} (cm³g⁻¹) ±0.01	d_{BET} (nm) ±1	d_{XRD} (nm) ±1	d_{TEM} (nm) ±1	$E_{g \cdot direct}$ (eV) ±0.01	$E_{g \cdot indirect}$ (eV) ±0.01
P25	50	-¹	0.03	31	18 ²–8 ³	-¹	3.45	3.05
TiO_2 pure	195	0.10	0.1	8	5	5	3.35	2.98
TiO_2/Zr0.7	205	0.11	0.11	8	4	6	3.36	3.03
TiO_2/Zr1.4	205	0.11	0.11	8	6	6	3.29	2.98
TiO_2/Zr2	210	0.12	0.11	7	6	5	3.26	2.90
TiO_2/Zr2.8	195	0.12	0.11	8	6	6	3.32	2.97
TiO_2/U4	270	0.24	0.16	6	6	6	3.34	3.05
TiO_2/U4/Zr0.7	260	0.28	0.15	6	7	6	3.35	3.07
TiO_2/U4/Zr1.4	280	0.27	0.17	5	4	5	3.26	2.98
TiO_2/U4/Zr2	280	0.34	0.17	5	6	6	3.32	3.04
TiO_2/U4/Zr2.8	200	0.28	0.12	8	6	7	3.32	3.07
TiO_2/N42	240	0.24	0.15	6	6	6	3.25	3.00
TiO_2/N42/Zr0.7	185	0.23	0.12	8	6	6	3.26	2.97
TiO_2/N42/Zr1.4	230	0.26	0.14	7	6	6	3.27	2.99
TiO_2/N42/Zr2	200	0.26	0.12	8	6	5	3.31	3.03
TiO_2/N42/Zr2.8	220	0.26	0.13	7	6	6	3.31	2.99

-¹ Not measured; ² value from anatase peak; ³ value from rutile peak; S_{BET}: specific surface area estimated by the Brunauer–Emmett–Teller (BET) theory; V_p: specific liquid volume adsorbed at saturation pressure of nitrogen; V_{DR}: microporous volume calculated thanks to the Dubinin–Raduskevitch theory; d_{BET}: mean diameter of TiO_2 nanoparticles obtained from S_{BET} values; d_{XRD}: mean diameter of TiO_2 crystallites calculated using the Scherrer equation; d_{TEM}: elementary TiO_2 particle diameter measured by TEM; E_g,direct: direct optical band gap value estimated with the transformed Kubelka–Munk function; E_g,indirect: indirect optical band gap values estimated with the transformed Kubelka–Munk function.

2.2. Composition of Samples

The Fourier transform infrared (FTIR) spectra of sample P25, pure TiO_2, TiO_2/U4, TiO_2/N42, TiO_2/U4/Zr2.8, and TiO_2/N42/Zr2.8 are shown in Figure 2. Generally, the spectra of aqueous samples are similar between each other, and few peaks are also similar to the commercial P25.

The strong broadband at 3200 cm⁻¹ as well as small peak at 1631 cm⁻¹ are due to vibrations of the –OH groups from the water adsorbed in the samples, as well as Ti-OH and Zr-OH groups in the Zr-doped samples. No additional peaks were observed for the Zr doping peaks, promoting efficient dispersion of zirconium [45].

The peak intensity in TiO_2/U4/Zr2.8 and TiO_2/N42/Zr2.8 powders relating to –OH groups are higher than those of pure TiO_2, indicating that nitrogen and zirconium doping increases the surface hydroxyl groups thus enhancing the photocatalytic activity.

The spectra undoubtedly reveal the presence of synthetic residues in addition to TiO_2. For pure TiO_2 and TiO_2/Zr2.8 samples, the doublet peaks located at 1553 cm⁻¹ and 1315 cm⁻¹ would seem to be attributed to nitrates [46–48], as well as the peak located at 1049 cm⁻¹ which returns to the free NO_3^- ions [49,50]. In fact, pure and single Zr-doped samples (prepared in the presence of HNO_3 during the synthesis) often contain residues of NO_3^- species [49], which is explained because these powders are not rinsed at the end of the synthesis, unlike the powders doped with nitrogen. It is observed that the intensity of these two peaks at 1553 cm⁻¹ and 1315 cm⁻¹ decreases with the doping in nitrogen source, explained by the washing of the powders when doped with urea or triethylamine as explained in the Section 3.5, as well as the disappearance of the peak at 1049 cm⁻¹.

The other doped samples have similar spectra.

The thermogravimetric analysis for the pure TiO_2, TiO_2/U4, TiO_2/N42, TiO_2/U4/Zr2.8, and TiO_2/N42/Zr2.8 samples is shown in Figure 3. The weight loss is comprised between 4 and 18%; the highest is reached with the pure TiO_2 sample. Indeed, this sample was not washed after synthesis. The losses occur between 100 °C and 400 °C. After 400 °C, no weight loss is observed.

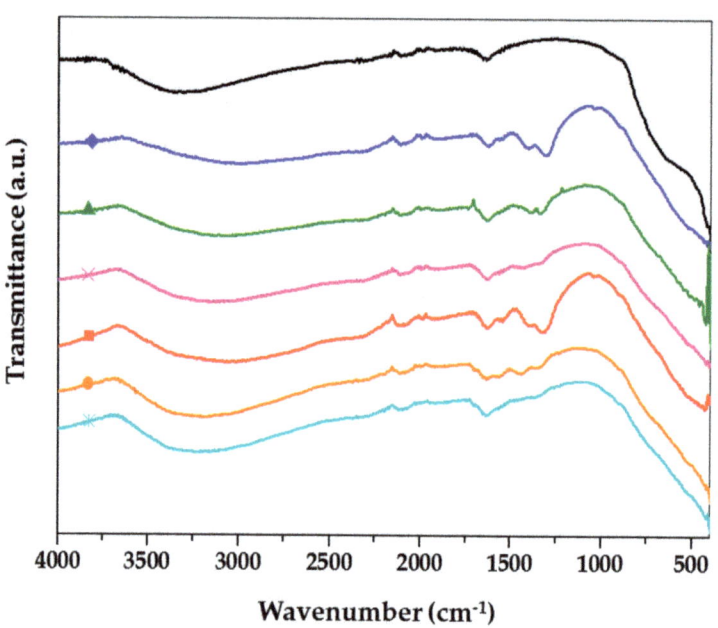

Figure 2. Fourier transformed infrared spectra (FTIR): (-) P25; (♦) pure TiO_2, (■) TiO_2/Zr2.8, (▲) TiO_2/U4, (●) TiO_2/U4/Zr2.8, (×) TiO_2/N42, (˙) TiO_2/N42/Zr2.8.

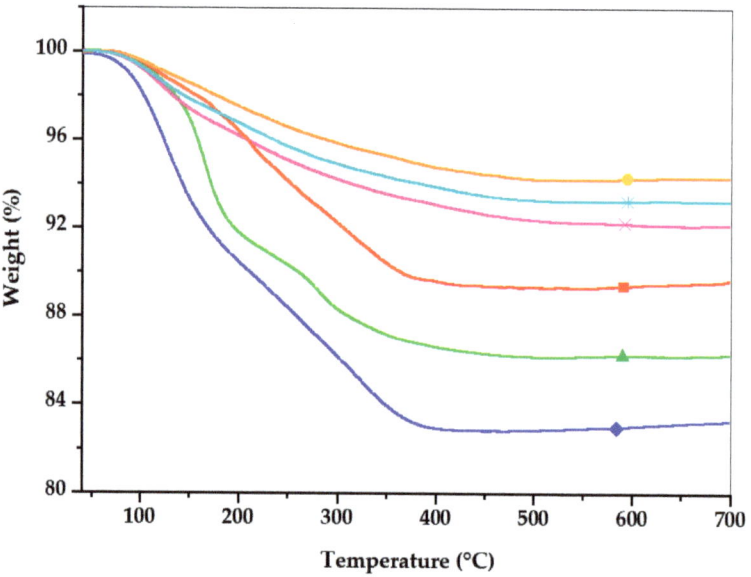

Figure 3. Thermogravimetric analysis of (♦) pure TiO_2, (■) TiO_2/Zr2.8, (▲) TiO_2/U4, (●) TiO_2/U4/Zr2.8, (×) TiO_2/N42, and (˙) TiO_2/N42/Zr2.8.

2.3. Textural Properties of Samples

Figure 4 shows the nitrogen adsorption-desorption isotherms for N/Zr doped and co-doped TiO$_2$ powders, with the corresponding pure TiO$_2$ sample as a reference.

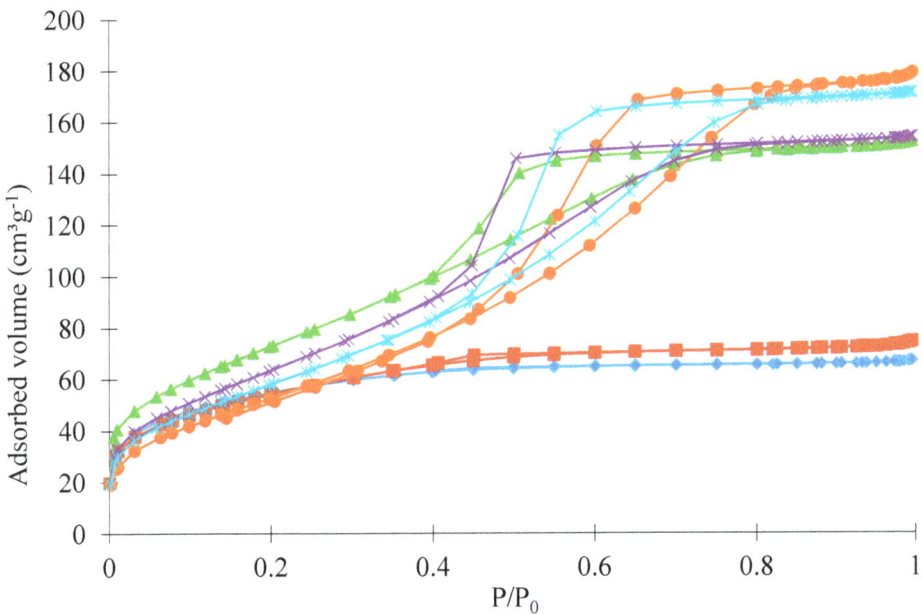

Figure 4. Nitrogen adsorption–desorption isotherms of samples: (♦) pure TiO$_2$, (■) TiO$_2$/Zr2.8, (▲) TiO$_2$/U4, (•) TiO$_2$/U4/Zr2.8, (×) TiO$_2$/N42, and (○) TiO$_2$/N42/Zr2.8.

For the pure TiO$_2$ sample and all samples only doped with Zr (TiO$_2$/Zr0.7, TiO$_2$/Zr1.4, TiO$_2$/Zr2 and TiO$_2$/Zr2.8), the isotherms are similar: a strong increase in adsorbed volume at low pressure, followed by a plateau, corresponding to a microporous solid (type I isotherm from the BDDT classification) [14,51]. The values of the specific surface area, S_{BET}, and microporous volume, V_{DR}, are similar for all samples, between 195 and 210 m^2 g^{-1} for S_{BET} and equal to 0.11 cm^3 g^{-1} for V_{DR} (Table 2). Furthermore, the V_{DR} and V_p (specific liquid volume adsorbed at saturation pressure of nitrogen) values are similar, a specific characteristic of microporous materials [51]. These textural properties are typical of TiO$_2$ samples doped with zirconium and prepared with the peptization-precipitation method [14]. This is due to the spherical shape of the nanoparticles in between which small voids (<2 nm) lie [14]. It is possible to see on TEM micrographs (see next section, Figure 5C) these small TiO$_2$ nanoparticles.

When nitrogen is incorporated inside the framework of TiO$_2$ (samples TiO$_2$/U4, TiO$_2$/U4/Zr2.8, TiO$_2$/N42, and TiO$_2$/N42/Zr2.8), the nitrogen adsorption–desorption isotherms evolve towards a mixture of the Type I to the Type IV [51]: (i) at low relative pressure, a sharp increase of the adsorbed volume is followed by a plateau which corresponds to type I isotherm, which is characteristic of microporous adsorbents; (ii) for relative pressure p/p_0 comprised between 0.4 and 0.8, a triangular hysteresis appears and is followed by a plateau, which is characteristic of mesoporous adsorbents. Furthermore, this type of hysteresis is characteristic of samples consisting of agglomerates (a few tens of nm), these agglomerates being themselves composed of elementary spherical TiO$_2$ particles. Finally, for samples doped with nitrogen and co-doped with nitrogen and zirconium, V_p values are higher than V_{DR} values (Table 2), meaning that these samples have a microporous volume and a mesoporous volume. In Figure 5A–D (TEM micrographs, see next section), it is possible to see that higher aggregates of TiO_2, nonexistent in Figure 5C.

From the specific surface area, S_{BET}, it is possible to estimate the elementary TiO_2 particle size, d_{BET}, with Equation (2) (see Section 3.7) by assuming elementary spherical and non-porous TiO_2 nanoparticles. The order of magnitude of d_{BET} is close to d_{XRD} values and d_{TEM} values (Table 2).

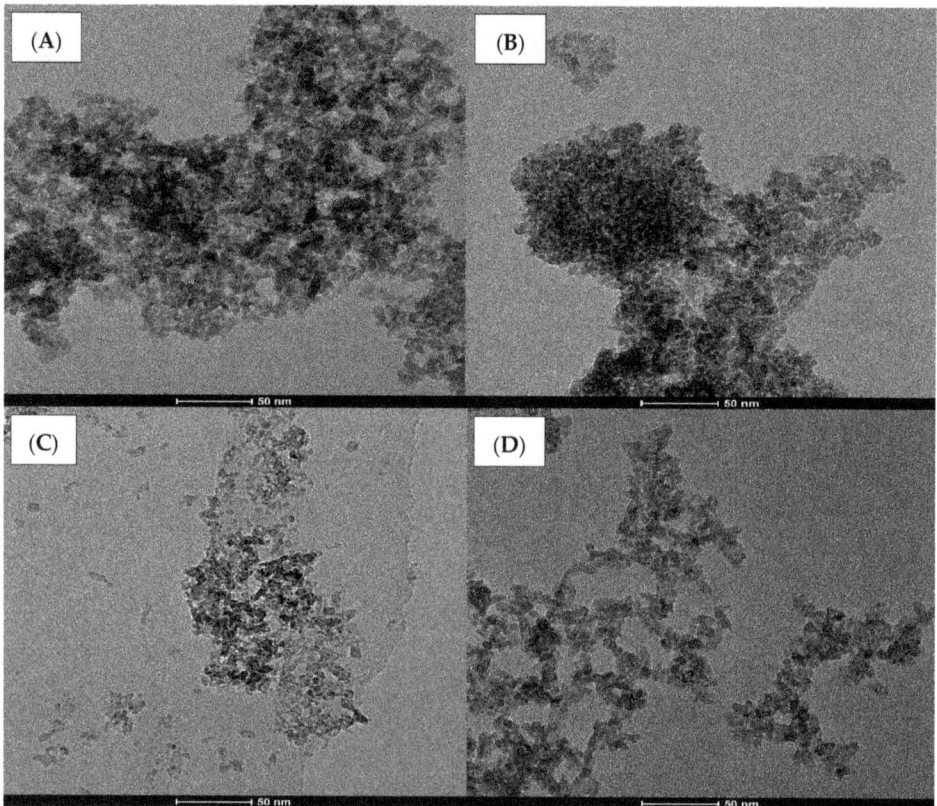

Figure 5. TEM micrographs of samples: (**A**) TiO_2/N42, (**B**) TiO_2/N42/Zr2.8, (**C**) TiO_2/Zr2.8, and (**D**) TiO_2/U4/Zr2.8.

2.4. Morphology of Samples

The morphology of the N/Zr doped and co-doped TiO_2 samples was visualized with a transmission electron microscope and is shown for four samples in Figure 5. The sample TiO_2/Zr2.8 (Figure 5C) presents a spherical shape and uniform distribution of the elementary TiO_2 nanoparticles. For the other samples (Figure 5A–D), higher aggregates (a few tens of nm) of TiO_2 particles are observed, these agglomerates being themselves composed of elementary spherical TiO_2 particles.

The size of elementary TiO_2 nanoparticles have been evaluated from TEM images on a series of fifty titania particles (Table 2). For all samples, the particle size range is similar to those found by XRD, d_{XRD}, and from nitrogen adsorption–desorption isotherms, d_{BET} (Table 2) [27].

2.5. Optical Properties of Samples

The normalized Kubelka–Munk function is shown in Figure 6 for pure TiO_2, $TiO_2/Zr2.8$, $TiO_2/U4/Zr2.8$, and $TiO_2/N42/Zr2.8$ samples.

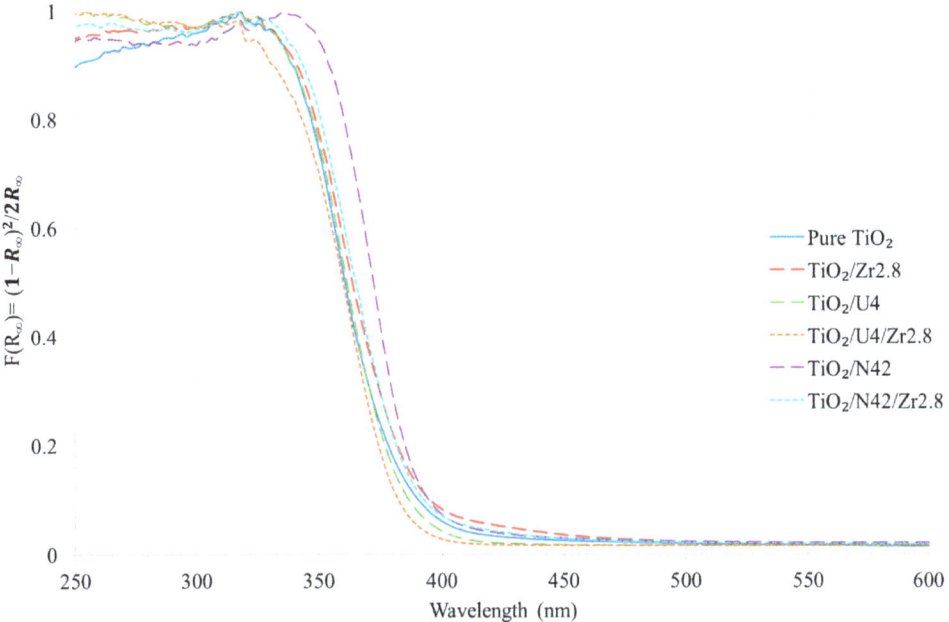

Figure 6. Normalized Kubelka–Munk function $F(R_\infty)$ calculated from DR-UV–Vis spectra for samples: pure TiO_2, $TiO_2/Zr2.8$, $TiO_2/U4$, $TiO_2/U4/Zr2.8$, $TiO_2/N42$, $TiO_2/N42/Zr2.8$ (caption in the figure).

The pure TiO_2 spectrum presents absorption around 365 nm. $TiO_2/N42$ sample presents a shift towards visible region compared to pure TiO_2, which can be explained by the insertion of N as already observed by Mahy et al. [27] using this synthesis method.

The other samples show absorption spectra close to the pure TiO_2. Nevertheless, the obtained spectra were all very slightly shifted to the visible range (Figure 6) and show slightly lower band gap values than pure TiO_2 (Table 2), especially if it is compared to commercial Evonik P25 photocatalyst. Indeed, N-doping occurred with aqueous sol–gel synthesis using HNO_3 as peptizing agent [14,27] even for pure TiO_2 (see Section 3.5), with as a consequence a positive effect on visible light activation.

2.6. XPS Analysis

Ti 2p, O 1s, N 1s, and Zr 3d XPS spectra are shown in Figure 7 for pure TiO_2 and $TiO_2/N42/Zr2.8$ samples shown as examples. Indeed, all the samples present similar XPS spectra.

On the Ti 2p spectra (Figure 7a), for both samples, the Ti $2p_{1/2}$ and Ti $2p_{3/2}$ peaks are observed at 464.2 eV and 458.5 eV, respectively. They correspond to Ti^{4+} species [27,52,53] and therefore the expected TiO_2 [14]. In Figure 7b, the peak at 530.1 eV is linked to Ti-O bonds and is present for both samples [14,54]. In the same figure, a shoulder is present above 530 eV, but this information is hard to exploit because of the presence of foreign oxygen from carbonaceous contamination [14].

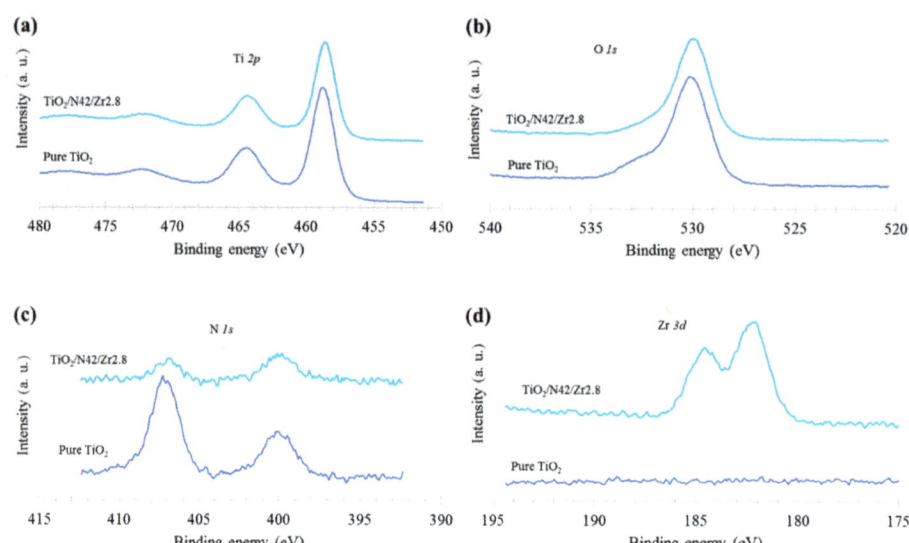

Figure 7. XPS spectra of pure TiO$_2$ and TiO$_2$/N42/Zr2.8 samples: (**a**) Ti *2p* region, (**b**) O *1s* region, (**c**) N *1s* region, and (**d**) Zr *3d* region.

For the N *1s* spectra (Figure 7c), two peaks are observed for both samples, at 400 eV and at 406.8 eV. The N *1s* peak at 400.1 eV is linked to interstitial Ti-O-N bonds [55,56]. These bonds help absorption of visible light [14]. This is coherent with the diffuse reflectance measurements (Figure 6) where the spectra of the samples are shifted towards visible compared to the Evonik P25. The other peak (406.8 eV) is probably due to nitrates originating from residual HNO$_3$ from the synthesis as explained in [14,27].

For the Zr *3d* spectrum of TiO$_2$/N42/Zr2.8 sample (Figure 7d), peaks are visible at 182.0 eV and 184.6 eV that correspond to Zr $3d_{3/2}$ and Zr $3d_{5/2}$ electronic states [14,53,57,58]. They confirm the presence of ZrO$_2$. Hybrid TiO$_2$-ZrO$_2$ nanoparticles could be present, as shown in [14,53,54,59]. Indeed, TEM images do not highlight different morphologies between pure and Zr-(co)-doped TiO$_2$ samples, increasing the likeliness of a TiO$_2$-ZrO$_2$ mixed structure [14].

The atomic ratios N/Ti and Zr/Ti estimated from XPS measurements for some samples are presented in Table 3. The variation of the N/Ti ratio is low across the tested samples, so that similar amounts of nitrogen are present, at least at the surface. This corroborates the absorption spectra shown in Table 2 and Figure 6, in which the similarity between the samples is obvious. Nevertheless, the photoactivity of the samples with N doping (Figures 8 and 9) increases with the doping, showing that the samples are different. It may be possible that the repartition of nitrogen is not homogeneous along the samples and, as previously observed [60,61], that some nitrogen signals are not detected by XPS as it is a surface analysis.

Table 3. Dopant molar ratios in TiO$_2$-based samples.

Sample	N/Ti$_{XPS}$	Zr/Ti$_{XPS}$	Zr/Ti$_{ICP}$
TiO$_2$/N42/Zr1.4	0.043	0.028	0.017
TiO$_2$/N42/Zr2	0.034	0.035	0.021
TiO$_2$/N42/Zr2.8	0.037	0.051	0.034
TiO$_2$/U4/Zr2.8	0.044	0.048	0.029

N/Ti$_{XPS}$: atomic ratio of nitrogen over titanium calculated from XPS measurements; Zr/Ti$_{XPS}$: atomic ratio of zirconium over titanium calculated from XPS measurements; Zr/Ti$_{ICP}$: molar ratio of zirconium over titanium calculated from ICP measurements.

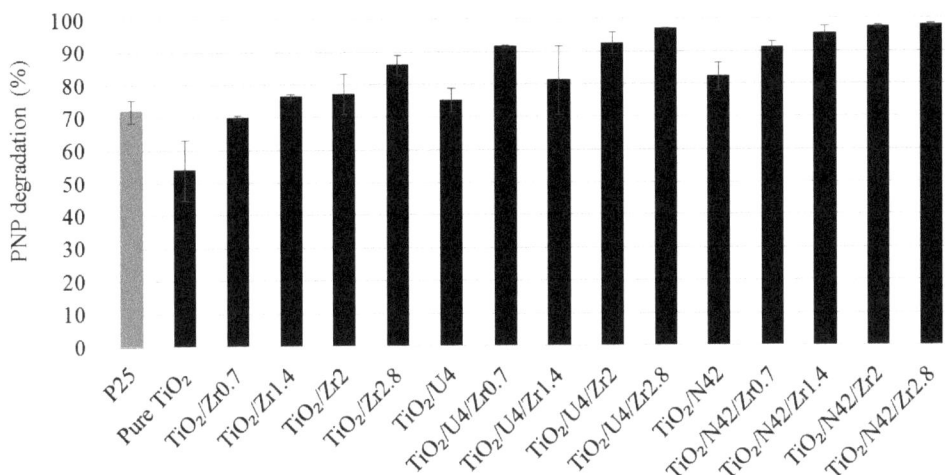

Figure 8. PNP degradation percentage for all the samples under UV/visible light after 8 h of illumination.

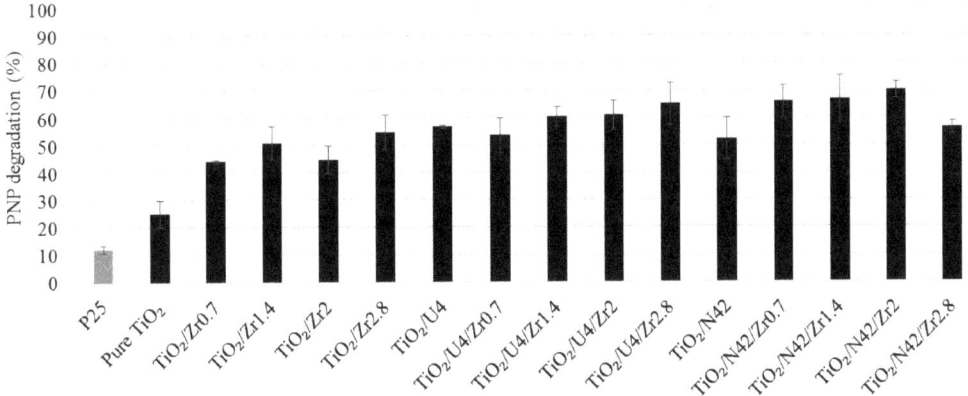

Figure 9. PNP degradation percentage for all the samples under visible light after 24 h of illumination.

The Zr/Ti ratio increases with the Zr doping percentage, as expected. This ratio was also obtained from ICP-AES results (Table 1). The values match each other relatively well, despite slightly lower values for the ICP results. Indeed, it seems that less ZrO_2 is present in the bulk of the samples, as previously observed with this type of synthesis [14]. It may be assumed that the titanium (IV) tetraisopropoxide (TTIP-TiO_2 precursor) is more reactive than the zirconium tert-butoxide (ZrO_2 precursor) in the conditions of the synthesis, leading to higher proportions of ZrO_2 on the surface.

2.7. Photocatalytic Activity

2.7.1. Experiments under UV–Visible Light

For all samples, Figure 8 shows the photocatalytic degradation of PNP under UV/visible light after 8 h of illumination. In the dark, no adsorption of PNP on the samples occurred, as observed in previous studies [14,27,61]. All doped and co-doped samples showed a higher degradation efficiency than pure TiO_2 sample. The degradation efficiency of PNP increases with increasing Zr content in the sample: from 68% for the TiO_2/Zr0.7 sample, to 84% for the TiO_2/Zr2.8 sample. PNP degradation percentage also increases with increasing Zr content in the TiO_2/U4/Zr series, starting from 90% for TiO_2/U4/Zr0.7 sample to

96% for the highest doping (TiO$_2$/U4/ Zr2.8 sample). For the TiO$_2$/N42/Zr series, the yield also increases from 90% for TiO$_2$/N42/Zr0.7 sample to 96% for TiO$_2$/N42/Zr2.8 sample.

The increase in performance under UV–visible light with Zr/N co-doping can be linked to the modification in crystallinity between pure, doped, and co-doped samples (Table 1) due to the introduction of dopants. Indeed, the distribution between anatase, brookite, and amorphous phase changes with the introduction of dopant. This difference in phase distribution can lead to compositions where a synergetic effect can enhance the photoactivity as for Evonik P25 commercial catalyst [27]. Moreover, it was also shown that the formation of a mixed oxide TiO$_2$-ZrO$_2$ could increase the lifetime of "e$^-$h$^+$" pairs [14]. In this study, the XPS spectra show the formation of ZrO$_2$ at the surface of the TiO$_2$ lattice (Figure 7d). When the Zr-doped TiO$_2$ photocatalysts are illuminated with UV/visible light, some photogenerated charges ("e$^-$h$^+$" pairs) can be delocalized on the ZrO$_2$, increasing the lifetime of the pair [14].

As the photocatalysts are illuminated by UV light, the N doping has a slight influence on the PNP degradation activity. However, the co-doped N/Zr TiO$_2$ photocatalysts present higher values for PNP degradation (80–98%) than the commercial Evonik P25 (70%).

2.7.2. Experiments under Visible Light

Figure 9 shows the photocatalytic activity of the samples after 24 h of illumination under visible light (>400 nm). The commercial Evonik P25 photocatalysts shows only 10% of PNP degradation against 25% for pure TiO$_2$ sample. TiO$_2$ samples doped with urea (TiO$_2$/U4 sample) and triethylamine (TiO$_2$/N42 sample) present a PNP degradation of 56% and 52%, respectively. As all aqueous sol–gel samples are doped with nitrogen (Figure 7), this doping leads to an increased visible light absorption compared to Evonik P25 sample (Figure 6), and so a higher photoactivity under illumination by visible light. Similarly, the degradation for Zr single doping increases from 44% for TiO$_2$/Zr0.7 sample to 54% for TiO$_2$/Zr2.8 samples. Indeed, as for UV/visible experiments, the Zr-doping can modifiy the crystallinity of the samples to produce synergetic compositions with higher photoefficiencies, and the formation of TiO$_2$-ZrO$_2$ mixed oxide can enhance the charge separation and so, the photoefficiency.

The best photocatalytic activities for PNP degradation are obtained with the co-doped samples. Indeed, TiO$_2$/U$_4$/Zr0.7 sample gives a PNP degradation of 52%, while it increases up to 64% for TiO$_2$/U4/Zr2.8 sample. Similarly, TiO$_2$/N42/Zr0.7 sample presents a photocatalytic activity of 64%, and this activity reaches 68% for TiO$_2$/N42/Zr2 sample.

Therefore, the doping of TiO$_2$ with Zr and N atoms could be used to increase the photocatalytic activity of TiO$_2$ in the visible range [41]. Indeed, the sensitivity to visible light of N/Zr/TiO$_2$ powders is caused by (i) N-doping from urea and triethylamine precursors by forming an intermediate energy level, and by (ii) Zr-doping, which increases the lifetime of "e$^-$h$^+$" pairs.

Concerning N-doping, when the interstitial nitrogen doping model is assumed, the nitrogen atoms are bonded to one or more oxygen atoms and are thus in any of the oxidation states corresponding to either NO$^-$, NO$_2$$^-$, or NO$_3$$^-$. The uncoupled electrons are distributed around the N and O atoms in these moieties. As a result, the incorporation of nitrogen into the TiO$_2$ lattice leads to the formation of a new energy state, i.e., the N 2p band above the O 2p valence band, which shifts the optical absorption of TiO$_2$ to the visible light region [43]. Furthermore, in this work, in addition to the N-doping from urea and triethylamine precursor, the activity under visible light for pure TiO$_2$ is probably due to the N-doping with the use of nitric acid for synthesis as shown in [14,27]. A ligand to metal charge transfer complex between the organic moieties and Ti(IV) ions cannot be excluded and could account for the improvement of photocatalytic properties of TiO$_2$/U4 samples [62].

In this study, Zr-doping is useful to improve the lifetime of electrons and holes produced by photoactivation [43]. Indeed, Zr^{4+} sites trap electrons more efficiently than Ti^{4+} sites, and the presence of vacant oxygen sites facilitates the transport of charge carriers to the reactive surface sites [41]. In addition, the valence band of TiO$_2$ is more stabilized with a Zr^{4+} doping because Zr^{4+} is more electropositive than Ti^{4+} [43].

The three precursors used in this work (urea, triethylamine, and zirconium tert-butoxide) show that the photocatalysts obtained at low temperatures, appear to be efficient photocatalysts under visible light and may offer promising prospects for the clean-up of water pollution and the degradation of organic pollutants.

2.7.3. Recyclability under Visible Light

The recyclability under visible light was evaluated for six samples: pure TiO$_2$, TiO$_2$/Zr2.8, TiO$_2$/U4, TiO$_2$/U4/Zr2.8, TiO$_2$/N42, and TiO$_2$/N42/Zr2. The samples with the best photoactivity from each series were chosen. The mean activity after 3 recycling cycles (96 h of illumination) is represented on Figure 10. The stability of the activity is maintained for all samples as already observed previously with titania made by aqueous sol–gel process.

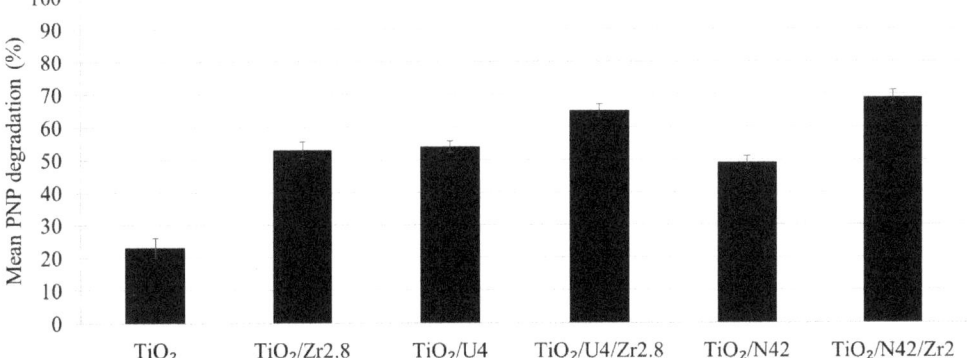

Figure 10. Mean PNP degradation percentages under visible light after 96 h of illumination (3 recycling experiments).

2.8. Photoluminescence Study

Photoluminescence (PL) spectra are represented on Figure 11 for the six samples representative of this study and for the commercial Evonik P25. The spectra were obtained after an excitation at 325 nm and 410 nm.

At 410 nm (Figure 11a), the excitation of 3.02 eV is mostly below TiO$_2$'s band gap. The peaks at 610 nm and 570 nm correspond to relaxation of trapped electrons to the valence band [63]. The oxygen vacancies are responsible for this trapping process [64]. Comparing the pure and Zr-doped sample shows that a lot of those vacancies are created by the dopant.

When the sample is excited at 325 nm (Figure 11b), the energy is sufficient to clear the band gap. The numerous peaks between 450 and 500 nm should correspond to relaxation of electrons in shallow states [63]. Finally, the shoulder at around 440 nm is present only for N-doped samples, indicating that a different localized energy state exists. The energy corresponding to 440 nm, i.e., 2.82 eV, indicates that this state is a few tenths of eV above the valence band. Figure 11c represents the different possible electronic transitions in TiO$_2$ samples

The role of Zr in the increase of photocatalytic activity could be to trap the electrons, thus increasing the lifetime of holes. The same phenomenon explains why the peaks between 450 nm and 490 nm decrease in height when doped with Zr: the e$^-$ traps decrease the number of electrons available in the shallow states between the conduction band, responsible for these peaks. On the other hand, doping with nitrogen decreases the band gap. The number of vacancies is reduced by the use of urea, but is unchanged in TiO$_2$/N42 samples compared to pure TiO$_2$. This could explain why the TiO$_2$/N42 samples perform slightly better than the TiO$_2$/U4 ones. However, the difference is too small to draw a definitive conclusion of the difference between those mechanisms.

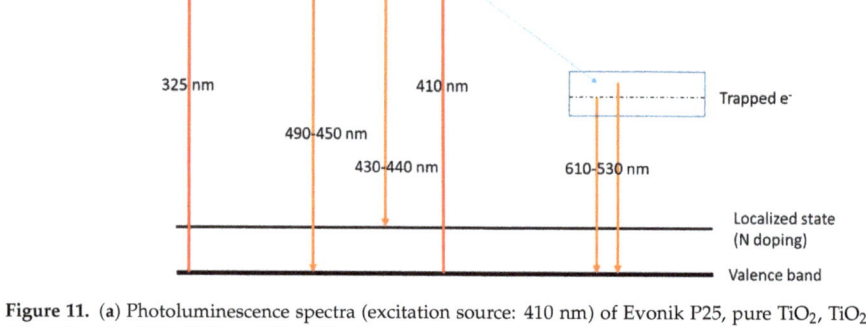

Figure 11. (a) Photoluminescence spectra (excitation source: 410 nm) of Evonik P25, pure TiO$_2$, TiO$_2$/Zr2.8, TiO$_2$/U4, TiO$_2$/U4/Zr2.8, TiO$_2$/N42, and TiO$_2$/N42/Zr2.8 samples (caption in the figure). (b) Photoluminescence spectra (excitation source: 325 nm) of pure TiO$_2$, TiO$_2$/Zr2.8, TiO$_2$/U4, TiO$_2$/U4/Zr2.8, TiO$_2$/N42, and TiO$_2$/N42/Zr2.8 samples (caption in the figure). (c) Proposed energy states for N-(co-)doped TiO$_2$, inspired from the work in [63].

3. Materials and Methods

3.1. Pure TiO$_2$ Synthesis

The TiO$_2$ powder is prepared by the sol–gel method, in which 250 mL of deionized water acidified with nitric acid (HNO$_3$, 65%, Merck, Darmstadt, Germany) at pH of 1 is introduced into a flask stirred at 700 rpm in a bath thermostatically controlled at 80 °C. 36.08 mL of titanium (IV) tetraisopropoxide (TTIP, >97%, Sigma-Aldrich, St. Louis, MO, USA), is added to 15.1 mL of isopropanol (IsoP, 99.5%, Acros, Hull, Belgium) at 25 °C and stirred for 30 min, then the TTIP-IsoP mixture is added dropwise to the thermostatically controlled flask, and left under stirring for 24 h in the closed flask. After this reaction time, the obtained sol is dried in ambient air; the obtained powder is crushed and used as such for further processing.

3.2. Urea-TiO$_2$ Powder Synthesis

The urea-doped powder is prepared using the same method as pure TiO$_2$ synthesis except that a mass (proportional to the desired molar concentration) of urea (NH$_2$-C(O)-NH$_2$ 98%, Sigma-Aldrich, St. Louis, MO, USA) is also included in the 250 mL of deionized water under stirring before acidifying to *pH* of 1. The molar ratio between TTIP and urea is equal to 4. It corresponds to the $TiO_2/U4$ sample described by Mahy et al. in [27].

3.3. Triethylamine-TiO_2 Doped Powder Synthesis

The powder doped with triethylamine is prepared with the same method as pure TiO_2 synthesis except that, once the sol is obtained after 24 h of stirring, 699 mL of triethylamine is added (in excess) to the pure TiO_2 suspension. The whole mixture is then left under magnetic stirring for 24 h. The samples containing triethylamine are denoted with "N42", corresponding to the molar ratio between triethylamine and TTIP [27].

3.4. Zr-TiO_2 Powder Syntheses

The powders doped with the zirconium precursor are prepared using also the same method as pure TiO_2 synthesis except that a mass (proportional to the desired molar concentration) of zirconium tert-butoxide (98%, Sigma Aldrich, St. Louis, MO, USA) is also added in the TTIP-isopropanol mixture. After 24 h of stirring in the thermostatically controlled flask, the samples are dried in the same way as the pure TiO_2 powder. Four Zr-doping molar percentages are studied: 0.7, 1.4, 2, and 2.8 mol%. The samples are designated as TiO_2/ZrX, where X is the amount of dopant.

3.5. Urea/Zr/TiO_2 Co-Doped Powder Synthesis

The co-doping of TiO_2 with urea and the zirconium precursor is done by a combination of the two doping methods applied by Mahy et al. [14,27]. Urea (28.4 g) is introduced into 250 mL of distilled water then acidified with HNO_3 to a pH of 1. At the same time, the calculated quantity of zirconium tert-butoxide (98%, Sigma Aldrich, St. Louis, MO, USA) is added to the TTiP-Isop mixture and left under stirring for 30 min until homogeneous mixing. At the end, the solution containing Isop-TTIP-Zr is added to the deionized urea water, followed by 24 h of stirring at 80 °C and 700 rpm. A white sol is obtained and dried in ambient air until a white powder is recovered. This powder is crushed then dried for 1 h under vacuum at 100 mbar, then rinsed with distilled water and centrifuged for 15 min. This rinsing is repeated 3 times and the obtained pellets are finally dried under vacuum at 100 °C and 100 mbar. The molar ratio between TTIP and urea is equal to 4 [27], and the zirconium precursor molar percentages are varied between 0.7, 1.4, 2, and 2.8 mol%.

3.6. Triethylamine/Zr/TiO_2 Co-Doped Powder Synthesis

The co-doping of TiO_2 with triethylamine and the zirconium precursor is done by a combination of the two doping methods applied by Mahy et al. [14,27]. Like the urea co-doping, the triethylamine co-doping is adapted by saturating each sol of the zirconium precursor (0.7, 1.4, 2, and 2.8 mol%) with triethylamine as described above. The molar ratio between triethylamine and TTIP is equal to 42 [27]. The rinsing and drying is done in the same way as the urea/Zr/TiO_2 powder.

3.7. Material Characterization

The sample composition is determined by inductively coupled plasma-atomic emission spectroscopy (ICP-AES), equipped with an ICAP 6500 THERMO Scientific device (Waltham, MA, USA).). The analysis protocol is fully detailed in [14].

The crystallographic properties are studied through X-ray diffraction (XRD) patterns from 10° to 70° with a Bruker D8 Twin-Twin powder diffractometer using Cu-Kα radiation (Bruker, Billerica, MA, USA). The Scherrer formula (Equation (1)) is used to determine the size of TiO_2 crystallites, d_{XRD}:

$$d_{XRD} = 0.9 \frac{\lambda}{\beta \, Cos(\theta)} \qquad (1)$$

where d_{XRD} is the crystallite size (nm), B is the total width of the peak at half its maximum value after correction for instrumental broadening (rad), λ is the X-ray wavelength (0.154 nm), and θ is the Bragg angle (rad).

The TOPAS software [65] was used to fit the diffractograms in order to estimate the percentages of crystalline phases. The structure parameters for the anatase and brookite phases were taken from the PDF 04-007-0701 and 04-007-0758 references (ICCD PDF4+ database), and the fundamental parameters approach [65] was used to model the instrumental contribution to the reflection profiles. The amount of amorphous TiO_2 was estimated through the same procedure using an internal standard of CaF_2 (calcium fluoride, Sigma-Aldrich, anhydrous powder, 99.99% trace metals).

The textural properties of the samples are characterized by nitrogen adsorption–desorption isotherms in an ASAP 2420 multi-sampler device from Micromeritics. From these isotherms, the microporous volume is calculated using Dubinin–Radushkevich theory (V_{DR}) [51]. The specific surface area is evaluated using the theory of Brunauer, Emmett, and Teller (S_{BET}) [51]. The mean particle size, d_{BET}, can be estimated from the S_{BET} values by assuming spherical and non-porous nanoparticles of anatase TiO_2 using the following formula [66]:

$$\frac{d_{BET}}{6} = \frac{\frac{1}{\rho_{anatase}}}{S_{BET}} \qquad (2)$$

where $\rho_{anatase}$ is the apparent density of anatase TiO_2 estimated as 3.89 g cm^{-3}.

X-ray photoelectron spectra are obtained with a SSI-X-probe (SSX-100/206) spectrometer equipped with a monochromatized microfocused Al X-ray source (1486.6 eV), operating at 10 kV and 20 mA. Samples are placed in the analysis chamber where the residual pressure was about 10^{-8} Pa. The following sequence of spectra is recorded: survey spectrum, C 1s, O 1s, N 1s, Ti 2p, Zr 3d, and again C 1s to check the stability of charge compensation with time and absence of degradation of the samples [14].

The C- (C, H) component of the carbon C 1s peak is fixed at 284.8 eV to calibrate the scale in binding energy. Three other components of the carbon peak (C-(O, N), C=O or O-C-O, and O-C=O) have been resolved, notably to determine the amount of oxygen involved in the carbon contamination [14]. Data processing is carried out with the CasaXPS program (Casa Software Ltd., Teignmouth, UK). The spectra are decomposed using the Gaussian and Lorentzian function product model (least squares fitting) after subtraction of a nonlinear Shirley baseline [14,67].

The optical properties of the sample are evaluated using diffuse reflection spectroscopy measurements in the 250–600 nm region with a Perkin Elmer Lambda 1050 S UV/VIS/NIR spectrophotometer, equipped with a spectralon coated integrating sphere (150 mm InGaAs Int. Sphere from PerkinElmer. Waltham, MA, USA) and using Al_2O_3 as reference. UV–Vis spectra, recorded in diffuse reflectance mode (R sample), are transformed using Kubelka–Munk function [27,68,69] to produce a signal, normalized for comparison between samples, and thus to calculate the band gap (direct and indirect ones). The details of this processing method are described in more details in [27,66].

Transmission electron microscopy (TEM) images were obtained with a FEI TEM-LaB6 TECNAI G2 microscope with a tungsten filament electron gun operating at 200 kV. The powders are dispersed in deionized water and sonicated for 15 min. Then, a drop of the dispersion is placed on a copper grid (Formvar/Carbon 200 Mesh Cu from Agar Scientific, Essex, UK) for observation.

The photoluminescence measurements were performed at room temperature using Shimadzu RF-6000 fluorimeter equipped with xenon excitation source. All emission spectra are obtained with a good spectral resolution of the detector.

Fourier Transform Infrared Spectroscopy (FTIR) was carried out using a 630 Cary infrared spectrometer (400 to 4000 cm^{-1}, Agilent, Santa Clara, CA, USA) in order to characterize the surface functional groups of the powders.

Thermogravimetric analyses were performed on a thermal analyzer SETARAM LabSysEvo1600 (KEP Technologies, Mougins, France) until 700 °C under air.

3.8. Photocatalytic Experiments

The photocatalytic activity of powders is evaluated by following the degradation of p-nitrophenol (PNP) after 8 h under UV–visible light and after 24 h under visible light, in triplicate, in an aqueous medium. For each test, the degradation percentage of PNP, (D_{PNPi}), is given by Equation (3) [27]:

$$D_{PNPi}\ (\%) = \left(1 - \frac{[PNP]_i}{[PNP]_0}\right) \times 100\% \quad (3)$$

where $[PNP]_i$ represents the residual concentration of PNP at time t = i h and $[PNP]_0$ represents the initial concentration of PNP at time t = 0 h [14].

The experimental set-up is described in a previous study [14,27]. The D_{PNP8} is evaluated after 8 h under UV/visible light, thanks to a halogen lamp with a continuous spectrum from 300 to 800 nm (300 W, 220 V) measured with a Mini-Spectrometer TMUV/Vis C10082MD from Hamamatsu [14]. The PNP degradation (D_{PNP24}) under visible light is evaluated after 24 h with the same halogen lamp covered by an UV filter that removes wavelengths shorter than 390 nm [14]. The residual concentration of PNP is measured by UV/Vis spectroscopy (GENESYS 10S UV–Vis from Thermo Scientific, Waltham, MA, USA) at 318 nm [14]. For each tested catalyst, three flasks containing the catalytic powder are exposed to light to calculate the PNP degradation, and one is kept in the dark to evaluate PNP adsorption on the sample. In each flask, the initial concentrations of catalyst (if present) and PNP are equal to 1 g/L and 10^{-4} M, respectively [14]. The volume of each flask is equal to 10 mL, and the flasks are agitated by a magnetic stirrer. Experiments are conducted in test tubes closed with a sealing cap. These tubes are placed in a cylindrical glass reactor with the halogen lamp in the center. The reactor is maintained at constant temperature (20 °C) by a cooling system with recirculating water. The lamp is also cooled by a similar system. Aluminum foil covers the outer wall of the reactor to prevent any interaction with the room lighting [11,14].

To test the stability and recyclability of the photoactivity of samples, photocatalytic recycling tests under visible light are made on six samples: pure TiO_2, TiO_2/Zr2.8, TiO_2/U4, TiO_2/U4/Zr2.8, TiO_2/N42, and TiO_2/N42/Zr2. The same protocol as explained in the above paragraph is performed on these catalysts [11]. After this, the samples are recovered by centrifugation (10,000 rpm for 1 h) followed by drying at 120 °C for 24 h [11]. A second and third cycle of photocatalytic tests as described above are applied to the reused catalysts. So, each tested catalyst undergoes four catalytic tests, and a mean PNP degradation on the three recycling tests is then calculated [11].

4. Conclusions

In this work, an aqueous sol–gel process was successfully applied to produce Zr/N single doped and co-doped TiO_2 photocatalysts at low temperature without any calcination step. The N/Ti molar ratio was set at 4 for urea and 42 for triethylamine. Different molar ratios were tested for Zr (0.7, 1.4, 2 and 2.8 mol%). Pure TiO_2 was also synthesized by the same aqueous sol–gel process for comparison. The photocatalyst Evonik P25 was also used as a reference commercial material.

Physico-chemical characterizations confirmed that in all the synthesized powders, anatase-brookite TiO_2 nanoparticles were present. The TiO_2 particle diameters estimated by three different techniques (XRD, BET, and TEM) were consistent and in the same range (~4–6 nm). From TEM micrographs, all TiO_2-based samples were composed of spheroidal nanoparticles arranged in agglomerates. Furthermore, all the samples synthesized by peptization were micro-mesoporous, with specific surface area values reaching 280 m^2/g.

The results of the XPS and ICP analyses showed that TiO_2 was successfully doped with nitrogen and zirconia. Indeed, the incorporation of nitrogen in TiO_2 materials with the presence of Ti-O-N bonds was observed, which allowed the absorption of light in the visible range and the enhancement of photoactivity in this wavelength range. The photocatalytic activity of TiO_2 was also improved by Zr-doping through the formation of TiO_2-ZrO_2 mixed oxide materials, which increased the lifetime of the photogenerated charges. In this way, all the samples of this work showed a higher photocatalytic efficiency for *p*-nitrophenol degradation compared to pure TiO_2 and commercial Evonik P25.

The highest PNP degradation percentages were obtained with co-doped samples: under UV/visible light, the best sample (TiO_2/N42/Zr2.8) reached 96% of PNP degradation after 8 h of irradiation, and under visible light, 68% of PNP degradation was reached with the best sample (TiO_2/N42/Zr2) after 24 h of illumination. These results confirmed the positive influence of the N and Zr dopants on the shift of TiO_2 photoactivity towards the visible region. Finally, it should be noted that these N/Zr single doped and co-doped TiO_2 samples were prepared with an environmentally friendly synthesis. Indeed, water was used as solvent, the synthesis of photocatalysts was done at ambient temperature and pressure, and no calcination step was required to obtain the crystallization of TiO_2. This study proposes very efficient photocatalysts under visible light offering promising prospects for the clean-up of water pollution and the degradation of organic pollutants.

Author Contributions: Conceptualization, methodology, investigation, analysis, and writing, H.B.-B., J.G.M., C.W., B.V., D.P., P.E., S.H., M.B., A.S., S.B.-B. and S.D.L.; writing—original draft preparation, H.B.-B. and J.G.M.; supervision, funding acquisition, and project administration, S.H., S.B.-B. and S.D.L. All the authors corrected the paper before submission and during the revision process. All authors have read and agreed to the published version of the manuscript.

Funding: This research received no external funding.

Institutional Review Board Statement: Not applicable.

Informed Consent Statement: Not applicable.

Data Availability Statement: The raw/processed data required to reproduce these findings cannot be shared at this time as the data also forms part of an ongoing study.

Acknowledgments: S.D.L. and S.H. thank the Belgian National Funds for Scientific Research (F.R.S.-FNRS) for their Associate Researcher and Research Director position, respectively. The authors acknowledge the Ministère de la Région Wallonne Direction Générale des Technologies, de la Recherche et de l'Energie and the Fonds de Bay. J.G.M. and S.H. also thank Innoviris Brussels for financial support through the Bridge project—COLORES.

Conflicts of Interest: The authors declare no conflict of interest.

References

1. Hieu, C.; Nguyen, H.; Fu, C.; Lu, Y.; Juang, R. Roles of adsorption and photocatalysis in removing organic pollutants from water by activated carbon À supported titania composites: Kinetic aspects. *J. Taiwan Inst. Chem. Eng.* **2020**, *109*, 51–61.
2. Helali, S.; Puzenat, E.; Perol, N.; Safi, M.; Guillard, C. Methylamine and dimethylamine photocatalytic degradation—Adsorption isotherms and kinetics. *Appl. Catal. A Gen.* **2011**, *402*, 201–207. [CrossRef]
3. Basavarajappa, P.S.; Patil, S.B.; Ganganagappa, N.; Raghava, K.; Raghu, A.V.; Venkata, C. Recent progress in metal-doped TiO_2, non-metal doped/codoped TiO_2 and TiO_2 nanostructured hybrids for enhanced photocatalysis. *Int. J. Hydrogen Energy* **2020**, *45*, 7764–7778. [CrossRef]
4. Mahy, J.G.; Hermans, S.; Lambert, D. Influence of nucleating agent addition on the textural and photo-Fenton properties of Fe(III)/SiO_2 catalysts. *J. Phys. Chem. Solids* **2020**, *144*, 109502. [CrossRef]
5. Oturan, M.A.; Aaron, J.J. Advanced oxidation processes in water/wastewater treatment: Principles and applications. A review. *Crit. Rev. Environ. Sci. Technol.* **2014**, *44*, 2577–2641. [CrossRef]
6. Belet, A.; Wolfs, C.; Mahy, J.G.; Poelman, D.; Vreuls, C. Sol-gel Syntheses of Photocatalysts for the Removal of Pharmaceutical Products in Water. *Nanomaterials* **2019**, *9*, 126. [CrossRef]
7. Levchuk, I.; Fern, P.; Sillanp, M.; Jos, J. A critical review on application of photocatalysis for toxicity reduction of real wastewaters. *J. Clean. Prod.* **2020**, *258*, 120694.
8. Zaleska, A. Doped-TiO_2: A Review. *Recent Patents Eng.* **2008**, *2*, 157–164. [CrossRef]
9. Gohin, M.; Allain, E.; Chemin, N.; Maurin, I.; Gacoin, T.; Boilot, J. Sol—Gel nanoparticulate mesoporous films with enhanced self-cleaning properties. *J. Photochem. Photobiol. A Chem.* **2010**, *216*, 142–148. [CrossRef]
10. Fujishima, A.; Hashimoto, K.; Watanabe, T. *TiO_2 Photocatalysis: Fundamentals and Applications*; Bkc: Tokyo, Japan, 1999.
11. Mahy, J.G.; Lambert, S.D.; Léonard, G.L.-M.; Zubiaur, A.; Olu, P.-Y.; Mahmoud, A.; Boschini, F.; Heinrichs, B. Towards a large scale aqueous sol-gel synthesis of doped TiO_2: Study of various metallic dopings for the photocatalytic degradation of p-nitrophenol. *J. Photochem. Photobiol. A Chem.* **2016**, *329*, 189–202. [CrossRef]

12. Yao, X.; Wang, X.; Su, L.; Yan, H.; Yao, M. Band structure and photocatalytic properties of N/Zr co-doped anatase TiO_2 from first-principles study. *J. Mol. Catal. A Chem.* **2011**, *351*, 11–16. [CrossRef]
13. Luciani, G.; Imparato, C.; Vitiello, G. Photosensitive Hybrid Nanostructured Materials: The Big Challenges for Sunlight Capture. *Materials* **2020**, *10*, 103. [CrossRef]
14. Mahy, J.G.; Lambert, S.D.; Tilkin, R.G.; Poelman, D.; Wolfs, C.; Devred, F.; Gaigneaux, E.M.; Douven, S. Ambient temperature ZrO_2-doped TiO_2 crystalline photocatalysts: Highly efficient powders and films for water depollution. *Mater. Today Energy* **2019**, *13*, 312–322. [CrossRef]
15. Garzon-Roman, A.; Zuñiga-islas, C.; Quiroga-gonzález, E. Immobilization of doped TiO_2 nanostructures with Cu or In inside of macroporous silicon using the solvothermal method: Morphological, structural, optical and functional properties. *Ceram. Int.* **2020**, *46*, 1137–1147. [CrossRef]
16. Siddiqa, A.; Masih, D.; Anjum, D.; Siddiq, M. Cobalt and sulfur co-doped nano-size TiO_2 for photodegradation of various dyes and phenol. *J. Environ. Sci.* **2015**, *37*, 100–109. [CrossRef]
17. Surendra, B.; Raju, B.M.; Noel, K.; Onesimus, S.; Choudhary, G.L.; Paul, P.F.; Vangalapati, M. Synthesis and characterization of Ni doped TiO_2 nanoparticles and its application for the degradation of malathion. *Mater. Today Proc.* **2020**, *26*, 1091–1095. [CrossRef]
18. Jemaa, I.B.; Chaabouni, F.; Ranguis, A. Cr doping effect on the structural, optoelectrical and photocatalytic properties of RF sputtered TiO_2 thin films from a powder target. *J. Alloys Compd.* **2020**, *825*, 153988. [CrossRef]
19. Bharati, B.; Mishra, N.C.; Sinha, A.S.K.; Rath, C. Unusual structural transformation and photocatalytic activity of Mn doped TiO_2 nanoparticles under sunlight. *Mater. Res. Bull.* **2020**, *123*, 110710. [CrossRef]
20. Manojkumar, P.; Lokeshkumar, E.; Saikiran, A.; Govardhanan, B.; Ashok, M. Visible light photocatalytic activity of metal (Mo/V/W) doped porous TiO_2 coating fabricated on Cp-Ti by plasma electrolytic oxidation. *J. Alloys Compd.* **2020**, *825*, 154092. [CrossRef]
21. Saito, K.; Yi, E.; Laine, R.M.; Sugahara, Y. Preparation of Nb-doped TiO_2 nanopowder by liquid-feed spray pyrolysis followed by ammonia annealing for tunable visible-light absorption and inhibition of photocatalytic activity. *Ceram. Int.* **2020**, *46*, 1314–1322. [CrossRef]
22. Ravishankar, T.N.; Vaz, M.D.O.; Ramakrishnappa, T.; Teixeira, S.R.; Dupont, J. Ionic liquid e assisted hydrothermal synthesis of Nb/ TiO_2 nanocomposites for efficient photocatalytic hydrogen production and photodecolorization of Rhodamine B under UV-visible and visible light illuminations. *Mater. Today Chem.* **2019**, *12*, 373–385. [CrossRef]
23. Thakur, I.; Örmeci, B. Inactivation of E. coli in water employing Fe-TiO_2 composite incorporating in-situ dual process of photocatalysis and photo-Fenton in fixed-mode. *J. Water Process Eng.* **2020**, *33*, 101085. [CrossRef]
24. Jiang, G.; Geng, K.; Wu, Y.; Han, Y.; Shen, X. High photocatalytic performance of ruthenium complexes sensitizing g-C_3N_4/TiO_2 hybrid in visible light irradiation. *Appl. Catal. B Environ.* **2018**, *227*, 366–375. [CrossRef]
25. Li, Y.; Cao, S.; Zhang, A.; Zhang, C.; Qu, T.; Zhao, Y.; Chen, A. Carbon and nitrogen co-doped bowl-like Au/TiO_2 nanostructures with tunable size for enhanced visible-light-driven photocatalysis. *Appl. Surf. Sci.* **2018**, *445*, 350–358. [CrossRef]
26. Onkani, S.P.; Diagboya, P.N.; Mtunzi, F.M.; Klink, M.J.; Olu-owolabi, B.I.; Pakade, V. Comparative study of the photocatalytic degradation of 2–chlorophenol under UV irradiation using pristine and Ag-doped species of TiO_2, ZnO and ZnS photocatalysts. *J. Environ. Manag.* **2020**, *260*, 110145. [CrossRef]
27. Mahy, J.G.; Cerfontaine, V.; Poelman, D.; Devred, F.; Gaigneaux, E.M.; Heinrichs, B.; Lambert, S.D. Highly efficient low-temperature N-doped TiO_2 catalysts for visible light photocatalytic applications. *Materials* **2018**, *11*, 584. [CrossRef]
28. Ohno, T.; Mitsui, Á.T.; Matsumura, M. Photocatalytic Activity of S-doped TiO_2 Photocatalyst under Visible Light. *Chem. Lett.* **2003**, *32*, 364–365. [CrossRef]
29. Payormhorm, J.; Idem, R. Synthesis of C-doped TiO_2 by sol-microwave method for photocatalytic conversion of glycerol to value-added chemicals under visible light. *Appl. Catal. A Gen.* **2020**, *590*, 117362. [CrossRef]
30. Yadav, V.; Verma, P.; Sharma, H.; Tripathy, S.; Saini, V.K. Photodegradation of 4-nitrophenol over B-doped TiO_2 nanostructure: Effect of dopant concentration, kinetics, and mechanism. *Environ. Sci. Pollut. Res.* **2020**, *27*, 10966–10980. [CrossRef] [PubMed]
31. Bodson, C.J.; Heinrichs, B.; Tasseroul, L.; Bied, C.; Mahy, J.G.; Man, M.W.C.; Lambert, S.D. Efficient P- and Ag-doped titania for the photocatalytic degradation of waste water organic pollutants. *J. Alloys Compd.* **2016**, *682*, 144–153. [CrossRef]
32. Tian, L.; Xing, L.; Shen, X.; Li, Q.; Ge, S.; Liu, B.; Jie, L. Visible light enhanced Fe–I–TiO_2 photocatalysts for the degradation of gaseous benzene. *Atmos. Pollut. Res.* **2020**, *11*, 179–185. [CrossRef]
33. Bayan, E.M.; Lupeiko, T.G.; Pustovaya, L.E.; Volkova, M.G.; Butova, V.V.; Guda, A.A. Zn-F co-doped TiO_2 nanomaterials: Synthesis, structure and photocatalytic activity. *J. Alloys Compd.* **2020**, *822*, 153662. [CrossRef]
34. Qin, Y.; Li, H.; Lu, J.; Meng, F.; Ma, C.; Yan, Y. Nitrogen-doped hydrogenated TiO_2 modified with CdS nanorods with enhanced optical absorption, charge separation and photocatalytic hydrogen evolution. *Chem. Eng. J.* **2020**, *384*, 123275. [CrossRef]
35. Mahy, J.G.; Paez, C.A.; Carcel, C.; Bied, C.; Tatton, A.S.; Damblon, C.; Heinrichs, B.; Man, M.W.C.; Lambert, S.D. Porphyrin-based hybrid silica-titania as a visible-light photocatalyst. *J. Photochem. Photobiol. A Chem.* **2019**, *373*, 66–76. [CrossRef]
36. Abbad, S.; Guergouri, K.; Gazaout, S.; Djebabra, S.; Zertal, A.; Barille, R.; Zaabat, M. Effect of silver doping on the photocatalytic activity of TiO_2 nanopowders synthesized by the sol-gel route. *J. Environ. Chem. Eng.* **2020**, *8*, 103718. [CrossRef]
37. Rathore, N.; Kulshreshtha, A.; Kumar, R.; Sharma, D. Study on morphological, structural and dielectric properties of sol-gel derived TiO_2 nanocrystals annealed at different temperatures. *Phys. B Phys. Condens. Matter.* **2020**, *582*, 411969. [CrossRef]

38. Mahy, J.G.; Claude, V.; Sacco, L.; Lambert, S.D. Ethylene polymerization and hydrodechlorination of 1,2-dichloroethane mediated by nickel (II) covalently anchored to silica xerogels. *J. Sol-Gel Sci. Technol.* **2017**, *81*, 59–68. [CrossRef]
39. Sato, S. Photocatalytic Activity of NOx-doped TiO_2 in the Visible Light Region. *Chem. Phys. Lett.* **1986**, *123*, 126–128. [CrossRef]
40. Asahi, R.; Morikawa, T.; Ohwaki, T.; Aoki, K.; Taga, Y. Visible-Light Photocatalysis in Nitrogen-Doped Titanium Oxides. *Science* **2001**, *293*, 269–271. [CrossRef]
41. Pouretedal, H.R. Visible photocatalytic activity of co-doped TiO_2/Zr, N nanoparticles in wastewater treatment of nitrotoluene samples. *J. Alloys Compd.* **2018**, *735*, 2507–2511. [CrossRef]
42. Park, J.-Y.; Lee, K.-H.; Kim, B.-S.; Kim, C.S.; Lee, S.-E.; Okuyama, K.; Jang, H.-D.; Kim, T.-O. Enhancement of dye-sensitized solar cells using Zr/N-doped TiO_2 composites as photoelectrodes. *RSC Adv.* **2014**, *4*, 9946–9952. [CrossRef]
43. Liu, H.; Liu, G.; Shi, X. N/Zr-codoped TiO_2 nanotube arrays: Fabrication, characterization, and enhanced photocatalytic activity. *Colloids Surf. A Physicochem. Eng. Asp.* **2010**, *363*, 35–40. [CrossRef]
44. Cha, J.; An, S.; Jang, H.; Kim, C.; Song, D.; Kim, T. Synthesis and photocatalytic activity of N-doped TiO_2/ZrO_2 visible-light photocatalysts. *Adv. Powder Technol.* **2012**, *23*, 717–723. [CrossRef]
45. Bineesh, K.V.; Kim, D.-K.; Park, D.-W. Synthesis and characterization of zirconium-doped mesoporous nano-crystalline TiO_2. *Nanoscale* **2010**, *2*, 1222–1228. [CrossRef] [PubMed]
46. Rubasinghege, G.; Grassian, V.H. Role(s) of adsorbed water in the surface chemistry of environmental interfaces. *Chem. Commun.* **2013**, *49*, 3071–3094. [CrossRef]
47. Luo, S.X.; Wang, F.M.; Shi, Z.S.; Xin, F. Preparation and photocatalytic activity of Zr doped TiO_2. *Mater. Res. Innov.* **2009**, *13*, 64–69. [CrossRef]
48. Nishino, N.; Finlayson-pitts, B.J. Thermal and photochemical reactions of NO_2 on chromium (III) oxide surfaces at atmospheric pressure. *Phys. Chem. Chem. Phys.* **2012**, *14*, 15840–15848. [CrossRef] [PubMed]
49. Burch, R.; Breen, J.P.; Meunier, F.C. A review of the selective reduction of NO x with hydrocarbons under lean-burn conditions with non-zeolitic oxide and platinum group metal catalysts. *Appl. Catal. B Environ.* **2002**, *39*, 283–303. [CrossRef]
50. Bollino, F.; Tranquillo, E. Zirconia/Hydroxyapatite Composites Synthesized Via Sol-Gel: Influence of Hydroxyapatite Content and. *Materials* **2017**, *10*, 757. [CrossRef]
51. Lecloux, A.J. Texture of catalysts. *Catal. Sci. Technol.* **1981**, *2*, 171.
52. Maver, K.; Štangar, U.L.; Černigoj, U.; Gross, S.; Cerc Korošec, R. Low-temperature synthesis and characterization of TiO_2 and TiO_2-ZrO_2 photocatalytically active thin films. *Photochem. Photobiol. Sci.* **2009**, *8*, 657–662. [CrossRef]
53. Thejaswini, T.V.L.; Prabhakaran, D.; Maheswari, M.A. Synthesis of mesoporous worm-like ZrO_2–TiO_2 monoliths and their photocatalytic applications towards organic dye degradation. *J. Photochem. Photobiol. A Chem.* **2017**, *344*, 212–222. [CrossRef]
54. Li, M.; Li, X.; Jiang, G.; He, G. Hierarchically macro—Mesoporous ZrO_2—TiO_2 composites with enhanced photocatalytic activity. *Ceram. Int.* **2015**, *41*, 5749–5757. [CrossRef]
55. Azouani, R.; Tieng, S.; Chhor, K.; Bocquet, J.F.; Eloy, P.; Gaigneaux, E.M.; Klementiev, K.; Kanaev, A. V TiO_2 doping by hydroxyurea at the nucleation stage: Towards a new photocatalyst in the visible spectral range. *Phys. Chem. Chem. Phys.* **2010**, *12*, 1–10. [CrossRef] [PubMed]
56. Bittencourt, C.; Rutar, M.; Umek, P.; Mrzel, A.; Vozel, K.; Arcon, D.; Henzler, K.; Krüger, P.; Guttmann, P. Molecular nitrogen in N-doped TiO_2 nanoribbons. *RSC Adv.* **2015**, *5*, 23350–23356. [CrossRef]
57. Mbiri, A.; Ta, D.H.; Gatebe, E.; Wark, M. Zirconium doped mesoporous TiO_2 multilayer thin films: Influence of the zirconium content on the photodegradation of organic pollutants. *Catal. Today* **2019**, *238*, 71–78. [CrossRef]
58. Qian, J.; Hu, Q.; Hou, X.; Qian, F.; Dong, L.; Li, B. Study of Different Ti/Zr Ratios on the Physicochemical Properties and Catalytic Activities for CuO/Ti−Zr−O Composites. *Ind. Eng. Chem. Res.* **2018**, *57*, 12792–12800. [CrossRef]
59. Tian, J.; Shao, Q.; Zhao, J.; Pan, D.; Dong, M.; Jia, C.; Ding, T. Microwave solvothermal carboxymethyl chitosan templated synthesis of TiO_2/ZrO_2 composites toward enhanced photocatalytic degradation of Rhodamine, B.J. *Colloid Interface Sci.* **2019**, *541*, 18–29. [CrossRef]
60. Livraghi, S.; Chierotti, M.R.; Giamello, E.; Magnacca, G.; Paganini, M.C.; Cappelletti, G.; Bianchi, C.L. Nitrogen-Doped Titanium Dioxide Active in Photocatalytic Reactions with Visible Light: A Multi-Technique Characterization of Differently Prepared Materials. *J. Phys. Chem. C* **2008**, *112*, 17244–17252. [CrossRef]
61. Douven, S.; Mahy, J.G.; Wolfs, C.; Reyserhove, C.; Poelman, D.; Devred, F.; Gaigneaux, E.M.; Lambert, S.D. Efficient N, Fe Co-Doped TiO_2 Active under Cost-Effective Visible LED Light: From Powders to Films. *Catalysts* **2020**, *10*, 547. [CrossRef]
62. Vitiello, G.; Pezzella, A.; Calcagno, V.; Silvestri, B.; Raiola, L.; Errico, G.D.; Costantini, A.; Branda, F.; Luciani, G. 5,6-Dihydroxyindole-2-carboxylic Acid−TiO_2 Charge Transfer Complexes in the Radical Polymerization of Melanogenic Precursor(s). *J. Phys. Chem. C* **2016**, *120*, 6262–6268. [CrossRef]
63. Pallotti, D.K.; Passoni, L.; Maddalena, P.; Di Fonzo, F.; Lettieri, S. Photoluminescence Mechanisms in Anatase and Rutile TiO_2. *J. Phys. Chem. C* **2017**, *121*, 9011–9021. [CrossRef]
64. Komaraiah, D.; Radha, E.; Kalarikkal, N.; Sivakumar, J.; Reddy, M.V.R.; Sayanna, R. Structural, optical and photoluminescence studies of sol-gel synthesized pure and iron doped TiO_2 photocatalysts. *Ceram. Int.* **2019**, *45*, 25060–25068. [CrossRef]
65. Cheary, B.Y.R.W.; Coelho, A. A Fundamental Parameters Approach to X-ray Line-Profile Fitting. *J. Appl. Crystallogr.* **1992**, *25*, 109–121. [CrossRef]

66. Malengreaux, C.M.; Douven, S.; Poelman, D.; Heinrichs, B.; Bartlett, J.R. An ambient temperature aqueous sol–gel processing of efficient nanocrystalline doped TiO_2-based photocatalysts for the degradation of organic pollutants. *J. Sol-Gel Sci. Technol.* **2014**, *71*, 557–570. [CrossRef]
67. Shirley, D.A. High-Resolution X-Ray Photoemission Spectrum of the Valence Bands of Gold. *Phys. Rev. B* **1972**, *5*, 4709–4714. [CrossRef]
68. Kubelka, P. Ein Beitrag zur Optik der Farban striche. *Z. Tech. Phys.* **1931**, *12*, 593–603.
69. Kubelka, P. New contributions to the optics of intensely light-scattering materials. *J. Opt. Soc. Am.* **1948**, *38*, 448–457. [CrossRef] [PubMed]

Article

Photocatalytic Reactivity of Carbon–Nitrogen–Sulfur-Doped TiO₂ Upconversion Phosphor Composites

Seong-Rak Eun [1], Shielah Mavengere [1], Bumrae Cho [2] and Jung-Sik Kim [1,*]

1. Department of Materials Science and Engineering, University of Seoul, Seoul 02504, Korea; dmstjdfkr@naver.com (S.-R.E.); lashiema@gmail.com (S.M.)
2. Department of Materials Engineering, Keimyung University, Daegu 42403, Korea; chobr@kmu.ac.kr
* Correspondence: jskim@uos.ac.kr

Received: 14 September 2020; Accepted: 12 October 2020; Published: 15 October 2020

Abstract: Sol–gel synthesized N-doped and carbon–nitrogen–sulfur (CNS)-doped TiO$_2$ solutions were deposited on upconversion phosphor using a dip coating method. Scanning electron microscopy (SEM) imaging showed that there was a change in the morphology of TiO$_2$ coated on NaYF$_4$:Yb,Er from spherical to nanorods caused by additional urea and thiourea doping reagents. Fourier transform infrared (FTIR) spectroscopy further verified the existence of nitrate–hyponitrite, carboxylate, and SO$_4^{2-}$ because of the doping effect. NaYF$_4$:Yb,Er composites coated with N- and CNS-doped TiO$_2$ exhibited a slight shift of UV-Vis spectra towards the visible light region. Photodecomposition of methylene blue (MB) was evaluated under 254 nm germicidal lamps and a 300 W Xe lamp with UV/Vis cut off filters. The photodegradation of toluene was evaluated on TiO$_2$/NaYF$_4$:Yb,Er and CNS-doped TiO$_2$/NaYF$_4$:Yb,Er samples under UV light illumination. The photocatalytic reactivity with CNS-doped TiO$_2$/NaYF$_4$:Yb,Er surpassed that of the undoped TiO$_2$/NaYF$_4$:Yb,Er for the MB solution and toluene. Photocatalytic activity is increased by CNS doping of TiO$_2$, which improves light sensitization as a result of band gap narrowing due to impurity sites.

Keywords: CNS-doped TiO$_2$; NaYF$_4$:Yb,Er; upconversion phosphor; UV-Vis–NIR photocatalysis

1. Introduction

Titanium dioxide (TiO$_2$) inorganic semiconductors have emerged as trending materials for photocatalysis applications [1–3]. The band gap of 3.2 eV necessitates ultraviolet (UV) light absorption to cause electron movement from the valence band to the conduction band to progress the photocatalytic reaction. However, these photoreactions only proceed within the UV portion of the solar spectrum, which leaves the broad portions of visible and near infrared (NIR) spectra undetected. Reversing the visible and NIR scattering phenomenon using TiO$_2$ remains an arduous task for researchers. The methods for improving spectral absorbance include doping with metals and non-metals [4–7] and coupling with other compounds to form nanocomposites [8,9].

Doping TiO$_2$ with elements such as B [10,11], C, N, and S [12–14] is among the extensively studied techniques used to narrow the band gap. C and N are organic elements that possess remarkable electronic structures and lower toxicity than metallic elements. Thus, non-metallic doping is advantageous because of the smaller ionic radii that can occupy the interstitial sites of TiO$_2$ [15–17]. The extra energy levels imparted by dopants in TiO$_2$ promote absorption of visible light photons [18]. To date, several studies have reported alternative mechanisms for electron–hole reaction pathways to facilitate photocatalysis. Specifically, N-doped TiO$_2$ forms a unique band structure as a substitution at vacancy or interstitial sites in TiO$_2$. As a result of this, the TiO$_2$ band gap of 3.2 eV is narrowed to between 2.8 and 3.06 eV. The doped TiO_2 then sensitizes visible light through the low-energy sites occupied by N for photocatalytic reactions. Consequently, N-doped TiO_2 is reported to perform better in photodegradation reactions due to suppressed electron recombination [16,19]. However, in tri-doped CNS-TiO_2, C, N, and S elements substitute at O sites in TiO_2. In other words, the 2p orbitals of C, N, and S interact with O orbitals inside TiO_2's conduction band. Therefore, new energy levels from the CNS doping elements lower the overall band structure of TiO_2. The great advantage of multi-element doping is that the sensitization sites increase several times as compared to singly doped element [14,20,21]. However, utilizing N- or CNS-doped TiO_2 nanoparticles in aquatic and air purification methods requires a stable substrate to immobilize and prevent release of TiO_2 into the environment. There are several immobilizing substrates for practical application of TiO_2 nanoparticles, which include stainless steel [22] and glass [23]. However, these substrates only offer support. Their chemical composition has minimal effects in terms of improving the light absorption, which is essential for photocatalytic activity progression. Thus, there is a need to stabilize the nanoparticles in micro-sized compounds such as $NaYF_4$:Yb,Er phosphors, which possess light-harvesting properties in the NIR region.

Although upconversion phosphors have short photoluminescence lifetimes, if coupled with TiO_2 they promote light harvesting or photocatalysis, even under visible light photons, due to the heterojunction effect [24,25]. The heterojunctions that exist at interfacial peripheries of the TiO_2 catalyst and phosphor support material are associated with modified electronic structure due to defects [24,26]. This phenomenon distinguishes $NaYF_4$:Yb,Er from other compounds, since both the heterojunctions and the light upconversion effects simultaneously improve the optical and photocatalytic performance of the composites.

Coupling TiO_2 with $NaYF_4$:Yb,Er phosphor has been reported as a promising approach for utilizing low-energy photons in the NIR region and emitting UV-visible light [27,28]. Over the past decades, $NaYF_4$:Yb,Er has been utilized to convert NIR 980 nm photons to emitted photons at 525–550 nm [26,29,30], and even at lower wavelengths such as 390–420 nm [31]. However, only the TiO_2 coating on phosphor has been reported to cause improved photocatalytic efficiencies after a long photoreaction time above 10 h [32]. There is a need to evaluate photocatalysis in consideration of the amount of catalyst in the reactor, the light source intensity, and the concentration of pollutant in an effort to complete the photoreaction rapidly. The effect of N doping [33] or carbon–nitrogen–sulfur (CNS) doping [21] has been reported as a method to promote visible light activation of TiO_2, but few studies have been reported that compare the effect of coupling N-doped or CNS-doped TiO_2 with $NaYF_4$:Yb,Er phosphor.

This study focuses on investigating the effects of N doping and CNS doping of TiO_2 and coupling with $NaYF_4$:Yb,Er upconversion phosphor. The stability of the organic–inorganic molecular bonding was characterized to confirm the existence of dopants in the TiO_2/$NaYF_4$:Yb,Er composites. Photocatalytic properties were evaluated with aqueous methylene blue (MB) and toluene pollutant mediums. The photocatalyst samples of N- or CNS-doped TiO_2/$NaYF_4$:Yb,Er were activated by UV, visible, and NIR light illumination of the solar spectrum.

2. Results

Figure 1 shows XRD spectra of $NaYF_4$:Yb,Er phosphor composites. The crystallinity is indexed for the phosphor in reference to hexagonal sodium, yttrium, ytterbium, erbium, and fluoride (JCPDS 00–028–1192). The $NaYF_4$:Yb,Er phosphor peaks were centered at (100), (110), (101), (200), (111), (201), (211), and (311). However, doping elements and TiO_2 crystal were undetectable in the composites owing to the relatively small amounts coated on the sub-micron phosphor matrix. The XRD peaks were invariant in terms of peak broadening, which normally confirms the effect of N or CNS doping in TiO_2.

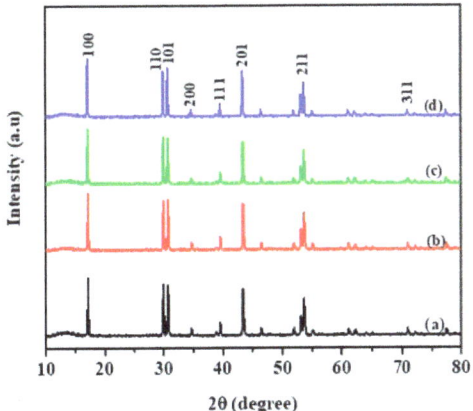

Figure 1. XRD spectra of (**a**) NaYF$_4$:Yb,Er phosphor and NaYF$_4$:Yb,Er coated with (**b**) TiO$_2$, (**c**) N-TiO$_2$, and (**d**) CNS-TiO$_2$.

Figure 2 shows SEM images of NaYF$_4$:Yb,Er (Figure 2a), TiO$_2$ (Figure 2b), TiO$_2$/NaYF$_4$:Yb,Er (Figure 2c), N-TiO$_2$/NaYF$_4$:Yb,Er (Figure 2d), and CNS-TiO$_2$/NaYF$_4$:Yb,Er (Figure 2e). The phosphor particles in Figure 2a are in the ~10 micrometer range, while the undoped TiO$_2$ particles in Figure 2b are agglomerates with a spherical morphology measuring 20–50 nm. Table 1 shows the elemental compositions of F, Na, Y, Er, and Yb. When the TiO$_2$ sol was coated on the phosphor, its spherical morphology was maintained (Figure 2c). Table 2 shows the elemental compositions of O, F, Na, Ti, Y, Er, and Yb. However, with N and CNS doping in TiO$_2$, the TiO$_2$ morphology changed to clustered nanorods on the surfaces of phosphor particles (Figure 2d,e). Table 3 shows the elemental compositions and confirms the presence of N, O, F, Na, Ti, Y, Er, Yb, and the N-doped TiO$_2$ at 0.17 at.% N. The dense nanorods' TiO$_2$ morphology is presumed to be because of the change in pH after addition of the urea (N) or thiourea carbon-nitrogen-sulfur (CNS) doping reagent, as described in the Materials section. Table 4 shows the elemental compositions of C, N, O, F, Na, S, Ti, Y, Er, and Yb. The EDS spectra are shown in Figures S1, S3, S5, and S7. The EDS elemental mapping data are shown in Figures S2, S4, S6, and S8. Interestingly, the CNS-doped TiO$_2$ shows values of C 46.30 at.%, N 0.79 at.%, and 0.15 at.%. Therefore, the thiourea doping reagent is a rich source for CNS elements.

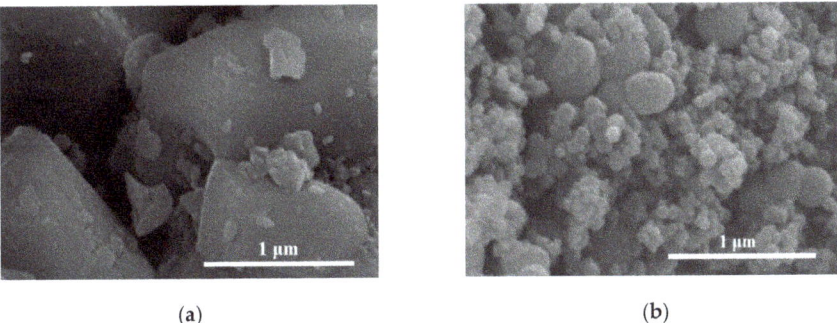

(**a**)   (**b**)

Figure 2. *Cont.*

Figure 2. SEM images of (**a**) NaYF$_4$:Yb,Er, (**b**) TiO$_2$ particles, and NaYF$_4$:Yb,Er phosphors coated with (**c**) TiO$_2$, (**d**) N-TiO$_2$, and (**e**) CNS-TiO$_2$.

Table 1. Elemental analysis for NaYF$_4$:Yb,Er.

Element	Mass%	Atom%
F	29.22	60.86
Na	10.19	17.54
Y	35.63	15.86
Er	5.11	1.21
Yb	19.84	4.54
Totals	100	-

Table 2. Elemental analysis for TiO$_2$/NaYF$_4$:Yb,Er.

Element	Mass%	Atom%
O	4.15	8.98
F	30.85	56.25
Na	10.66	16.05
Ti	4.66	3.37
Y	28.35	11.05
Er	4.42	0.92
Yb	16.91	3.38
Totals	100	-

Table 3. Elemental analysis for N-TiO$_2$/NaYF$_4$:Yb,Er.

Element	Mass%	Atom%
N	0.07	0.17
O	7.33	16.30
F	25.62	48.01
Na	7.59	11.75
Ti	11.12	8.27
Y	28.43	11.38
Er	4.27	0.91
Yb	15.57	3.20
Totals	100	-

Table 4. Elemental analysis for CNS-TiO$_2$/NaYF$_4$:Yb,Er.

Element	Mass%	Atom%
C	25.97	46.30
N	0.52	0.79
O	12.79	17.12
F	19.41	21.89
Na	5.41	5.04
S	0.23	0.15
Ti	7.94	3.55
Y	14.60	3.52
Er	2.78	0.36
Yb	10.36	1.28
Totals	100	-

FTIR spectroscopy results for N-TiO$_2$/NaYF$_4$:Yb,Er and CNS-TiO$_2$/NaYF$_4$:Yb,Er are shown in Figure 3a,b, respectively. The stretching vibrations at 3410 and 1624 cm^{-1} are assigned to the -OH groups bonded to Ti- atoms and H$_2$O bending, respectively; while the stretching vibrations at 1369 and 1130 cm^{-1} arise from carboxylate and S = O bonds, respectively, because of surface-adsorbed SO$_4^{2-}$ species. The stretching vibrations at 1044 and 616 cm^{-1} are because of hyponitrite and the Ti-O groups, respectively [34].

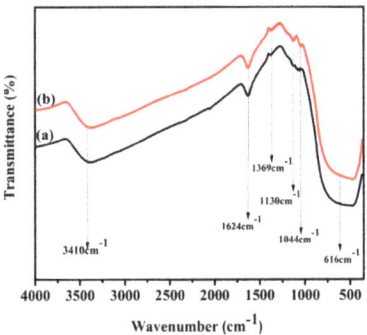

Figure 3. Fourier transform infrared spectra of NaYF$_4$:Yb,Er coated with (**a**) N-TiO$_2$ and (**b**) CNS-TiO$_2$.

Figure 4 exhibits UV-Vis absorption spectra of NaYF$_4$:Yb,Er coated with undoped, N-doped, and CNS-doped TiO$_2$. The NaYF$_4$:Yb,Er phosphor absorbs UV light from 200 nm with an edge at 250 nm (due to the NaYF$_4$ host), and in Figure S9 absorption peaks were centered at 524 and 654 nm due to the Er^{3+} co-activator [35]. However, coating TiO$_2$ on phosphor broadened the absorption spectra. Specifically, the TiO$_2$/NaYF$_4$:Yb,Er composites exhibited UV-Vis absorption between 200 and 400 nm, with peak absorption at 300 nm. The additional N-doped and CNS-doped TiO$_2$ phosphor composites show higher absorption intensities and a slight shift towards the visible region; the red-shift has been referenced by other researchers as being because of the lower energy levels in TiO_2 imparted by N or CNS doping elements [20,36].

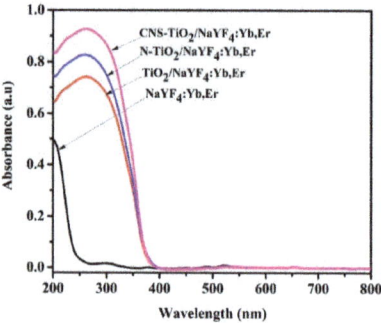

Figure 4. UV-Vis absorption spectra of $NaYF_4$:Yb,Er coated with TiO_2, N-TiO_2, and CNS-TiO_2.

Figure 5a shows photoluminescence (PL) spectra as obtained after 980 nm NIR irradiation of $NaYF_4$:Yb,Er upconversion phosphor composites. Firstly, the full emission spectra (UV-Vis–NIR emission) in Figure 5a exhibit emissions of visible light photons at 520, 527, 540, and 548 nm. Thus, as a result of simultaneous energy transfer occurring in the excited phosphor, the energy losses resulted in lower energy photons at 652 and 658 nm. The $^2H_{11/2}$-to-$^4I_{15/2}$ transition is assigned to the emission at 527 nm, while the $^4S_{3/2}$ to $^4I_{15/2}$ is assigned to the emission at 540 nm. Additionally, the transitions emitting low-energy photons at 652 nm are from $^4F_{9/2}$ to $^4I_{15/2}$ [25]. The overall reduction in emission peaks in the TiO_2-coated $NaYF_4$:Yb,Er samples was also observed in previous research [32]. Thus, TiO_2 nanoparticles on phosphor act as barriers to emitted light.

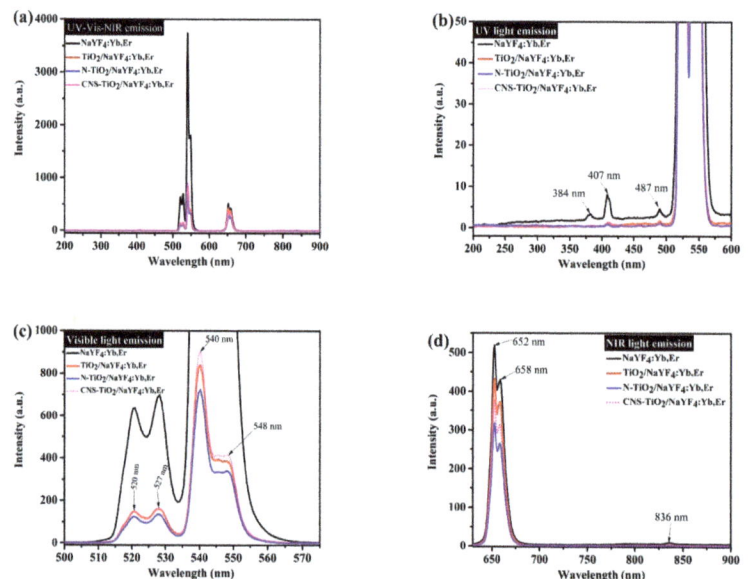

Figure 5. Photoluminescence spectra of $NaYF_4$:Yb,Er coated with TiO_2, N-TiO_2 and CNS-TiO_2. (**a**) UV-Vis–NIR emissions and the enlarged spectra for (**b**) UV light, (**c**) visible light, and (**d**) NIR light emissions.

Figure 5a was enlarged and labeled (Figure 5b–d) to clearly observe peak variations. Figure 5b shows UV light emission at 384 nm and visible light photons at 407 and 484 nm in the NaYF$_4$:Yb,Er phosphor. However, all of these peaks were suppressed in TiO$_2$/NaYF$_4$:Yb,Er and N/CNS-doped TiO$_2$/NaYF$_4$:Yb,Er composites. The effect of doping on photoluminescence was observed in the enlarged peaks in Figure 5c,d. Specifically, at 520 and 527 nm in Figure 5c, the TiO$_2$/NaYF$_4$:Yb,Er and CNS-doped TiO$_2$/NaYF$_4$:Yb,Er show unchanged emission intensities, while only N-doped TiO$_2$/NaYF$_4$:Yb,Er exhibits further peak suppression. However, at the peaks located at 540 and 548 nm, the CNS-doped TiO$_2$/NaYF$_4$:Yb,Er exhibits the highest photoluminescence, followed by TiO$_2$/NaYF$_4$:Yb,Er and N-doped TiO$_2$/NaYF$_4$:Yb,Er.

Figure 5d shows the enlarged PL emission spectra to clearly show peaks in the NIR region. At 652 and 658 nm, the peak intensity decreases in the order of NaYF$_4$:Yb,Er > TiO$_2$/NaYF$_4$:Yb,Er TiO$_2$ > CNS-doped TiO$_2$/NaYF$_4$:Yb,Er > N-doped TiO$_2$/NaYF$_4$:Yb,Er. The peak suppression and enhancement phenomena represent a multi-energy transfer process. However, some researchers have highlighted that the pH level in urea (the doping reagent in phosphor) tunes the PL emissions for visible and NIR emissions in Y$_2$O$_3$:Yb,Er nanophosphors [37]. Thus, in our work we confirmed the tuning of visible and NIR emissions. In the N-doped TiO$_2$/NaYF$_4$:Yb,Er (urea additive: pH 5.79) and in CNS-doped TiO$_2$/NaYF$_4$:Yb,Er (thiourea additive: pH 5.73) as compared to undoped TiO$_2$/NaYF$_4$:Yb,Er (pH 6.39), the pH values are almost the same, however the most acidic CNS-doped phosphor exhibits the highest PL intensity (at 540 and 548 nm), followed by the TiO$_2$ phosphor and the doped TiO$_2$-coated samples. Upconversion phosphors are characterized by having short lifetimes, as exhibited in the decay curves in Figure S10.

Figure 6 shows the photobleaching variation of the MB solution for NaYF$_4$:Yb,Er phosphor composites under UV light illumination. The maximum peak at 664 nm is the characteristic MB absorbance that is monitored to evaluate intensity variations as a measure of the degradation of the organic compound. In a 4 h reaction period, the CNS-TiO$_2$/NaYF$_4$:Yb,Er absorbance spectra with the lowest intensities at seen at 4 h. Therefore, CNS-TiO$_2$/NaYF$_4$:Yb,Er shows the highest photocatalytic reactivity among the three samples. However, the TiO$_2$ and N-TiO$_2$ upconversion phosphor composites require more time to completely degrade MB solutions (Figure 6b,c).

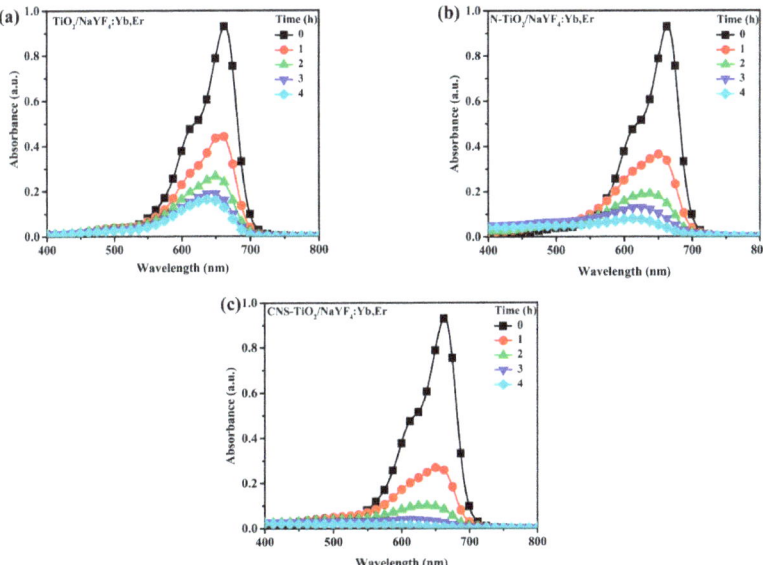

Figure 6. Methylene blue (MB) absorbance variation for NaYF$_4$:Yb,Er phosphor coated with (**a**) TiO$_2$, (**b**) N-TiO$_2$, and (**c**) CNS-TiO$_2$ under UV light activation.

Figure 7a shows the peak absorbance variations for photodegradation of the MB solution under UV illumination. The photodegradation of the MB solution proceeded to 40% efficiency for the photoreaction mixture with TiO_2 only. However, N-doped TiO_2 and CNS-doped TiO_2 improves the efficiency to 80%. This is owing to the light absorption property related to N- and CNS-doped TiO_2. Supporting TiO_2 on the $NaYF_4$:Yb,Er upconversion phosphor improved the photocatalytic efficiency up to 90%. Moreover, with N-TiO_2 or CNS-TiO_2 supported on $NaYF_4$:Yb,Er, the photocatalytic efficiencies were enhanced to completion (100% in 4 h). Thus, N-TiO_2 or CNS-TiO_2 coupling with $NaYF_4$:Yb,Er phosphor improves the catalytic activity due to the improvement in light absorption by phosphor and N or CNS doping elements. In detail, the $NaYF_4$:Yb,Er phosphor support has light absorption properties due to the $NaYF_4$ host (as exhibited in Figure 4), which coincides with the TiO_2 absorption band. Additionally, light absorption for the composite is improved with additional N and CNS doping.

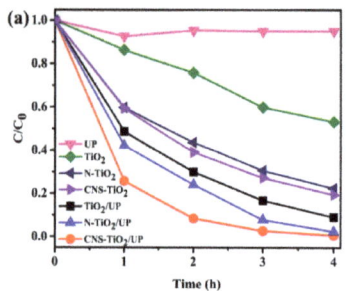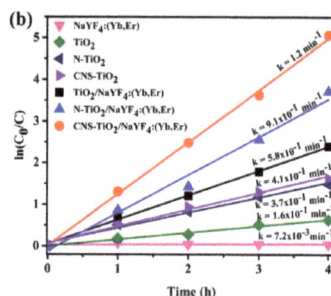

Figure 7. (a) MB peak absorbance variation as degraded with 254 nm UV illumination. Here, UP is the $NaYF_4$:Yb,Er upconversion phosphor and (b) correlated rate of kinetics.

Both CNS doping and N doping of TiO_2 further cause improvements in photocatalytic efficiencies. The N or CNS doping of TiO_2 incorporates new energy levels in the interstitial and substitution sites. Thus, lower energy levels lower the TiO_2 absorption band, which promotes the flow of electrons into the conduction band. As follows, the photocatalytic activity achieves completion in a 4 h cycle for phosphor coated with doped CNS or N-TiO_2. Figure 7b illustrates the rate of kinetics for the samples in Figure 7a. As shown in Figure 7b, the reactions have a linear and typical relationship for first-order kinetics. The rate constants are 0.16 min^{-1} for TiO_2, 0.37 min^{-1} for N-TiO_2, 0.41 min^{-1} for CNS-TiO_2, 0.0072 min^{-1} for $NaYF_4$:Yb,Er, 0.58 min^{-1} for $TiO_2/NaYF_4$:Yb,Er, 0.91 min^{-1} for N-$TiO_2/NaYF_4$:Yb,Er, and 1.2 min^{-1} for CNS-$TiO_2/NaYF_4$:Yb,Er. This result shows that MB photodegradation with CNS-$TiO_2/NaYF_4$:Yb,Er is the fastest reaction by 7.5 times for TiO_2, by 3.2 times for N-TiO_2, by 2.9 times for CNS-TiO_2, by 2.1 times for $TiO_2/NaYF_4$:Yb,Er, and by 1.3 times for N-$TiO_2/NaYF_4$:Yb,Er. Therefore, CNS doping of TiO_2 and its support on $NaYF_4$:Yb,Er phosphor improves the photocatalytic performances.

Figure 8 exhibits the MB peak absorbance against time for visible light activation. The N-doped TiO_2 and CNS-doped TiO_2 show improvements in visible light photocatalytic efficiencies as compared to TiO_2 only. Furthermore, coupling TiO_2 with phosphor and doping the $TiO_2/NaYF_4$:Yb,Er causes enhancements of photocatalytic efficiencies. Precisely, $TiO_2/NaYF_4$:Yb,Er, N-doped $TiO_2/NaYF_4$:Yb,Er, and CNS-doped $TiO_2/NaYF_4$:Yb,Er showed 50%, 60%, and 70% efficiencies after 120 min of light irradiation. The undoped $TiO_2/NaYF_4$:Yb,Er composite showed significant photocatalytic efficiency as compared to TiO_2, N-TiO_2, and CNS-TiO_2, mainly due to the heterojunction effect that exists between phosphor and TiO_2. The CNS-doped $TiO_2/NaYF_4$:Yb,Er shows the highest photocatalytic reactivity over visible light illumination. As a result, the effect of doping with N-$TiO_2/NaYF_4$:Yb,Er or CNS-$TiO_2/NaYF_4$:Yb,Er is clearly exhibited by 10% and 20% efficiency enhancements, respectively, compared with undoped-$TiO_2/NaYF_4$:Yb,Er. Visible light sensitization is conceptualized by the low-energy states induced by doping with N or CNS elements. Thus, tri-element doping in TiO_2 imparts more impurities in the TiO_2 band than N doping only. Figure 8b shows the rate of kinetics for the samples in Figure 8a. The CNS-doped TiO$_2$/NaYF$_4$:Yb,Er exhibited the swiftest reaction with a 9.7 × 10^{-3} min^{-1} rate constant. In comparison with other photocatalysts, the CNS-doped TiO$_2$/NaYF$_4$:Yb,Er reaction is 9.7 times greater for TiO$_2$ only, 4.4 times greater for N-TiO$_2$, 2.8 times greater for CNS-TiO$_2$, 1.8 times greater for TiO$_2$/NaYF$_4$:Yb,Er, and 1.2 times greater for N-TiO$_2$/NaYF$_4$:Yb,Er. Therefore, the photocatalytic efficiencies are improved by tri-doping TiO$_2$ and its support on the NaYF$_4$:Yb,Er upconversion phosphor.

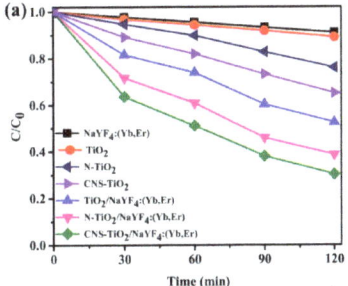 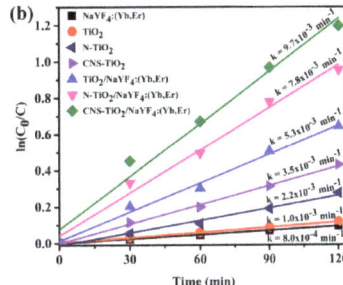

Figure 8. (a) MB peak absorbance against the visible light activation time and (b) correlated rate of kinetics.

Figure 9 shows the MB peak absorbance against time for NIR light activation. Undoped TiO$_2$–phosphor and N-doped TiO$_2$–phosphor composites only show ~15% efficiency with NIR illuminations, but CNS-doped TiO$_2$/NaYF$_4$:Yb,Er exhibits a ~25% improvement. Therefore, NIR irradiations in CNS-doped TiO$_2$/NaYF$_4$:Yb,Er have the highest photocatalytic activity. Under NIR, the Yb^{3+} in NaYF$_4$:Yb,Er phosphor sensitizes the NIR photons and emits UV-Vis–NIR photons through the Er^{3+} emission center, as discussed in Figure 5. Visible light photons are sensitized through the low energy levels imparted by N- or CNS doping and electrons are injected into the TiO$_2$ conduction band for photocatalysis. Additionally, the doping elements promote electron–hole generation efficiencies to facilitate photocatalysis. However, the overall photocatalytic efficiencies were low due to the mono-wavelength absorption properties of Yb^{3+} ions at 980 nm.

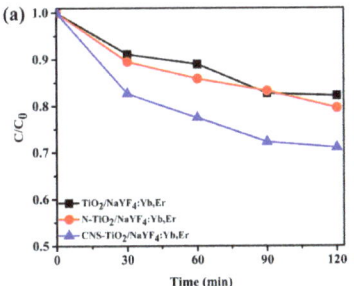 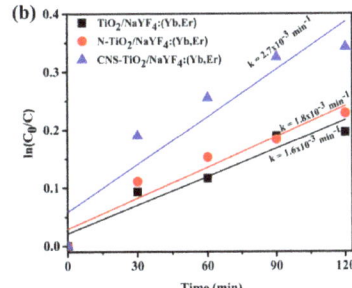

Figure 9. (a) MB peak absorbance against the NIR light activation time and (b) correlated rate of kinetics.

Figure 10 exhibits the concentration variations of toluene during photodegradation with TiO$_2$-NaYF$_4$:Yb,Er and CNS/TiO$_2$-NaYF$_4$:Yb,Er powders under UV light illumination. The toluene concentration in the Y-axis corresponds to the concentration of toluene remaining in the Teflon bag after 1 h sampling under UV light illumination. Through the first 1 h sampling cycle, the CNS/TiO$_2$-NaYF$_4$:Yb,Er photocatalyst has 1.5 ppm of toluene less than the undoped TiO$_2$-NaYF$_4$:Yb,Er sample. The 4 h toluene degradation is above 95% for the CNS/TiO$_2$-NaYF$_4$:Yb,Er.

This result indicates that the CNS/TiO$_2$-NaYF$_4$:Yb,Er has exceptionally superior photoactivity compared with the TiO$_2$-NaYF$_4$:Yb,Er sample. Thus, CNS doping is essential for improving photocatalytic activity.

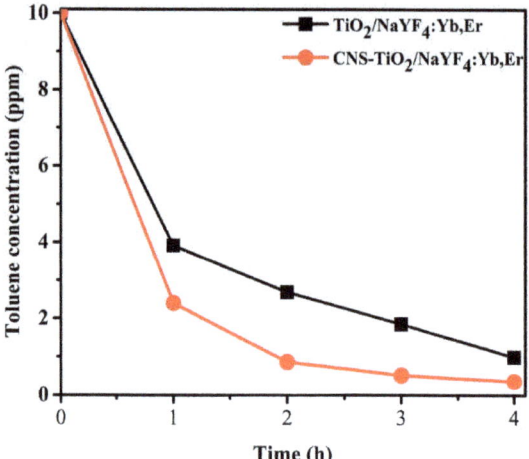

Figure 10. Photodegradation for toluene with CNS-TiO$_2$/NaYF$_4$:Yb,Er and TiO$_2$/NaYF$_4$:Yb,Er composites under UV light illumination.

3. Discussions

Figure 11 is a schematic diagram depicting the synthesis and photocatalytic test performances. The pH of the coating sol varied due to the interaction of the TiO$_2$ precursor and ethanol without pH modifiers. Thus, the TiO$_2$ sol without the doping reagent showed slightly alkaline conditions at pH 6.39. As a result of this slight alkalinity, the TiO$_2$ morphology existed as spherical particles on phosphor. However, with additional thiourea and urea as doping reagents for N and CNS in the TiO$_2$ sol, the pH decreased to acidic conditions. For instance, in the N-doped TiO$_2$ sol the pH was 5.79, while in the CNS-doped TiO$_2$ sol the pH was 5.73. As a result, the morphologies were changed to nanorods. Hence, the drop in pH promoted the formation of nanorods. The photoluminescence emission peaks were also observed to be tuned by urea or thiourea reagents owing to the drop in pH.

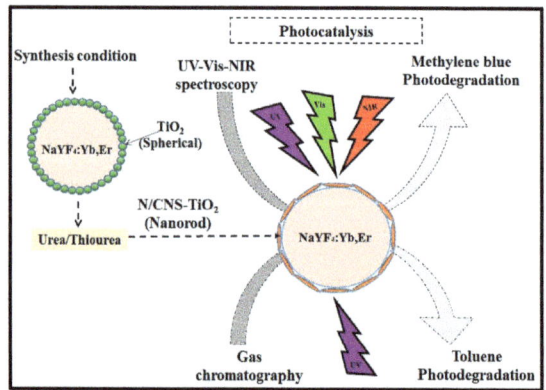

Figure 11. Schematic diagram for synthesis and photocatalysis tests.

Figure 12 shows the proposed photocatalysis mechanism for the NaYF$_4$:Yb,Er phosphor coated with TiO$_2$ under different doping conditions. The three conditions for TiO$_2$ coating on NaYF$_4$:Yb,Er are illustrated in Figure 12 as TiO_2, $N-TiO_2$, and $CNS-TiO_2$. The first condition is TiO_2 supported on $NaYF_4$:Yb,Er phosphor, where only the TiO_2 energy band is present. The second condition is N-doped TiO_2 supported on $NaYF_4$:Yb,Er, where the continuous solid line inside the TiO_2 represents N energy levels. The third condition is CNS-doped TiO_2 supported on $NaYF_4$:Yb,Er, where the dotted line in the TiO_2 band represents discrete energy levels imparted by C, N, or S elements. As illustrated, the upconversion phosphor as the support material contains Yb^{3+} as the NIR sensitizer in the $^2F_{7/2}$ state, which transfers energy to the co-activator or Er^{3+} through the $^4F_{7/2}$ and $^4I_{11/2}$ states [35,38,39]. Simultaneously, the Er^{3+} emits UV-Vis–NIR photons of less than 980 nm. Hence, the upconversion phosphor converts low-energy NIR photons at 980 nm to high-energy photons (Figure 5). Peak wavelengths were observed at 384, 407, 520, 527, 540, 548, 652, 658, and 836 nm. Thus, $NaYF_4$:Yb,Er upconversion phosphor emits UV-Vis–NIR photons.

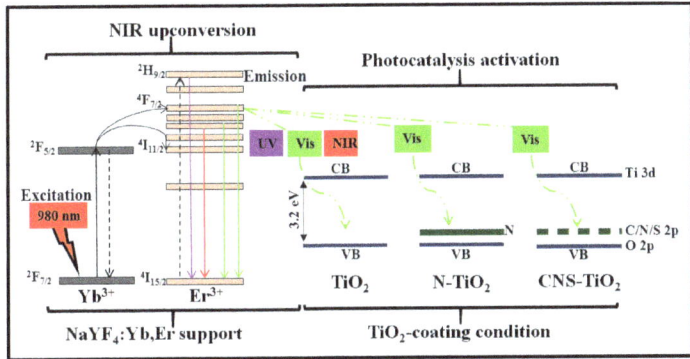

Figure 12. Proposed photocatalysis mechanism for the $NaYF_4$:Yb,Er phosphor coated with TiO_2 under different doping conditions.

Photocatalytic activity proceeds under UV-Vis–NIR irradiation. The reactions are related to light sensitization centers [40]. For instance, under UV light activation, the TiO_2-$NaYF_4$:Yb,Er composite absorbs light and facilitates photocatalysis. It is noteworthy that the $NaYF_4$ phosphor host also sensitizes UV light photons (Figure 4), which enhances the overall light absorption by the TiO_2-$NaYF_4$:Yb,Er composite. Under visible light, N-TiO_2-$NaYF_4$:Yb,Er or CNS-TiO_2/$NaYF_4$:Yb,Er light is sensitized through the lower energy levels in the N or CNS, which inject electrons into the TiO_2 conduction band for photoreaction at the nanorod surface. Under NIR irradiation, the light is sensitized through the Yb^{3+} and energy is transferred through Er^{3+}. Then, the emitted visible light is absorbed through low-energy impurities in N or CNS elements and through the low energy levels due to heterojunctions between TiO_2 and $NaYF_4$:Yb,Er phosphor. Consequently, after sensitization of UV-Vis–NIR light, electrons are injected into the TiO_2 conduction band, leaving holes in the valence band. At the surface of TiO_2, electrons are adsorbed by O_2 molecules while holes are adsorbed by H_2O molecules to form superoxide and hydroxyl radicals, respectively. After several intermediate reactions, superoxide or hydroxyl radicals attack and photodegrade the MB or toluene pollutants. The CNS-doped TiO_2/$NaYF_4$:Yb,Er outperformed the N-doped TiO_2-$NaYF_4$:Yb,Er and TiO_2-$NaYF_4$:Yb,Er in UV-Vis–NIR photocatalytic activities in methylene blue due to the existence of low-energy CNS doping elements with improved light absorption properties [20].

This work mainly focused on adding 2.5% urea or thiourea to TiO_2 to study the effects of doping with N or CNS. Since the photoluminescence spectra exhibited interesting visible and NIR light tuning, further investigations into the effects of varying urea or thiourea (N or CNS, respectively) doping of TiO_2/$NaYF_4$:Yb,Er is recommended. Specifically, the photoluminescence emissions at 540 and 548 nm in CNS-doped TiO_2/$NaYF_4$:Yb,Er were enhanced (Figure 5b) compared to the TiO_2-$NaYF_4$:Yb,Er and N-TiO_2-$NaYF_4$:Yb,Er. This insinuates the possibility that CNS doping improves the emission of visible light. Additionally, the photoluminescence under NIR is limited due to the narrow absorption of Yb^{3+} ions to the 980 nm wavelengths [39,40]. It is also recommended to evaluate the effects of adding 2 co-sensitizers to improve the absorption of NIR light to achieve significant photocatalytic efficiency.

4. Materials and Methods

4.1. Experimental

The $NaYF_4$:Yb,Er upconversion phosphor was synthesized from a stoichiometric amount of Y:Yb:Er at 77:20:3 mol% from yttrium, ytterbium, and erbium oxides (99.9%, Sigma Aldrich, St. Louis, MO, USA) by mixing in a nitric acid solution. The mixture solution was constantly stirred at 120 °C for 30 min to form a transparent sol. In a separate beaker, urea, ammonium hydrogen fluoride, and sodium silicon fluoride (Duksan Pure Chemicals Co., Ansan-si, Kyunggi-do, Korea) were dissolved in distilled H_2O by magnetic stirring at 80 °C for 2 h. Then, the multi-component $NaYF_4$:Yb,Er solution was poured into a closed crucible for combustion at 650 °C for 5 min in a box furnace (SK1700-B30, Thermotechno Co., Siheung-si, Gyeonggi-do, Korea).

A flow chart of the TiO_2 sol preparation and coating process on $NaYF_4$:Yb,Er is shown in Figure 13. The TiO_2 sols were prepared from 10 mL titanium(IV)-butylate (99%, Acros Organics) with anhydrous ethyl alcohol (99.9%, Duksan Pure Chemicals Co., Ansan-si, Kyunggi-do, Korea) and distilled H_2O at a volumetric ratio of 5:1. After a 10 min mixing procedure, the 2.5 mol% urea or 2.5 mol% thiourea doping reagents (Daejung Chemicals Siheung-si, Gyeonggi-do, Korea) for N or CNS were added to separate beakers with TiO_2 sol. The N- or CNS-doped TiO_2 sols were further magnetically stirred at 50 °C for 2 h. For comparison, undoped TiO_2 sol was also prepared using similar conditions. The pH for the as-synthesized coating solutions was measured using a Hanna Instruments portable digital meter (HI 8424, Smithfield, RI, USA). The pH readings for the TiO_2 sol, N-doped TiO_2 sol, and CNS-doped TiO_2 sol were 6.39, 5.79, and 5.73, respectively. Thus, the pH values dropped with urea and thiourea reagents.

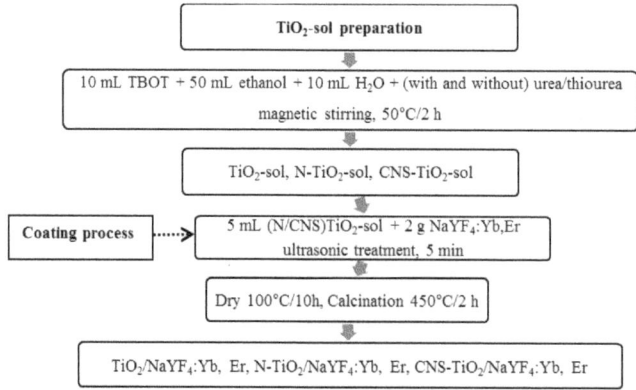

Figure 13. Flow chart of TiO_2 sol preparation and its coating process on $NaYF_4$:Yb,Er.

The sol–gel coating process was proceeded by dip coating 2 g $NaYF_4$:Yb,Er upconversion phosphor powder samples in 5 mL of TiO_2 sol, N-TiO_2 sol, or CNS-TiO_2 sol. The dispersed $NaYF_4$:Yb,Er phosphor was treated in an ultrasonic bath for 5 min followed by removal of excess ethanol solution through Advantec filter paper (Toyo Roshi Kaisha, Ltd., Tokyo, Japan). Then, the coated phosphors were placed on a quartz petri dish for drying at 100 °C for 10 h. Finally, the undoped TiO_2, N-doped TiO_2, and CNS-doped-TiO_2 coated $NaYF_4$:Yb,Er samples were placed in alumina crucibles with lids for a 2 h calcination process at 450 °C. Only N-doped TiO_2 and CNS-doped TiO_2 were also calcined using a similar process for the purpose of acting as photocatalytic test controls.

4.2. Characterizations

Crystallinity and morphology were evaluated by X-ray diffractometry (Bruker AXS8 Advanced, D8 Discover, Bruker AXS GmbH, Karlsruhe, Germany) and scanning electron microscopy (SEM, Hitachi S-4300, Hitachi Ltd., Tokyo, Japan). The organic and inorganic molecular bonds were examined by Fourier transform infrared (FTIR; Bruker Vertex 70, Karlsruhe, Germany) spectroscopy at a 50:1 wt.% ratio of KBr/phosphor. UV-Vis diffuse reflectance spectra (DRS) for the synthesized composites were evaluated by UV-Vis–NIR spectroscopy (UV-3150 Shimadzu, Kyoto, Japan) in fast scan speed mode through a 30 nm slit. Photoluminescence (PL) characteristics were examined using a fluorescence spectrophotometer (Hitachi F-4500, Tokyo, Japan). The PL powder samples were excited with a 40 mW, 980 nm infrared dot laser (ILaser Lab Co. Ltd., Seong Dong-gu, Seoul, Korea), and the emitted photons were detected through a 10 nm slit.

Photocatalytic activity was evaluated under three conditions of light illumination. First, UV light from 254 nm photons (G6T5 Sankyo Denki 2-fluorescent lamps, Sankyo Denki Co, Kanagawa, Japan) in a protective chamber was illuminated over a mixture of 100 mg catalyst and 100 mL MB 5 ppm solution. Photodegraded 2 mL samples were collected and labeled with respective light irradiation times throughout the 3 light illumination conditions. Second, visible light (10,000 lux, 300 W Xe Solar Simulators Luzchem SolSim2) was irradiated over a 410 nm cut-off filter to the 100 mg catalyst in 50 mL MB 5 ppm solution. Third, NIR photons (15,000 lux, Luzchem SolSim2, Ottawa, Ontario, Canada) were illuminated on a photoreaction of 100 mg catalyst–50 mL MB 2 ppm solution with 410 and 760 nm cut-off filters. The absorbance variation of UV-Vis–NIR photodegraded MB solutions was analyzed in quartz cell holders of a UV-Vis–NIR spectroscope (UV-3150 Shimadzu, Kyoto, Japan). Photocatalytic activity control experiments were also conducted on N-doped TiO_2 only and CNS-doped TiO_2 using the same UV and visible light conditions.

Toluene photodegradation was evaluated on 6 g TiO_2-$NaYF_4$:Yb,Er and CNS/TiO_2-$NaYF_4$:Yb,Er powders. The photocatalytic powders were sparsely dispersed on a borosilicate Petri dish in a Teflon sampling bag. The light illumination source was an 8 W Philips TUV G6T5. At 1 h intervals, 1 mL toluene vapor was withdrawn from the sampling bag and injected into an Agilent Technologies gas chromatography (GC) system (7890 A, Agilent Technologies Inc., California, USA) to determine the extent of degradation in relation to the UV light illumination time.

5. Conclusions

Upconversion phosphor was coupled with undoped TiO_2 and N- and CNS-doped TiO_2 using a sol–gel coating method. The TiO_2 nanocrystalline morphologies on $NaYF_4$:Yb,Er upconversion phosphor changed from spherical to nanorods, most probably due to the use of urea and thiourea doping reagents. N doping and CNS doping in TiO_2 were successful, as confirmed by EDS and FTIR through broad molecular bonds of SO_4^{2-}, carboxylate, and NO_3^-. The UV-Vis DRS of the photocatalyst powder samples exhibited a slight red-shift with N doping and CNS doping. The MB photocatalytic degradation efficiencies under UV-Vis–NIR illumination sources are enhanced with doping with N or CNS elements. Specifically, N-TiO_2 and CNS-TiO_2 photocatalysts exhibited lower UV-Vis photocatalytic efficiencies as compared to the N-TiO_2 and CNS-TiO_2 supported on $NaYF_4$:Yb,Er phosphor. Moreover, toluene degradation efficiencies were improved by CNS doping on TiO_2/$NaYF_4$:Yb,Er. The CNS-doped-TiO_2/$NaYF_4$:Yb,Er photocatalyst is a plausible candidate for pollutant remediation.

Supplementary Materials: The following are available online at http://www.mdpi.com/2073-4344/10/10/1188/s1, Figure S1: Elemental mapping for $NaYF_4$:Yb,Er, Figure S2: Elemental spectra for $NaYF_4$:Yb,Er, Figure S3: Elemental mapping for TiO_2/$NaYF_4$:Yb,Er, Figure S4: Elemental spectra for TiO_2/$NaYF_4$:Yb,Er, Figure S5: Elemental mapping for N-TiO_2/$NaYF_4$:Yb,Er, Figure S6: Elemental spectra for N-TiO_2/$NaYF_4$:Yb,Er, Figure S7: Elemental mapping for CNS-TiO_2/$NaYF_4$:Yb,Er, Figure S8: Elemental spectra for CNS-TiO_2/$NaYF_4$:Yb,Er, Figure S9: UV-Vis absorption spectra enlarged along the absorbance axis from Figure 4, Figure S10: Decay curve for $NaYF_4$:Yb,Er coated with TiO_2, N-TiO_2 and CNS-TiO_2 monitored at 540 nm under 980 nm excitation.

Author Contributions: Supervision, J.-S.K.; methodology and experimentation, S.-R.E.; writing—original draft preparation, S.M.; writing—review and editing, B.C. All authors have read and agreed to the published version of the manuscript.

Funding: This research was funded by the Ministry of Land, Infrastructure, and Transport of Korea, grant number 20CTAP-C157721-01, under the Infrastructure and Transportation Technology Promotion Research Program.

Conflicts of Interest: The authors declare no conflict of interest.

References

1. Xiong, Z.; Lei, Z.; Li, Y.; Dong, L.; Zhao, Y.; Zhang, J. A review on modification of facet-engineered TiO_2 for photocatalytic CO_2 reduction. *J. Photochem. Photobiol. C Photochem. Rev.* **2018**, *36*, 24–47. [CrossRef]
2. Shayegan, Z.; Lee, C.-S.; Haghighat, F. TiO_2 photocatalyst for removal of volatile organic compounds in gas phase—A review. *Chem. Eng. J.* **2018**, *334*, 2408–2439. [CrossRef]
3. Fujishima, A.; Rao, T.N.; Tryk, D.A. Titanium dioxide photocatalysis. *J. Photochem. Photobiol. C Photochem. Rev.* **2000**, *1*, 1–21. [CrossRef]
4. Akpan, U.G.; Hameed, B.H. The advancements in sol-gel method of doped-TiO_2 photocatalysts. *Appl. Catal. A Gen.* **2010**, *375*, 1–11. [CrossRef]
5. Kumaravel, V.; Mathew, S.; Bartlett, J.; Pillai, S.C. Photocatalytic hydrogen production using metal doped TiO_2: A review of recent advances. *Appl. Catal. B Environ.* **2019**, *244*, 1021–1064. [CrossRef]
6. Yadav, H.M.; Kolekar, T.V.; Pawar, S.H.; Kim, J.-S. Enhanced photocatalytic inactivation of bacteria on Fe-containing TiO_2 nanoparticles under fluorescent light. *J. Mater. Sci. Mater. Med.* **2016**, *27*, 57. [CrossRef]
7. Koli, V.B.; Delekar, S.D.; Pawar, S.H. Photoinactivation of bacteria by using Fe-doped TiO_2-MWCNTs nanocomposites. *J. Mater. Sci. Mater. Med.* **2016**, *27*, 177. [CrossRef]
8. Muzakki, A.; Shabrany, H.; Saleh, R. Synthesis of ZnO/CuO and TiO_2/CuO nanocomposites for light and ultrasound assisted degradation of a textile dye in aqueous solution. *AIP Conf. Proc.* **2016**, *1725*, 020051. [CrossRef]
9. Balayeva, N.O.; Fleisch, M.; Bahnemann, D.W. Surface-grafted WO_3/TiO_2 photocatalysts: Enhanced visible-light activity towards indoor air purification. *Catal. Today* **2018**, *313*, 63–71. [CrossRef]
10. Xue, X.; Wang, Y.; Yang, H. Preparation and characterization of boron-doped titania nano-materials with antibacterial activity. *Appl. Surf. Sci.* **2013**, *264*, 94–99. [CrossRef]
11. Koli, V.B.; Mavengere, S.; Kim, J.-S. Boron-doped TiO_2–CNTs nanocomposites for photocatalytic application. *J. Mater. Sci. Mater. Electron.* **2018**. [CrossRef]
12. Chen, X.; Burda, C. The Electronic Origin of the Visible-Light Absorption Properties of C-, N- and S-Doped TiO_2 Nanomaterials. *J. Am. Chem. Soc.* **2008**, *130*, 5018–5019. [CrossRef] [PubMed]
13. Yang, J.; Bai, H.; Jiang, Q.; Lian, J. Visible-light photocatalysis in nitrogen–carbon-doped TiO_2 films obtained by heating TiO_2 gel–film in an ionized N_2 gas. *Thin Solid Films* **2008**, *516*, 1736–1742. [CrossRef]
14. Cheng, X.; Yu, X.; Xing, Z. One-step synthesis of visible active CNS-tridoped TiO_2 photocatalyst from biomolecule cystine. *Appl. Surf. Sci.* **2012**, *258*, 7644–7650. [CrossRef]
15. Cheng, X.; Yu, X.; Xing, Z.; Yang, L. Synthesis and characterization of N-doped TiO_2 and its enhanced visible-light photocatalytic activity. *Arab. J. Chem.* **2016**, *9*, S1706–S1711. [CrossRef]
16. Khore, S.K.; Tellabati, N.V.; Apte, S.K.; Naik, S.D.; Ojha, P.; Kale, B.B.; Sonawane, R.S. Green sol-gel route for selective growth of 1D rutile N-TiO_2: A highly active photocatalyst for H_2 generation and environmental remediation under natural sunlight. *RSC Adv.* **2017**, *7*, 33029–33042. [CrossRef]
17. Dozzi, M.V.; Selli, E. Doping TiO_2 with p-block elements: Effects on photocatalytic activity. *J. Photochem. Photobiol. C Photochem. Rev.* **2013**, *14*, 13–28. [CrossRef]
18. Zhou, M.; Yu, J. Preparation and enhanced daylight-induced photocatalytic activity of C,N,S-tridoped titanium dioxide powders. *J. Hazard. Mater.* **2008**, *152*, 1229–1236. [CrossRef]
19. Liu, X.; Liu, Z.; Zheng, J.; Yan, X.; Li, D.; Chen, S.; Chu, W. Characteristics of N-doped TiO_2 nanotube arrays by N_2-plasma for visible light-driven photocatalysis. *J. Alloys Compd.* **2011**, *509*, 9970–9976. [CrossRef]
20. Malini, B.; Allen Gnana Raj, G. C,N and S-doped TiO_2-characterization and photocatalytic performance for rose bengal dye degradation under day light. *J. Environ. Chem. Eng.* **2018**, *6*, 5763–5770. [CrossRef]
21. Wang, F.; Ma, Z.; Ban, P.; Xu, X. C,N and S codoped rutile TiO_2 nanorods for enhanced visible-light photocatalytic activity. *Mater. Lett.* **2017**, *195*, 143–146. [CrossRef]

22. Patil, U.M.; Kulkarni, S.B.; Deshmukh, P.R.; Salunkhe, R.R.; Lokhande, C.D. Photosensitive nanostructured TiO$_2$ grown at room temperature by novel "bottom-up" approached CBD method. *J. Alloys Compd.* **2011**, *509*, 6196–6199. [CrossRef]
23. Okunaka, S.; Tokudome, H.; Hitomi, Y.; Abe, R. Facile preparation of stable aqueous titania sols for fabrication of highly active TiO$_2$ photocatalyst films. *J. Mater. Chem. A* **2015**, *3*, 1688–1695. [CrossRef]
24. Nie, Z.; Ke, X.; Li, D.; Zhao, Y.; Zhu, L.; Qiao, R.; Zhang, X.L. NaYF$_4$:Yb,Er,Nd@NaYF$_4$:Nd Upconversion Nanocrystals Capped with Mn:TiO$_2$ for 808 nm NIR-Triggered Photocatalytic Applications. *J. Phys. Chem. C* **2019**, *123*, 22959–22970. [CrossRef]
25. Liu, Y.; Xia, Y.; Jiang, Y.; Zhang, M.; Sun, W.; Zhao, X.-Z. Coupling effects of Au-decorated core-shell β-NaYF$_4$:Er/Yb@SiO$_2$ microprisms in dye-sensitized solar cells: Plasmon resonance versus upconversion. *Electrochim. Acta* **2015**, *180*, 394–400. [CrossRef]
26. Mavengere, S.; Jung, S.C.; Kim, J.S. Visible light photocatalytic activity of NaYF$_4$:(Yb,Er)-CuO/TiO$_2$ composite. *Catalysts* **2018**, *8*, 521. [CrossRef]
27. Wang, W.; Li, Y.; Kang, Z.; Wang, F.; Yu, J.C. A NIR-driven photocatalyst based on α-NaYF$_4$: YB,Tm@TiO$_2$ core-shell structure supported on reduced graphene oxide. *Appl. Catal. B Environ.* **2016**, *182*, 184–192. [CrossRef]
28. Mavengere, S.; Kim, J.-S. UV–visible light photocatalytic properties of NaYF$_4$:(Gd, Si)/TiO$_2$ composites. *Appl. Surf. Sci.* **2018**, *444*, 491–496. [CrossRef]
29. Wu, X.; Yin, S.; Dong, Q.; Liu, B.; Wang, Y.; Sekino, T.; Lee, S.W.; Sato, T. UV, visible and near-infrared lights induced NOx destruction activity of (Yb,Er)-NaYF$_4$/C-TiO$_2$ composite. *Sci. Rep.* **2013**, *3*, 1–8. [CrossRef]
30. Wang, Z.; Wang, C.; Han, Q.; Wang, G.; Zhang, M.; Zhang, J.; Gao, W.; Zheng, H. Metal-enhanced upconversion luminescence of NaYF$_4$:Yb/Er with Ag nanoparticles. *Mater. Res. Bull.* **2017**, *88*, 182–187. [CrossRef]
31. Wu, X.; Yin, S.; Dong, Q.; Sato, T. Blue/green/red colour emitting up-conversion phosphors coupled C-TiO$_2$ composites with UV, visible and NIR responsive photocatalytic performance. *Appl. Catal. B Environ.* **2014**, *156–157*, 257–264. [CrossRef]
32. Mavengere, S.; Yadav, H.M.; Kim, J.-S. Photocatalytic properties of nanocrystalline TiO$_2$ coupled with up-conversion phosphors. *J. Ceram. Sci. Technol.* **2017**, *8*, 67–72. [CrossRef]
33. Marques, J.; Gomes, T.D.; Forte, M.A.; Silva, R.F.; Tavares, C.J. A new route for the synthesis of highly-active N-doped TiO$_2$ nanoparticles for visible light photocatalysis using urea as nitrogen precursor. *Catal. Today* **2019**, *326*, 36–45. [CrossRef]
34. Anil Kumar Reddy, P.; Venkata Laxma Reddy, P.; Maitrey Sharma, V.; Srinivas, B.; Kumari, V.D.; Subrahmanyam, M. Photocatalytic Degradation of Isoproturon Pesticide on C, N and S Doped TiO$_2$. *J. Water Resour. Prot.* **2010**, *2*, 235–244. [CrossRef]
35. Maurya, S.K.; Tiwari, S.P.; Kumar, A.; Kumar, K. Plasmonic enhancement of upconversion emission in Ag@NaYF$_4$:Er^{3+}/Yb^{3+} phosphor. *J. Rare Earths* **2018**, *36*, 903–910. [CrossRef]
36. Vaiano, V.; Sacco, O.; Iervolino, G.; Sannino, D.; Ciambelli, P.; Liguori, R.; Bezzeccheri, E.; Rubino, A. Enhanced visible light photocatalytic activity by up-conversion phosphors modified N-doped TiO$_2$. *Appl. Catal. B Environ.* **2015**, *176*, 594–600. [CrossRef]
37. Leng, J.; Tang, J.; Xie, W.; Li, J.; Chen, L. Impact of pH and urea content on size and luminescence of upconverting Y$_2$O$_3$:Yb, Er nanophosphors. *Mater. Res. Bull.* **2018**, *100*, 171–177. [CrossRef]
38. Van Sark, W.G.J.H.M.; de Wild, J.; Rath, J.K.; Meijerink, A.; Schropp, R.E.I. Upconversion in solar cells. *Nanoscale Res. Lett.* **2013**, *8*, 81. [CrossRef]
39. Shan, S.-N.; Wang, X.-Y.; Jia, N.-Q. Synthesis of NaYF$_4$:Yb^{3+}, Er^{3+} upconversion nanoparticles in normal microemulsions. *Nanoscale Res. Lett.* **2011**, *6*, 539. [CrossRef]
40. Wang, W.; Ding, M.; Lu, C.; Ni, Y.; Xu, Z. A study on upconversion UV–vis–NIR responsive photocatalytic activity and mechanisms of hexagonal phase NaYF$_4$:Yb^{3+},Tm^{3+}@TiO$_2$ core-shell structured photocatalyst. *Appl. Catal. B Environ.* **2014**, *144*, 379–385. [CrossRef]

Publisher's Note: MDPI stays neutral with regard to jurisdictional claims in published maps and institutional affiliations.

© 2020 by the authors. Licensee MDPI, Basel, Switzerland. This article is an open access article distributed under the terms and conditions of the Creative Commons Attribution (CC BY) license (http://creativecommons.org/licenses/by/4.0/).

Article

Activation Treatments and SiO₂/Pd Modification of Sol-gel TiO₂ Photocatalysts for Enhanced Photoactivity under UV Radiation

Julien G. Mahy [1,2,*], **Valériane Sotrez** [2], **Ludivine Tasseroul** [2], **Sophie Hermans** [1] and **Stéphanie D. Lambert** [2]

1. Institute of Condensed Matter and Nanosciences (IMCN), Université catholique de Louvain, Place Louis Pasteur 1, 1348 Louvain-la-Neuve, Belgium; sophie.hermans@uclouvain.be
2. Department of Chemical Engineering—Nanomaterials, Catalysis & Electrochemistry, University of Liège, B6a, Quartier Agora, Allée du six Août 11, 4000 Liège, Belgium; valeriane.sotrez@gmail.com (V.S.); ludivine.tasseroul@gmail.com (L.T.); stephanie.lambert@uliege.be (S.D.L.)
* Correspondence: julien.mahy@uclouvain.be; Tel.: +32-4-3664771

Received: 3 September 2020; Accepted: 13 October 2020; Published: 14 October 2020

Abstract: The objective of this work is to improve the efficiency of TiO_2 photocatalysts by activation treatments and by modification with palladium nanoparticles and doping with SiO_2. The influence of the additive loading was explored, and two activation treatments were performed: UV exposition and H_2 reduction. $TiO_2/SiO_2/Pd$ photocatalysts were synthesized by an original cogelation method: a modified silicon alkoxide, i.e., [3-(2-aminoethyl)aminopropyl]trimethoxysilane (EDAS), was used to complex the palladium ions, thanks to the ethylenediamine group, while the alkoxide groups reacted with TiO_2 precursors. Pure TiO_2 was also synthesized by the sol-gel process for comparison. X-ray diffraction evidenced that the crystallographic structure of TiO_2 was anatase and that Pd was present, either in its oxidized form after calcination, or in its reduced form after reduction. The specific surface area of the samples varied from 5 to 145 $m^2\,g^{-1}$. Transmission electron microscopy allowed us to observe the homogeneous dispersion and nanometric size of Pd particles in the reduced samples. The width of the band gap for pure TiO_2 sample, measured by UV/Visible diffuse reflectance spectroscopy at approximately 3.2 eV, corresponded to that of anatase. The band gap for the $TiO_2/SiO_2/Pd$ composite samples could not be calculated, due to their high absorption in visible range. The photocatalytic activity of the various catalysts was evaluated by the degradation of a methylene blue solution under UV radiation. The results showed that the photocatalytic activity of the catalysts was inversely proportional to the content of silica present in the matrix. A small amount of silica improved the photocatalytic activity, as compared to the pure TiO_2 sample. By contrast, a high amount of silica delayed the crystallization of TiO_2 in its anatase form. The activation treatment under UV had little influence on photocatalytic efficiency. The introduction of Pd species increased the photocatalytic activity of the samples because it allowed for a decrease in the rate of electron–hole recombinations in TiO_2. The reduction treatment improved the activity of photocatalysts, whatever the palladium content, thanks to the reduction of Ti^{4+} into Ti^{3+}, and the formation of defects in the crystallographic structure of anatase.

Keywords: TiO_2; sol-gel process; SiO_2; Pd-modification; activation treatments; photocatalysis

1. Introduction

For decades, different processes have been developed to reduce the current pollution in water, air and soil [1,2]. Among these various methods, photocatalysis had emerged as a potential technique to degrade organic pollutants [3,4]. Its principle lies on the use of a semiconductor as photocatalyst, which is activated by light illumination. Its activation produces a series of redox reactions leading to radicals' formation. These radicals can attack organic pollutants and degrade them in CO_2 and H_2O if the degradation is complete [3,4].

The most used photocatalyst is TiO_2 [5–7]. This material is activated under UV light due to its band gap around 3.2 eV for the anatase phase [4]. TiO_2 presents the advantage of being easily available, inexpensive and non-toxic. The use of TiO_2 as photocatalyst has two main limitations [4]: (i) the fast charge recombination and (ii) the large band gap value. Indeed, if the recombination of the photo-generated species (electrons, e^-, and holes, h^+) is fast, the production of radicals is low and the degradation is less effective. Furthermore, if the band gap is large, the energy required for the electron transfer is high and only UV radiation can be used. This study focused on the improvement of the charge recombination phenomenon.

Regarding the improvement of the recombination time, the major modification of TiO_2 materials is the addition of metallic nanoparticles such as Ag [8–11], Au [12], Pt [12,13] or Pd [8,14]. In this case, the metallic nanoparticle plays the role of electron trap, allowing us to increase the recombination time [9]. Combination with other semiconductors has also been investigated to increase the recombination time [4]. In this case, synergetic effects can be observed leading to a better charge separation or an increased photostability. Different combinations of semiconductors were tested, such as TiO_2/ZnO [15], TiO_2/CdS [4] or TiO_2/ZrO_2 [16]. The introduction of SiO_2 in the TiO_2 matrix can also enhance the photoactivity due to surface acidity modification, the introduction of defects or increased surface area [9,17]. In order to optimize the modification efficiency, the additive needs to be highly dispersed in the TiO_2 matrix. Previous works have reported that a high dispersion of metallic nanoparticles into a SiO_2 and TiO_2 matrix can be achieved through the use of a complexing agent, e.g., ethylenediamine [9,18,19]. Indeed, by using modified silicon alkoxide, small amounts of Ag, Cu or Ni nanoparticles are dispersed in either TiO_2 or SiO_2. The use of this silicon alkoxide introduces a SiO_2 doping in the TiO_2 matrix, and this doping is referenced as effective to increase the photoactivity [9,17,20].

In this study, TiO_2 photocatalyst was modified with Pd species, in order to increase the photocatalytic property under UV light for the degradation of organic dye. To finely disperse these Pd species, a modified silicon alkoxide (3-(2-aminoethyl) aminopropyl]trimethoxysilane, EDAS) was cogelled with Ti alkoxide. Indeed, this silicon alkoxide possesses an ethylenediamine function able to complex metallic ions; SiO_2 was also introduced into the materials. Different amounts were added to study the influence of SiO_2 addition on TiO_2 properties. This original preparation method allows the modification in one step. Four different samples were prepared with four Pd loadings; one pure TiO_2 and one TiO_2/SiO_2 were also prepared for comparison. To further increase the photoefficiency, two activation treatments were applied to the sample: UV activation and H_2 reduction. The physicochemical properties of the resulting materials were determined thanks to nitrogen adsorption–desorption, X-ray diffraction, inductively coupled plasma optical emission spectroscopy, diffuse reflectance and transmission electron microscopy. The photocatalytic activity was followed on the degradation of methylene blue (MB), under UV-A illumination, during 6 h of reaction.

2. Results and Discussion

Six samples were prepared: one pure titania (named TiO_2), one titania doped with silica (named TiO_2/SiO_2) and four titania samples doped with silica and modified with Pd named $TiO_2/SiO_2/PdX$, where X corresponds to the theoretical wt.% of Pd and can vary between 1, 5, 12 and 22 wt.% of Pd. Table 1 summarizes the amounts of reactants for each sample. The weight amount of SiO_2 is in the same range as Pd. The UV-treated samples are denoted with "- UV", while those reduced under H_2 are denoted with "-H_2".

Table 1. Synthesis operating variables and additive loadings determined by ICP.

Sample	$n_{Solvent}$ (mmol)	n_{Pd} (mmol)	n_{EDAS} (mmol)	n_{TIPT} (mmol)	n_{Water} (mmol)	Gelification Time (min)	Theoretical Loading SiO_2/Pd (wt.%)	Actual Loading SiO_2/Pd (wt.%)
Pure TiO_2	1830	-	-	91.5	183	9	-	-
TiO_2/SiO_2	1830	-	6.88	84.6	180	8	5.8/0	4.9/0
$TiO_2/SiO_2/Pd1$	1830	0.69	1.38	90.1	182	>60	0.98/1	1.2/1.1
$TiO_2/SiO_2/Pd5$	1830	3.44	6.88	84.6	180	40	4.9/5	4.7/4.2
$TiO_2/SiO_2/Pd12$	1830	8.60	17.20	74.3	174	>60	11.76/12	12.9/12.2
$TiO_2/SiO_2/Pd22$	1830	17.4	34.90	56.6	166	>60	21.56/22	24.6/21.8

$n_{Solvent}$ = amount of 2-methoxyethanol; n_{Pd} = amount of palladium acetylacetate; n_{EDAS} = amount of EDAS; n_{TIPT} = amount of TIPT; n_{Water} = amount of water.

2.1. Colors and Compositions of Photocatalytic Samples

The pure and TiO_2/SiO_2 samples had a white color, characteristic of TiO_2 and SiO_2 materials. The Pd-modified TiO_2/SiO_2 samples were brown after calcination and darker when the amount of Pd increased. The brown color is characteristic of PdO material. The samples, after UV activation, kept their initial coloration, while after H_2 reduction, the Pd-modified ones became black, which is characteristic of metallic palladium. The pictures of the samples are shown in Supplementary Materials Figure S1.

The actual loading (measured by ICP) is given in Table 1 and was close to the theoretical one, calculated on the basis of the engaged amounts. In each composite sample, the amount of SiO_2 is similar to the amount of Pd, as expected.

2.2. Crystallinity

The XRD patterns of all calcined samples are presented in Figure 1a. For pure TiO_2 sample, the characteristic peaks of TiO_2 anatase were detected (JCPDS 21-1272: 25.3°, 37.9°, 48.0°, 54.6°, 62.8° and 68.8°). When additive was added, the intensity of anatase peaks decreased and even disappeared for $TiO_2/SiO_2/Pd22$ sample. For the samples modified with Pd, additional peaks appeared at 34°, 42°, 54.9° and 60.3° (JCPDS 75-584), which corresponded to palladium oxide. The intensity of these peaks increased with the amount of Pd (Figure 1a).

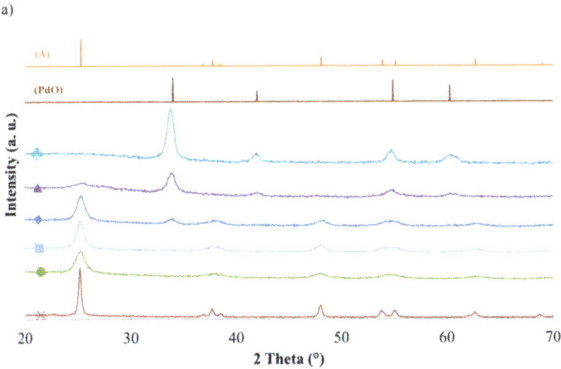

Figure 1. *Cont.*

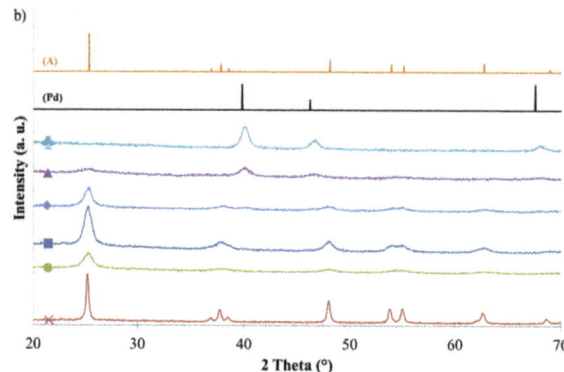

Figure 1. (a) XRD patterns of calcined samples: (×) pure TiO_2, (●) TiO_2/SiO_2, (■) $TiO_2/SiO_2/Pd1$, (♦) $TiO_2/SiO_2/Pd5$, (▲) $TiO_2/SiO_2/Pd12$, and (♣) $TiO_2/SiO_2/Pd22$; and (b) XRD patterns of reduced samples: (×) pure TiO_2—H_2, (●) TiO_2/SiO_2—H_2, (■) $TiO_2/SiO_2/Pd1$—H_2, (♦) $TiO_2/SiO_2/Pd5$—H_2, (▲) $TiO_2/SiO_2/Pd12$—H_2 and (♣) $TiO_2/SiO_2/Pd22$—H_2. (A) Reference pattern of anatase, (PdO) reference pattern of palladium oxide and (Pd) reference pattern of palladium.

After UV activation, no modification of the crystallinity was observed (XRD patterns identical to Figure 1a, not shown). Contrarily, after H_2 reduction (Figure 1b), the peaks of metallic palladium (JCPDS 87-0638: 39.8°, 46.2° and 67.6°) were observed replacing those of palladium oxide, as expected.

Concerning the crystallite size evaluated by using the Scherrer equation, the TiO_2 crystallites (d_{XRDTi}) decreased with the additive content: from 34 nm for pure TiO_2 to 7 nm for $TiO_2/SiO_2/Pd12$ calcined sample (Table 2). For the PdO crystallites (d_{XRDPdO}, for the calcined samples), they increased with the amount of Pd: from 8 nm for $TiO_2/SiO_2/Pd5$ to 25 nm for $TiO_2/SiO_2/Pd22$ sample. The metallic Pd nanoparticles (d_{XRDPd} for the reduced samples) had a slightly smaller size than PdO crystallites (Table 2).

Table 2. Samples' crystalline, textural and optical properties.

Sample	d_{XRDTi} (nm) ±1	d_{XRDPdO} (nm) ±1	d_{XRDPd} (nm) ±1	S_{BET} ($m^2\,g^{-1}$) ±5	V_{DR} ($cm^3\,g^{-1}$) ±0.01	V_P ($cm^3\,g^{-1}$) ±0.1	d_{TEMTi} (nm) ±5	d_{TEMPdO} (nm) ±4	d_{TEMPd} (nm) ±4	E_g (eV) ±0.01
Pure TiO_2	34	-b	-b	<5	-	0.04	25	-b	-b	3.17
Pure TiO_2-H_2	34	-b	-b	<5	-	0.04	-a	-b	-b	3.16
Pure TiO_2-UV	34	-b	-b	<5	-	0.04	25	-b	-b	3.25
TiO_2/SiO_2	8	-b	-b	100	0.04	0.2	10	-b	-b	3.25
TiO_2/SiO_2-H_2	10	-b	-b	145	0.06	0.4	-a	-b	-b	3.23
TiO_2/SiO_2-UV	8	-b	-b	100	0.04	0.2	10	-b	-b	3.27
$TiO_2/SiO_2/Pd1$	19	-c	-b	40	0.02	0.1	18	-c	-b	-d
$TiO_2/SiO_2/Pd1$-H_2	25	-b	-c	60	0.02	0.2	-a	-b	-c	-d
$TiO_2/SiO_2/Pd1$-UV	19	-c	-b	40	0.02	0.1	18	-c	-b	-d
$TiO_2/SiO_2/Pd5$	11	8	-b	45	0.02	0.2	12	8	-b	-d
$TiO_2/SiO_2/Pd5$-H_2	11	-b	17	60	0.02	0.2	-a	-b	10	-d
$TiO_2/SiO_2/Pd5$-UV	10	8	-b	45	0.02	0.2	12	9	-b	-d
$TiO_2/SiO_2/Pd12$	7	17	-b	20	0.01	0.6	10	12	-b	-d
$TiO_2/SiO_2/Pd12$-H_2	6	-b	9	25	0.01	0.4	-a	-b	11	-d
$TiO_2/SiO_2/Pd12$-UV	7	17	-b	20	0.01	0.6	10	12	-b	-d
$TiO_2/SiO_2/Pd22$	-b	25	-b	45	0.02	0.4	-b	14	-b	-d
$TiO_2/SiO_2/Pd22$-H_2	-b	-b	12	100	0.05	0.1	-b	-b	11	-d
$TiO_2/SiO_2/Pd22$-UV	-b	25	-b	45	0.02	0.4	-b	14	-b	-d

-a, not measured; -b, not present; -c, not detectable; -d, unmeasurable; d_{XRDTi}, mean diameter of TiO_2 crystallites measured by the Scherrer method; d_{XRDPdO}, mean diameter of PdO crystallites measured by the Scherrer method; d_{XRDPd}, mean diameter of Pd crystallites measured by the Scherrer method; S_{BET}, specific surface area determined by the BET method; V_{DR}, specific micropore volume determined by Dubinin–Raduskevitch theory; V_P, specific liquid volume adsorbed at the saturation pressure of nitrogen; d_{TEMTi}, mean diameter of TiO_2 nanoparticles measured by TEM; d_{TEMPdO}, mean diameter of PdO nanoparticles measured by TEM; d_{TEMPd}, mean diameter of Pd nanoparticles measured by TEM; E_g, optical band-gap values calculated by using the transformed Kubelka–Munk function.

The decrease of the TiO$_2$ crystallinity can be explained by the introduction of SiO$_2$ in the TiO$_2$ matrix. Indeed, silica was introduced by the EDAS, which was used to disperse Pd in the TiO$_2$ matrix. The introduction of silica delayed the crystallization of TiO$_2$ [9,21,22], leading to amorphous TiO$_2$ for the sample with the highest content of SiO$_2$ (TiO$_2$/SiO$_2$/Pd22 sample).

There was no modification of the crystallite sizes, both for TiO$_2$ and PdO, for all UV-activated samples (Table 2). For the H$_2$ reduced samples, as explained above, the PdO species are convert in Pd species and the evolution of TiO$_2$ crystallite follows the evolution of the non-reduced samples.

2.3. Morphology

The nitrogen adsorption–desorption isotherms are presented in Figure 2a for all calcined samples. Pure TiO$_2$ and all Pd-modified TiO$_2$ samples were non-porous solids with a very low surface area (S$_{BET}$ < 50 m^2 g^{-1}, Table 2). Only TiO$_2$/SiO$_2$ sample was a mesoporous solid (type IV isotherm [23]) with a hysteresis between 0.4 and 0.6 in relative pressure. Indeed, the introduction of sol-gel silica increases in the TiO$_2$ leads to more porous solid [22]. The Pd modification seemed to have low influence on the texture of the samples.

After H$_2$ reduction (Figure 2b), the S$_{BET}$ increased for all samples (Table 2). Indeed, some remaining organic contaminants could be removed by this thermal treatment, leading to an increase of the microporosity (V$_{DR}$ increased in the majority of the reduced samples, Table 2).

The UV activation had no influence on the textural properties (nitrogen adsorption–desorption isotherms identical to Figure 2a, not shown).

The samples were observed by TEM, as shown in Figure 3, for four representative samples. The estimation of the different sizes of particles (TiO$_2$, PdO and Pd) are given in Table 2. The TiO$_2$ nanoparticles (d$_{TEMTi}$) were observed in the size range of 10–18 nm for all samples doped with SiO$_2$ (Figure 3a–d). For TiO$_2$/SiO$_2$/Pd22 sample, no TiO$_2$ crystallite was observed in the matrix, confirming the amorphous TiO$_2$/SiO$_2$ materials. For pure TiO$_2$, the size was around 25 nm. This result was coherent with the XRD patterns (Figure 1) where crystallization delays were obtained with SiO$_2$ doping [9,21]. For the Pd-modified ones, darker nanoparticles highly dispersed in the TiO$_2$/SiO$_2$ matrix were observed which corresponded to the PdO (calcined samples) or Pd (reduced samples) nanoparticles (Figure 3b–d). The Pd species were highly dispersed in the oxide matrix, thanks to the use of EDAS [24,25]. When the amount of Pd became too high, the size of the nanoparticles increased (d$_{TEMPdO}$ and d$_{TEMPd}$, Table 2). Indeed, the palladium species were closer to each other, leading to sintering when the samples were heated. The size of PdO and metallic Pd was quite similar for all samples, before and after H$_2$ reduction (Table 2). The UV activation treatment had no influence on the size of the materials.

The XRD crystallite sizes and the nanoparticle sizes estimated with TEM, for the Pd and PdO species, were in the same range. The small differences result from the fact that the crystallites sizes determined by XRD correspond to the average diameter in volume of crystallites. Therefore, XRD gives more statistical significance to large particles, compared to TEM.

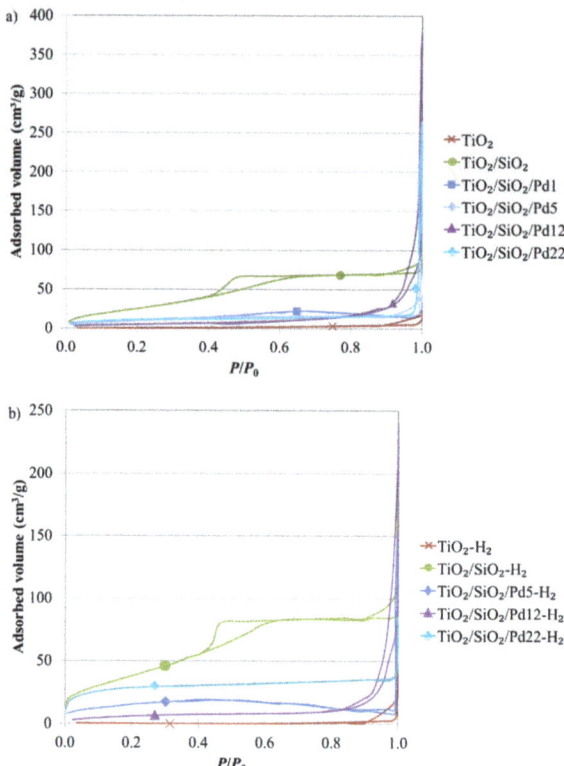

Figure 2. Nitrogen adsorption–desorption isotherms of (**a**) calcined samples and (**b**) reduced samples.

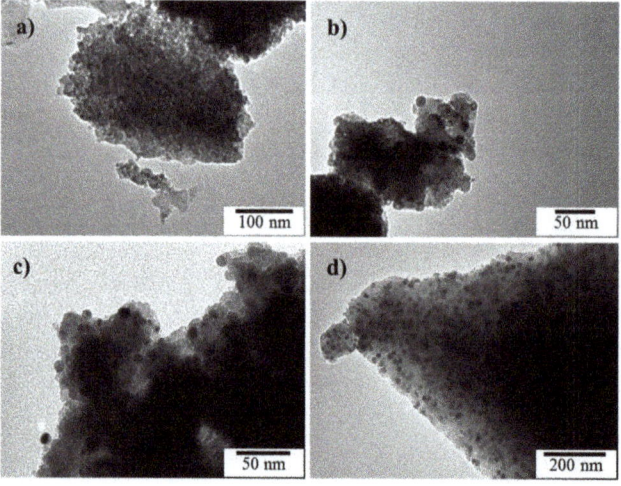

Figure 3. TEM micrographs of calcined and reduced samples: (**a**) TiO_2/SiO_2, (**b**) $TiO_2/SiO_2/Pd5$—H_2, (**c**) $TiO_2/SiO_2/Pd12$-H_2 and (**d**) $TiO_2/SiO_2/Pd22$.

2.4. Optical Properties

The diffuse reflectance UV/Visible spectra of the calcined samples are presented in Figure 4. The pure TiO_2 and TiO_2/SiO_2 samples had the characteristic spectrum of titania materials with absorption in UV range from around 400 nm. The Pd-modified samples had a larger band of absorption in the UV/Visible region due to the presence of the PdO nanoparticles. As the reduced samples were black, they were not analyzed with this technique. The band gap can be calculated only for the pure TiO_2 and TiO_2/SiO_2 samples, and is given in Table 2. For both samples, values around 3.2 eV were found in agreement with titania band gap [26].

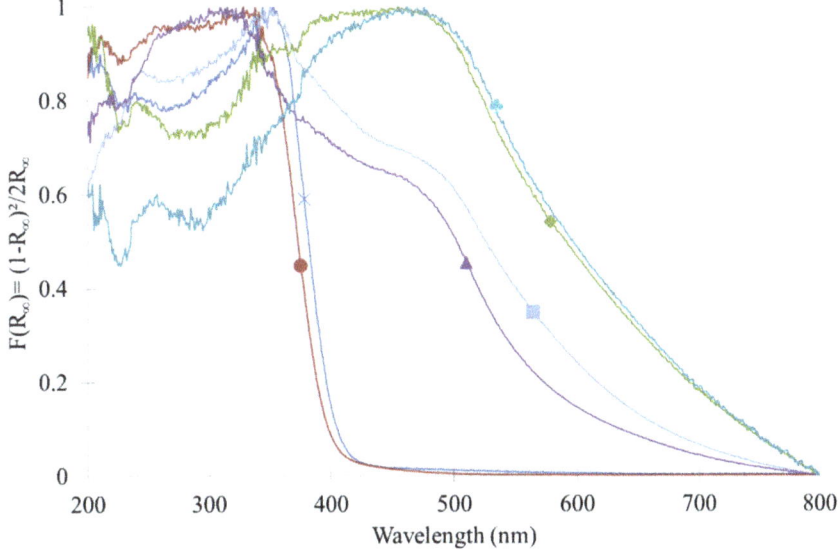

Figure 4. Normalized diffuse reflectance UV/Visible spectra of all calcined samples: (×) pure TiO_2, (●) TiO_2/SiO_2, (■) $TiO_2/SiO_2/Pd1$, (♦) $TiO_2/SiO_2/Pd5$, (▲) $TiO_2/SiO_2/Pd12$ and (♣) $TiO_2/SiO_2/Pd22$.

2.5. Photocatalytic Activity

The evolution of the MB concentration during the photocatalytic experiments is given in Figure 5 for all samples. For each sample, curves obtained with samples treated by both UV and H_2 are also represented. No photolysis of MB was observed under illumination without catalyst (blank experiment). Thanks to the dark experiments, the adsorption of MB on the samples without illumination was deduced from the observed decrease of MB concentration, under illumination, to measure only the MB degradation.

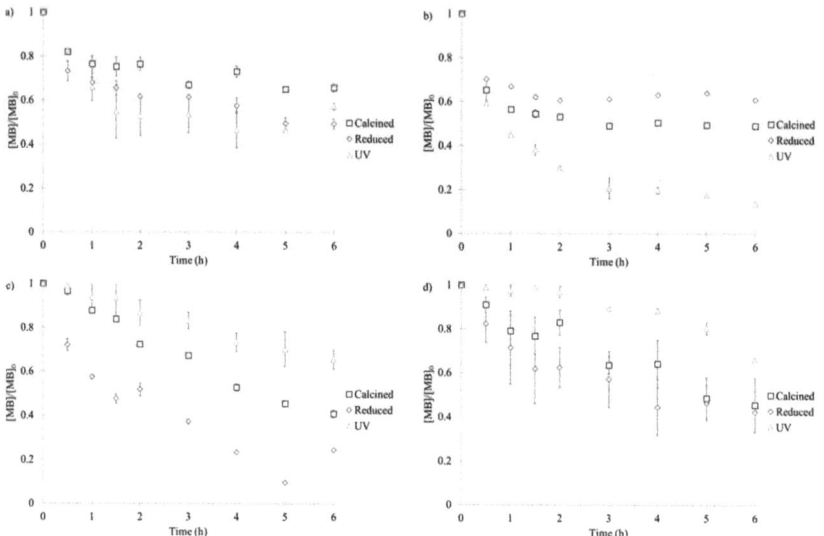

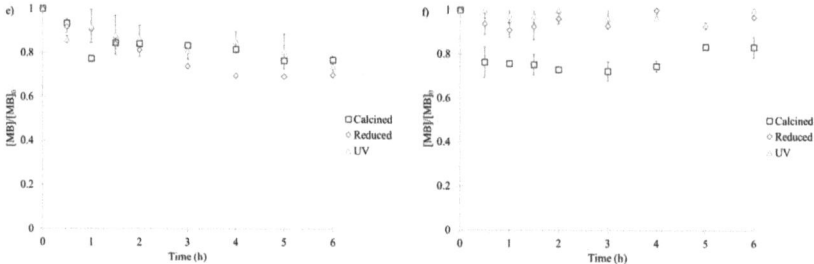

Figure 5. *Cont.*

Figure 5. Methylene blue (MB) concentration evolution ([MB]/[MB]$_0$ ratio) over time (6 h) for all sample series under UV-A illumination: (**a**) pure TiO$_2$, (**b**) TiO$_2$/SiO$_2$, (**c**) TiO$_2$/SiO$_2$/Pd1, (**d**) TiO$_2$/SiO$_2$/Pd5, (**e**) TiO$_2$/SiO$_2$/Pd12 and (**f**) TiO$_2$/SiO$_2$/Pd22. The contribution of MB adsorption was subtracted for all samples.

For samples without palladium (pure TiO$_2$ and TiO$_2$/SiO$_2$), the H$_2$ reduction treatment did not seem to have much influence on the percentage of MB degradation, while the activation treatment under UV seemed to improve it slightly for the pure TiO$_2$ sample, and even more for TiO$_2$/SiO$_2$ sample. For samples with low palladium content (TiO$_2$/SiO$_2$/Pd1 and TiO$_2$/SiO$_2$/Pd5 samples), the opposite phenomenon was observed: Activation under UV seemed to decrease the photoactivity, while H$_2$ reduction greatly improved it for the TiO$_2$/SiO$_2$/Pd1 sample and slightly more for the TiO$_2$/SiO$_2$/Pd5 and TiO$_2$/SiO$_2$/Pd12 samples. The TiO$_2$/SiO$_2$/Pd22 sample had a low photocatalytic activity, regardless of the treatment, due to its amorphous nature (Figure 1). The MB conversion for all the samples is given in Table 3, together with the percentage of MB adsorption.

Table 3. Photocatalytic experiments for MB degradation after 6 h, under UV-A illumination.

Sample	MB Adsorption after 6 h (Blank Test—%) ±5	MB Conversion after 6 h (Blank Deduced—%) ±5	Kinetic Constant k (h^{-1})
Pure TiO_2	0	35	0.069
Pure TiO_2-H_2	5	50	0.116
Pure TiO_2-UV	0	45	0.092
TiO_2/SiO_2	50	50	0.119
TiO_2/SiO_2-H_2	60	40	0.082
TiO_2/SiO_2-UV	15	85	0.333
TiO_2/SiO_2/Pd1	0	60	0.147
TiO_2/SiO_2/Pd1-H_2	15	75	0.233
TiO_2/SiO_2/Pd1-UV	0	35	0.070
TiO_2/SiO_2/Pd5	0	55	0.131
TiO_2/SiO_2/Pd5-H_2	5	60	0.147
TiO_2/SiO_2/Pd5-UV	0	35	0.069
TiO_2/SiO_2/Pd12	20	25	0.039
TiO_2/SiO_2/Pd12-H_2	15	30	0.059
TiO_2/SiO_2/Pd12-UV	15	25	0.050
TiO_2/SiO_2/Pd22	10	15	0.030
TiO_2/SiO_2/Pd22-H_2	10	0	0.005
TiO_2/SiO_2/Pd22-UV	10	0	0.000

The pure TiO_2 samples had very low MB adsorption by opposition to TiO_2/SiO_2 samples. This difference could come from the textural properties of the TiO_2/SiO_2 samples, which presented higher S_{BET} values, as compared to pure TiO_2 samples (< 5 vs. 100–145 $m^2 g^{-1}$, Table 2). The Pd-modified samples presented S_{BET} values between pure TiO_2 and TiO_2/SiO_2 samples (~ 40 $m^2 g^{-1}$, Table 2), so their MB adsorption behavior was intermediate with a small adsorption (~10%), except for the TiO_2/SiO_2/Pd12 series, which adsorbed around 20%. The MB adsorption was therefore directly linked to the surface properties of the samples [27].

On the one hand, the photoactivity of TiO_2 was increased by the introduction of a small amount of silica into its network [9,28]. The TiO_2/SiO_2/Pd1 sample degraded 60% of MB against 35% for the pure TiO_2 sample (Table 3). On the other hand, when the quantity of silica introduced became too large, the crystallization of TiO_2 in its anatase form was delayed [21]. Thus, in the samples containing a large amount of palladium (TiO_2/SiO_2/Pd5, TiO_2/SiO_2/Pd12 and TiO_2/SiO_2/Pd22), and therefore a large amount of silica, photocatalytic efficiency decreased: We obtained 60% MB degradation for the TiO_2/SiO_2/Pd1 sample against 15% MB degradation for the TiO_2/SiO_2/Pd22 sample (Table 3). In all cases, except for the TiO_2/SiO_2/Pd22 sample, which contains only amorphous TiO_2, the H_2 reduction treatment had the effect of increasing the photocatalytic activity. Indeed, H_2 reduction allowed us to decrease the rate of recombination of the "electron–hole" pairs present in TiO_2 via two distinct phenomena: (i) The first was the possible creation of oxygen vacancies on the surface of TiO_2 by reduction of Ti^{4+} ions to Ti^{3+} [29,30], and (ii) the second was the reduction of PdO particles to metallic palladium acting as an electron trap and thus limiting the recombination's of the "electron–hole" pairs [31,32]. Activation processing under UV, on the other hand, has little influence on the photocatalytic activity of calcined samples.

The optimal sample is TiO_2/SiO_2/Pd1, which combines the advantages of both additives (SiO_2 and Pd) with a weight loading percentage of 1% for each. Indeed, the introduction of SiO_2 allows us to increase the specific surface area, as compared to pure TiO_2 (Table 2), while the crystallinity of anatase is maintained (Figure 1). Moreover, Pd introduction allows an absorption in the visible range (Figure 4).

Assuming that the photocatalyst is homogeneously dispersed in a perfectly agitated medium, the derivative of the material balance in a semi-continuous reactor can be expressed as follows [33]:

$$\frac{dC}{dt} = -r_V \frac{m_{cata,0}}{V_0} \tag{1}$$

where C is the concentration of methylene blue in the solution (kmol/m^3), t is the time (h), r_V is the specific rate of the degradation of methylene blue (kmol/kg$_{catalyst}$·h), $m_{cata,0}$ is the initial catalyst mass homogeneously dispersed in the solution (kg of catalyst) and V_0 is the initial solution volume (m^3).

In general, the degradation curves of methylene blue in photocatalysis had an exponential appearance. The degradation kinetics of methylene blue could then be modeled by assuming that the reaction was of order 1 under the operating conditions used. The kinetic equation could thus be expressed as follows:

$$\frac{dC}{dt} = -k \times C \times \frac{m_{cata,0}}{V_0} \qquad (2)$$

where k is the kinetic constant (m^3/kg$_{catalyst}$ h).

As part of this work, the photocatalytic activity of the samples was evaluated in media containing 50 mg of catalysts dispersed in 50 mL of MB solution at 20 µmol/L concentration. The $m_{cata,0}/V_0$ ratio was thus equal to 1. The kinetic equation could then be developed as follows:

$$\frac{dC}{dt} = -k \times C \qquad (3)$$

$$\frac{dC}{C} = -k \times dt \qquad (4)$$

$$\ln\left(\frac{C}{C_0}\right) = -k \times t \qquad (5)$$

$$C = C_0 \times e^{-k \times t} \qquad (6)$$

where C_0 is the initial MB concentration in the solution (kmol/m^3).

Figure 6 compares the theoretical kinetic model to the experimental values for the TiO$_2$/SiO$_2$/Pd1 sample. The two curves superimposed perfectly, justifying the hypothesis of an order 1 kinetics of degradation.

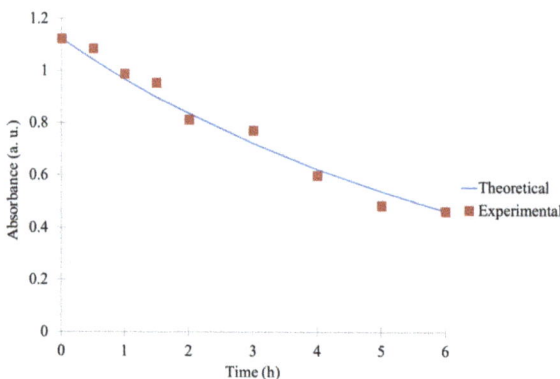

Figure 6. Kinetic analysis of the MB degradation for TiO$_2$/SiO$_2$/Pd1 sample: (-) theoretical evolution and (■) experimental points.

Equation (6) is commonly found in the literature [34,35] for photocatalytic process. Knowing the evolution over time of the concentration of methylene blue in the solutions, the kinetic constant, k, could be calculated for all samples. The values obtained are listed in Table 3.

The evolution of the kinetic constants values corresponds to the observations made previously. Indeed, the kinetic constant, k, was increased by the introduction of a small quantity of silica into the TiO$_2$ network: The TiO$_2$/SiO$_2$/Pd1 sample had a kinetic constant equal to 0.147 h^{-1} versus 0.069 h^{-1} for the pure TiO$_2$ sample (Table 3). On the other hand, when the quantity of silica introduced became too large, the crystallization of TiO$_2$ in its anatase form was delayed [9,21]. Thus, in samples containing a large amount of palladium (TiO$_2$/SiO$_2$/Pd5, TiO$_2$/SiO$_2$/Pd12 and TiO$_2$/SiO$_2$/Pd22 samples), and therefore a large quantity of silica, the kinetic constant decreased, i.e., 0.147 h^{-1} for the TiO$_2$/SiO$_2$/Pd1 sample, as compared to 0.030 h^{-1} for the TiO$_2$/SiO$_2$/Pd22 sample. In all cases (except for the sample TiO$_2$/SiO$_2$/Pd22, which contained only amorphous TiO$_2$), the H$_2$ reduction treatment had the consequence of increasing the kinetic constant of the samples. Activation processing under UV had little influence on the kinetic constant of the calcined samples.

With the combination of Pd/SiO$_2$ modification and the reduction of the photocatalyst, the photoactivity for the degradation of MB can be greatly increase compared to the pure material. Moreover, the use of modified silicon alkoxide allowed a good distribution of the metal in the titania matrix. This cogelation synthesis method appears as an easy process for homogeneous doping in titania photocatalysts.

3. Materials and Methods

3.1. Photocatalyst Synthesis

The pure and modified TiO$_2$ samples were prepared by a sol-gel cogelation method inspired from Reference [21]. The steps were the following: Palladium acetylacetonate (Pd(acac)$_2$, Sigma Aldrich, 99%) was dissolved in 2-methoxyethanol (Sigma Aldrich, 99.8%); then 3-(2-aminoethyl) aminopropyl]trimethoxysilane (EDAS, Merck, 97%) was added to the solution and stirred for 1 h, for complexation. The resulting solution became yellow. Then titanium isopropoxide (TIPT, Sigma Aldrich, 98%) was added, under stirring. In a second vial, distilled water was mixed with 2-methoxyethanol. This second solution was added, dropwise, to the first one, in 5 min, under stirring. The resulting suspension was placed in an oven, for gelation, at 80 °C, for 72 h. Then, the gel was dried under vacuum at 80 °C for 48 h, until the pressure fell at 1000 Pa. At this pressure, the temperature was increased to 150 °C for 24 h. The dried powder was crushed with a mortar and calcined at 550 °C for 5 h.

Six samples were prepared with the following protocol: one pure TiO$_2$ (without Pd(acac)$_2$ and EDAS), one TiO$_2$/SiO$_2$ (without Pd(acac)$_2$) and four TiO$_2$/SiO$_2$/Pd samples (1, 5, 12 and 22 theoretical wt.% of Pd). Table 1 summarizes the amounts of reactants for each sample: the weight amount of SiO$_2$ is in the same range as Pd in all cases.

3.2. Activation Treatments

In order to increase the photoactivity of the samples, they underwent two different activation treatments after calcination: UV activation and H$_2$ reduction.

Concerning the UV activation, the calcined powder was mixed in 50 mL of distilled water, in a Petri dish that was illuminated for 24 h, under stirring, at 25 °C. The lamp (Osram Sylvania, Blacklight-Bleu Lamp, F18W/BLB-T8) emitted UV-A radiation considered to be monochromatic, with a wavelength of 365 nm and an intensity of 1.2 mW/cm^2. After the UV treatment, the photocatalyst was filtered and dried for 24 h, at 120 °C. The UV-treated samples are denoted with "- UV".

Concerning the H$_2$ reduction, the calcined powder was heated under H$_2$, for 5 h, at 450 °C. The reduced samples are denoted with "- H$_2$".

3.3. Characterizations

The nitrogen adsorption–desorption isotherms were obtained thanks to a Fisons Sorptomatic 1900 device. The specific surface area, S_{BET}, the microporous volume, V_{DR}, and the porous volume, V_P, were obtained from the isotherms, thanks to BET and Dubinin–Raduskevic theories [23].

The crystallographic phases were determined with a Siemens D5000 device, by measurements between 20° and 70°, with a step size of 0.05° and a step time of 5 s. The crystallite size was estimated thanks to the Scherrer equation [36].

The atomic composition of the samples was determined by inductively coupled plasma atomic emission spectrometry (ICP-AES), with an ICAP 6500 device from Thermo Scientific. Before introducing the sample in the device, it was mineralized as follows: 75 mg of sample was mixed with 1.6 g of $Li_2B_4O_7$ and 0.4 g of Li_3BO_3 in a graphitic crucible. Then it was heated to 1000 °C for 5 min. After cooling, the pearl was dissolved, under stirring, in hot HNO_3 2 M.

The sample suspensions were observed with transmission electron microscopy (TEM), using a Phillips CM 100 device (accelerating voltage 200 kV). The average nanoparticle size was estimated by measuring approximatively 100 nanoparticles with the TEM software.

Optical properties of the sample materials in the range of 200–800 nm region were obtained with diffuse reflectance measurements on a Varian Cary 500 UV–Vis-NIR spectrophotometer equipped with an integrating sphere (Varian External DRA-2500), and using $BaSO_4$ as reference. The absorbance spectra were transformed by using the Kubelka–Munk function [37–39] to produce a signal, normalized for comparison between samples, allowing us to calculate the band gap (E_g). The details of this treatment method are widely described elsewhere [40–42].

3.4. Photocatalytic Activity on Methylene Blue Degradation

The photoactivity of the samples was determined by following the methylene blue (MB) degradation, under UV-A illumination. The lamp (Osram Sylvania, Blacklight-Bleu Lamp, F18W/BLB-T8) emitted UV-A radiation considered as monochromatic, with a wavelength of 365 nm and an intensity of 1.2 mW/cm^2 (same as for the UV activation, Section 3.2). A scheme of similar installation was presented in Reference [43].

Then, 50 mL of MB solution with a concentration of 20 µmol/L was mixed with the photocatalyst (1 g/L). The suspension was illuminated for 6 h, at a constant temperature of 25 °C, under UV-A illumination. The concentration of MB was followed by recording UV/VIS spectrum between 300 and 800 nm with GENESYS 105 UV-VIS spectrophotometer from Thermo Scientific. A calibration curve was made to link the absorbance to the concentration. Aliquots were taken every 30 min during the first 2 h of the experiment, and then each hour, until a total of 6 h. Each sample was tested 3 times to obtain mean degradation values; the variation on the sample concentration was around 5%. Blank experiment without photocatalyst was also performed to ensure that MB did not undergo photobleaching under UV-A illumination. Dark tests without illumination were performed to assess the MB adsorption on the samples.

4. Conclusions

In this study, TiO_2 photocatalysts were cogelled with silicon alkoxide, in order to finely disperse Pd species in TiO_2 matrix and dope them with SiO_2; these modifications were made in one step during the synthesis. This Pd modification was made in order to increase the photoefficiency on methylene blue degradation. Two activation treatments were also performed in the same goal: UV activation and H_2 reduction. A pure and a mixed TiO_2/SiO_2 sample were also synthesized as reference.

Results showed that the crystallographic structure of TiO_2 is the anatase phase and that Pd is present either in its oxidized form after the calcination stage, or in its reduced form after the reduction stage. The introduction of SiO_2 delayed the crystallization of TiO_2, leading to amorphous TiO_2 material when the highest Pd loading (22 wt.%) was produced. The specific surface area of the samples, determined by nitrogen adsorption–desorption, varied from 5 to 145 m^2 g^{-1}, with the highest S_{BET} value obtained after H_2 reduction treatment. Transmission electron microscopy allowed us to observe the homogeneous dispersion and nanometric size of Pd particles in the reduced samples. The size of Pd crystallites increased with the Pd amount. The band gap measured by UV/Visible diffuse reflectance spectroscopy corresponded to that of the TiO_2 anatase, approximately 3.2 eV for the pure and SiO_2 doped TiO_2. The band gaps for Pd-modified samples were not determined due to the dark color of the samples and the high absorption in the UV/Visible range. The UV activation did not modify the physicochemical properties of the samples, and it had little effect on photocatalytic activity of calcined samples.

The photoactivity of TiO_2 was increased by the introduction of a small amount of silica into its network. However, when the quantity of silica introduced became too large, the crystallization of TiO_2 in its anatase form was delayed, and the photocatalytic activity of samples decreased. The introduction of Pd in the samples also increased the photoactivity. Moreover, the H_2 reduction treatment further increased the photocatalytic activity of the samples, because this treatment allowed us to decrease the rate of electron–hole recombination in TiO_2 via two distinct phenomena: (i) the creation of oxygen vacancies on the surface of TiO_2 by reduction of Ti^{4+} ions to Ti^{3+}; (ii) the reduction of PdO particles to palladium metallic nanoparticles acting as an electron trap and thus limiting the electron–hole recombination. The optimal loading is 1 wt.% for each (SiO_2 and Pd) additive.

Supplementary Materials: The following are available online at http://www.mdpi.com/2073-4344/10/10/1184/s1, Figure S1. Pictures of the calcined samples: from left to right: pure TiO_2, TiO_2/SiO_2, $TiO_2/SiO_2/Pd1$, $TiO_2/SiO_2/Pd5$, $TiO_2/SiO_2/Pd12$ and $TiO_2/SiO_2/Pd22$.

Author Contributions: Conceptualization, methodology, investigation, analysis and writing, V.S., L.T., J.G.M. and S.D.L.; writing—original draft preparation, J.G.M., S.H. and S.D.L.; supervision, funding acquisition and project administration, S.H. and S.D.L. All the authors corrected the paper before submission and during the revision process. All authors have read and agreed to the published version of the manuscript.

Funding: This research received no external funding.

Acknowledgments: S.D.L. thanks the Belgian National Funds for Scientific Research (F.R.S.-FNRS) for her Associate Researcher position. The authors acknowledge the Ministère de la Région Wallonne Direction Générale des Technologies, de la Recherche et de l'Energie (DGO6), the Fonds de Bay and the Fonds de Recherche Fondamentale et Collective for financial support. J.G.M. and S.H. also thank Innoviris Brussels for financial support through the Bridge project—COLORES.

Conflicts of Interest: The authors declare no conflict of interest.

References

1. Pignatello, J.J.; Oliveros, E.; MacKay, A. Advanced oxidation processes for organic contaminant destruction based on the fenton reaction and related chemistry. *Crit. Rev. Environ. Sci. Technol.* **2006**, *36*, 1–84. [CrossRef]
2. Kuyukina, M.S.; Ivshina, I.B. Application of *Rhodococcus* in Bioremediation of Contaminated Environments. In *Biology of Rhodococcus*; Alvarez, H., Ed.; Springer: Berlin/Heidelberg, Germany, 2010; pp. 231–262. ISBN 9783642129377.
3. Linsebigler, A.L.; Lu, G.; Yates, J.T. Photocatalysis on TiO_2 Surfaces: Principles, Mechanisms, and Selected Results. *Chem. Rev.* **1995**, *95*, 735–758. [CrossRef]
4. Pelaez, M.; Nolan, N.T.; Pillai, S.C.; Seery, M.K.; Falaras, P.; Kontos, A.G.; Dunlop, P.S.M.; Hamilton, J.W.J.; Byrne, J.A.; O'Shea, K.; et al. A review on the visible light active titanium dioxide photocatalysts for environmental applications. *Appl. Catal. B Environ.* **2012**, *125*, 331–349. [CrossRef]
5. Oseghe, E.O.; Ofomaja, A.E. Study on light emission diode/carbon modified TiO_2 system for tetracycline hydrochloride degradation. *J. Photochem. Photobiol. A Chem.* **2018**, *360*, 242–248. [CrossRef]
6. Mahy, J.G.; Paez, C.A.; Carcel, C.; Bied, C.; Tatton, A.S.; Damblon, C.; Heinrichs, B.; Wong Chi Man, M.; Lambert, S.D. Porphyrin-based hybrid silica-titania as a visible-light photocatalyst. *J. Photochem. Photobiol. A Chem.* **2019**, *373*, 66–76. [CrossRef]
7. Banerjee, S.; Dionysiou, D.D.; Pillai, S.C. Self-cleaning applications of TiO_2 by photo-induced hydrophilicity and photocatalysis. *Appl. Catal. B Environ.* **2015**, *176*, 396–428. [CrossRef]
8. Espino-Estévez, M.R.; Fernández-Rodríguez, C.; González-Díaz, O.M. Effect of TiO_2—Pd and TiO_2—Ag on the photocatalytic oxidation of diclofenac, isoproturon and phenol. *Chem. Eng. J.* **2016**, *298*, 82–95. [CrossRef]
9. Léonard, G.L.-M.; Malengreaux, C.M.; Mélotte, Q.; Lambert, S.D.; Bruneel, E.; Van Driessche, I.; Heinrichs, B. Doped sol-gel films vs. powders TiO_2: On the positive effect induced by the presence of a substrate. *J. Environ. Chem. Eng.* **2016**, *4*, 449–459. [CrossRef]
10. Tunc, I. The effect of the presence of Ag nanoparticles on the photocatalytic degradation of oxalic acid adsorbed on TiO_2 nanoparticles monitored by ATR-FTIR. *Mater. Chem. Phys.* **2014**, *144*, 444–450. [CrossRef]

11. Bodson, C.J.; Heinrichs, B.; Tasseroul, L.; Bied, C.; Mahy, J.G.; Wong Chi Man, M.; Lambert, S.D. Efficient P- and Ag-doped titania for the photocatalytic degradation of waste water organic pollutants. *J. Alloy. Compd.* **2016**, *682*, 144–153. [CrossRef]
12. Vaiano, V.; Iervolino, G.; Sannino, D.; Murcia, J.J.; Hidalgo, M.C.; Ciambelli, P.; Navío, J.A. Photocatalytic removal of patent blue V dye on Au—TiO_2 and Pt—TiO_2 catalysts. *Appl. Catal. B Environ.* **2016**, *188*, 134–146. [CrossRef]
13. Borzyszkowska, A.F.; Stepnowski, P.; Ofiarska, A.; Pieczynska, A.; Siedlecka, E.M. Pt—TiO_2-assisted photocatalytic degradation of the cytostatic drugs ifosfamide and cyclophosphamide under artificial sunlight. *Chem. Eng. J.* **2016**, *285*, 417–427.
14. Abdelaal, M.Y.; Mohamed, R.M. Novel Pd/TiO_2 nanocomposite prepared by modified sol-gel method for photocatalytic degradation of methylene blue dye under visible light irradiation. *J. Alloy. Compd.* **2013**, *576*, 201–207. [CrossRef]
15. Léonard, G.L.-M.; Pàez, C.A.; Ramírez, A.E.; Mahy, J.G.; Heinrichs, B. Interactions between Zn^{2+} or ZnO with TiO_2 to produce an efficient photocatalytic, superhydrophilic and aesthetic glass. *J. Photochem. Photobiol. A Chem.* **2018**, *350*, 32–43. [CrossRef]
16. Mahy, J.G.; Lambert, S.D.; Tilkin, R.G.; Poelman, D.; Wolfs, C.; Devred, F.; Gaigneaux, E.M.; Douven, S. Ambient temperature ZrO_2-doped TiO_2 crystalline photocatalysts: Highly efficient powders and films for water depollution. *Mater. Today Energy* **2019**, *13*, 312–322. [CrossRef]
17. Chen, Q.; Shi, W.; Xu, Y.; Wu, D.; Sun, Y.; Si, A. Visible-light-responsive Ag—Si codoped anatase TiO_2 photocatalyst with enhanced thermal stability. *Mater. Chem. Phys.* **2011**, *125*, 825–832. [CrossRef]
18. Pirard, S.L.; Mahy, J.G.; Pirard, J.-P.; Heinrichs, B.; Raskinet, L.; Lambert, S.D. Development by the sol-gel process of highly dispersed $Ni—Cu/SiO_2$ xerogel catalysts for selective 1,2-dichloroethane hydrodechlorination into ethylene. *Microporous Mesoporous Mater.* **2015**, *209*, 197–207. [CrossRef]
19. Belet, A.; Wolfs, C.; Mahy, J.G.; Poelman, D.; Vreuls, C. Sol-Gel Syntheses of Photocatalysts for the Removal of Pharmaceutical Products in Water. *Nanomaterials* **2019**, *9*, 126. [CrossRef]
20. Yang, J.; Xu, X.; Liu, Y.; Gao, Y.; Chen, H.; Li, H. Preparation of $SiO_2@TiO_2$ composite nanosheets and their application in photocatalytic degradation of malachite green at emulsion interface. *Colloids Surf. A Physicochem. Eng. Asp.* **2019**, *582*, 123858. [CrossRef]
21. Braconnier, B.; Páez, C.A.; Lambert, S.; Alié, C.; Henrist, C.; Poelman, D.; Pirard, J.P.; Cloots, R.; Heinrichs, B. Ag- and SiO_2-doped porous TiO_2 with enhanced thermal stability. *Microporous Mesoporous Mater.* **2009**, *122*, 247–254. [CrossRef]
22. Bodson, C.J.; Lambert, S.D.; Alié, C.; Cattoën, X.; Pirard, J.; Bied, C.; Wong Chi Man, M.; Heinrichs, B. Effects of additives and solvents on the gel formation rate and on the texture of P- and Si-doped TiO_2 materials. *Microporous Mesoporous Mater.* **2010**, *134*, 157–164. [CrossRef]
23. Lecloux, A.J. Texture of Catalysts. In *Catalysis: Science and Technology*; Anderson, J.R., Boudart, M., Eds.; Springer: Berlin/Heidelberg, Germany, 1981; Volume 2, pp. 171–230.
24. Mahy, J.G.; Claude, V.; Sacco, L.; Lambert, S.D. Ethylene polymerization and hydrodechlorination of 1,2-dichloroethane mediated by nickel(II) covalently anchored to silica xerogels. *J. Sol-Gel Sci. Technol.* **2017**, *81*, 59–68. [CrossRef]
25. Lambert, S.; Cellier, C.; Grange, P.; Pirard, J.P.; Heinrichs, B. Synthesis of Pd/SiO_2, Ag/SiO_2, and Cu/SiO_2 cogelled xerogel catalysts: Study of metal dispersion and catalytic activity. *J. Catal.* **2004**, *221*, 335–346. [CrossRef]
26. Khaki, M.R.D.; Shafeeyan, M.S.; Raman, A.A.A.; Daud, W.M.A.W. Application of doped photocatalysts for organic pollutant degradation—A review. *J. Environ. Manag.* **2017**, *198*, 78–94. [CrossRef]
27. Yukselen, Y.; Kaya, A. Suitability of the methylene blue test for surface area, cation exchange capacity and swell potential determination of clayey soils. *Eng. Geol.* **2008**, *102*, 38–45. [CrossRef]
28. Calleja, G.; Serrano, D.P.; Sanz, R.; Pizarro, P. Mesostructured SiO_2-doped TiO_2 with enhanced thermal stability prepared by a soft-templating sol-gel route. *Microporous Mesoporous Mater.* **2008**, *111*, 429–440. [CrossRef]
29. Zhang, Z.; Long, J.; Xie, X.; Zhuang, H.; Zhou, Y.; Lin, H.; Yuan, R.; Dai, W.; Ding, Z.; Wang, X.; et al. Controlling the synergistic effect of oxygen vacancies and N dopants to enhance photocatalytic activity of N-doped TiO_2 by H_2 reduction. *Appl. Catal. A Gen.* **2012**, *425–426*, 117–124. [CrossRef]

30. Páez, C.A.; Lambert, S.D.; Poelman, D.; Pirard, J.P.; Heinrichs, B. Improvement in the methylene blue adsorption capacity and photocatalytic activity of H_2-reduced rutile-TiO_2 caused by Ni(II)porphyrin preadsorption. *Appl. Catal. B Environ.* **2011**, *106*, 220–227.
31. Liu, H.; Ma, H.T.; Li, X.Z.; Li, W.Z.; Wu, M.; Bao, X.H. The enhancement of TiO_2 photocatalytic activity by hydrogen thermal treatment. *Chemosphere* **2003**, *50*, 39–46. [CrossRef]
32. Tong, H.-X.; Chen, Q.-Y.; Yin, Z.-L.; Hu, H.-P.; Wu, D.-X.; Yang, Y.-H. Preparation, characterization and photo-catalytic behavior of WO_3-TiO_2 catalysts with oxygen vacancies. *Trans. Nonferrous Met. Soc. China* **2009**, *19*, 1483–1488. [CrossRef]
33. Pirard, S.L.; Malengreaux, C.M.; Toye, D.; Heinrichs, B. How to correctly determine the kinetics of a photocatalytic degradation reaction? *Chem. Eng. J.* **2014**, *249*, 1–5. [CrossRef]
34. Arabatzis, I.M.; Stergiopoulos, T.; Andreeva, D.; Kitova, S.; Neophytides, S.G.; Falaras, P. Characterization and photocatalytic activity of Au/TiO_2 thin films for azo-dye degradation. *J. Catal.* **2003**, *220*, 127–135. [CrossRef]
35. Fisher, M.B.; Keane, D.A.; Fernández-Ibáñez, P.; Colreavy, J.; Hinder, S.J.; McGuigan, K.G.; Pillai, S.C. Nitrogen and copper doped solar light active TiO_2 photocatalysts for water decontamination. *Appl. Catal. B Environ.* **2013**, *130–131*, 8–13. [CrossRef]
36. Patterson, A.L. The Scherrer Formula for X-Ray Particle Size Determination. *Phys. Rev.* **1939**, *56*, 978–982. [CrossRef]
37. Malengreaux, C.M.; Douven, S.; Poelman, D.; Heinrichs, B.; Bartlett, J.R. An ambient temperature aqueous sol-gel processing of efficient nanocrystalline doped TiO_2-based photocatalysts for the degradation of organic pollutants. *J. Sol-Gel Sci. Technol.* **2014**, *71*, 557–570. [CrossRef]
38. Kubelka, P. Ein Beitrag zur Optik der Farban striche. *Z Tech. Phys.* **1931**, *12*, 593–601.
39. Kubelka, P. New contributions to the optics of intensely light-scattering materials. *J. Opt. Soc. Am.* **1948**, *38*, 448–457. [CrossRef]
40. Mahy, J.G.; Lambert, S.D.; Léonard, G.L.-M.; Zubiaur, A.; Olu, P.-Y.; Mahmoud, A.; Boschini, F.; Heinrichs, B. Towards a large scale aqueous sol-gel synthesis of doped TiO_2: Study of various metallic dopings for the photocatalytic degradation of p-nitrophenol. *J. Photochem. Photobiol. A Chem.* **2016**, *329*, 189–202. [CrossRef]
41. Mahy, J.G.; Cerfontaine, V.; Poelman, D.; Devred, F.; Gaigneaux, E.M.; Heinrichs, B.; Lambert, S.D. Highly efficient low-temperature N-doped TiO_2 catalysts for visible light photocatalytic applications. *Materials* **2018**, *11*, 584. [CrossRef]
42. Malengreaux, C.M.; Pirard, S.L.; Léonard, G.; Mahy, J.G.; Herlitschke, M.; Klobes, B.; Hermann, R.; Heinrichs, B.; Bartlett, J.R. Study of the photocatalytic activity of Fe^{3+}, Cr^{3+}, La^{3+} and Eu^{3+} single-doped and co-doped TiO_2 catalysts produced by aqueous sol-gel processing. *J. Alloy. Compd.* **2017**, *691*, 726–738. [CrossRef]
43. Tasseroul, L.; Pirard, S.L.; Lambert, S.D.; Páez, C.A.; Poelman, D.; Pirard, J.P.; Heinrichs, B. Kinetic study of p-nitrophenol photodegradation with modified TiO_2 xerogels. *Chem. Eng. J.* **2012**, *191*, 441–450. [CrossRef]

Publisher's Note: MDPI stays neutral with regard to jurisdictional claims in published maps and institutional affiliations.

© 2020 by the authors. Licensee MDPI, Basel, Switzerland. This article is an open access article distributed under the terms and conditions of the Creative Commons Attribution (CC BY) license (http://creativecommons.org/licenses/by/4.0/).

Review

Eco-Friendly Colloidal Aqueous Sol-Gel Process for TiO$_2$ Synthesis: The Peptization Method to Obtain Crystalline and Photoactive Materials at Low Temperature

Julien G. Mahy [1,*], Louise Lejeune [1], Tommy Haynes [1], Stéphanie D. Lambert [2], Raphael Henrique Marques Marcilli [3], Charles-André Fustin [3] and Sophie Hermans [1,*]

1. Molecular Chemistry, Materials and Catalysis (MOST), Institute of Condensed Matter and Nanosciences (IMCN), Université Catholique de Louvain, Place Louis Pasteur 1, B-1348 Louvain-la-Neuve, Belgium; l.lejeune@student.uclouvain.be (L.L.); tommy.haynes@uclouvain.be (T.H.)
2. Department of Chemical Engineering—Nanomaterials, Catalysis, Electrochemistry, B6a, University of Liège, B-4000 Liège, Belgium; stephanie.lambert@uliege.be
3. Bio and Soft Matter Division (BSMA), Institute of Condensed Matter and Nanosciences (IMCN), Université Catholique de Louvain, Place Louis Pasteur 1, B-1348 Louvain-la-Neuve, Belgium; raphael.marques@uclouvain.be (R.H.M.M.); charles-andre.fustin@uclouvain.be (C.-A.F.)
* Correspondence: julien.mahy@uclouvain.be (J.G.M.); sophie.hermans@uclouvain.be (S.H.); Tel.: +32-10-47-28-10 (S.H.)

Abstract: This work reviews an eco-friendly process for producing TiO$_2$ via colloidal aqueous sol–gel synthesis, resulting in crystalline materials without a calcination step. Three types of colloidal aqueous TiO$_2$ are reviewed: the as-synthesized type obtained directly after synthesis, without any specific treatment; the calcined, obtained after a subsequent calcination step; and the hydrothermal, obtained after a specific autoclave treatment. This eco-friendly process is based on the hydrolysis of a Ti precursor in excess of water, followed by the peptization of the precipitated TiO$_2$. Compared to classical TiO$_2$ synthesis, this method results in crystalline TiO$_2$ nanoparticles without any thermal treatment and uses only small amounts of organic chemicals. Depending on the synthesis parameters, the three crystalline phases of TiO$_2$ (anatase, brookite, and rutile) can be obtained. The morphology of the nanoparticles can also be tailored by the synthesis parameters. The most important parameter is the peptizing agent. Indeed, depending on its acidic or basic character and also on its amount, it can modulate the crystallinity and morphology of TiO$_2$. Colloidal aqueous TiO$_2$ photocatalysts are mainly being used in various photocatalytic reactions for organic pollutant degradation. The as-synthesized materials seem to have equivalent photocatalytic efficiency to the photocatalysts post-treated with thermal treatments and the commercial Evonik Aeroxide P25, which is produced by a high-temperature process. Indeed, as-prepared, the TiO$_2$ photocatalysts present a high specific surface area and crystalline phases. Emerging applications are also referenced, such as elaborating catalysts for fuel cells, nanocomposite drug delivery systems, or the inkjet printing of microstructures. Only a few works have explored these new properties, giving a lot of potential avenues for studying this eco-friendly TiO$_2$ synthesis method for innovative implementations.

Keywords: TiO$_2$; photocatalysis; sol–gel synthesis; peptization; doping; pollutant degradation; mild temperature

1. Introduction

Photocatalysis is a well-established process for the effective and sustainable removal of a large range of organic pollutants, both in liquid and gaseous media [1]. This phenomenon consists of a set of oxidation-reduction (redox) reactions between the organic compounds (pollutants) and the active species formed at the surface of an illuminated photocatalyst (usually a photoactivable semiconductor solid). Generally, when the solid photocatalyst is illuminated (Figure 1), electrons from the valence band are promoted to the conduction band. This results in electron–hole pairs, which can react with O_2 and H_2O, adsorbed at the surface of the photocatalyst, to produce hydroxyl ($^\bullet$OH) and superoxide ($O_2^{-\bullet}$) radicals. These radicals can attack organic molecules and induce their degradation in CO_2 and H_2O, if the degradation is complete [2].

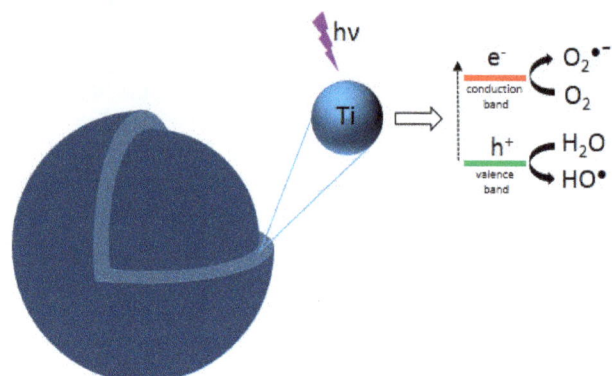

Figure 1. Schematic representation of photocatalytic TiO_2 NP: photogenerated charges (electron and hole) upon absorption of radiation.

Various semi-conductors can be used as photocatalysts, such as NiO [3], ZnO [4], CeO_2 [5], MnO_2 [6], or TiO_2 [7]. The most widely used solid photocatalyst is TiO_2 [7,8], which is a non-toxic and cheap semiconductor sensitive to UV radiation [8]. TiO_2 exists in three different crystallographic structures: anatase (tetragonal structure with a band gap of 3.2 eV), brookite (orthorhombic structure with a band gap >3.2 eV), and rutile (tetragonal structure with a band gap of 3.0 eV) [7]. The best phase for photocatalytic applications is anatase [7]. However, the use of TiO_2 as a photocatalyst has two main limitations [7]: (i) the fast charge recombination, and (ii) the high band gap value which calls for UV light for activation. Therefore, the amount of energy required to activate anatase TiO_2 is high. Indeed, its band gap width (3.2 eV) corresponds to light with a wavelength inferior or equal to 388 nm [7] and so, in the case of illumination by natural light, only the most energetic light will be used for activation, which corresponds to 5–8% of the solar spectrum [8]. To prevent these limitations, several studies have been conducted [9–12] to increase the recombination time and extend the activity towards the visible range. Most works consisted in modifying TiO_2 materials by doping or modification with a large range of different elements, such as Ag [9], P [13], N [14], Fe [11,12], porphyrin [15,16], etc. Therefore, the synthesis process of TiO_2 must be easily adjustable to incorporate such dopants/additives when needed, depending on the targeted application.

Several processes exist to produce TiO_2 photocatalysts, the main methods being chemical or physical vapor deposition [17,18], aerosol process [19], microwave [20], reverse micelle [21], hydrothermal [22], and laser pyrolysis [23]. These processes often use severe synthesis conditions, such as high pressure, high-temperature, or complex protocols. Another possible synthesis pathway is the sol–gel method [24], which has proven to be effective for the synthesis of TiO_2 in the form of powders or films, with control of the nanostructure and surface properties [25–29]. The sol–gel process is classified among "soft chemistry" protocols because reactions occur at low temperature and low pressure. The titanium precursor, usually an alkoxide, undergoes two main reactions: hydrolysis and condensation ((1)–(3) from Figure 2) [24,30,31]. The condensation gives the Ti-O-Ti network formation.

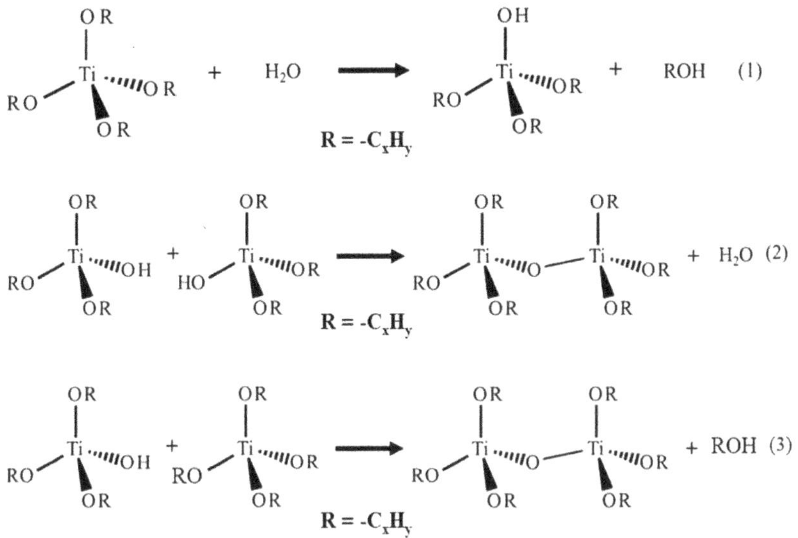

Figure 2. Hydrolysis and condensation reactions of the sol–gel process with Ti alkoxide precursor.

By controlling the rate of the hydrolysis and condensation reactions, a liquid sol or a solid gel is obtained. In order to produce TiO_2 by sol–gel processes, an organic solvent is often used. This organic solvent, such as 2-methoxyethanol, is able to complex the titanium precursor (for example, titanium tetraisopropoxide, TTIP, Ti-$(OC_3H_7)_4$) to control its reactivity. A stoichiometric amount of water is added to avoid fast precipitation [24,31]. The material then undergoes drying and calcination steps to remove residual organic molecules and to crystallize amorphous TiO_2 in anatase, brookite, or rutile phases [32]. In the last decade, attempts at reducing the use of large amounts of organic solvent have been heavily investigated, in order to develop greener syntheses. The use of water as the main solvent was made possible by the use of a peptizing agent. By definition, a peptizing agent (PA) is a substance that, even in small amounts, prevents the agglomeration/flocculation of particles and a decrease in viscosity through enhancing the dispersion in aqueous media [33]. The PA allows crystallization at low temperature, even if the titanium precursor has precipitated. The synthesis of high crystalline TiO_2 nanoparticles, through colloidal aqueous sol–gel in presence of PA, has been successfully reported in the literature [34] and is the main subject of this review.

This synthesis path was first referenced at the end of the 1980s [35–37]. Water is present in a large excess compared to the Ti precursor, and peptizing agents are used to form small TiO_2-crystalline nanoparticles from various Ti precursors at low temperature (<100 °C) [8,38,39], resulting in the formation of a crystalline colloid. Although it is seldom used in the development of TiO_2 synthesis processes, since organic solvents are preferred to better control the Ti precursor reactivity, this preparation method presents a lot of advantages and fulfills the principles of green chemistry that are currently being promoted: (i) the synthesis conditions are soft as it is a sol–gel process; (ii) easy protocol with no risky conditions; (iii) low use of organic reagents, as water is the main solvent; and (iv) crystalline materials are obtained without thermal treatment. Additionally, this synthesis has other advantages, such as: (i) very stable colloids are obtained, allowing the elaboration of coatings very easily by classical deposition techniques (spray-, dip-, spin-, or bar-coating); (ii) protocol easily modified to introduce dopants or additives; and (iii) production at larger scale, up to 20 L.

The goal of this review is to evaluate the state of the art of the research into this not very well-known eco-friendly process for producing TiO_2 via colloidal aqueous sol–gel synthesis, resulting in crystalline materials without a calcination step. A literature review allowed us to find about 115 articles making use of this synthesis process to produce TiO$_2$ materials, spanning from 1987 to 2020. Figure 3 represents the year distribution of these 115 articles. The number of articles over the past 30 years was quite low, due to several reasons: (i) the hydrolysis of the Ti precursor is much easier to control in alcohol solvent and (ii) very fast in water, (iii) the use of water to replace organic solvents for greener processes is a quite recent requirement in chemical processes. Nevertheless, the development of this process has become more and more important over the last ten years.

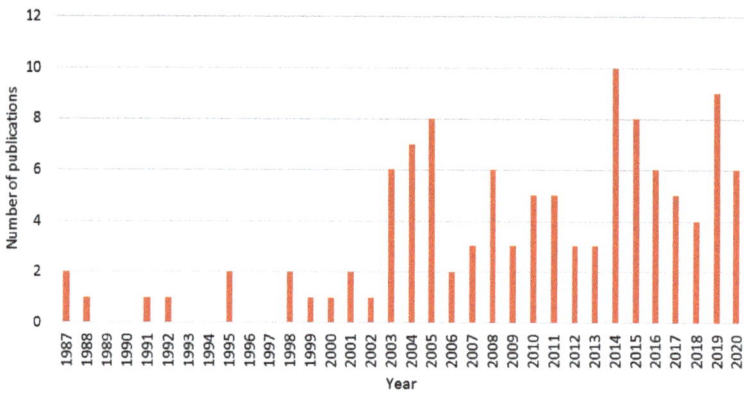

Figure 3. Number of publications per year about colloidal aqueous sol–gel synthesis of TiO$_2$ materials collected for this review.

An increase of interest in this topic in the past ten years is clearly observed. Throughout this review article, the synthesis protocol will be detailed with a focus on the most important parameters, in order to template the resulting TiO$_2$ material. Indeed, by changing synthesis parameters, the three different phases of TiO$_2$ can be obtained, without any thermal treatments. Moreover, specific morphologies can also be produced. In some of the selected articles, thermal post-treatments (calcination or hydrothermal treatment) are applied to the as-synthesized materials, therefore their impact on the crystallinity and morphology of the resulting TiO$_2$ materials will also be reviewed in this paper.

Finally, the photocatalytic properties of these aqueous TiO$_2$ materials will be also reviewed and linked to their physico-chemical characteristics. In the end, new emerging applications will be highlighted.

2. Synthesis of TiO$_2$ with PA in Water

The synthesis uses three main components: the Ti precursor, the peptizing agent, and water. Two operations will take place during the synthesis: the precipitation and the peptization. Indeed, usually the Ti precursor is very reactive on contact with water, resulting in its rapid hydrolysis and condensation. It produces a precipitate of mainly amorphous TiO$_2$. Then, the addition of the peptizing agent will induce the peptization, i.e., the slow dissolution of the TiO$_2$ precipitate and its crystallization into small TiO$_2$ crystallites (<10 nm). Indeed, the introduction of peptizing agent modifies the pH of the solution and increases the solubility of the amorphous titania [39]. The heating of the solution further increases the dissolution of this amorphous TiO$_2$ and accelerates the crystallization [40]. The high concentration of hydroxylated titanium leads to a rapid crystallization, with high nucleation rate [40]. Due to this rapid nucleation rate, metastable polymorphs (i.e., anatase and brookite phases) are favored. When the crystallization is slower, the stable rutile phase is produced [39,40].

Figure 4 presents the general scheme of the synthesis. Usually, the reaction medium can be heated up to 95 °C during peptization.

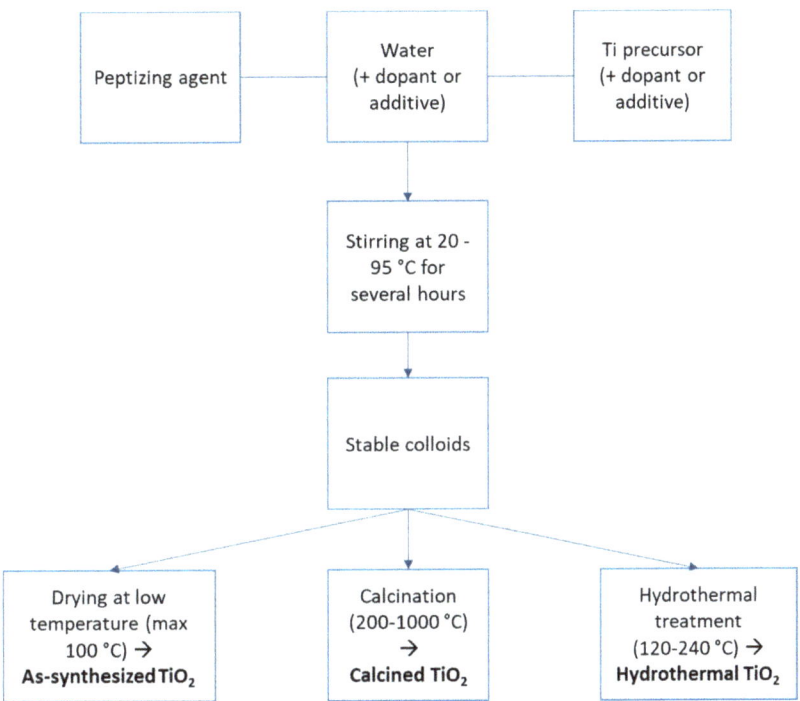

Figure 4. General scheme of the sol–gel TiO$_2$ colloidal aqueous synthesis.

The resulting colloids are very stable (up to years [41]) due to the surface charges of the nanoparticles and can be composed of different crystalline phases and morphologies, depending on the synthesis parameters. The parameters that can be varied are: the type and amount of peptizing agent, the temperature and duration of peptization, and the type of Ti precursor.

Numerous variants of these synthetic parameters have been collected and summarized in Table 1. In addition to the above-mentioned components, possible dopants, applied post-treatments, and shapings are also listed. From this summary, it appears clear that the most used Ti precursor is titanium isopropoxide (TTIP), used in 75 out of the 115 considered studies, due to its relatively low cost; while the peptizing agent is mainly nitric acid (in 71 out of 115 works). The reaction mixture is often heated to reduce the reaction time. When doping is performed, mainly metallic or nitrogen species are used, as they are the main dopants that are known to enhance TiO$_2$ photoactivity. Each author tries to keep the synthesis protocol easy and eco-friendly by reducing the amount of additive/dopant used during the synthesis process. Some organic solvents can be added to stabilize the Ti precursor during the synthesis, but only in very small quantities (less than 10% in volume). With the obtained colloids, it is easy to produce materials with different shapes, such as coatings on various substrates, powders by just drying the colloids, or as colloids directly. The study of Douven et al. [42] refers to the possibility of easily synthesizing colloidal aqueous TiO$_2$ at larger scale, up to 10 L batches. This shows the potential for scaling-up towards industrial scale.

Table 1. Main TiO$_2$ synthesis parameters.

Synthesis Parameters	Corresponding Parameters Collected in the Literature (Variants)
Ti precursor	Ti isopropoxide [8,16,34,35,38,41–95], Ti ethoxide [39,96], Ti butoxide [37,97–103], Ti trichloride [104,105], Ti tetrachloride [106–113], Titanyl sulfate and disulfate [114–117], Titanium(IV) bis(acetylacetonate) diisopropoxide [118], metatitanic acid [119–122], Ti propoxide [96].
Peptizing agent	Nitric acid [8,16,36,39,41–43,45,46,53,56,59,61,62,64,66,67,70,73,76–79,82,86,93,96,100,106,107,109,114,117,119,120,123–131], acetic acid [37,44,87,125,132,133], hydrochloric acid [39,49,54,68,72,87,104,108,121,134–139], malonic acid [125], sulfuric acid [39,53,89,107], tetramethylammonium hydroxide [50,101,140], sodium hydroxide [52,54], phosphoric acid [54,107], perchloric acid [83,141], ammonium hydroxide [38,58,91], hydrogen peroxide [105,116], lactic acid [71], citric acid [138], boric acid [85]
Temperature range of reaction	20–95 °C
Trace of organic solvent	Isopropanol, ethanol, methanol
Additive or dopant	Other metallic alkoxides, metallic salts, carbon materials, nitrogen compounds
Thermal treatment	Ambient drying, calcination in the range 200–1000 °C, hydrothermal treatment
Shaping	Powder, coating, colloid

3. Crystallinity

One of the main advantages of this colloidal aqueous TiO$_2$ synthesis method is to produce crystalline materials without any thermal treatment. Nevertheless, some studies performed post-synthetic hydrothermal and/or calcination steps in order to obtain specific physico-chemical properties. The following sections detail the crystalline properties obtained depending on these three possibilities: as-synthesized, after calcination, or hydrothermal treatments.

3.1. As-Synthesized Aqueous TiO$_2$

As mentioned, after the synthesis, a stable TiO$_2$ colloid in water is obtained. This suspension can be dried under ambient air or precipitated by a pH change to recover the as-synthesized powder. This powder can be easily redispersed in acidic water [41]. In the majority of the reviewed studies, the powders are characterized by XRD in order to evaluate their crystalline phases.

Concerning the crystallinity, the peptizing agent seems to play a very important role. Indeed, the three different TiO$_2$ phases, namely anatase/brookite/rutile, can be obtained by merely changing the amount of peptizing agent, its acid-basic character, or the nature of the counter ion [82]. In all these studies, the crystallite size remains in the same range, between 3 and 10 nm [47,78].

3.1.1. Acid Peptizing Agent

With the most used peptizing agent, HNO$_3$, when it is used without any other additive, a mixture of anatase/brookite is often produced, [39,76,87,97,142]; with a higher proportion of anatase, as presented in Figure 5a. Only the peak at 30.8° is observed for brookite. An increase in the amount of HNO$_3$ during peptization (from pH of 2 up to pH of 0.5) induces the formation of rutile phase, as show in [39,118,143]. A mixture of crystalline phases is often reported. When additives that cause a shift in pH value are used, the distribution can be modified. In the works of Burda et al. [133] and Chen et al. [95], only anatase is produced when amine is added with HNO$_3$ at the beginning of the synthesis.

With HCl, which is the second most common peptizing agent used, anatase or anatase/brookite is also mainly reported [39,69,121,135,138,139]. A mixture of anatase/rutile is also produced when the amount of HCl increases [104,107,139]. At very high concentration, such as a Ti/H$^+$ molar ratio of 0.08, rutile alone is even observed [108]. Moreover, when different types of acids are used in the same concentration, different phase distributions can be obtained. As examples, Vinogradov et al. [87] used a Ti/H$^+$ ratio of 0.5, and obtained anatase/brookite mixtures with HNO$_3$ or HCl, while only anatase was produced with acetic acid, and an anatase/Ti sulfate mixture with H$_2$SO$_4$. This suggests that the counter ion (Bronsted conjugate base) also plays an important role in the preferential crystalline phase formed [87]. In Kanna et al. [107], with a similar acid amount (not specified), anatase is produced with H$_2$SO$_4$ and H$_3$PO$_4$, and an anatase/rutile mixture with HNO$_3$, HCl, or acetic acid. With carboxylic acids such as acetic, lactic, malonic, or citric acid, anatase is the main phase reported [57,74,76,87,125,133,138], as shown in Figure 5b. Only Kanna et al. [107] report an anatase/rutile mixture.

Globally, when inorganic acid is used, anatase and/or brookite phases are produced, but when the amount of acid leads to a pH smaller than 1, rutile phase is also produced. With organic acids, only anatase phase is formed. The different distributions of phases will impact the resulting surface area. Indeed, anatase and brookite phases lead to a higher specific surface area than rutile [63].

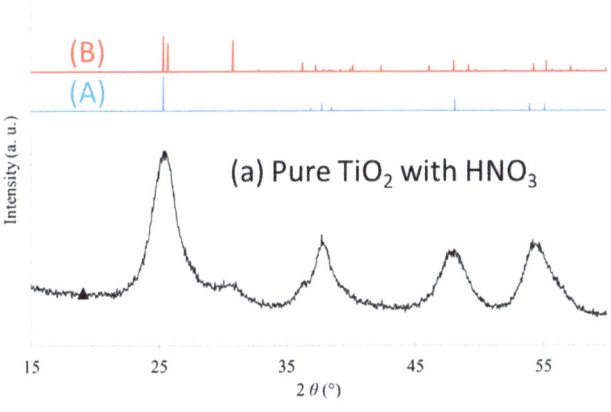

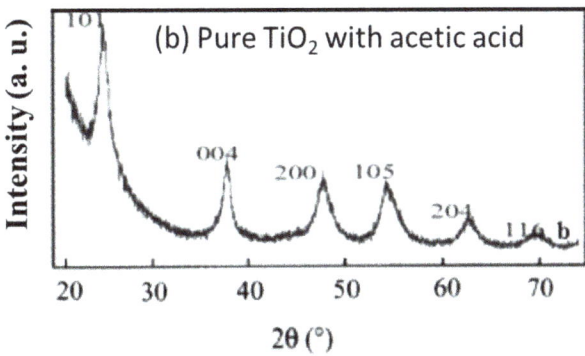

Figure 5. XRD patterns of pure TiO$_2$ material obtained with (**a**) HNO$_3$, where A stands for anatase phase and B for brookite phase from [142] (reproduced with permission from J. G. Mahy et al., AIMS Materials Science; published by AIMS Press, 2018, open access); and (**b**) acetic acid peptizing agents, from [133] (reproduced with permission from J. L. Gole et al., The Journal of Physical Chemistry B; published by The American Chemical Society, 2004).

3.1.2. Basic Peptizing Agent

The basic peptization is far less common (about 8 out of 115 references considered in this review), but some studies still reference it. In Mashid et al. [38], NH$_4$OH is used to synthesize anatase/brookite mixture, as illustrated in Figure 6 for pH 8 and 9. Similarly, with NaOH anatase/brookite is reported in Mutuma et al. [70]. In Yu et al. [91], only anatase is observed with NH$_4$OH peptizing agent at high pH. Zhang et al. [113] report an anatase/rutile mixture with NH$_4$OH at neutral pH. To conclude, the nature (acidic or basic) of the peptizing agent and the amount used will impact the resulting phases, but the type of phase is difficult to predict.

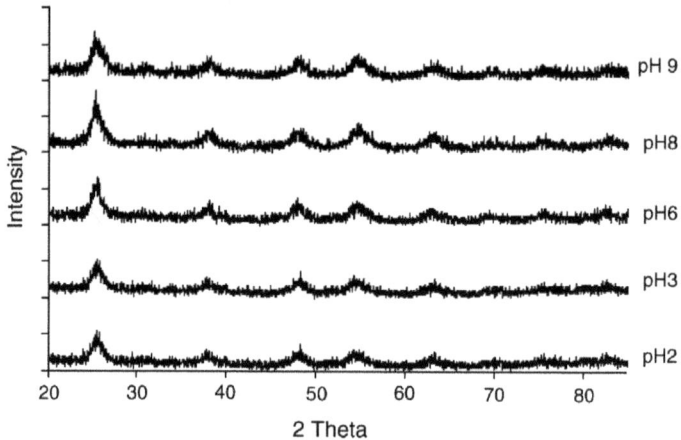

Figure 6. XRD patterns of pure TiO$_2$ obtained at different pH values, with HNO$_3$ (pH < 7) or NH$_4$OH (pH > 7) peptizing agent, from [38] (reproduced with permission from S. Mahshid et al., Journal of Materials Processing Technology; published by Elsevier, 2007).

As shown in the above paragraphs, both acidic or basic PA lead to crystalline TiO$_2$ materials. It is worth mentioning that the resulting TiO$_2$ materials are not 100% crystalline, as is the case when thermal treatments such as calcination or hydrothermal treatment (next paragraphs) are applied. Nevertheless, it was shown [63,65] that the crystalline fraction can be quite high (up to 85–90%) and that this fraction can be optimized by playing with the synthesis parameters, such as the time of reaction or the amount of PA.

3.2. Aqueous TiO$_2$ after a Calcination Treatment

Even if a crystalline material is already obtained right after the synthesis, often composed of two or three TiO$_2$ crystalline phases, as shown in the previous section, a large range of studies perform a calcination step to further crystallize the TiO$_2$ materials, also leading to an increase in the crystallite sizes. When the calcination temperature is high (>600 °C), rutile is often produced, as it is the most stable phase at high-temperature, as represented in Figure 7 [144]. Nie et al. [144] present a study of a structural dependence in function of the temperature and pressure on the calcination post-treatment of TiO$_2$, Figure 7. For temperatures below T <200 °C and pressure lower than 2 GPa the preferential crystalline phase is anatase, for calcinations in the same range of temperatures but with pressures higher than 2 GPa, the preferred crystalline phase formed is srilankite. On the other hand, for calcination performed at a temperature higher than 600 °C a preferential rutile phase is normally observed, independent of the applied pressure, Figure 7. Additionally, a phase anatase–rutile transition is often observed around 500 °C.

The phase transition from brookite to anatase or rutile has been less studied and no phase diagram is found in the literature. Nevertheless, some authors claimed that brookite evolves to anatase then rutile when the calcination temperature increases, [145,146], while others claim than brookite evolves directly to rutile [147,148].

In the considered studies, the temperature of calcination varies between 200 and 1000 °C. In all of these cases, the crystallite size increases, from 3–10 nm in the as-synthesized TiO$_2$ materials, to a range of 20–100 nm, depending on the calcination temperature [45,46,86,149]. Obviously, the higher the temperature, the higher the obtained size.

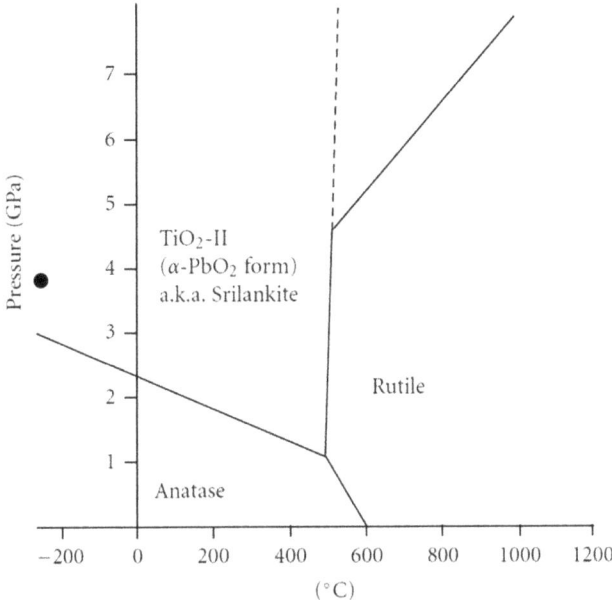

Figure 7. TiO$_2$ phase transition diagram from [144] (reproduced with permission from X. Nie et al., International Journal of Photoenergy; published by Hindawi Publishing Corporation, 2009, open access).

3.2.1. Calcination after Acidic Peptization

In Borlaf et al. [93], a HNO$_3$ peptized TiO$_2$ colloid is calcined between 200 and 1000 °C, and the crystalline phases are compared at various temperatures. As-synthesized, the TiO$_2$ material is composed of an anatase/brookite mixture, whose crystallite size increases, while keeping the same crystalline mixture until 500 °C. From 600 °C to 800 °C, the mixture is composed of anatase/rutile, with the proportion of rutile increasing with the temperature. From 800 to 1000 °C, only rutile is present. This is illustrated in Figure 8.

In [38,43,45,49,53,60,69,70,80,89,112,130,134], similar evolutions are obtained when using HNO$_3$ or HCl peptizer followed by a calcination from 300 to 900 °C. The anatase/brookite mixture is converted into a anatase/brookite/rutile mixture around 500 °C and becomes only rutile around 700 °C. Globally, the colloidal aqueous TiO$_2$ synthesis allows keeping anatase/brookite phase until 500–700 °C during calcination [58,71,73,85,91,98,114,129,130], which is coherent with the anatase-to-rutile transition temperature (Figure 7).

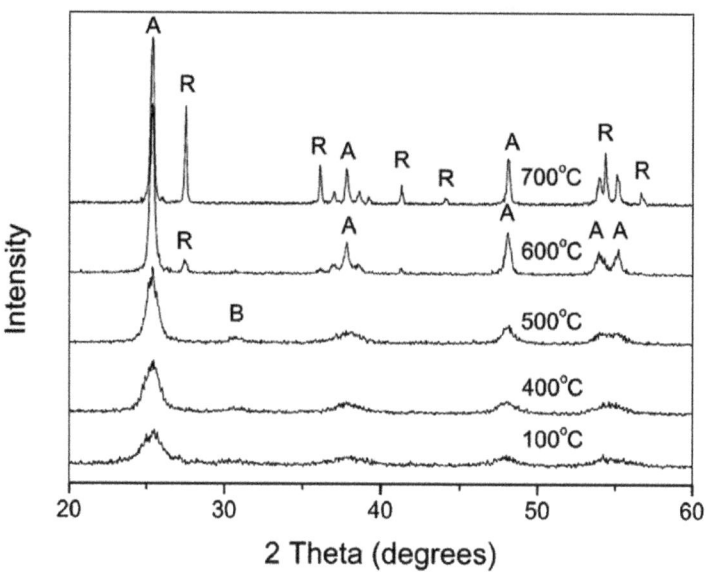

Figure 8. Evolution of the XRD pattern of a TiO$_2$ sample peptized with HNO$_3$ and calcined at different temperatures, from [91]. The A, B, and R labels stand for anatase, brookite, and rutile phases, respectively (reproduced with permission from J. Yu et al., Journal of Catalysis; published by Elsevier, 2003).

3.2.2. Basic Peptization Followed by Calcination

The same trends are globally observed in the case of the basic peptizers, even if these are less studied: an increase in anatase or anatase/brookite content is observed until a calcination temperature around 500–700 °C [52,58,70,91], then rutile becomes the main phase, as illustrated in Figure 9 [58,70,91].

3.3. Aqueous TiO$_2$ after Hydrothermal Treatment

This treatment consists in placing the precursor suspension in water in an autoclave under pressure, and heated at a controlled temperature. Similarly to calcination, a hydrothermal treatment allows the increase of the crystallinity of the as-synthesized samples thanks to the Ostwald ripening mechanism [50]. The temperature of such a treatment is usually between 170 and 240 °C. The crystallite size increases compared to the as-synthesized TiO$_2$ crystallite, in the range of 5 to 70 nm. When the treatment is very long (i.e., several days), a phase change may occur towards rutile (thermodynamically the most stable). A calcination step can be also applied after the hydrothermal treatment, and this will further increase the crystallite size of the phase present after the hydrothermal treatment [55,90,103,109,140,150], until the temperature of anatase-to-rutile transition is reached, where only rutile crystallites continue to grow [55,150]. For both types of peptizers, acid or basic, similar evolutions are observed.

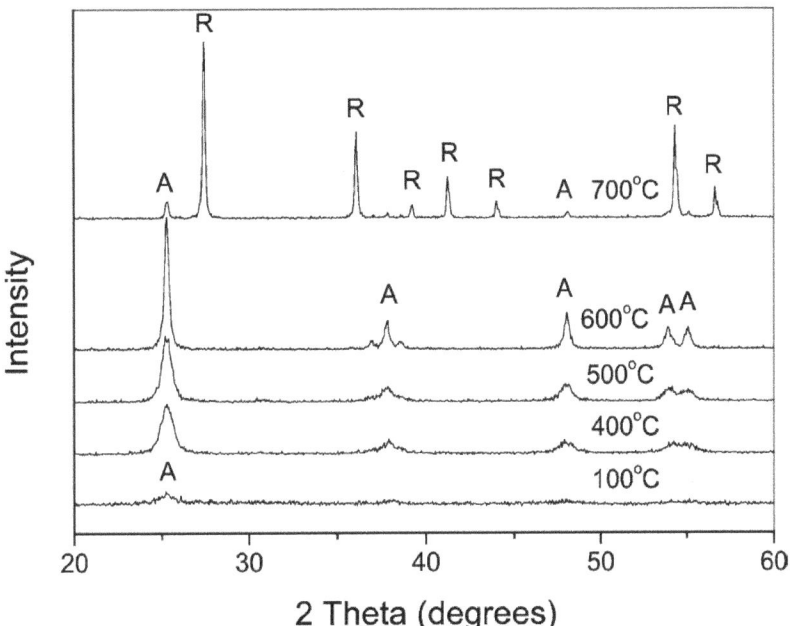

Figure 9. Evolution of the XRD pattern of a TiO_2 sample peptized with NH_4OH and calcined at different temperatures, from [91]. The A and R labels stand for anatase and rutile phases, respectively (reproduced with permission from J. Yu et al., Journal of Catalysis; published by Elsevier, 2003).

In [50,54,59,90,103,116,140,151], the hydrothermal treatment allows the increase of the crystallite size of the crystalline phase present in the as-synthesized sample. An increase of the duration, or temperature, of the hydrothermal treatment leads to larger crystallite size [50,55]. An as-synthesized anatase phase can also be converted into the rutile phase if the temperature or duration is sufficient, as illustrated in Figure 10, while an as-synthesized anatase/brookite mixture is converted to rutile phase after hydrothermal treatment at 200 °C or 240 °C for 2 h.

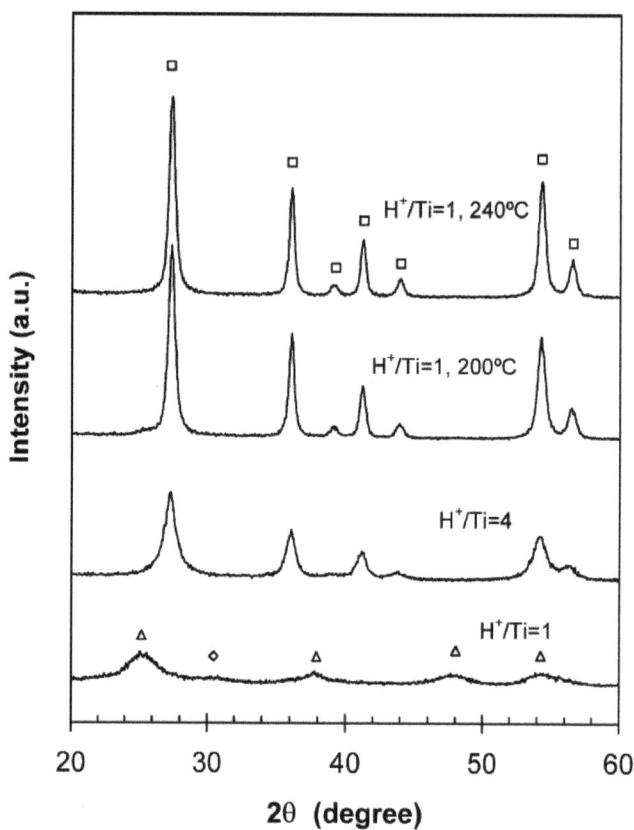

Figure 10. Evolution of the XRD pattern of a TiO$_2$ sample peptized with HNO$_3$ and hydrothermally treated at different temperatures, from [90]. (Δ) anatase, (◊), brookite, and (□) rutile phases (reproduced with permission from J. Yang et al., Journal of Colloid and Interface Science; published by Elsevier, 2005).

4. Morphology

Besides the crystallite formation at low temperature, colloidal aqueous TiO$_2$ synthesis allows the production of specific morphologies, depending on the synthesis conditions and the post-treatments applied. The following sections detail the TiO$_2$ morphologies obtained, depending on the same three synthetic steps: as-synthesized, and after calcination or hydrothermal treatments. The morphology is linked to the crystalline phase produced. The morphology depends on the crystalline phases produced during the synthesis. Indeed, anatase and brookite phases mainly lead to spherical nanoparticles, while rutile gives rod-like nanoparticles [104].

A particularity of this synthesis method using peptization is that the crystallite size and the nanoparticle size are the same. Indeed, it was shown in many studies [8,38,41,60,61,63,66,80] that one particle is made of one crystallite, thanks to comparisons made between XRD (crystallite size estimated by Scherrer formula) and TEM imaging.

4.1. Morphology of As-Synthesized Aqueous TiO$_2$

As-synthesized TiO$_2$ materials are stable colloids that are composed of nanoparticles in the range of 3–10 nm [96,100]. For the materials composed of anatase or an anatase/brookite mixture, all studies report similar spherical nanoparticles below 10 nm, as shown in

Figure 11a as an illustrative example [61]. When rutile phase is present, the morphology of rutile crystallites corresponds to nanorods, as depicted in Figure 11b [104]. Therefore, two main morphologies are observed, depending on the crystalline phases.

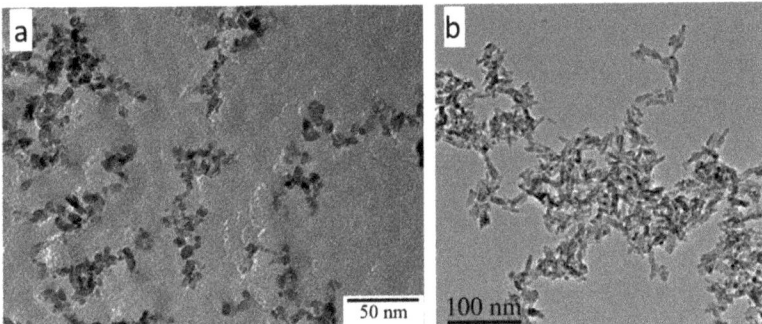

Figure 11. TEM micrographs of (**a**) TiO$_2$ anatase/brookite spherical nanoparticles, from [61] (reproduced with permission from J. G. Mahy et al., Journal of Photochemistry and Photobiology A: Chemistry; published by Elsevier, 2016) and (**b**) TiO$_2$ rutile nanorods, from [104] (reproduced with permission from S. Cassaignon et al., Journal of Physics and Chemistry of Solids; published by Elsevier, 2007).

The effect of PA on the final morphology of TiO$_2$ will depend on the crystalline phase that is formed during the synthesis. Indeed, when anatase and/or brookite phases are formed, spherical nanoparticles are produced. Basic or acidic PA can lead to anatase/brookite phases, and thus basic or acidic PA can lead to spherical nanoparticles. When organic acid PA is used, spherical nanoparticles are produced because only anatase phase is formed. When rutile is produced, a nanorod morphology is obtained and, globally, it is when a large amount of acidic PA is used that this is the case. Therefore, in conclusion, it is difficult to state that one type of PA (acidic or basic) will produce a specific type of morphology, but it is rather linked to the resulting crystalline phase.

4.2. Morphology of Aqueous TiO$_2$ after Calcination Treatment

As explained above, calcination permits further crystallizing the as-synthesized TiO$_2$ materials, yielding an increase in the crystallite size. Therefore, as for the as-synthesized materials, two morphologies (sphere [88] and nanorod [73]) are observed depending on the crystalline phases, but the size range of the nanoparticles is larger than the as-synthesized (10–70 nm vs. 2–10 nm). Figure 12 presents the spherical [43] and nanorod [73] morphologies obtained after calcination at 500 °C.

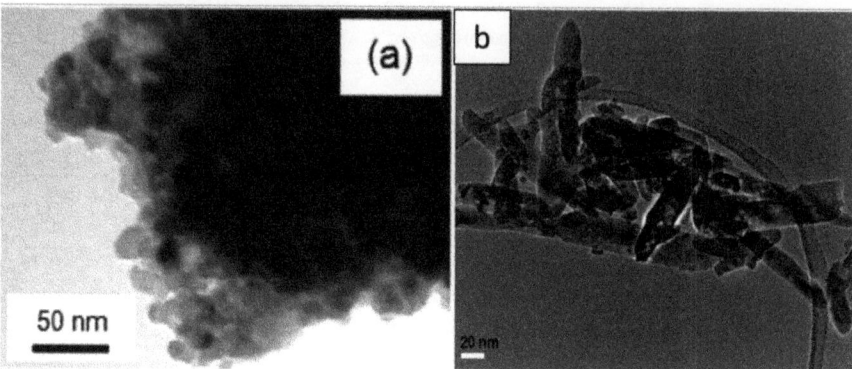

Figure 12. TEM micrographs of (**a**) TiO$_2$ anatase spherical sample calcined at 500 °C from [43] (reproduced with permission from F;.R. Cesconeto et al., Ceramics International; published by Elsevier, 2018) and (**b**) TiO$_2$ rutile nanorod sample calcined at 500 °C from [73] (reproduced with permission from P. Periyat et al., Materials Science in Semiconductor Processing; published by Elsevier, 2015).

4.3. Morphology of Aqueous TiO$_2$ after Hydrothermal Treatment

As for the calcination, the hydrothermal treatment allows the increase of the crystallite size (comprised between 10 and 80 nm), while keeping the morphology of the as-synthesized materials (sphere or nanorod) [83,152]. Figure 13 gives an example of spheres [50] and nanorods [141] obtained by hydrothermal treatment.

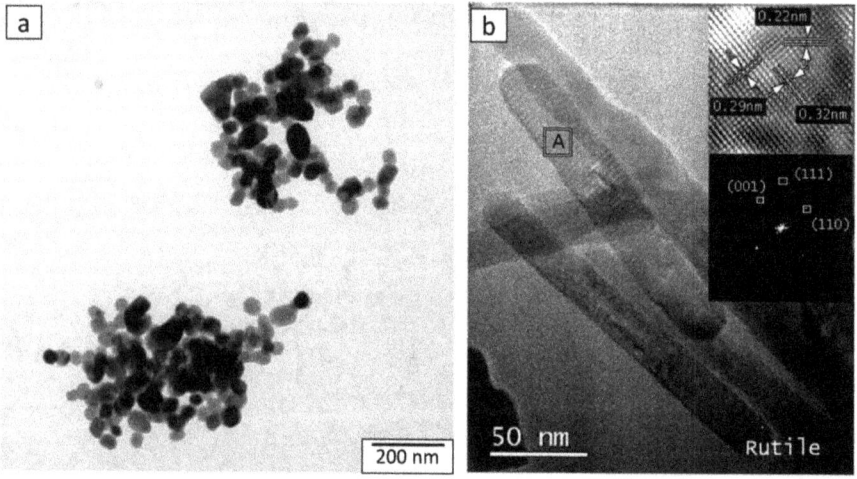

Figure 13. TEM micrographs of (**a**) TiO$_2$ anatase spherical sample hydrothermally treated at 230 °C from [50] (reproduced with permission from S. Hore et al., Journal of Materials Chemistry; published by RSC, 2005) and (**b**) TiO$_2$ rutile nanorod sample hydrothermally treated at 200 °C from [141] (reproduced with permission from H. Li et al., Materials Research Bulletin; published by Elsevier, 2011).

5. Doping and Additives

As mentioned in the introduction, the two intrinsic limitations of TiO$_2$ as a photocatalyst are (i) the fast charge recombination, and (ii) the high band gap value, which calls for UV light for activation [7]. Therefore, the doping and/or modification of colloidal aqueous TiO$_2$ are also described in the literature to prevent these limitations. Throughout the literature, four main modification strategies of aqueous TiO_2 were found: doping with (i) metallic or (ii) non-metallic species, (iii) a combination with other semiconductors, and (iv) sensitization with dye molecules.

The modification of TiO_2 with metallic species introduces metallic ions or metallic nanoparticles into the material. Metallic ions can produce intermediate levels of energy between the valence and conduction bands of TiO_2, leading to a reduction of the energy necessary for electron photoexcitation. As a consequence, near-visible light can activate the photocatalytic process. These metallic ions can also act as photoelectron-hole traps, increasing the recombination time and enhancing the electron–hole separation. Metallic nanoparticles dispersed in the TiO_2 matrix also act as electron traps due to their conductive nature. The metallic species listed are Ag [84,99,149], Fe [42,61,73], Cu [8,61], Rh [93], Pd, Ca [43], Cr [61,66], Pt [51], Zn [8,61,124], Nd [110,111], Tb [132], Ce [44,109,120], Eu [117,126], and W [123].

The doping with non-metallic elements is usually conducted with N, P, or S, and can reduce the band gap by creating an intermediate band for the electrons between the conduction band and the valence band. This doping allows the use of less energetic light to activate TiO_2. Here, we mainly found N-doping (around 5 mol%), due to the frequent use of HNO_3 as a peptizing agent, even in the materials referenced as pure TiO_2. Supplementary sources of N were also used: mainly amine as trimethylamine [63,95,127,133], urea [54,63], melamine [116], hydrazine [133], ethylene diamine [63,75], etc. Many studies reported photoactivity under near-visible range illumination (see Section 6). The combination with other semiconductors in heterojunction is also reported: with ZrO_2 [65,67], $g-C_3N_4$ [135], SnO [131], and Bi_2O_3 [77]. This modification produces a heterojunction at the interface of the two materials, which enhances the electron–hole separation due to the difference in energy levels of the conduction and valence bands of the two photocatalysts.

The introduction of dyes is reported in Mahy et al. [16]. In this case, the grafting of the porphyrin molecule at the surface allows the TiO_2 activation in the visible range, due to the transfer of electrons from the dye by its excitation under visible illumination [16]. One study reports the production of composites made of aqueous TiO_2 with carbon nanotubes [56]. In this case, the role of the carbon materials is similar to the introduction of metallic nanoparticles. As a carbon nanotube is a conductive material, it can trap the photo-generated electrons and decrease the recombination process.

6. Photocatalytic Properties

It is shown in the above paragraphs that colloidal aqueous TiO_2 synthesis can produce crystalline TiO_2 materials with specific morphologies, even without any thermal treatment. These crystalline materials are mainly being used for pollutant degradation. This section will summarize the photocatalytic activity of these aqueous TiO_2 materials identified in the literature. A fraction of the articles dealing with aqueous TiO_2 do not explore its photocatalytic properties and are limited to the description of the physico-chemical properties. This represents 47 out of 115 articles, but in 10 cases, another application is also explored (see Section 7).

6.1. Photoactivity of As-Synthesized Aqueous TiO₂

Table 2 lists the parameters of the photocatalytic experiments in the studies using as-synthesized TiO_2 materials. The most tested molecule as a model "pollutant" is methylene blue (MB) [95,126,127,133,143], but 16 other molecules, such as methyl orange [125], p-nitrophenol [42], and rhodamine B [42,84,87], have also been tested, showing the versatility of this material. The majority of these "pollutants" are model molecules (dyes); photocatalytic degradations of real wastewater or mixed pollutant solutions are very rare. The pollutant concentration is kept low as the photocatalysis process is a finishing water treatment step to remove residual pollution if still present, for example, after a classical wastewater treatment plant. Concerning the illumination, the information is often not very complete. Indeed, sometimes the wavelength and/or the intensity are not given. Globally,

UV-A light or visible light (~350–500 nm range) is used in most of the cases, as it corresponds to the band gap of TiO_2. The time of irradiation can vary from minutes [106,126] to hours [42,120], up to 24 h [42], and depends on the power of the lamp.

Various dopants or additives are added at the beginning of the reaction to increase the photodegradation and/or the adsorption spectrum. Classically, metallic dopants such as Ag [84,99] or Fe [42] are added to enhance the electron–hole separation. As explained above, N-doping allows the increase of the light absorption in the visible range, and thus increases the photoactivity in the visible range [63,127].

Different shapes of photocatalysts can be used: powder [106,126,138], film deposited on various substrates [97,119,135], or even fabric [74]. Numerous studies [42,65,95,133,138] compare their photocatalysts to the most famous commercial TiO_2, Evonik Aeroxide P25, which is produced by high-temperature process. Usually, similar or better activities are obtained with the aqueous TiO_2. A direct comparison between all studies is very complicated, as the experimental conditions are different from one paper to another. Indeed, the lamp, illumination duration, concentration of photocatalyst or pollutant, and type of pollutant are the major parameters which differ from study to study (Table 2). Nevertheless, the high specific surface area obtained with the aqueous sol–gel process is referred to in most studies as the main reason for the increased photocatalytic activity compared to Evonik P25 (250 $m^2\ g^{-1}$ for aqueous sol–gel samples vs. 50 $m^2\ g^{-1}$ for P25). Therefore, the specific structure made of small nanoparticles (<10 nm, see Figure 10 from [8]) highly dispersed in water medium seems to play the most important role in its photocatalytic properties for pollutant removal in water.

Table 2 demonstrates that it is possible to obtain a very efficient TiO_2 material with an eco-friendly and easy synthesis without any additional high-temperature treatment. Indeed, the anatase phase, which is known to be the most efficient photocatalytic phase of TiO_2, due to its better charge separation efficiency, is easily produced.

Table 2. Parameters of photocatalytic experiments in studies using as-synthesized TiO_2 materials.

Paper	Photocatalyst and Shape (Concentration)	Pollutant (Concentration)	Illumination and Time	Best Degradation Results
Bazrafshan et al., 2015 [106]	• Pure TiO_2 • Powder (0.5 g/L)	Reactive orange dye (200 ppm)	Xenon lamp—40 min	100%
Belet et al., 2019 [124]	• Pure TiO_2, TiO_2/Zn • Film on glass	• Methylene blue (MB) (5×10^{-5} M) • pharma products (lorazepam, tramadol, alprazolam, ibuprofen, and metformin. 10 µg/L each)	254 nm—4 h	• 60% on MB • 10–50% on different pharma products
Bergamonti et al., 2014 [125]	• Pure TiO_2 • Powder (9.22 mM)	• Methyl orange (MO) (0.03 mM) • MB (0.03 mM)	365 nm—160 min	100% on both
Borlaf et al., 2014 [126]	• Pure TiO_2, TiO_2/Eu • Powder (0.33×10^{-2} M)	MB (0.33×10^{-2} M)	254 or 312 or 365 nm—40 min	Only kinetic constants given

Table 2. Cont.

Paper	Photocatalyst and Shape (Concentration)	Pollutant (Concentration)	Illumination and Time	Best Degradation Results
Gole et al., 2004 [133]	• N/TiO$_2$ • Powder (5 g/L)	MB (–)	• 390 nm—600 min • 540 nm—600 min	• 80% at 390 nm • 23% at 540 nm
Chen et al., 2005 [95]	• N/TiO$_2$ • Powder (5 g/L)	MB (–)	• 390 nm—600 min • 540 nm—600 min • 780 nm—600 min	• 80% • 25% • 5%
Douven et al., 2020 [42]	• Pure TiO$_2$, N, Fe doping • Powder (1 g/L) • Film on steel	• p-nitrophenol (PNP) (10^{-4} M) • Rhodamine B (RB) (2.5×10^{-6} M)	• Visible (400–800)—24 h • 395 nm (LED)—120 min	• 65% • 95%
Hu et al., 2005 [97]	• Pure TiO$_2$ • Film on quartz	Reactive brilliant red dye XB3 (50 mg/L)	365 nm—120 min	100%
Hu et al., 2014 [127]	• Pure TiO$_2$, • N/TiO$_2$ • Powder (0.5 g/L)	MB (20 µM)	• UV—90 min • Visible (>420 nm)—300 min	• 75% (UV) • 65% (visible)
Huang et al., 2019 [135]	• gC3N4/TiO$_2$ • Composite film	NO$_x$ (gas phase- 400 ppb)	Visible—cycle of 30 min	25% for one cycle
Kanna et al., 2008 [107]	• Pure TiO$_2$ • Powder (0.5 g/L)	• MB (2.5×10^{-5} M) • Cristal violet (CV) (2.5×10^{-5} M) • Congo red (CR) (2.5×10^{-5} M)	366 nm—3 h	• 90% • 95% • 100%
Léonard et al., 2016 [56]	• TiO$_2$/Nanotube • Film on glass	PNP (10^{-4} M)	• 365 nm—24 h • Visible (400–800 nm)—24 h	• 55% • 0%
Li et al., 2014 [115]	• Composite TiO$_2$/PSS or PEI • Powder (1 g/L)	• MB (10 mg/L) • RB (10 mg/L)	365 nm—280 or 400 min	• 95% • 97%
Liu et al., 2008 [119]	• Pure TiO$_2$ • Powder (0.5 g/L) • Film on aluminum and film on glass	• RB (liquid phase- 10 mg/L) • CH3SH (gas phase—100 ppmv) • HCHO (gas phase—5.5 ppmv)	• 50 min—365 nm • 25 min—365 nm • 3 h—365 nm	• 95% • 97% • 85%

Table 2. Cont.

Paper	Photocatalyst and Shape (Concentration)	Pollutant (Concentration)	Illumination and Time	Best Degradation Results
Liu et al., 2010 [120]	• Pure TiO$_2$, TiO$_2$/Ce^{3+} • Powder (1 g/L) • Film on filter paper	• MB (10 mg/L) • 2,3-dichloriphenol (10 mg/L) • Benzene (gas phase 5.5 ppmv)	• UV-A (365 nm) and visible (>420 nm) for liquid—50–180 min • 365,405,430,540,580 nm for gas—7–10 h	• 95–70% • 100–70% • 70–15%
Mahy et al. [16,41,61,62,64,65]	• Pure TiO$_2$, various doping (N, metallic ions, Zr, Pt, porphyrin) • Powder (1 g/L) • Film on pre-painted steel	• PNP (10^{-4} M) • MB (2×10^{-5} M)	• UV-visible (300–800 nm)—8 h • Visible (400–800)—24 h • 365 nm—17 h	• 95% • 70% • 80%
Malengreaux et al. [8,66]	• Pure TiO$_2$, various doping (metallic ions) • Powder (1 g/L)	PNP (10^{-4} M)	UV-visible (300–800 nm)—7 h	75%
Qi et al., 2010 [74]	• Pure TiO$_2$ • Film on cotton fabric	Neolan Blue 2G (0.2 g/L)	365 nm—2 h	70%
Sharma et al., 2020 [138]	• Pure TiO$_2$ • Powder (0.01—0.35 M)	Solophenyl green (3.15 g/L)	365 nm—350 min	70%
Suligoj et al., 2016 [121]	• Pure TiO$_2$ • Composite film with SiO$_2$ on glass	Toluene (gas phase 49 ppmv)	365 nm—100 min	100%
Sung-Suh et al., 2004 [84]	• Pure TiO$_2$, TiO$_2$/Ag • Powder (0.4—4 g/L)	RB (10^{-5} M)	• UV—1 h • Visible—4 h	• 95% • 90%
Vinogradov et al., 2014 [87]	• Pure TiO$_2$ • Film on glass	RB (40 mg/L)	UV—120 min	95%
Wang et al., 2009 [99]	• Pure TiO$_2$, TiO$_2$/Ag • Powder (1 g/L)	MB (30 μM)	UV—90 min	55%
Wang et al., 2005 [143]	• Pure TiO$_2$ • Powder (0.09 M)	MB (0.016 g/L)	UV—25 min	45%

Table 2. Cont.

Paper	Photocatalyst and Shape (Concentration)	Pollutant (Concentration)	Illumination and Time	Best Degradation Results
Xie et al., 2005 [110]	• Pure TiO$_2$, TiO$_2$/Nd^{3+} • Powder (1 g/L)	X3B (100 mg/L)	400–800 nm—120 min	90%
Yan et al., 2013 [131]	• Pure TiO$_2$, TiO$_2$/Sn • Powder (0.28 g/L)	MB (16 mg/L)	Visible (>420 nm)—100 min	45%
Yun et al., 2004 [92]	• Pure TiO$_2$ • Film on glass	Ethanol (gas phase 450 ppmv)	UV—50 min	100%
Zhang et al., 2001 [122]	• Pure TiO$_2$ • Powder (0.8 g/L)	sodium benzenesulfate (12 mM)	UV—4 h	100%

6.2. Photoactivity of Aqueous TiO$_2$ after a Calcination Treatment

Table 3 summarizes the parameters of the photocatalytic experiments for the studies using calcined aqueous TiO$_2$ materials. The observations are similar to Section 6.1 above: numerous pollutants can be degraded (but mainly model pollutants are studied, such as methylene blue), several efficient dopants are used to increase photo-degradation, and the various experimental conditions do not allow a direct comparison of the results. Nevertheless, the photoactivity of the calcined materials does not seem to be better than the as-synthesized materials. Indeed, similar degradation rates are obtained with similar illumination times (compare Table 3 vs. Table 2).

Table 3. Parameters of photocatalytic experiments for studies using calcined aqueous TiO$_2$ materials.

Paper	Photocatalyst and Shape (Concentration)	Pollutant (Concentration)	Illumination and Time	Best Degradation Results
Al-Maliki et al., 2017 [132]	• Pure TiO$_2$, TiO$_2$/Tb • Film	KMnO$_4$ (2 × 10^{-5} M)	• UV (200–400 nm)—75 min • 400–600 nm—75 min	• 65% • 50%
Borlaf et al., 2012 [93]	• Pure TiO$_2$, TiO$_2$/Rh^{3+} • Powder (0.33 × 10^{-2} M)	MB (0.33 × 10^{-2} M)	254 or 312 or 365 nm—40 min	Only kinetic constants given
Cano-Franco et al., 2019 [44]	• Pure TiO$_2$, TiO$_2$/Ce • Powder (1 g/L)	MB (400 ppm)	Solar lamp (Xe lamp)—150 min	98%
Cesconeto et al., 2018 [43]	• Pure TiO$_2$, TiO$_2$/Ca • Powder (0.1 g/L)	MB (1.25 × 10^{-3} M)	254 or 312 or 365 nm—40 min	Only kinetic constants given
Chung et al., 2016 [134]	• Pure TiO$_2$ • Powder (0.1 g/L)	Dye reactive orange 16 (RO16) (25 ppm)	UV—120 min	100%

Table 3. *Cont.*

Paper	Photocatalyst and Shape (Concentration)	Pollutant (Concentration)	Illumination and Time	Best Degradation Results
Haque et al., 2017 [49]	• Pure TiO_2 • Powder (0.5 g/L)	MB and MO (–)	Visible—120 min	70%
Ibrahim et al., 2010 [52]	• Pure TiO_2 • Powder (0.1 g)	MO (30 ppm)	UV—5 h	100%
Kattoor et al., 2014 [114]	• Pure TiO_2 • Powder (0.03 g)	MB (10^{-5} M)	UV-A—100 min	85%
Khan et al., 2017 [129]	• Pure TiO_2 • Powder (0.063 g/L)	PNP (0.02 g/L)	254 nm—30 min	65%
Ma et al., 2012 [117]	• Pure TiO_2, TiO_2/Eu • Powder (1 g/L)	Salicylic acid (50 mg/L)	Visible (>420 nm)—300 min	88%
Mahmoud et al., 2018 [34]	• Pure TiO_2 • Powder (1 g/L)	• MB (10 ppm) • PNP • CV	UV—120 min	100%
Mao et al., 2005 [130]	• Pure TiO_2 • Powder (0.3 g/L)	X3B (30 mg/L)	UV—40 min	100%
Maver et al., 2009 [67]	• Pure TiO_2, TiO_2/Zr • Film on glass and silicon	PlasmocorinthB (40 mg/L)	UV-A—3000 s	70%
Molea et al., 2014 [105]	• Pure TiO_2 • Powder (0.1 g/L)	MB (2.75×10^{-3} g/L)	300–400 nm + 400–700 nm—300 min	47%
Mutuma et al., 2015 [70]	• Pure TiO_2 • Powder (0.6 g/L)	MB (32 mg/L)	UV—70 min	95%
Periyat et al., 2015 [73]	• Pure TiO_2, TiO_2/Fe • Powder (1.2 g/L)	R6G (5×10^{-6} M)	420–800 nm—20 min	100%
Qiu et al., 2007 [75]	• Pure TiO_2, TiO_2/N • Powder (11 mg/L)	MB (–)	Visible (>400 nm)—350 min	85%
Quintero et al., 2020 [76]	• Pure TiO_2 • Powder (1 g/L)	MB (5 ppm)	365 nm—250 min	90%

Table 3. Cont.

Paper	Photocatalyst and Shape (Concentration)	Pollutant (Concentration)	Illumination and Time	Best Degradation Results
Ropero-Vega et al., 2019 [77]	• Pure TiO$_2$, TiO$_2$/Bi2O3 • Film on glass	Salicylic acid (0.1 mM)	UV-Visible (325–650 nm) —1 h	10%
Su et al., 2004 [98]	• Pure TiO$_2$ • Powder (–)	Salicylic acid (4×10^{-4} M)	254 nm—250 min	65%
Tobaldi et al., 2014 [85]	• Pure TiO$_2$ • Powder (0.25 g/L) • Film on petri dishes	MB (liquid phase—5 mg/L)NO$_x$ (gas phases—0.5 ppmv)	Solar light—7 hSolar light—40 min	100%60%
Xie et al., 2005 [111]	• Pure TiO$_2$, TiO$_2$/Nd • Powder (1 g/L)	X3B (100 mg/L)	365 nm + 400–800 nm—120 min400–800 nm—120 min	95%35%
Yamazaki et al., 2001 [89]	• Pure TiO$_2$ • Powder (0.2 g)	Ethylene (gas phase 160 ppmv)	4W fluorescence black light bulbs—2 h	100%
Yu et al., 2003 [91]	• Pure TiO$_2$ • Film on petri dishes (0.3 g)	Acetone (gas phase—400 ppm)	365 nm—60 min	Only kinetic constants given

6.3. Photoactivity of Aqueous TiO$_2$ after Hydrothermal Treatment

Table 4 summarizes the parameters of the photocatalytic experiments for the studies using aqueous TiO$_2$ materials after a hydrothermal treatment. As for the calcined TiO$_2$ materials, the photoactivity does not seem to be improved compared to the as-synthesized materials (compare Table 4 vs. Table 2). In terms of photoactivity, it can be deduced that a thermal treatment (calcination or hydrothermal) is not necessary to obtain an efficient photocatalyst with this type of synthesis method. Indeed, before thermal treatment, crystalline materials are already present with a high specific surface area. The thermal treatment increases the crystallite size and allows a 100% crystalline material to be obtained, but reduces the specific surface area, hence it is not advantageous because photocatalysis occurs at the surface.

One study [151] tested the photo efficiency of their catalysts on real wastewater, where multiple pollutants were present as pharmaceutical products, pesticides, and various organic chemicals. This study showed the effectiveness of the TiO$_2$ photocatalysts for the degradation of these molecules.

Table 4. Parameters of photocatalytic experiments for studies using aqueous TiO$_2$ materials after hydrothermal treatment.

Paper	Photocatalyst and Shape (Concentration)	Pollutant (Concentration)	Illumination and Time	Best Degradation Results
Fallet et al., 2006 [150]	• Pure TiO$_2$ • Film on Si wafer	Malic acid (3.7 × 10^{-4} M)	UV (>340 nm)—3 h	90%
Jiang et al., 2011 [128]	• Pure TiO$_2$ • Powder (1 g/L)	MO (10 mg/L)	Visible (>400 nm)—100 min	35%
Kaplan et al., 2016 [54]	• Pure TiO$_2$ • Powder (0–125 mg/L)	Bisphenol A (BPA) (10 mg/L)	365 nm—60 min	100%
Liu et al., 2014 [116]	• Pure TiO$_2$, TiO$_2$/N • Film on glass	HCHO (gas phase—0.32 mg/m^3)	Visible ()—24 h	95%
Mahata et al., 2012 [59]	• Pure TiO$_2$ • Powder (–)	MO (–)	UV Visible—120 min	85%
Saif et al., 2012 [151]	• Pure TiO$_2$ • Powder (–)	Real wastewater	Solar light—3 h	57% mineralization
Xie et al., 2003 [109]	• Pure TiO$_2$, TiO$_2$/Ce • Powder (1 g/L)	X3B (100 mg/L)	400–800 nm—120 min	95%

7. Addition Features for Aqueous Sol–Gel TiO$_2$

Some other studies used colloidal aqueous TiO$_2$ materials in applications other than photocatalytic pollutant degradation. All these applications used the other properties of titania, such as its hydrophilicity, its high refractive index, or its semi-conducting property. In Alcober et al. [123], aqueous TiO$_2$ material is utilized to produce photochromic coatings with tungsten doping. In Antonello et al. [139], high refractive index coatings are produced from aqueous TiO$_2$ suspensions. In Bugakova et al. [94], TiO$_2$ inks, based on aqueous TiO$_2$ colloids, are used for applications derived from the inkjet printing of microstructures for electronic devices. In Haq et al. [48] and Lin et al. [153], aqueous TiO$_2$ suspensions give adsorbent materials for heavy metals and dye adsorption. Indeed, as aqueous synthesis of TiO$_2$ suspensions produces TiO$_2$ nanoparticles, the specific surface area of these materials is high compared to titania obtained by high-temperature synthesis. In Hore et al. [50] and Kashyout et al. [55], aqueous TiO$_2$ materials are used in solar cell fabrication. In Papiya et al. [72], a cathode catalyst for microbial fuel cells is produced with aqueous TiO$_2$ materials. In Salahuddin et al. [79], aqueous TiO$_2$ is mixed with PLA to design a nanocomposite system for Norfloxacin drug delivery. Hydrophilic surfaces are also produced with aqueous TiO$_2$ [62,138]. The use of photocatalyst materials such as aqueous TiO$_2$ can be also implemented in energy related fields, such as the production of H$_2$ by photocatalyzed decomposition of water [154]. The possibility of integrating heterogeneous photocatalysis with electrochemical processes to exploit their synergistic actions can be also envisaged [155]. Numerous further studies can be imagined to explore fully the properties of this green TiO$_2$ synthesis pathway.

8. Conclusions and Outlook

The aim of this review was to establish the state of the art of the research in the area of the little known eco-friendly process of producing TiO_2 via colloidal aqueous sol–gel synthesis, resulting in a crystalline material without a calcination step. From 1987 to 2020, about 115 articles were found dealing with colloidal aqueous sol–gel TiO_2 preparation, taking into account three types of aqueous TiO_2: the as-synthesized type obtained directly after synthesis, without any specific treatment; the calcined, obtained after a subsequent calcination step; and the hydrothermal, obtained after this specific autoclave treatment.

This eco-friendly process is based on the hydrolysis of a Ti precursor in excess of water, followed by the peptization of the precipitated TiO_2. Compared to classical TiO_2 synthesis, this colloidal aqueous sol–gel method results in crystalline TiO_2 nanoparticles without a thermal treatment, and it is a green synthesis method because it uses small amounts of chemicals, water as a solvent, and a low temperature for crystallization. Moreover, some works have shown that this synthesis method can be easily upscaled to 20 L.

Depending on the synthesis parameters, the three crystalline phases of TiO_2 (anatase, brookite, rutile) can be obtained. The morphology of the nanoparticles can also be tailored by the synthesis parameters. The most important parameter is the peptizing agent. Indeed, depending on its acidic or basic character and also on its amount, it can modulate the crystallinity, and so, the morphology of the material. HNO_3 seems to be the most versatile PA. Indeed, it allows obtaining the three different phases of TiO_2 and the corresponding morphologies (nanosphere or nanorod) just by changing its quantity during the synthesis.

The exact mechanism of the TiO_2 material formation and the exact influence of the PA on the resulting TiO_2 materials needs deeper studies, to understand clearly the formation of the different crystalline phases and morphologies. For example, the use of in-situ XRD or FTIR to probe the exact formation mechanism of PA-assisted sol–gel synthesis of TiO_2 could be a path to explore. Moreover, machine learning and big data analysis will open a new avenue in this TiO_2 material research. Indeed, they could help to find a correlation between the many different experimental parameters and their ability to produce highly crystalline TiO_2.

Even if crystalline TiO_2 materials are obtained after aqueous sol–gel synthesis, some studies apply a thermal post-treatment, calcination, or hydrothermal to further crystallize the materials. These treatments can also increase the crystallite size of the as-synthesized material and modify its morphology. Moreover, the surface area will decrease during the calcination due to particle growth with the phase change. Furthermore, the increase in the calcination temperature causes the particles to coalesce, creating tightly connected agglomerates, blocking the entry of N_2 gas during the BET analysis.

The aqueous TiO_2 photocatalysts are mainly used in various photocatalytic reactions for organic pollutant degradation. More than 20 different molecules have been reported to be degraded with these materials, but mainly model pollutants. Experiments on real wastewater are lacking in the literature for this type of material. The numerous experimental conditions make it difficult to compare the performance of catalysts. Nevertheless, the as-synthesized materials seem to have an equivalent photocatalytic efficiency to the photocatalysts post-treated with thermal treatments. Indeed, as-prepared, the TiO_2 photocatalysts are crystalline and present a high specific surface area. Thermal treatments do not seem to be necessary from a photocatalytic point of view. Moreover, studies showed that aqueous TiO_2 presents better photoactivity than commercial Evonik Aeroxide P25, which is produced by high-temperature process.

Emerging applications are also referenced, such as elaborating catalysts for fuel cells, nanocomposite drug delivery systems, or the inkjet printing of microstructures. As the development of alternative energy sources is very prominent in current research activities, the use of this kind of photocatalyst to produce H_2 from the photocatalyzed decomposition of water also seems a promising path to explore. Moreover, the development of electrophotocatalytic devices for various applications, in water pollution treatment for example, will be realized in the next few years. However, only a few works have explored these other properties, giving a lot of potential avenues for studying this eco-friendly TiO_2 synthesis method for innovative implementations.

Author Contributions: Writing—original draft preparation, J.G.M., L.L., T.H., S.D.L., R.H.M.M., C.-A.F. and S.H.; writing—review and editing, J.G.M., L.L., T.H., S.D.L., R.H.M.M., C.-A.F. and S.H. All authors have read and agreed to the published version of the manuscript.

Funding: This research was funded by INNOVIRIS Brussels (Institute for Research and Innovation) through the Bridge project platform—as part of COLORES project.

Data Availability Statement: All data were taken from the articles of the bibliography section.

Acknowledgments: S.D.L. and S.H. are grateful to F.R.S.-F.N.R.S. for their Senior Research Associate position. J.G.M., R.H.M.M., C.A.F. and S.H. also thank INNOVIRIS Brussels for financial support through the Bridge project—COLORES.

Conflicts of Interest: The authors declare no conflict of interest.

References

1. Oturan, M.A.; Aaron, J.-J. Advanced Oxidation Processes in Water/Wastewater Treatment: Principles and Applications. A Review. *Crit. Rev. Environ. Sci. Technol.* **2014**, *44*, 2577–2641. [CrossRef]
2. Nakata, K.; Fujishima, A. TiO_2 photocatalysis: Design and applications. *J. Photochem. Photobiol. C Photochem. Rev.* **2012**, *13*, 169–189. [CrossRef]
3. Hermawan, A.; Hanindriyo, A.T.; Ramadhan, E.R.; Asakura, Y.; Hasegawa, T.; Hongo, K.; Inada, M.; Maezono, R.; Yin, S. Octahedral morphology of NiO with (111) facet synthesized from the transformation of NiOHCl for the NOx detection and degradation: Experiment and DFT calculation. *Inorg. Chem. Front.* **2020**, *7*, 3431–3442. [CrossRef]
4. Ong, C.B.; Ng, L.Y.; Mohammad, A.W. A review of ZnO nanoparticles as solar photocatalysts: Synthesis, mechanisms and applications. *Renew. Sustain. Energy Rev.* **2018**, *81*, 536–551. [CrossRef]
5. Ma, R.; Zhang, S.; Wen, T.; Gu, P.; Li, L.; Zhao, G.; Niu, F.; Huang, Q.; Tang, Z.; Wang, X. A critical review on visible-light-response CeO2-based photocatalysts with enhanced photooxidation of organic pollutants. *Catal. Today* **2019**, *335*, 20–30. [CrossRef]
6. Chiam, S.-L.; Pung, S.-Y.; Yeoh, F.-Y. Recent developments in MnO2-based photocatalysts for organic dye removal: A review. *Environ. Sci. Pollut. Res.* **2020**, *27*, 5759–5778. [CrossRef] [PubMed]
7. Pelaez, M.; Nolan, N.T.; Pillai, S.C.; Seery, M.; Falaras, P.; Kontos, A.G.; Dunlop, P.S.; Hamilton, J.W.; Byrne, J.; O'Shea, K.; et al. A review on the visible light active titanium dioxide photocatalysts for environmental applications. *Appl. Catal. B Environ.* **2012**, *125*, 331–349. [CrossRef]
8. Malengreaux, C.M.; Douven, S.; Poelman, D.; Heinrichs, B.; Bartlett, J.R. An ambient temperature aqueous sol–gel processing of efficient nanocrystalline doped TiO_2-based photocatalysts for the degradation of organic pollutants. *J. Sol Gel Sci. Technol.* **2014**, *71*, 557–570. [CrossRef]
9. Espino-Estévez, M.; Fernández-Rodríguez, C.; González-Díaz, O.M.; Araña, J.; Espinós, J.; Ortega-Méndez, J.; Doña-Rodríguez, J.M. Effect of TiO_2–Pd and TiO_2–Ag on the photocatalytic oxidation of diclofenac, isoproturon and phenol. *Chem. Eng. J.* **2016**, *298*, 82–95. [CrossRef]
10. Vaiano, V.; Iervolino, G.; Sannino, D.; Murcia, J.J.; Hidalgo, M.C.; Ciambelli, P.; Navío, J.A. Photocatalytic removal of patent blue V dye on Au-TiO_2 and Pt-TiO_2 catalysts. *Appl. Catal. B Environ.* **2016**, *188*, 134–146. [CrossRef]
11. Di Paola, A.; Marci, G.; Palmisano, L.; Schiavello, M.; Uosaki, K.; Ikeda, A.S.; Ohtani, B. Preparation of Polycrystalline TiO_2 Photocatalysts Impregnated with Various Transition Metal Ions: Characterization and Photocatalytic Activity for the Degradation of 4-Nitrophenol. *J. Phys. Chem. B* **2002**, *106*, 637–645. [CrossRef]
12. Rauf, M.; Meetani, M.; Hisaindee, S. An overview on the photocatalytic degradation of azo dyes in the presence of TiO_2 doped with selective transition metals. *Desalination* **2011**, *276*, 13–27. [CrossRef]
13. Bodson, C.J.; Heinrichs, B.; Tasseroul, L.; Bied, C.; Mahy, J.G.; Man, M.W.C.; Lambert, S.D. Efficient P- and Ag-doped titania for the photocatalytic degradation of waste water organic pollutants. *J. Alloys Compd.* **2016**, *682*, 144–153. [CrossRef]
14. Di Valentin, C.; Pacchioni, G.; Selloni, A.; Livraghi, S.; Giamello, E. Characterization of Paramagnetic Species in N-Doped TiO_2 Powders by EPR Spectroscopy and DFT Calculations. *J. Phys. Chem. B* **2005**, *109*, 11414–11419. [CrossRef]
15. Gilma, G.O.; Carlos, A.P.M.; Fernando, M.O.; Edgar, A.P.-M. Photocatalytic degradation of phenol on TiO_2 and TiO_2/Pt sensitized with metallophthalocyanines. *Catal. Today* **2005**, *107–108*, 589–594. [CrossRef]
16. Mahy, J.G.; Paez, C.A.; Carcel, C.; Bied, C.; Tatton, A.S.; Damblon, C.; Heinrichs, B.; Man, M.W.C.; Lambert, S.D. Porphyrin-based hybrid silica-titania as a visible-light photocatalyst. *J. Photochem. Photobiol. A Chem.* **2019**, *373*, 66–76. [CrossRef]
17. Xie, H.; Gao, G.; Tian, Z.; Bing, N.; Wang, L. Synthesis of TiO_2 nanoparticles by propane/air turbulent flame CVD process. *Particuology* **2009**, *7*, 204–210. [CrossRef]
18. Djenadic, R.; Winterer, M. Chemical Vapor Synthesis of Nanocrystalline Oxides. In *2D Nanoelectronics*; Springer Science and Business Media LLC: Berlin/Heidelberg, Germany, 2012; pp. 49–76.

19. Inturi, S.N.R.; Boningari, T.; Suidan, M.; Smirniotis, P.G. Flame Aerosol Synthesized Cr Incorporated TiO$_2$ for Visible Light Photodegradation of Gas Phase Acetonitrile. *J. Phys. Chem. C* **2013**, *118*, 231–242. [CrossRef]
20. Dar, M.I.; Chandiran, A.K.; Graetzel, M.; Nazeeruddin, M.K.; Shivashankar, S.A. Controlled synthesis of TiO$_2$ nanoparticles and nanospheres using a microwave assisted approach for their application in dye-sensitized solar cells. *J. Mater. Chem. A* **2013**, *2*, 1662–1667. [CrossRef]
21. Zhang, D.; Qi, L.; Ma, J.; Cheng, H. Formation of crystalline nanosized titania in reverse micelles at room temperature. *J. Mater. Chem.* **2002**, *12*, 3677–3680. [CrossRef]
22. Nian, J.-N.; Teng, H. Hydrothermal Synthesis of Single-Crystalline Anatase TiO$_2$ Nanorods with Nanotubes as the Precursor. *J. Phys. Chem. B* **2006**, *110*, 4193–4198. [CrossRef]
23. Simon, P.; Pignon, B.; Miao, B.; Coste-Leconte, S.; Leconte, Y.; Marguet, S.; Jegou, P.; Bouchet-Fabre, B.; Reynaud, C.; Herlin-Boime, N. N-Doped Titanium Monoxide Nanoparticles with TiO Rock-Salt Structure, Low Energy Band Gap, and Visible Light Activity. *Chem. Mater.* **2010**, *22*, 3704–3711. [CrossRef]
24. Gratzel, M. Sol-Gel Processed TiO$_2$ Films for Photovoltaic Applications. *J. Sol Gel Sci. Technol.* **2001**, *22*, 7–13. [CrossRef]
25. Carp, O. Photoinduced reactivity of titanium dioxide. *Prog. Solid State Chem.* **2004**, *32*, 33–177. [CrossRef]
26. Huang, T.; Huang, W.; Zhou, C.; Situ, Y.; Huang, H. Superhydrophilicity of TiO$_2$/SiO$_2$ thin films: Synergistic effect of SiO$_2$ and phase-separation-induced porous structure. *Surf. Coat. Technol.* **2012**, *213*, 126–132. [CrossRef]
27. Guan, K. Relationship between photocatalytic activity, hydrophilicity and self-cleaning effect of TiO$_2$/SiO$_2$ films. *Surf. Coat. Technol.* **2005**, *191*, 155–160. [CrossRef]
28. Antonelli, D.M.; Ying, J. Synthesis of Hexagonally Packed Mesoporous TiO$_2$ by a Modified Sol–Gel Method. *Angew. Chem. Int. Ed.* **1995**, *34*, 2014–2017. [CrossRef]
29. Braconnier, B.; Páez, C.A.; Lambert, S.; Alié, C.; Henrist, C.; Poelman, D.; Pirard, J.-P.; Cloots, R.; Heinrichs, B. Ag- and SiO$_2$-doped porous TiO$_2$ with enhanced thermal stability. *Microporous Mesoporous Mater.* **2009**, *122*, 247–254. [CrossRef]
30. Anderson, C.; Bard, A.J. An Improved Photocatalyst of TiO$_2$/SiO$_2$ Prepared by a Sol-Gel Synthesis. *J. Phys. Chem.* **1995**, *99*, 9882–9885. [CrossRef]
31. Brinker, G.W.; Jeffrey, C.S. Sol-gel science. In *The Physics and Chemistry of Sol-Gel Processing*; Academic Press: Cambridge, MA, USA, 2013.
32. Schubert, U. Chemical modification of titanium alkoxides for sol–gel processing. *J. Mater. Chem.* **2005**, *15*, 3701–3715. [CrossRef]
33. Jan, W.G. Encyclopedic Dictionary of Polymers. *Encycl. Dict. Polym.* **2011**. [CrossRef]
34. Mahmoud, H.A.; Narasimharao, K.; Ali, T.T.; Khalil, K.M.S. Acidic Peptizing Agent Effect on Anatase-Rutile Ratio and Photocatalytic Performance of TiO$_2$ Nanoparticles. *Nanoscale Res. Lett.* **2018**, *13*, 48. [CrossRef] [PubMed]
35. Yamanaka, S.; Nishihara, T.; Hattori, M.; Suzuki, Y. Preparation and properties of titania pillared clay. *Mater. Chem. Phys.* **1987**, *17*, 87–101. [CrossRef]
36. Anderson, M.A.; Gieselmann, M.J.; Xu, Q. Titania and alumina ceramic membranes. *J. Membr. Sci.* **1988**, *39*, 243–258. [CrossRef]
37. Doeuff, S.; Henry, M.; Sanchez, C.; Livage, J. Hydrolysis of titanium alkoxides: Modification of the molecular precursor by acetic acid. *J. Non Cryst. Solids* **1987**, *89*, 206–216. [CrossRef]
38. Mahshid, S.; Askari, M.; Ghamsari, M.S. Synthesis of TiO$_2$ nanoparticles by hydrolysis and peptization of titanium isopropoxide solution. *J. Mater. Process. Technol.* **2007**, *189*, 296–300. [CrossRef]
39. Bischoff, B.L.; Anderson, M.A. Peptization Process in the Sol-Gel Preparation of Porous Anatase (TiO$_2$). *Chem. Mater.* **1995**, *7*, 1772–1778. [CrossRef]
40. Matijevic, E. Monodispersed metal (hydrous) oxides—A fascinating field of colloid science. *Acc. Chem. Res.* **1981**, *14*, 22–29. [CrossRef]
41. Mahy, J.G.; Deschamps, F.; Collard, V.; Jérôme, C.; Bartlett, J.; Lambert, S.D.; Heinrichs, B. Acid acting as redispersing agent to form stable colloids from photoactive crystalline aqueous sol–gel TiO$_2$ powder. *J. Sol Gel Sci. Technol.* **2018**, *87*, 568–583. [CrossRef]
42. Douven, S.; Mahy, J.G.; Wolfs, C.; Reyserhove, C.; Poelman, D.; Devred, F.; Gaigneaux, E.M.; Lambert, S.D. Efficient N, Fe Co-Doped TiO$_2$ Active under Cost-Effective Visible LED Light: From Powders to Films. *Catalysts* **2020**, *10*, 547. [CrossRef]
43. Cesconeto, F.R.; Borlaf, M.; Nieto, M.I.; de Oliveira, A.P.N.; Moreno, R. Synthesis of CaTiO3 and CaTiO3/TiO$_2$ nanoparticulate compounds through Ca^{2+}/TiO$_2$ colloidal sols: Structural and photocatalytic characterization. *Ceram. Int.* **2018**, *44*, 301–309. [CrossRef]
44. Cano-Franco, J.C.; Álvarez-Láinez, M. Effect of CeO2 content in morphology and optoelectronic properties of TiO$_2$-CeO2 nanoparticles in visible light organic degradation. *Mater. Sci. Semicond. Process.* **2019**, *90*, 190–197. [CrossRef]
45. Colomer, M.T.; Guzmán, J.; Moreno, R. Determination of Peptization Time of Particulate Sols Using Optical Techniques: Titania As a Case Study. *Chem. Mater.* **2008**, *20*, 4161–4165. [CrossRef]
46. Colomer, M.T.; Guzmán, J.; Moreno, R. Peptization of Nanoparticulate Titania Sols Prepared Under Different Water−Alkoxide Molar Ratios. *J. Am. Ceram. Soc.* **2009**, *93*, 59–64. [CrossRef]
47. Ghamsari, M.S.; Gaeeni, M.R.; Han, W.; Park, H.-H. Highly stable colloidal TiO$_2$ nanocrystals with strong violet-blue emission. *J. Lumin.* **2016**, *178*, 89–93. [CrossRef]
48. Haq, S.; Rehman, W.; Waseem, M. Adsorption Efficiency of Anatase TiO$_2$ Nanoparticles Against Cadmium Ions. *J. Inorg. Organomet. Polym. Mater.* **2018**, *29*, 651–658. [CrossRef]
49. Haque, F.Z.; Nandanwar, R.; Singh, P. Evaluating photodegradation properties of anatase and rutile TiO$_2$ nanoparticles for organic compounds. *Optik* **2017**, *128*, 191–200. [CrossRef]

50. Hore, S.; Palomares, E.; Smit, H.; Bakker, N.J.; Comte, P.; Liska, P.; Thampi, K.R.; Kroon, J.M.; Hinsch, A.; Durrant, J.R. Acid versus base peptization of mesoporous nanocrystalline TiO$_2$ films: Functional studies in dye sensitized solar cells. *J. Mater. Chem.* **2004**, *15*, 412–418. [CrossRef]
51. Huang, B.-S.; Tseng, H.-H.; Su, E.-C.; Chiu, I.-C.; Wey, M.-Y. Characterization and photoactivity of Pt/N-doped TiO$_2$ synthesized through a sol–gel process at room temperature. *J. Nanoparticle Res.* **2015**, *17*, 282. [CrossRef]
52. Ibrahim, S.A.; Sreekantan, S. Effect of pH on TiO$_2$ Nanoparticles via Sol-Gel Method. *Adv. Mater. Res.* **2010**, *173*, 184–189. [CrossRef]
53. Khalil, K.M.; El-Khatib, R.M.; Ali, T.T.; Mahmoud, H.A.; Elsamahy, A.A. Titania nanoparticles by acidic peptization of xerogel formed by hydrolysis of titanium(IV) isopropoxide under atmospheric humidity conditions. *Powder Technol.* **2013**, *245*, 156–162. [CrossRef]
54. Kaplan, R.; Erjavec, B.; Dražić, G.; Grdadolnik, J.; Pintar, A. Simple synthesis of anatase/rutile/brookite TiO$_2$ nanocomposite with superior mineralization potential for photocatalytic degradation of water pollutants. *Appl. Catal. B Environ.* **2016**, *181*, 465–474. [CrossRef]
55. Kashyout, A.; Soliman, M.; Fathy, M. Effect of preparation parameters on the properties of TiO$_2$ nanoparticles for dye sensitized solar cells. *Renew. Energy* **2010**, *35*, 2914–2920. [CrossRef]
56. Léonard, G.L.-M.; Remy, S.; Heinrichs, B. Doping TiO$_2$ films with carbon nanotubes to simultaneously optimise antistatic, photocatalytic and superhydrophilic properties. *J. Sol Gel Sci. Technol.* **2016**, *79*, 413–425. [CrossRef]
57. Leyva-Porras, C.; Toxqui-Teran, A.; Vega-Becerra, O.; Miki-Yoshida, M.; Rojas-Villalobos, M.; García-Guaderrama, M.; Aguilar-Martínez, J. Low-temperature synthesis and characterization of anatase TiO$_2$ nanoparticles by an acid assisted sol–gel method. *J. Alloys Compd.* **2015**, *647*, 627–636. [CrossRef]
58. Lim, C.S. Effect of pH on the Microstructural Morphology and Phase Transformation of TiO$_2$ Nanopowders Prepared by Sol-Gel Method. *Asian J. Chem.* **2014**, *26*, 1843–1847. [CrossRef]
59. Mahata, S.; Mahato, S.S.; Nandi, M.M.; Mondal, B. Synthesis of TiO[sub 2] nanoparticles by hydrolysis and peptization of titanium isopropoxide solution. *AIP Conf. Proc.* **2011**, *1461*, 225–228. [CrossRef]
60. Mahshid, S.; Askari, M.; Ghamsari, M.S.; Afshar, N.; Lahuti, S. Mixed-phase TiO$_2$ nanoparticles preparation using sol–gel method. *J. Alloys Compd.* **2009**, *478*, 586–589. [CrossRef]
61. Mahy, J.G.; Lambert, S.D.; Léonard, G.L.-M.; Zubiaur, A.; Olu, P.-Y.; Mahmoud, A.; Boschini, F.; Heinrichs, B. Towards a large scale aqueous sol-gel synthesis of doped TiO$_2$: Study of various metallic dopings for the photocatalytic degradation of p-nitrophenol. *J. Photochem. Photobiol. A Chem.* **2016**, *329*, 189–202. [CrossRef]
62. Mahy, J.G.; Léonard, G.L.-M.; Pirard, S.; Wicky, D.; Daniel, A.; Archambeau, C.; Liquet, D.; Heinrichs, B. Aqueous sol–gel synthesis and film deposition methods for the large-scale manufacture of coated steel with self-cleaning properties. *J. Sol Gel Sci. Technol.* **2017**, *81*, 27–35. [CrossRef]
63. Mahy, J.G.; Cerfontaine, V.; Poelman, D.; Devred, F.; Gaigneaux, E.M.; Heinrichs, B.; Lambert, S.D. Highly Efficient Low-Temperature N-Doped TiO$_2$ Catalysts for Visible Light Photocatalytic Applications. *Materials* **2018**, *11*, 584. [CrossRef]
64. Mahy, J.G.; Tilkin, R.G.; Douven, S.; Lambert, S.D. TiO$_2$ nanocrystallites photocatalysts modified with metallic species: Comparison between Cu and Pt doping. *Surf. Interfaces* **2019**, *17*, 100366. [CrossRef]
65. Mahy, J.G.; Lambert, S.D.; Tilkin, R.G.; Wolfs, C.; Poelman, D.; Devred, F.; Gaigneaux, E.M.; Douven, S. Ambient temperature ZrO2-doped TiO$_2$ crystalline photocatalysts: Highly efficient powders and films for water depollution. *Mater. Today Energy* **2019**, *13*, 312–322. [CrossRef]
66. Malengreaux, C.M.; Pirard, S.L.; Léonard, G.; Mahy, J.G.; Herlitschke, M.; Klobes, B.; Hermann, R.; Heinrichs, B.; Bartlett, J.R. Study of the photocatalytic activity of Fe3+, Cr3+, La3+ and Eu3+ single-doped and co-doped TiO$_2$ catalysts produced by aqueous sol-gel processing. *J. Alloys Compd.* **2017**, *691*, 726–738. [CrossRef]
67. Maver, K.; Štangar, U.L.; Černigoj, U.; Gross, S.; Korošec, R.C. Low-temperature synthesis and characterization of TiO$_2$ and TiO$_2$–ZrO2 photocatalytically active thin films. *Photochem. Photobiol. Sci.* **2009**, *8*, 657–662. [CrossRef]
68. Mohammadi, M.; Fray, D.; Mohammadi, A. Sol–gel nanostructured titanium dioxide: Controlling the crystal structure, crystallite size, phase transformation, packing and ordering. *Microporous Mesoporous Mater.* **2008**, *112*, 392–402. [CrossRef]
69. Mohammadi, M.R.; Cordero-Cabrera, M.C.; Ghorbani, M.; Fray, D.J. Synthesis of high surface area nanocrystalline anatase-TiO$_2$ powders derived from particulate sol-gel route by tailoring processing parameters. *J. Sol Gel Sci. Technol.* **2006**, *40*, 15–23. [CrossRef]
70. Mutuma, B.K.; Shao, G.; Kim, W.D.; Kim, H.T. Sol–gel synthesis of mesoporous anatase–brookite and anatase–brookite–rutile TiO$_2$ nanoparticles and their photocatalytic properties. *J. Colloid Interface Sci.* **2015**, *442*, 1–7. [CrossRef]
71. Okunaka, S.; Tokudome, H.; Hitomi, Y.; Abe, R. Facile preparation of stable aqueous titania sols for fabrication of highly active TiO 2 photocatalyst films. *J. Mater. Chem. A* **2014**, *3*, 1688–1695. [CrossRef]
72. Papiya, F.; Pattanayak, P.; Kumar, V.; Das, S.; Kundu, P.P. Sulfonated graphene oxide and titanium dioxide coated with nanostructured polyaniline nanocomposites as an efficient cathode catalyst in microbial fuel cells. *Mater. Sci. Eng. C* **2020**, *108*, 110498. [CrossRef]
73. Periyat, P.; Saeed, P.; Ullattil, S. Anatase titania nanorods by pseudo-inorganic templating. *Mater. Sci. Semicond. Process.* **2015**, *31*, 658–665. [CrossRef]

74. Qi, K.; Xin, J.H. Room-Temperature Synthesis of Single-Phase Anatase TiO$_2$ by Aging and its Self-Cleaning Properties. *ACS Appl. Mater. Interfaces* **2010**, *2*, 3479–3485. [CrossRef]
75. Qiu, X.; Zhao, Y.; Burda, C. Synthesis and Characterization of Nitrogen-Doped Group IVB Visible-Light-Photoactive Metal Oxide Nanoparticles. *Adv. Mater.* **2007**, *19*, 3995–3999. [CrossRef]
76. Quintero, Y.; Mosquera, E.; Diosa, J.; García, A. Ultrasonic-assisted sol–gel synthesis of TiO$_2$ nanostructures: Influence of synthesis parameters on morphology, crystallinity, and photocatalytic performance. *J. Sol Gel Sci. Technol.* **2020**, *94*, 477–485. [CrossRef]
77. Ropero-Vega, J.L.; Candal, R.J.; Pedraza-Avella, J.A.; Niño-Gómez, M.E.; Bilmes, S.A. Enhanced visible light photoelectrochemical performance of β-Bi2O3-TiO$_2$/ITO thin films prepared by aqueous sol-gel. *J. Solid State Electrochem.* **2019**, *23*, 1757–1765. [CrossRef]
78. Ryu, D.H.; Kim, S.C.; Koo, S.M.; Kim, D.P. Deposition of Titania Nanoparticles on Spherical Silica. *J. Sol Gel Sci. Technol.* **2003**, *26*, 489–493. [CrossRef]
79. Salahuddin, N.; Abdelwahab, M.; Gaber, M.; Elneanaey, S. Synthesis and Design of Norfloxacin drug delivery system based on PLA/TiO$_2$ nanocomposites: Antibacterial and antitumor activities. *Mater. Sci. Eng. C* **2020**, *108*, 110337. [CrossRef]
80. Ghamsari, M.S.; Radiman, S.; Hamid, M.A.A.; Mahshid, S.; Rahmani, S. Room temperature synthesis of highly crystalline TiO$_2$ nanoparticles. *Mater. Lett.* **2013**, *92*, 287–290. [CrossRef]
81. Shinozaki, K.; Zack, J.W.; Richards, R.M.; Pivovar, B.S.; Kocha, S.S. Oxygen Reduction Reaction Measurements on Platinum Electrocatalysts Utilizing Rotating Disk Electrode Technique. *J. Electrochem. Soc.* **2015**, *162*, F1144–F1158. [CrossRef]
82. Shin, H.; Jung, H.S.; Hong, K.S.; Lee, J.-K. Crystallization Process of TiO$_2$ Nanoparticles in an Acidic Solution. *Chem. Lett.* **2004**, *33*, 1382–1383. [CrossRef]
83. Sugimoto, T.; Zhou, X.; Muramatsu, A. Synthesis of uniform anatase TiO$_2$ nanoparticles by gel–sol method 4. Shape control. *J. Colloid Interface Sci.* **2003**, *259*, 53–61. [CrossRef]
84. Sung-Suh, H.M.; Choi, J.R.; Hah, H.J.; Koo, S.M.; Bae, Y.C. Comparison of Ag deposition effects on the photocatalytic activity of nanoparticulate TiO$_2$ under visible and UV light irradiation. *J. Photochem. Photobiol. A Chem.* **2004**, *163*, 37–44. [CrossRef]
85. Tobaldi, D.M.; Pullar, R.; Binions, R.; Jorge, A.B.; McMillan, P.F.; Saeli, M.; Seabra, M.P.; Labrincha, J.A. Influence of sol counter-ions on the visible light induced photocatalytic behaviour of TiO$_2$ nanoparticles. *Catal. Sci. Technol.* **2014**, *4*, 2134–2146. [CrossRef]
86. Uchiyama, H.; Bando, T.; Kozuka, H. Effect of the amount of H2O and HNO3 in Ti(OC3H7)4 solutions on the crystallization of sol-gel-derived TiO$_2$ films. *Thin Solid Film* **2019**, *669*, 157–161. [CrossRef]
87. Vinogradov, A.V.; Vinogradov, V.V. Effect of Acidic Peptization on Formation of Highly Photoactive TiO$_2$ Films Prepared without Heat Treatment. *J. Am. Ceram. Soc.* **2014**, *97*, 290–294. [CrossRef]
88. Xu, Q.; Anderson, M.A. Synthesis of porosity controlled ceramic membranes. *J. Mater. Res.* **1991**, *6*, 1073–1081. [CrossRef]
89. Yamazaki, S.; Fujinaga, N.; Araki, K. Effect of sulfate ions for sol–gel synthesis of Titania photocatalyst. *Appl. Catal. A Gen.* **2001**, *210*, 97–102. [CrossRef]
90. Yang, J.; Mei, S.; Ferreira, J.M.; Norby, P.; Quaresmâ, S. Fabrication of rutile rod-like particle by hydrothermal method: An insight into HNO3 peptization. *J. Colloid Interface Sci.* **2005**, *283*, 102–106. [CrossRef]
91. Yu, J.; Leung, M.K.-P.; Ho, W.; Cheng, B.; Zhao, X. Effects of acidic and basic hydrolysis catalysts on the photocatalytic activity and microstructures of bimodal mesoporous Titania. *J. Catal.* **2003**, *220*, 69–78. [CrossRef]
92. Yun, Y.J.; Chung, J.S.; Kim, S.; Hahn, S.H.; Kim, E.J. Low-temperature coating of sol–gel anatase thin films. *Mater. Lett.* **2004**, *58*, 3703–3706. [CrossRef]
93. Borlaf, M.; Poveda, J.M.; Moreno, R.; Colomer, M.T. Synthesis and characterization of TiO$_2$/Rh^{3+} nanoparticulate sols, xerogels and cryogels for photocatalytic applications. *J. Sol Gel Sci. Technol.* **2012**, *63*, 408–415. [CrossRef]
94. Bugakova, D.; Slabov, V.; Sergeeva, E.; Zhukov, M.; Vinogradov, A. Comprehensive characterization of TiO$_2$ inks and their application for inkjet printing of microstructures. *Colloids Surf. A Physicochem. Eng. Asp.* **2020**, *586*, 124146. [CrossRef]
95. Chen, X.; Lou, Y.; Samia, A.C.S.; Burda, C.; Gole, J.L. Formation of Oxynitride as the Photocatalytic Enhancing Site in Nitrogen-Doped Titania Nanocatalysts: Comparison to a Commercial Nanopowder. *Adv. Funct. Mater.* **2005**, *15*, 41–49. [CrossRef]
96. Vorkapic, D.; Matsoukas, T. Effect of Temperature and Alcohols in the Preparation of Titania Nanoparticles from Alkoxides. *J. Am. Ceram. Soc.* **2005**, *81*, 2815–2820. [CrossRef]
97. Hu, Y.; Yuan, C. Low-temperature preparation of photocatalytic TiO$_2$ thin films from anatase sols. *J. Cryst. Growth* **2005**, *274*, 563–568. [CrossRef]
98. Su, C.; Hong, B.-Y.; Tseng, C.-M. Sol–gel preparation and photocatalysis of titanium dioxide. *Catal. Today* **2004**, *96*, 119–126. [CrossRef]
99. Wang, J.; Zhao, H.; Liu, X.; Li, X.; Xu, P.; Han, X. Formation of Ag nanoparticles on water-soluble anatase TiO$_2$ clusters and the activation of photocatalysis. *Catal. Commun.* **2009**, *10*, 1052–1056. [CrossRef]
100. Wang, J.; Han, X.; Liu, C.; Zhang, W.; Cai, R.; Liu, Z. Adjusting the Crystal Phase and Morphology of Titania via a Soft Chemical Process. *Cryst. Growth Des.* **2010**, *10*, 2185–2191. [CrossRef]
101. Yang, J.; Mei, S.; Ferreira, J.M. In situ preparation of weakly flocculated aqueous anatase suspensions by a hydrothermal technique. *J. Colloid Interface Sci.* **2003**, *260*, 82–88. [CrossRef]
102. Yang, J.; Mei, S.; Ferreira, J.M.F. Hydrothermal Synthesis of Nanosized Titania Powders: Influence of Peptization and Peptizing Agents on the Crystalline Phases and Phase Transitions. *J. Am. Ceram. Soc.* **2000**, *83*, 1361–1368. [CrossRef]

103. Yang, J.; Mei, S.; Ferreira, J.M.F. Hydrothermal Synthesis of Nanosized Titania Powders: Influence of Tetraalkyl Ammonium Hydroxides on Particle Characteristics. *J. Am. Ceram. Soc.* **2004**, *84*, 1696–1702. [CrossRef]
104. Cassaignon, S.; Koelsch, M.; Jolivet, J.-P. From TiCl3 to TiO_2 nanoparticles (anatase, brookite and rutile): Thermohydrolysis and oxidation in aqueous medium. *J. Phys. Chem. Solids* **2007**, *68*, 695–700. [CrossRef]
105. Molea, A.; Popescu, V.; Rowson, N.; Dinescu, A.M. Influence of pH on the formulation of TiO_2 nano-crystalline powders with high photocatalytic activity. *Powder Technol.* **2014**, *253*, 22–28. [CrossRef]
106. Bazrafshan, H.; Tesieh, Z.A.; Dabirnia, S.; Naderifar, A. Low Temperature Synthesis of TiO 2 Nanoparticles with High Photocatalytic Activity and Photoelectrochemical Properties through Sol-Gel Method. *Mater. Manuf. Process.* **2015**, *31*, 119–125. [CrossRef]
107. Kanna, M.; Wongnawa, S. Mixed amorphous and nanocrystalline TiO_2 powders prepared by sol–gel method: Characterization and photocatalytic study. *Mater. Chem. Phys.* **2008**, *110*, 166–175. [CrossRef]
108. Lee, J.H.; Yang, Y.S. Effect of HCl concentration and reaction time on the change in the crystalline state of TiO_2 prepared from aqueous TiCl4 solution by precipitation. *J. Eur. Ceram. Soc.* **2005**, *25*, 3573–3578. [CrossRef]
109. Xie, Y.; Yuan, C. Visible-light responsive cerium ion modified Titania sol and nanocrystallites for X-3B dye photodegradation. *Appl. Catal. B Environ.* **2003**, *46*, 251–259. [CrossRef]
110. Xie, Y.; Yuan, C.; Li, X. Photocatalytic degradation of X-3B dye by visible light using lanthanide ion modified titanium dioxide hydrosol system. *Colloids Surf. A Physicochem. Eng. Asp.* **2005**, *252*, 87–94. [CrossRef]
111. Xie, Y.; Yuan, C. Photocatalytic and photoelectrochemical performance of crystallized titanium dioxide sol with neodymium ion modification. *J. Chem. Technol. Biotechnol.* **2005**, *80*, 954–963. [CrossRef]
112. Zeng, T.; Qiu, Y.; Chen, L.; Song, X. Microstructure and phase evolution of TiO_2 precursors prepared by peptization-hydrolysis method using polycarboxylic acid as peptizing agent. *Mater. Chem. Phys.* **1998**, *56*, 163–170. [CrossRef]
113. Zhang, Q.-H.; Gao, L.; Guo, J.-K. Preparation and characterization of nanosized TiO_2 powders from aqueous TiCl4 solution. *Nanostruct. Mater.* **1999**, *11*, 1293–1300. [CrossRef]
114. Kattoor, V.; Smitha, V.S.; Mohamed, A.P.; Hareesh, U.N.S.; Warrier, K.G. Temperature assisted acid catalyzed peptization of TiO_2; facile sol–gel approach for thermally stable anatase phase. *RSC Adv.* **2014**, *4*, 21664–21671. [CrossRef]
115. Li, Y.; Qin, Z.; Guo, H.; Yang, H.; Zhang, G.; Ji, S.; Zeng, T. Low-Temperature Synthesis of Anatase TiO_2 Nanoparticles with Tunable Surface Charges for Enhancing Photocatalytic Activity. *PLoS ONE* **2014**, *9*, e114638. [CrossRef] [PubMed]
116. Liu, W.-X.; Jiang, P.; Shao, W.-N.; Zhang, J.; Cao, W.-B. A novel approach for the synthesis of visible-light-active nanocrystalline N-doped TiO_2 photocatalytic hydrosol. *Solid State Sci.* **2014**, *33*, 45–48. [CrossRef]
117. Ma, Y.; Zhang, J.; Tian, B.; Chen, F.; Bao, S.; Anpo, M. Synthesis of visible light-driven Eu, N co-doped TiO_2 and the mechanism of the degradation of salicylic acid. *Res. Chem. Intermed.* **2012**, *38*, 1947–1960. [CrossRef]
118. Kim, Y.T.; Park, Y.S.; Myung, H.; Chae, H.K. A chelate-assisted route to anatase TiO_2 nanoparticles in acidic aqueous media. *Colloids Surf. A Physicochem. Eng. Asp.* **2008**, *313-314*, 260–263. [CrossRef]
119. Liu, T.-X.; Li, F.-B.; Li, X.-Z. Effects of peptizing conditions on nanometer properties and photocatalytic activity of TiO_2 hydrosols prepared by H2TiO3. *J. Hazard. Mater.* **2008**, *155*, 90–99. [CrossRef]
120. Liu, T.-X.; Li, X.-Z.; Li, F.-B. Enhanced photocatalytic activity of Ce^{3+}-TiO_2 hydrosols in aqueous and gaseous phases. *Chem. Eng. J.* **2010**, *157*, 475–482. [CrossRef]
121. Šuligoj, A.; Štangar, U.L.; Ristić, A.; Mazaj, M.; Verhovšek, D.; Tušar, N.N. TiO_2–SiO_2 films from organic-free colloidal TiO_2 anatase nanoparticles as photocatalyst for removal of volatile organic compounds from indoor air. *Appl. Catal. B Environ.* **2016**, *184*, 119–131. [CrossRef]
122. Zhang, R.; Gao, L. Effect of peptization on phase transformation of TiO_2 nanoparticles. *Mater. Res. Bull.* **2001**, *36*, 1957–1965. [CrossRef]
123. Alcober, C.; Alvarez, F.; Bilmes, S.A.; Candal, R.J. Photochromic W-TiO_2 membranes. *J. Mater. Sci. Lett.* **2002**, *21*, 501–504. [CrossRef]
124. Belet, A.; Wolfs, C.; Mahy, J.G.; Poelman, D.; Vreuls, C.; Gillard, N.; Lambert, S.D. Sol-gel Syntheses of Photocatalysts for the Removal of Pharmaceutical Products in Water. *Nanomaterials* **2019**, *9*, 126. [CrossRef]
125. Bergamonti, L.; Alfieri, I.; Lorenzi, A.; Montenero, A.; Predieri, G.; Di Maggio, R.; Girardi, F.; Lazzarini, L.; Lottici, P.P. Characterization and photocatalytic activity of TiO_2 by sol–gel in acid and basic environments. *J. Sol Gel Sci. Technol.* **2014**, *73*, 91–102. [CrossRef]
126. Borlaf, M.; Moreno, R.; Ortiz, A.L.; Colomer, M.T. Synthesis and photocatalytic activity of Eu3+-doped nanoparticulate TiO_2 sols and thermal stability of the resulting xerogels. *Mater. Chem. Phys.* **2014**, *144*, 8–16. [CrossRef]
127. Hu, L.; Wang, J.; Zhang, J.; Zhang, Q.; Liu, Z. An N-doped anatase/rutile TiO_2hybrid from low-temperature direct nitridization: Enhanced photoactivity under UV-/visible-light. *RSC Adv.* **2014**, *4*, 420–427. [CrossRef]
128. Jiang, J.; Long, M.; Wu, D.; Cai, W. Alkoxyl-derived visible light activity of TiO_2 synthesized at low temperature. *J. Mol. Catal. A Chem.* **2011**, *335*, 97–104. [CrossRef]
129. Khan, H. Sol–gel synthesis of TiO_2 from TiOSO4: Characterization and UV photocatalytic activity for the degradation of 4-chlorophenol. *React. Kinet. Mech. Catal.* **2017**, *121*, 811–832. [CrossRef]
130. Mao, L.; Li, Q.; Dang, H.; Zhang, Z. Synthesis of nanocrystalline TiO_2 with high photoactivity and large specific surface area by sol–gel method. *Mater. Res. Bull.* **2005**, *40*, 201–208. [CrossRef]

131. Yan, Q.; Wang, J.; Han, X.; Liu, Z. Soft-chemical method for fabrication of SnO–TiO$_2$ nanocomposites with enhanced photocatalytic activity. *J. Mater. Res.* **2013**, *28*, 1862–1869. [CrossRef]
132. Al-Maliki, F.J.; Al-Lamey, N.H. Synthesis of Tb-doped titanium dioxide nanostructures by sol–gel method for environmental photocatalysis applications. *J. Sol Gel Sci. Technol.* **2016**, *81*, 276–283. [CrossRef]
133. Gole, J.L.; Stout, J.D.; Burda, C.; Lou, Y.; Chen, X. Highly Efficient Formation of Visible Light Tunable TiO$_{2-x}$N$_x$ Photocatalysts and Their Transformation at the Nanoscale. *J. Phys. Chem. B.* **2004**, *108*, 1230–1240. [CrossRef]
134. Chung, W.; Kim, S.; Chang, S. A Study of the Correlation Between the Physical Characteristics and Efficiency of TiO$_2$ Photocatalyst Prepared with the Sol–Gel Method. *J. Nanosci. Nanotechnol.* **2016**, *16*, 11040–11045. [CrossRef]
135. Huang, Y.; Wang, P.; Wang, Z.; Rao, Y.; Cao, J.-J.; Pu, S.; Ho, W.; Lee, S.-C. Protonated g-C3N4/Ti3+ self-doped TiO$_2$ nanocomposite films: Room-temperature preparation, hydrophilicity, and application for photocatalytic NO removal. *Appl. Catal. B Environ.* **2019**, *240*, 122–131. [CrossRef]
136. Look, J.L.; Zukoski, C.F. Alkoxide-Derived Titania Particles: Use of Electrolytes to Control Size and Agglomeration Levels. *J. Am. Ceram. Soc.* **1992**, *75*, 1587–1595. [CrossRef]
137. Look, J.-L.; Zukoski, C.F. Colloidal Stability and Titania Precipitate Morphology: Influence of Short-Range Repulsions. *J. Am. Ceram. Soc.* **1995**, *78*, 21–32. [CrossRef]
138. Sharma, B.; Agarwal, R.; Jassal, M.; Agrawal, A.K. Stabilizer-free low-acid rapid synthesis of highly stable transparent aqueous titania nano sol and its photocatalytic activity. *J. Mol. Liq.* **2020**, *305*, 112842. [CrossRef]
139. Antonello, A.; Brusatin, G.; Guglielmi, M.; Bello, V.; Mattei, G.; Zacco, G.; Martucci, A. Nanocomposites of titania and hybrid matrix with high refractive index. *J. Nanoparticle Res.* **2011**, *13*, 1697–1708. [CrossRef]
140. Bi-Tao, X.; Bao-Xue, Z.; Long-Hai, L.; Jun, C.; Yan-Biao, L.; Wei-Min, C. Preparation of nanocrystalline anatase TiO$_2$ using basic sol-gel method. *Chem. Pap.* **2008**, *62*, 382–387. [CrossRef]
141. Li, H.; Afanasiev, P. On the selective growth of titania polymorphs in acidic aqueous medium. *Mater. Res. Bull.* **2011**, *46*, 2506–2514. [CrossRef]
142. Mahy, J.G.; Lambert, S.D.; Geens, J.; Daniel, A.; Wicky, D.; Archambeau, C.; Heinrichs, B. Large scale production of photocatalytic TiO$_2$ coating for volatile organic compound (VOC) air remediation. *AIMS Mater. Sci.* **2018**, *5*, 945–956. [CrossRef]
143. Wang, J.-Y.; Yu, J.-X.; Liu, Z.-H.; He, Z.-K.; Cai, R.-X. A simple new way to prepare anatase TiO$_2$ hydrosol with high photocatalytic activity. *Semicon. Sci. Technol.* **2005**, *20*, L36–L39. [CrossRef]
144. Nie, X.; Zhuo, S.; Maeng, G.; Sohlberg, K. Doping ofTiO$_2$Polymorphs for Altered Optical and Photocatalytic Properties. *Int. J. Photoenergy* **2009**, *2009*, 294042. [CrossRef]
145. Xu, Q.; Zhang, J.; Feng, Z.; Ma, Y.; Wang, X.; Li, C. Surface Structural Transformation and the Phase Transition Kinetics of Brookite TiO$_2$. *Chem. Asian J.* **2010**, *5*, 2158–2161. [CrossRef] [PubMed]
146. Bakardjieva, S.; Štengl, V.; Szatmary, L.; Subrt, J.; Lukac, J.; Murafa, N.; Niznansky, D.; Cizek, K.; Jirkovsky, J.; Petrova, N. Transformation of brookite-type TiO$_2$ nanocrystals to rutile: Correlation between microstructure and photoactivity. *J. Mater. Chem.* **2006**, *16*, 1709–1716. [CrossRef]
147. Balaganapathi, T.; Kaniamuthan, B.; Vinoth, S.; Arun, T.; Thilakan, P. Controlled synthesis of brookite and combined brookite with rutile phases of titanium di-oxide and its characterization studies. *Ceram. Int.* **2017**, *43*, 2438–2440. [CrossRef]
148. Li, J.-G.; Ishigaki, T. Brookite rutile phase transformation of TiO$_2$ studied with monodispersed particles. *Acta Mater.* **2004**, *52*, 5143–5150. [CrossRef]
149. Lin, Y.; Cai, Y.; Qiu, M.; Drioli, E.; Fan, Y. Environment-benign preparation of Ag toughening TiO$_2$/Ti tight ultrafiltration membrane via aqueous sol–gel route. *J. Mater. Sci.* **2015**, *50*, 5307–5317. [CrossRef]
150. Fallet, M.; Permpoon, S.; Deschanvres, J.-L.; Langlet, M. Influence of physico-structural properties on the photocatalytic activity of sol-gel derived TiO$_2$ thin films. *J. Mater. Sci.* **2006**, *41*, 2915–2927. [CrossRef]
151. Mamadou, S.D.; Neil, A.F.; Myung, S.J. Nanotechnology for Sustainable Development. *Nanotechnol. Sustain. Dev.* **2014**, *14*, 101–111. [CrossRef]
152. Sugimoto, T.; Zhou, X.; Muramatsu, A. Synthesis of uniform anatase TiO$_2$ nanoparticles by gel–sol method 3. Formation process and size control. *J. Colloid Interface Sci.* **2003**, *259*, 43–52. [CrossRef]
153. Chang, C.-J.; Lin, C.-Y.; Hsu, M.-H. Enhanced photocatalytic activity of Ce-doped ZnO nanorods under UV and visible light. *J. Taiwan Inst. Chem. Eng.* **2014**, *45*, 1954–1963. [CrossRef]
154. Vaiano, V.; Lara, M.; Iervolino, G.; Matarangolo, M.; Navio, J.; Hidalgo, M.C. Photocatalytic H2 production from glycerol aqueous solutions over fluorinated Pt-TiO$_2$ with high {001} facet exposure. *J. Photochem. Photobiol. A Chem.* **2018**, *365*, 52–59. [CrossRef]
155. Cao, D.; Wang, Y.; Zhao, X. Combination of photocatalytic and electrochemical degradation of organic pollutants from water. *Curr. Opin. Green Sustain. Chem.* **2017**, *6*, 78–84. [CrossRef]

Article

Investigation of Photocatalysis by Mesoporous Titanium Dioxide Supported on Glass Fibers as an Integrated Technology for Water Remediation

Cristina De Ceglie [1,†], Sudipto Pal [2,*,†], Sapia Murgolo [1], Antonio Licciulli [2] and Giuseppe Mascolo [1,*]

1. Istituto di Ricerca sulle Acque, Consiglio Nazionale delle Ricerche, Via F. De Blasio 5, 70132 Bari, Italy; cristina.deceglie@ba.irsa.cnr.it (C.D.C.); sapia.murgolo@ba.irsa.cnr.it (S.M.)
2. Department of Innovation Engineering, University of Salento, Via Per Monteroni, 73100 Lecce, Italy; antonio.licciulli@unisalento.it
* Correspondence: sudipto.pal@unisalento.it (S.P.); giuseppe.mascolo@ba.irsa.cnr.it (G.M.)
† These authors contributed equally to this work.

Abstract: The photocatalytic efficiency of an innovative UV-light catalyst consisting of a mesoporous TiO_2 coating on glass fibers was investigated for the degradation of pharmaceuticals (PhACs) in wastewater effluents. Photocatalytic activity of the synthesized material was tested, for the first time, on a secondary wastewater effluent spiked with nine PhACs and the results were compared with the photolysis used as a benchmark treatment. Replicate experiments were performed in a flow reactor equipped with a UV radiation source emitting at 254 nm. Interestingly, the novel photocatalyst led to the increase of the degradation of carbamazepine and trimethoprim (about 2.2 times faster than the photolysis). Several transformation products (TPs) resulting from both the spiked PhACs and the compounds naturally occurring in the secondary wastewater effluent were identified through UPLC-QTOF/MS/MS. Some of them, produced mainly from carbamazepine and trimethoprim, were still present at the end of the photolytic treatment, while they were completely or partially removed by the photocatalytic treatment.

Keywords: mesoporous titania; glass fiber; photocatalysis; contaminants of emerging concern; high resolution mass spectrometry; transformation products

1. Introduction

The presence of compounds of emerging concern (CECs) such as pharmaceuticals, pesticides, personal care products, and surfactants in secondary wastewater effluents poses a threat to the receiving water bodies and, consequently, to wildlife and human health. This risk has led to the publication of guidelines by the World Health Organization (WHO), Food and Agriculture Organization of the United Nations (FAO), and Environmental Protection Agency (EPA), in which chemical and microbiological parameters of the wastewaters are considered [1]. The guidelines have been periodically updated since 1973. CECs and some of their transformation products (TPs) are poorly removed by the activated sludge process in conventional wastewater treatment plants (WWTPs) and the long-term exposure to them could cause reproductive and hormonal disorders as potential health problems [2] and the increase of bacterial antibiotic resistance [3].

Chlorination and UV-C light-assisted disinfection methods are the most commonly used treatment technologies in WWTPs and they also affect the removal of different classes of micropollutants [4–9]. However, transportation, handling, and chemical hazards are the main disadvantages of the chlorination process, whereas wastewater turbidity and adverse photoreactivation limit the use of UV treatment technology. In this regard, advanced oxidation processes (AOPs) represent a more efficient alternative due to their versatility toward degrading organic and inorganic contaminants in water and on solid phases [5,10–13].

Among the AOPs, heterogeneous photocatalysis is promising for water treatment due to its versatility and it is also eco-friendly since no chemicals are directly required during the process [14–16]. The working principle of photocatalytic oxidation relies on the generation of strong oxidant and highly reactive species like hydroxyl radicals (OH•), irradiating the surface of a semiconductor with a light source having energy greater than its bandgap [17,18]. In a controlled reaction, the OH• radicals can completely mineralize the organic contaminants in water matrices. The advantages of using titanium dioxide (TiO_2) as a photocatalyst for the removal of CECs in water treatment are widely reported in the literature [17–19], concerning its high photo-stability and inertness in the chemical environment, wide availability, low cost and non-toxicity.

One of the most interesting challenges in water treatment is the synthesis of novel and efficient catalysts based on TiO_2 nanoparticles for the degradation of CECs in real secondary wastewater effluents. Although the activity of TiO_2 is higher in its powder form while performing photocatalysis in suspension, the recovery of the catalyst at the end of the oxidation process is very complex and cost-effective especially in a scaled-up system limiting its applicability [20]. In this perspective, the immobilization of TiO_2 catalyst on inert materials such as glass, silica, activated carbon, polymeric materials is preferable to the suspended TiO_2 as it facilitates the recovery of the catalyst at the end of the treatment and minimizes the traces of photocatalyst nanoparticles in the final treated waters [21–23]. Achievement of highly available active surfaces, the selection of the best support in terms of thermal and mechanical stability, or chemical inertness towards the catalyst, as well as the choice of the synthesis procedure are factors of primary importance in the developed immobilization technique. Porous substrates, such as activated carbon, diatomaceous earth, nanoclays, hollow glass spheres, and polymeric materials, are often used as supported catalysts due to their higher adsorption capacity [24]. In this context, the application of mesoporous TiO_2 in water remediation has been also investigated [25,26] However, there is a serious issue of using these kinds of materials since a sintering process is necessary to fix nanostructured TiO_2 coating to the substrate, which causes deformation and sometimes the loss of the porous structure [27]. In this context, glass fiber mats could be a good alternative due to their high flexibility, lightweight, high aspect ratio of the fibers, thermal stability, low cost and ability to remain stable under oxidation atmosphere and UV light irradiation [17,28,29].

In this work, a sol–gel dip coating technique was employed to obtain nanostructured TiO_2 coating on glass fiber mats. A pore generating agent was introduced in the sol preparation stage to achieve a mesoporous structure since TiO_2 coating often suffers low photocatalytic activity due to its lower surface area, especially when a highly crystalline phase is obtained by sintering the coating at high temperature. The mesoporous structure with a higher surface area increases the number of active sites on the catalyst surface that enhances the rate of photo-oxidation reaction thus improving the overall efficiency [30].

The present study aims to evaluate the effectiveness of photocatalysis employing a novel nanostructured and mesoporous TiO_2 coating on glass fiber mats for the removal of nine PhACs (carbamazepine, cetirizine, clarithromycin, climbazole, diclofenac, irbesartan, lidocaine, torsemide, and trimethoprim) spiked in a real secondary wastewater effluent, beside the removal of the naturally occurring CECs, as an integrated technology for water remediation. In addition, the identification of the transformation products (TPs), formed during both photolytic and photocatalytic treatments was also performed.

2. Results and Discussion

2.1. Mesoporous TiO_2/Glass Fibers: Synthesis and Characterization

Figure 1 shows the XRD pattern of the mesoporous nanocrystalline TiO_2 film deposited on the glass fiber mat, which shows several well-resolved diffraction peaks corresponding to various reflection planes of the crystalline anatase phase of TiO_2 (JCPDS no. 84-1286) as indicated in the figure. The absence of any rutile peak indicates the formation of pure anatase nanocrystalline phase. The average particle size estimated from the (101) plane of the XRD spectra using the Scherrer's equation (D = kλ/βCosθ, where D is the average crystallite size, k is the Scherrer constant (0.9), λ is the wavelength of the radiation source, β is the full line width at half-maxima of the (101) diffraction peak and θ is the angle of the corresponding diffraction peak) was found to be about 15 nm.

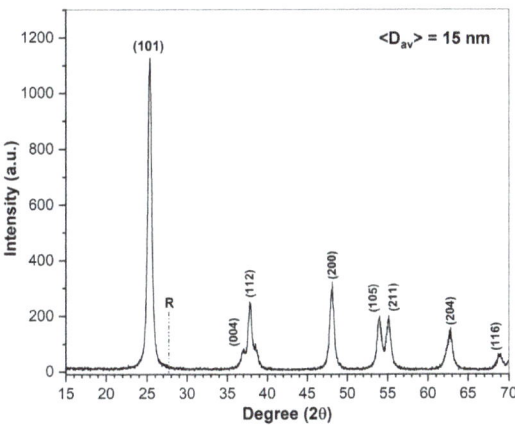

Figure 1. XRD pattern of the mesoporous TiO$_2$ coating on glass fibers.

The amount of TiO$_2$ loading on the glass fiber mats was estimated from the XRF spectra shown in Figure 2. As a comparison, the spectra of bare glass fiber is also shown. The appearance of a strong Ti peak indicates the presence of TiO$_2$ in the composite mat. The elemental distribution (Figure 2b) shows an almost homogeneous distribution of TiO$_2$ over the substrate. The average TiO$_2$ loading was estimated to be about 16% by weight.

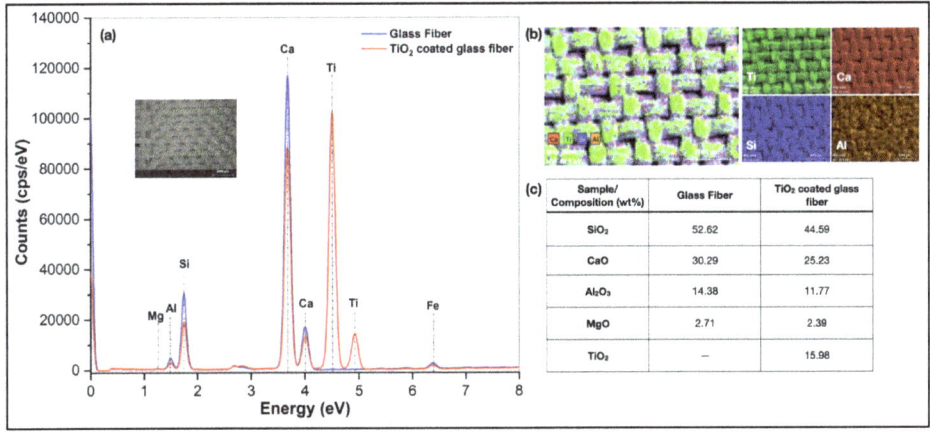

Figure 2. (**a**) XRF spectra of the TiO$_2$ coating on glass fiber. The inset shows real image of the TiO$_2$ coated glass fiber with the indicated area from where the spectrum was recorded; (**b**) elemental mapping of the major elements present; (**c**) estimation of TiO$_2$ loading on glass fiber.

The mesoporosity of the nanostructured film was analyzed by the BET surface area measurements. The specific surface area was calculated to be 61.2 m^2/g. Figure 3a shows the N$_2$ adsorption–desorption isotherm plot that corresponds to the typical type IV isotherm pattern with type H1 hysteresis. The adsorption branch shows a lower slope in the low relative pressure range (multilayer adsorption) followed by a sharp rise at a higher relative pressure (pore condensation in mesopore), whereas the desorption branch follows a narrow hysteresis loop where the desorption branch is parallel to the adsorption branch, indicating the formation of a narrow distribution of uniform mesopores. This is reflected in Figure 3b, which shows the BJH pore size distribution plot calculated from the desorption branch of the isotherm, where a narrow pore size distribution with an average pore diameter of 5.57 nm is observed. This data confirms the formation of mesoporous structure in the coating material [31].

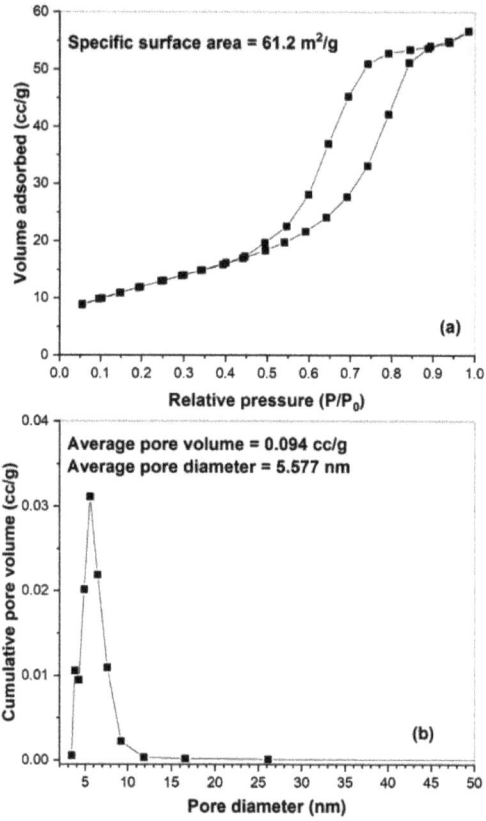

Figure 3. (a) N_2 adsorption-desorption isotherm and (b) BJH pore size distribution of the mesoporous TiO_2 coating. Specific surface area was calculated from the multilayer BET plot of the adsorption branch in the relative pressure of 0.05–0.30 range.

Morphological analysis was carried out to realize how TiO_2 nanoparticles are attached to the glass fiber surface. Figure 4a–d shows the FESEM micrographs of the mesoporous TiO_2 coating on the fibers at different magnifications. Every single fiber coated with TiO_2 nanoparticles is observed in Figure 4a, which is strongly supported by the presence of Ti in XRF spectra (Figure 2). More enlarged images are shown in Figure 4b–d, where the coating is clearly visible consisting of spherical TiO_2 nanoparticles. The broken part shows agglomerated nanoparticles, whereas the top smooth part shows a uniform coating with mesoporous nature, particularly in Figure 4d. This data confirms the strong attachment of the TiO_2 nanoparticles to the fiber surface. Moreover, due to the mesoporous structure, the coated surface became superhydrophilic showing water contact angle between 5 to 8° (data not shown) that would enhance the photocatalytic activity.

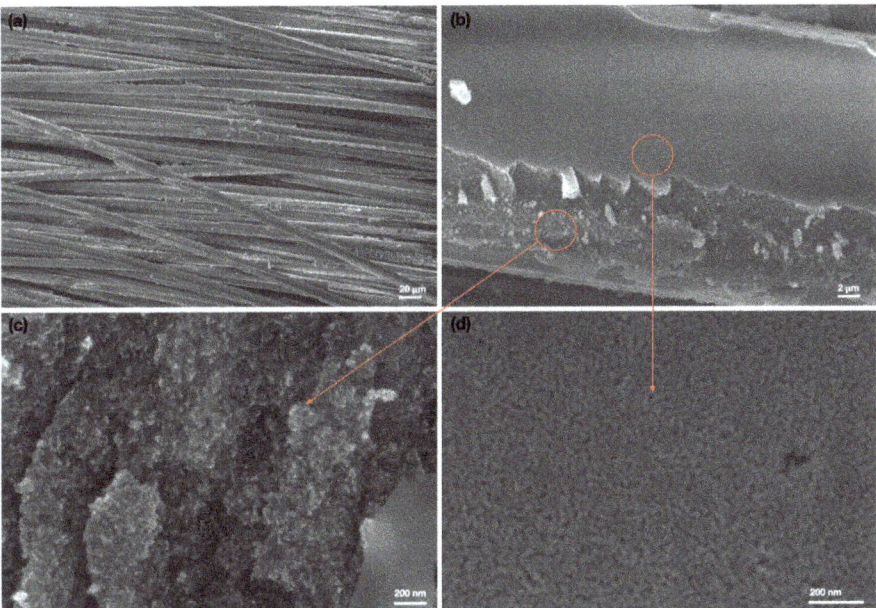

Figure 4. FESEM images of the mesoporous TiO$_2$ coating on glass fiber: (**a**) every single fiber coated with TiO$_2$ nanoparticles; (**b**–**d**) enlarged images showing the spherical TiO$_2$ nanoparticles.

2.2. Photocatalytic Degradation of Spiked PhACs

An up-flow reactor (0.5 L) equipped with a low-pressure mercury UV lamp emitting monochromatic UV radiation at a wavelength of 254 nm (40 W) was employed in recirculation mode for photolytic and photocatalytic experiments [32]. The volume of the treated solution was 1.2 L and each test was performed twice at a controlled temperature (30 °C). For the photocatalytic experiments, the catalyst fabric was wrapped around the quartz tube, which protects the UV lamp (Figure 5). Before starting the photocatalytic treatments, the catalyst fabric was exposed to a water up-flow of 6 L h^{-1} for 30 min in order to verify the adsorption of CECs on the supported catalyst. Photolysis experiments with only UV light irradiation were performed as benchmark treatments. The secondary wastewater effluent used for the investigation was spiked with the target compounds (at a concentration of about 200 µg L^{-1}). The structures of the drugs studied are shown in Figure 6.

After exposure to the UV light, some compounds such as diclofenac and cetirizine were completely degraded, after 20 and 60 min of reaction time, respectively (data not shown). The phototransformation rate is strictly dependent on the nature of the compound, in particular, the presence in the structures of groups absorbing UV energy as for example conjugated double bonds and hetero-atoms. Diclofenac is well-known to be susceptible to direct photolysis [33,34] and cetirizine, showing a very complex structure, is more susceptible to fast UV degradation due to the several routes of fragmentation. The photochemical behavior of the other investigated PhACs that exhibited slower removal kinetics, mainly CBZ, IBS and TMP, is different. In Table 1 the kinetic constants (k, min^{-1}) obtained for each contaminant during both photolytic and photocatalytic treatments (performed in duplicate) are listed. As far as diclofenac, it was quickly removed during the photolysis and therefore it was not possible to investigate the performance of the catalyst.

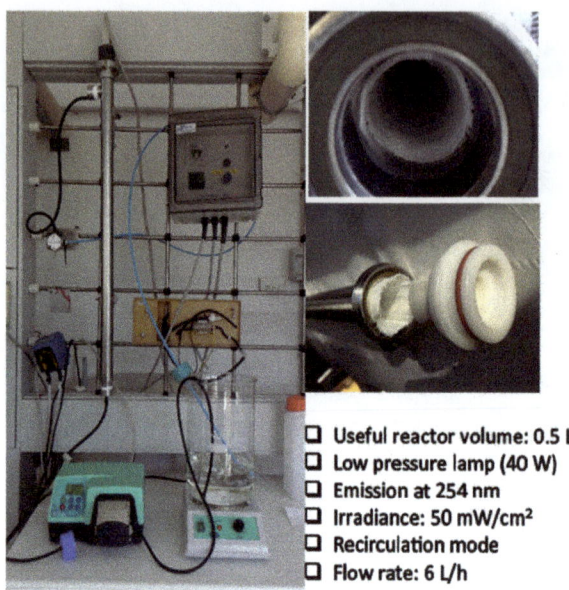

- Useful reactor volume: 0.5 L
- Low pressure lamp (40 W)
- Emission at 254 nm
- Irradiance: 50 mW/cm²
- Recirculation mode
- Flow rate: 6 L/h

Figure 5. Photocatalytic system equipped with UV lamp, quartz tube and mesoporous titanium dioxide supported on glass fiber mats.

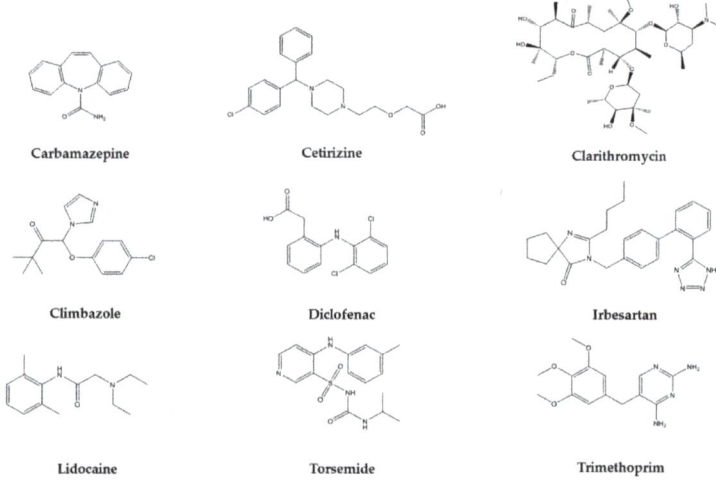

Figure 6. Chemical structures of the studied pharmaceuticals.

These results demonstrated that the compounds CBZ and TMP are the most recalcitrant to the photolytic treatment being the values of rate constants as the smallest ones (0.029 min^{-1} and 0.028 min^{-1}, respectively). CBZ and TMP are known for their slow photo-transformation rates [35] and consequently, their removal as well as the removal of formed TPs is important to increase water quality. In a recent paper, Paredes and colleagues investigated a novel catalyst based on immobilized TiO$_2$ on PVDF dual-layer hollow fiber membranes for the photo-transformation of eight target pharmaceuticals and they found that CBZ, TMP and metoprolol were more rapidly removed by photocatalysis compared to photolysis [36].

Table 1. First-order kinetic constants (k, min^{-1}) for spiked PhACs obtained during photolytic and photocatalytic treatments, treating secondary wastewater effluent.

Spiked PhACs	Elemental Composition	m/z (Da)	k (min^{-1})	
			Photolysis (Average ± SD)	Photocatalysis Mesoporous TiO$_2$ on Glass Fibers (Average ± SD)
CBZ	$C_{15}H_{12}N_2O$	237.1014	0.029 ± 0.001	0.068 ± 0.023
CTZ	$C_{21}H_{25}N_2O_3Cl$	389.1610	0.402 ± 0.066	0.289 ± 0.022
CLR	$C_{38}H_{69}NO_{13}$	748.4762	0.051 ± 0.004	0.061 ± 0.007
CLI	$C_{15}H_{17}N_2O_2Cl$	293.1050	0.233 ± 0.020	0.183 ± 0.011
DCF	$C_{14}H_{11}NO_2Cl_2$	296.0241	-	-
IBS	$C_{25}H_{28}N_6O$	429.2362	0.050 ± 0.001	0.059 ± 0.002
LDC	$C_{14}H_{22}N_2O$	235.1794	0.061 ± 0.002	0.065 ± 0.010
TOR	$C_{16}H_{20}N_4O_3S$	349.1306	0.081 ± 0.003	0.088 ± 0.011
TMP	$C_{14}H_{18}N_4O_3$	291.1439	0.028 ± 0.001	0.062 ± 0.014

According to Lian et al. [37], in wastewaters effluents, the photolabile species can be classified into five groups. IBS being included in the IV group (photochemically produced reactive intermediates combination-dominated group), was characterized by a slow degradation reaction with respect to the compounds belonging to the first three groups. CLI is considered moderately photo-susceptible while in the study of Kim et al. [38], clarithromycin is classified as a slow degrading pharmaceutical. It seems that the presence of dissolved organic matters promotes photodegradation of lidocaine and torsemide

The application of the mesoporous titanium dioxide supported on glass fibers allowed for increasing the photo-transformation rate of CBZ and TMP compared to the photolytic treatment of about 2.2 times. For the other tested PhACs the kinetic constants were slightly higher in presence of the novel photocatalyst except for CTZ and CLI for which a decrease in removal was observed during the photocatalytic process (Table 1).

2.3. Photocatalytic Degradation of Naturally Occurring PhACs

The performance of the novel catalyst was investigated for the removal of CECs naturally present in the secondary wastewater effluent. In this perspective, a targeted screening was performed employing the AB SCIEX software for both compound identification and trend detection. A group of eight additional PhACs was detected in the secondary wastewater effluent including candesartan, flecainide, gabapentin, irbesartan 446, lamotrigine, niflumic acid, telmisartan, and venlafaxine. The concentration of such CECs was measured between 0.5 and 5 µg/L. For five of these compounds (flecainide, gabapentin, irbesartan 446, lamotrigine and telmisartan), it was possible to measure the pseudo first-order kinetic constant during the UV-based treatments (with and without the photocatalyst) while candesartan, niflumic acid, and venlafaxine were quickly removed during the treatment by UV light alone (Table 2).

Average k values for all the detected substances (Table 2) were higher for photocatalysis, so the presence of the mesoporous TiO$_2$-based photocatalyst increased their photodegradation. In Figure 7, the main results relative to photolysis and photocatalysis for both the spiked and the naturally occurring contaminants, in terms of k (min^{-1}) values, are summarized. For each CEC, the average k value and the standard deviation are reported.

Table 2. First-order kinetic constants (k, min^{-1}) for naturally occurring PhACs obtained during photolytic and photocatalytic treatments, treating secondary wastewater effluent.

PhACs	Elemental Composition	m/z (Da)	k (min^{-1})	
			Photolysis (Avarage ± SD)	Photocatalysis Mesoporous TiO$_2$ on Glass Fibers (Avarage ± SD)
Flecainide	C$_{17}$H$_{20}$N$_2$O$_3$F$_6$	415.1440	0.032 ± 0.001	0.061 ± 0.016
Gabapentin	C$_9$H$_{17}$NO$_2$	172.1331	0.021 ± 0.001	0.052 ± 0.019
Irbesartan 446	C$_{25}$H$_{30}$N$_6$O$_2$	447.2486	0.049 ± 0.001	0.073 ± 0.011
Lamotrigine	C$_9$H$_7$N$_5$Cl$_2$	256.0150	0.028 ± 0.001	0.046 ± 0.014
Telmisartan	C$_{33}$H$_{30}$N$_4$O$_2$	515.2428	0.058 ± 0.002	0.081 ± 0.013

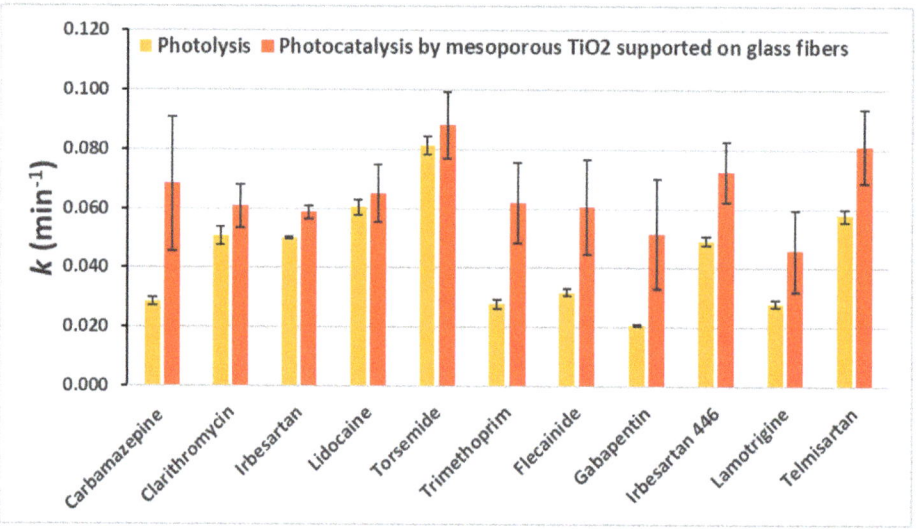

Figure 7. Identification of phototransformation products.

The software MetabolitePilot allowed for the identification of 26 transformation products formed from the spiked PhACs (Table S1). In the attempt to propose a chemical structure for the detected TPs, the data deriving from a deep bibliography research were combined with the accurate MS and MS/MS information achieved with the UPLC-QTOF/MS/MS analyses. Several detected TPs are already known in literature and the relative references (mentioned in Table S1) were used as a suggestion for structure elucidation [39,40]. The main difficulty in the attribution of the correct structure is the presence of numerous isomers for some TPs (for example CBZ-5 and CBZ-6). These isomers having the same accurate mass, show different retention times because they only differ for the position of a specific group, for example an OH group. If the analyte intensity is too low and MS/MS spectra are of poor quality, it is not possible to obtain a confident attribution. Among the six TPs of carbamazepine ([M+H]$^+$ 237.1014), five of them were discussed in detail by Calza et al., Franz et al. and Martinez-Piernas et al. [39–41]. Only CBZ-3, at m/z value of 241.0601, was considered as a new TP because, at present, its formation was not revealed in other scientific works. As far as clarithromycin ([M+H]$^+$ 748.4762) all the detected TPs were described by Calza et al. [39] and Buchicchio et al. [42]

Interestingly, for climbazole ([M+H]$^+$ 293.1050) the TP with m/z 167.1175 showed a MS/MS spectrum identical to that acquired by Castro et al. [43] and it results from the cleavage of the ether bond. According to the identification confidence levels of Schymanski et al. [44] the molecular structure of TP CLI-1 can be assigned as a probable structure (Level 2a). Two additional CLI-TPs were revealed for the first time in this work, the first one at m/z of 247.1448 (CLI-2) and the second one, at m/z 338.0887 (CLI-3, a nitro-derivative of climbazole).

The MS/MS spectrum of CLI-3 revealed a fragment at m/z 69.0456 that is present only in the MS/MS spectrum of climbazole and that corresponds to the protonated imidazole ($C_3H_5N_2^+$). The mass shift between the TP (m/z 338.0887) and the parent compound (m/z 293.1050) was 44.98 Da corresponding to nitration, so the empirical formula $C_{15}H_{16}N_3O_4Cl$ was attributed to this product, identified as nitro-climbazole. The observation of the protonated imidazole in the MS/MS spectrum could be proof of the nitration of the phenyl ring. This kind of modification was discussed by Nelieu et al. [45] during the photodegradation of monuron in an aqueous solution. In addition to IBS-1 and IBS-2 [46,47] another TP of irbesartan at m/z 445.2338 obtaining from the addition of one -OH group to the parent compound ([M+H]$^+$ 429.2362) was detected and its probable structure is illustrated in the Table S1.

For lidocaine ([M+H]$^+$ 235.1794) three new TPs were identified LDC-1, LDC-2, and LDC-3 at 233.1647 m/z, 283.1656 m/z, and 299.1603 m/z, respectively, and their chemical structures were attributed with a high confidence level. The proposed structures were suggested by the detailed study of Rayaroth et al. [48] about the degradation mechanism of lidocaine by photocatalysis that involves the hydroxyl radicals as major reactive species.

The transformation products of torsemide ([M+H]$^+$ 349.1306) in the aquatic environment were investigated in the recent work of Lege et al. [49]. The article focuses on the degradation products derived from various treatments included photo-transformation; 4 photolysis TPs (TP 364b, TP 362, TP 258 and TP 393) were identified by Lege and coworkers, the first three TPs with a confidence level 2 and the last one with a confidence level 4. For TP 364 three isomers (TP 364 a, b, c) with molecular formula $C_{16}H_{20}N_4O_4S$ were detected after the different degradation studies. Only TP 364b (exact monoisotopic mass of [M+H]$^+$: 365.1278 m/z) was present after each kind of treatment, thus revealing the importance of this TP in the degradation pathway of this drug. The three isomers represent the hydroxylation products of torsemide and they differ in the hydroxylation site [49].

In the present work, the molecular ion at 365.1264 m/z was detected after photodegradation experiments (TOR-2) and the MS/MS spectrum matches that of TP 364b. The molecular ion at 363.1112 m/z (TOR-1) was the same as TP 362, deriving from ketone formation. The formation of TP 258 after photolysis was of minor significance compared to the other TPs in the investigation of Lege et al. [49]; this type of by-product was not detected in the present study. Finally, the TP 393 (exact monoisotopic mass of [M+H]$^+$: 394.1180 m/z), attributed as nitro-torsemide, was also recognized in our reaction samples. Unlike Lege et al., we succeeded in identifying the TP with a confidence level of 2 (TOR-3) because of the good quality of the acquired MS/MS spectrum and the high intensity of the fragments.

For trimethoprim ([M+H]$^+$ 291.1439) the detected TPs were already identified by Paredes et al. [36]. No degradation products of cetirizine and diclofenac having a significant removal trend were detected in the present investigation. In Figure 8, the time profiles of the most representative transformation product for three (CBZ, CLR, TMP) spiked compounds were reported. Most of the detected TPs showed a bell-shape trend (a more comprehensive view of TPs time profiles is present in Figures S1–S7) and four different trends were noted: (i) the photolytic treatment generates a TP with higher intensity respect to photocatalytic treatment, with a delay in the removal of the compound (CBZ-5 and CLR-1, Figure 8); (ii) the TP increases during photolytic treatment and it seems to accumulate along reaction time while negligible formation is observed during photocatalysis (TMP-1, Figure 8); (iii) no differences are observed between photolysis and photocatalysis (IBS-3, Figure S4); (iv) at first, a higher amount of the TP is observed during the photocatalytic treatment but, then, the decrease is similar for both the treatments (CLR-3, Figure S2).

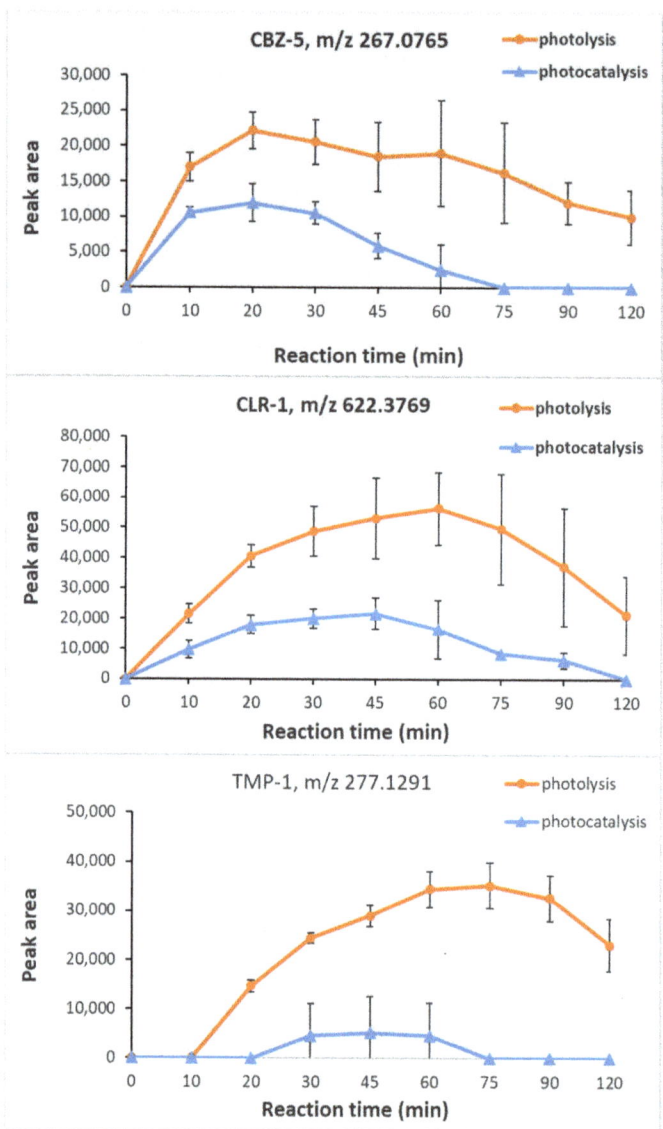

Figure 8. Time profiles of the TPs CBZ-5, CLR-1, and TMP-1 during photolysis and photocatalysis with mesoporous TiO_2 coated on glass fibers, using secondary wastewater effluent.

Among the previously described trends, it is worth noting the second one, in fact, in other works such as that of Paredes et al. [36], a similar behaviour was not highlighted. Considering the first trend typology, different TPs are still present in the reaction sample after the end of the photolysis while they were fast removed during photocatalysis: i.e., CBZ-5 and CBZ-6, CLR-1, CLR-4, CLI-2, TOR-1 and, interestingly, all the detected TPs of trimethoprim.

3. Materials and Methods

3.1. Selection of Pharmaceutical Compounds

Nine pharmaceutical compounds not completely removed in conventional wastewater treatment and usually detected in secondary wastewater effluents at trace concentrations (i.e., µg/L–ng/L) were selected as target contaminants to investigate the efficiency of the novel photocatalyst: carbamazepine (CBZ), cetirizine (CTZ), clarithromycin (CLR), climbazole (CLI), diclofenac (DCF), irbesartan (IBS), lidocaine (LDC), torsemide (TOR), and trimethoprim (TMP). All the listed compounds were spiked in a real secondary wastewater effluent at a final concentration ranging between 100 and 200 µg/L. The effluent was taken from a self-forming dynamic membrane bioreactor treating municipal wastewater and characterized according to standard methods. All chemicals were purchased from Sigma–Aldrich as well as the solvents used for chromatographic analyses and for preparing standard solutions, e.g., acetonitrile, methanol and formic acid (UPLC grade).

3.2. Synthesis and Characterization of the Mesoporous TiO_2 Coating on Glass Fiber

Mesoporous TiO_2 coating was deposited on the glass fiber mats by sol–gel method using the dip technique. The TiO_2 sol (containing 5% of TiO_2 by weight in ethanol) was prepared by hydrolysis-condensation of titanium isopropoxide (TTIP, Sigma–Aldrich, St. Louis, MO, USA, 97%) in presence of hydrochloric acid (HCl, Alfa-Aesar 37%) and deionized water (Millipore Milli Q). The triblock copolymer, Pluronic P123 ($PEO_{20}PPO_{70}PEO_{20}$, average M_{av}~5800, Sigma–Aldrich) was used as a mesopore generating template. The molar ratios of the reagents were TTIP:HCl:H_2O:P123 = 1:0.5:4:0.0125. At first, P123 was dissolved in the appropriate amount of ethanol followed by the addition of TTIP drop-wise to the above solution. After homogeneous mixing of the solution, the required amount of HCl was diluted with water and drop-wise added. The sol was left under stirring for several hours to complete the hydrolysis-condensation reaction. The as prepared sol aged for one day before the coating application. Commercially available glass fiber mats (approx. composition 52% SiO_2, 30% CaO, 14% Al_2O_3, 2% MgO) were used as the substrate to deposit the TiO_2 coating. The glass fiber mats were cut into 30 × 20 cm size and preliminary treated at 500 °C in the air for 1 h to decompose any organic binder present. The organic-free glass fiber mats were coated with the TiO_2 sol by dip-coating method at a withdrawal speed of 20 cm/min. After the coating deposition, the mats were kept in an oven at 65 °C for one night to gently dry the coated film and avoid any unwanted cracks. Sintering and template removal took place at 500 °C for 2 h in the air at the heating rate of 1 °C/min and maintaining a similar cooling rate as well, after which the mesoporous nanocrystalline TiO_2 coating was formed on the glass fiber mat.

3.3. Microstructural Characterization

Nanocrystalline phase formation of TiO_2 in the coated film was investigated by X-ray diffraction (XRD) analysis that performed with a Rigaku Ultima X-ray diffractometer using Cu Kα radiation (λ = 1.5406 Å) operating at 40 kV/30 mA with the step size of 0.02°. Morphological characterization of the coating was carried out on a Zeiss Sigma VP field emission scanning electron microscope (FESEM). The amount of TiO_2 loading on the glass fiber mat was determined by X-ray fluorescence spectroscopy (XRF). It was performed with a Bruker M4 Tornado (Bruker Nano Germany) X-ray fluorescence spectrometer operating at 50 kV/600 µA (30 W) equipped with X-Flash silicon drift detector. The measurements were performed in area mode (approx. 35 mm^2) and the elemental quantification was estimated from the average of five measurements.

3.4. PhACs Concentration Measurements

The concentration of the selected PhACs during photolytic and photocatalytic treatments was determined using a high-resolution mass spectrometer, TripleTOF 5600+ system (AB Sciex), coupled to a liquid chromatographic system, Ultimate 3000 (Thermo Fisher Scientific, Waltham, MA, USA), by means of a duo-spray ion source operated in positive electrospray (ESI) mode. All MS analyses were acquired with an acquisition method based on double experiments, i.e., full-scan survey TOF-MS and IDA (information dependent acquisition) experiment. 50 µL samples were injected and eluted with a binary gradient consisting of 0.1% formic acid in water (solvent A) and 0.1% formic acid in MeCN (solvent B), employing a Waters BEH C18 column (2.1 × 150 mm, 1.7 µm) operating at a flow of 0.200 mL/min. Before LC/MS analysis, carbamazepine D10 was added as an internal standard to each sample at a final concentration of 10 µg/L. AB Sciex software was used for data processing, i.e., SciexOS 1.2, LibraryView 1.0.2 and MetabolitePilot 1.5. ChemBioDraw Ultra 13.0 was used for TPs structure elucidation.

4. Conclusions

A novel photocatalytic system consisting of mesoporous titanium dioxide supported on glass fibers as the catalyst substrate in a UV reactor arrangement has been developed to remove the pharmaceutical contaminants. Highly crystalline and nanostructured mesoporous TiO_2 coatings with a high surface area were successfully fixed to the fiber surfaces after the sintering process. The microstructural analysis confirmed the formation of mesoporosity in the coating matrix. The large exposed area of the glass fiber mat allowed the fast photo-oxidation rate compared to the photolysis. The degradation study carried out with the new catalyst support showed that it was effective not only in the removal of most of the PhACs investigated, with particular reference to carbamazepine and trimethoprim, but also in the abatement of their TPs. The strong attachment of the TiO_2 coating to the fiber surfaces showed the reusability of the supported catalyst without spending the effort to recover it after the photocatalytic reaction. This simple coating strategy can be extended to either other suitable catalyst support or to modify the starting solution to make it visible light active (e.g., by doping with Ag, Cu, Au, Fe_2O_3, etc.).

Supplementary Materials: The following are available online at https://www.mdpi.com/article/10.3390/catal12010041/s1, Figure S1. Time profiles of CBZ TPs during photolysis and photocatalysis with mesoporous TiO_2 coated on glass fibers, using secondary wastewater effluent, Figure S2. Time profiles of CLR TPs during photolysis and photocatalysis with mesoporous TiO_2 coated on glass fibers, using secondary wastewater effluent, Figure S3. Time profiles of CLI TPs during photolysis and photocatalysis with mesoporous TiO_2 coated on glass fibers, using secondary wastewater effluent, Figure S4. Time profiles of IBS TPs during photolysis and photocatalysis with mesoporous TiO_2 coated on glass fibers, using secondary wastewater effluent, Figure S5. Time profiles of LDC TPs during photolysis and photocatalysis with mesoporous TiO_2 coated on glass fibers, using secondary wastewater effluent, Figure S6. Time profiles of TOR TPs during photolysis and photocatalysis with mesoporous TiO_2 coated on glass fibers, using secondary wastewater effluent, Figure S7. Time profiles of TMP TPs during photolysis and photocatalysis with mesoporous TiO_2 coated on glass fibers, using secondary wastewater effluent, Table S1. List of transformation products of the *spiked compounds* detected by suspect screening in photolytic and photocatalytic experiments (mesoporous TiO_2 supported on glass fibers), treating secondary wastewater effluent.

Author Contributions: Conceptualization, C.D.C. and S.P.; methodology, C.D.C., S.P. and S.M.; software, C.D.C., S.M.; validation, C.D.C., S.M. and G.M.; investigation, C.D.C., S.P. and S.M.; resources, C.D.C., S.P. and S.M.; data curation, C.D.C., S.M.; writing—original draft preparation, C.D.C., S.P. and S.M.; writing—review and editing, C.D.C., S.P., S.M., A.L. and G.M.; visualization, A.L., G.M.; supervision, A.L., G.M. All authors have read and agreed to the published version of the manuscript.

Funding: This research received no external funding.

Data Availability Statement: The data, either raw or processed required to reproduce these research works cannot be shared at this time as the data also form part of an ongoing study.

Acknowledgments: The authors thank to Donato Cannoletta for performing the XRD measurements and Fabio Marzo for providing the FESEM measurements.

Conflicts of Interest: The authors declare no conflict of interest.

References

1. Jaramillo, M.F.; Restrepo, I. Wastewater Reuse in Agriculture: A Review about Its Limitations and Benefits. *Sustainability* **2017**, *9*, 1734. [CrossRef]
2. Gonsioroski, A.; Mourikes, V.E.; Flaws, J.A. Endocrine Disruptors in Water and Their Effects on the Reproductive System. *Int. J. Mol. Sci.* **2020**, *21*, 1929. [CrossRef]
3. Webb, S.; Ternes, T.; Gibert, M.; Olejniczak, K. Indirect Human Exposure to Pharmaceuticals via Drinking Water. *Toxicol. Lett.* **2003**, *142*, 157–167. [CrossRef]
4. Mezzanotte, V.; Antonelli, M.; Citterio, S.; Nurizzo, C. Wastewater Disinfection Alternatives: Chlorine, Ozone, Peracetic Acid, and UV Light. *Water Environ. Res.* **2007**, *79*, 2373–2379. [CrossRef] [PubMed]
5. Patel, M.; Kumar, R.; Kishor, K.; Mlsna, T.; Pittman, C.U.; Mohan, D. Pharmaceuticals of Emerging Concern in Aquatic Systems: Chemistry, Occurrence, Effects, and Removal Methods. *Chem. Rev.* **2019**, *119*, 3510–3673. [CrossRef]
6. Lopez, A.; Mascolo, G.; Földényi, R.; Passino, R. Disinfection By-Products Formation during Hypochlorination of Isoproturon Contaminated Groundwater. *Water Sci. Technol.* **1996**, *34*, 351–358. [CrossRef]
7. Lopez, A.; Mascolo, G.; Tiravanti, G.; Passino, R. Degradation of Herbicides (Ametryn and Isoproturon) during Water Disinfection by Means of Two Oxidants (Hypochlorite and Chlorine Dioxide). *Water Sci. Technol.* **1997**, *35*, 129–136. [CrossRef]
8. Lopez, A.; Mascolo, G.; Detomaso, A.; Lovecchio, G.; Villani, G. Temperature Activated Degradation (Mineralization) of 4-Chloro-3-Methyl Phenol by Fenton's Reagent. *Chemosphere* **2005**, *59*, 397–403. [CrossRef]
9. Mascolo, G.; Lopez, A.; Passino, R.; Ricco, G.; Tiravanti, G. Degradation of Sulphur Containing S-Triazines during Water Chlorination. *Water Res.* **1994**, *28*, 2499–2506. [CrossRef]
10. Wang, S.; Wang, J. Activation of Peroxymonosulfate by Sludge-Derived Biochar for the Degradation of Triclosan in Water and Wastewater. *Chem. Eng. J.* **2019**, *356*, 350–358. [CrossRef]
11. Yang, X.; Sun, J.; Fu, W.; Shang, C.; Li, Y.; Chen, Y.; Gan, W.; Fang, J. PPCP Degradation by UV/Chlorine Treatment and Its Impact on DBP Formation Potential in Real Waters. *Water Res.* **2016**, *98*, 309–318. [CrossRef]
12. De la Cruz, N.; Esquius, L.; Grandjean, D.; Magnet, A.; Tungler, A.; de Alencastro, L.F.; Pulgarín, C. Degradation of Emergent Contaminants by UV, UV/H2O2 and Neutral Photo-Fenton at Pilot Scale in a Domestic Wastewater Treatment Plant. *Water Res.* **2013**, *47*, 5836–5845. [CrossRef] [PubMed]
13. Pizzigallo, M.D.R.; Ruggiero, P.; Crecchio, C.; Mascolo, G. Oxidation of Chloroanilines at Metal Oxide Surfaces. *J. Agric. Food Chem.* **1998**, *46*, 2049–2054. [CrossRef]
14. Ibhadon, A.O.; Fitzpatrick, P. Heterogeneous Photocatalysis: Recent Advances and Applications. *Catalysts* **2013**, *3*, 189–218. [CrossRef]
15. Emerging Contaminants from Industrial and Municipal Waste. Available online: https://www.springerprofessional.de/en/emerging-contaminants-from-industrial-and-municipal-waste/2873406 (accessed on 17 December 2021).
16. Raja, P.; Bozzi, A.; Jardim, W.F.; Mascolo, G.; Renganathan, R.; Kiwi, J. Reductive/Oxidative Treatment with Superior Performance Relative to Oxidative Treatment during the Degradation of 4-Chlorophenol. *Appl. Catal. B Environ.* **2005**, *59*, 249–257. [CrossRef]
17. Pasini, S.M.; Valério, A.; Yin, G.; Wang, J.; de Souza, S.M.A.G.U.; Hotza, D.; de Souza, A.A.U. An Overview on Nanostructured TiO_2-Containing Fibers for Photocatalytic Degradation of Organic Pollutants in Wastewater Treatment. *J. Water Process Eng.* **2021**, *40*, 101827. [CrossRef]
18. Yuan, R.; Zhu, Y.; Zhou, B.; Hu, J. Photocatalytic Oxidation of Sulfamethoxazole in the Presence of TiO_2: Effect of Matrix in Aqueous Solution on Decomposition Mechanisms. *Chem. Eng. J.* **2019**, *359*, 1527–1536. [CrossRef]
19. Murgolo, S.; De Ceglie, C.; Di Iaconi, C.; Mascolo, G. Novel TiO_2-Based Catalysts Employed in Photocatalysis and Photoelectrocatalysis for Effective Degradation of Pharmaceuticals (PhACs) in Water: A Short Review. *Curr. Opin. Green Sustain. Chem.* **2021**, *30*, 100473. [CrossRef]
20. Pal, S.; Laera, A.M.; Licciulli, A.; Catalano, M.; Taurino, A. Biphase TiO2 Microspheres with Enhanced Photocatalytic Activity. *Ind. Eng. Chem. Res.* **2014**, *53*, 7931–7938. [CrossRef]
21. Byrne, C.; Subramanian, G.; Pillai, S.C. Recent Advances in Photocatalysis for Environmental Applications. *J. Environ. Chem. Eng.* **2018**, *6*, 3531–3555. [CrossRef]
22. Rachel, A.; Subrahmanyam, M.; Boule, P. Comparison of Photocatalytic Efficiencies of TiO_2 in Suspended and Immobilised Form for the Photocatalytic Degradation of Nitrobenzenesulfonic Acids. *Appl. Catal. B Environ.* **2002**, *37*, 301–308. [CrossRef]
23. Kim, D.S.; Park, Y.S. Photocatalytic Decolorization of Rhodamine B by Immobilized TiO_2 onto Silicone Sealant. *Chem. Eng. J.* **2006**, *116*, 133–137. [CrossRef]
24. Padmanabhan, S.K.; Pal, S.; Ul Haq, E.; Licciulli, A. Nanocrystalline TiO_2–Diatomite Composite Catalysts: Effect of Crystallization on the Photocatalytic Degradation of Rhodamine B. *Appl. Catal. A Gen.* **2014**, *485*, 157–162. [CrossRef]
25. Yacou, C.; Smart, S.; Diniz da Costa, J.C. Mesoporous TiO_2 Based Membranes for Water Desalination and Brine Processing. *Sep. Purif. Technol.* **2015**, *147*, 166–171. [CrossRef]
26. Akhavan, O.; Ghaderi, E. Self-Accumulated Ag Nanoparticles on Mesoporous TiO2 Thin Film with High Bactericidal Activities. *Surf. Coat. Technol.* **2010**, *204*, 3676–3683. [CrossRef]

27. Hofstadler, K.; Bauer, R.; Novalic, S.; Heisler, G. New Reactor Design for Photocatalytic Wastewater Treatment with TiO_2 Immobilized on Fused-Silica Glass Fibers: Photomineralization of 4-Chlorophenol. *Environ. Sci. Technol.* **1994**, *28*, 670–674. [CrossRef] [PubMed]
28. Fukugaichi, S. Fixation of Titanium Dioxide Nanoparticles on Glass Fiber Cloths for Photocatalytic Degradation of Organic Dyes. *ACS Omega* **2019**, *4*, 15175–15180. [CrossRef] [PubMed]
29. Chen, L.; Yang, S.; Mäder, E.; Ma, P.-C. Controlled Synthesis of Hierarchical TiO_2 Nanoparticles on Glass Fibres and Their Photocatalytic Performance. *Dalton Trans.* **2014**, *43*, 12743–12753. [CrossRef]
30. Petronella, F.; Truppi, A.; Dell'Edera, M.; Agostiano, A.; Curri, M.L.; Comparelli, R. Scalable Synthesis of Mesoporous TiO_2 for Environmental Photocatalytic Applications. *Materials* **2019**, *12*, 1853. [CrossRef] [PubMed]
31. Licciulli, A.; Riccardis, A.D.; Pal, S.; Nisi, R.; Mele, G.; Cannoletta, D. Ethylene Photo-Oxidation on Copper Phthalocyanine Sensitized TiO_2 Films under Solar Radiation. *J. Photochem. Photobiol. A Chem.* **2017**, *346*, 523–529. [CrossRef]
32. Murgolo, S.; Yargeau, V.; Gerbasi, R.; Visentin, F.; El Habra, N.; Ricco, G.; Lacchetti, I.; Carere, M.; Curri, M.L.; Mascolo, G. A New Supported TiO_2 Film Deposited on Stainless Steel for the Photocatalytic Degradation of Contaminants of Emerging Concern. *Chem. Eng. J.* **2017**, *318*, 103–111. [CrossRef]
33. Carlson, J.C.; Stefan, M.I.; Parnis, J.M.; Metcalfe, C.D. Direct UV Photolysis of Selected Pharmaceuticals, Personal Care Products and Endocrine Disruptors in Aqueous Solution. *Water Res.* **2015**, *84*, 350–361. [CrossRef] [PubMed]
34. Homem, V.; Santos, L. Degradation and Removal Methods of Antibiotics from Aqueous Matrices—A Review. *J. Environ. Manag.* **2011**, *92*, 2304–2347. [CrossRef] [PubMed]
35. Afonso-Olivares, C.; Fernández-Rodríguez, C.; Ojeda-González, R.J.; Sosa-Ferrera, Z.; Santana-Rodríguez, J.J.; Rodríguez, J.M.D. Estimation of Kinetic Parameters and UV Doses Necessary to Remove Twenty-Three Pharmaceuticals from Pre-Treated Urban Wastewater by UV/H_2O_2. *J. Photochem. Photobiol. A Chem.* **2016**, *329*, 130–138. [CrossRef]
36. Paredes, L.; Murgolo, S.; Dzinun, H.; Dzarfan Othman, M.H.; Ismail, A.F.; Carballa, M.; Mascolo, G. Application of Immobilized TiO2 on PVDF Dual Layer Hollow Fibre Membrane to Improve the Photocatalytic Removal of Pharmaceuticals in Different Water Matrices. *Appl. Catal. B Environ.* **2019**, *240*, 9–18. [CrossRef]
37. Lian, L.; Yan, S.; Zhou, H.; Song, W. Overview of the Phototransformation of Wastewater Effluents by High-Resolution Mass Spectrometry. *Environ. Sci. Technol.* **2020**, *54*, 1816–1826. [CrossRef]
38. Kim, I.; Tanaka, H. Photodegradation Characteristics of PPCPs in Water with UV Treatment. *Environ. Int.* **2009**, *35*, 793–802. [CrossRef] [PubMed]
39. Calza, P.; Medana, C.; Padovano, E.; Giancotti, V.; Baiocchi, C. Identification of the Unknown Transformation Products Derived from Clarithromycin and Carbamazepine Using Liquid Chromatography/High-Resolution Mass Spectrometry. *Rapid Commun. Mass Spectrom.* **2012**, *26*, 1687–1704. [CrossRef]
40. Martínez-Piernas, A.B.; Nahim-Granados, S.; Polo-López, M.I.; Fernández-Ibáñez, P.; Murgolo, S.; Mascolo, G.; Agüera, A. Identification of Transformation Products of Carbamazepine in Lettuce Crops Irrigated with Ultraviolet-C Treated Water. *Environ. Pollut.* **2019**, *247*, 1009–1019. [CrossRef] [PubMed]
41. Franz, S.; Falletta, E.; Arab, H.; Murgolo, S.; Bestetti, M.; Mascolo, G. Degradation of Carbamazepine by Photo(Electro)Catalysis on Nanostructured TiO2 Meshes: Transformation Products and Reaction Pathways. *Catalysts* **2020**, *10*, 169. [CrossRef]
42. Buchicchio, A.; Bianco, G.; Sofo, A.; Masi, S.; Caniani, D. Biodegradation of Carbamazepine and Clarithromycin by Trichoderma Harzianum and Pleurotus Ostreatus Investigated by Liquid Chromatography—High-Resolution Tandem Mass Spectrometry (FTICR MS-IRMPD). *Sci. Total Environ.* **2016**, *557–558*, 733–739. [CrossRef]
43. Castro, G.; Casado, J.; Rodríguez, I.; Ramil, M.; Ferradás, A.; Cela, R. Time-of-Flight Mass Spectrometry Assessment of Fluconazole and Climbazole UV and $UV/H2O2$ Degradability: Kinetics Study and Transformation Products Elucidation. *Water Res.* **2016**, *88*, 681–690. [CrossRef]
44. Schymanski, E.L.; Jeon, J.; Gulde, R.; Fenner, K.; Ruff, M.; Singer, H.P.; Hollender, J. Identifying Small Molecules via High Resolution Mass Spectrometry: Communicating Confidence. *Environ. Sci. Technol.* **2014**, *48*, 2097–2098. [CrossRef] [PubMed]
45. Nélieu, S.; Shankar, M.V.; Kerhoas, L.; Einhorn, J. Phototransformation of Monuron Induced by Nitrate and Nitrite Ions in Water: Contribution of Photonitration. *J. Photochem. Photobiol. A Chem.* **2008**, *193*, 1–9. [CrossRef]
46. Boix, C.; Ibáñez, M.; Sancho, J.V.; Parsons, J.R.; de Voogt, P.; Hernández, F. Biotransformation of Pharmaceuticals in Surface Water and during Waste Water Treatment: Identification and Occurrence of Transformation Products. *J. Hazard. Mater.* **2016**, *302*, 175–187. [CrossRef] [PubMed]
47. Shah, R.P.; Sahu, A.; Singh, S. Identification and Characterization of Degradation Products of Irbesartan Using LC–MS/TOF, MSn, on-Line H/D Exchange and LC–NMR. *J. Pharm. Biomed. Anal.* **2010**, *51*, 1037–1046. [CrossRef] [PubMed]
48. Rayaroth, M.P.; Aravind, U.K.; Aravindakumar, C.T. Photocatalytic Degradation of Lignocaine in Aqueous Suspension of TiO_2 Nanoparticles: Mechanism of Degradation and Mineralization. *J. Environ. Chem. Eng.* **2018**, *6*, 3556–3564. [CrossRef]
49. Lege, S.; Sorwat, J.; Yanez Heras, J.E.; Zwiener, C. Abiotic and Biotic Transformation of Torasemide—Occurrence of Degradation Products in the Aquatic Environment. *Water Res.* **2020**, *177*, 115753. [CrossRef] [PubMed]

Article

Visible Light Induced Nano-Photocatalysis Trimetallic Cu$_{0.5}$Zn$_{0.5}$-Fe: Synthesis, Characterization and Application as Alcohols Oxidation Catalyst

Asma Ghazzy [1], Lina Yousef [2] and Afnan Al-Hunaiti [2,*]

[1] Faculty of Pharmacy, Al-Ahliyya Amman University, Amman 19328, Jordan; a.alghazzy@ammanu.edu.jo
[2] Department of Chemistry, The University of Jordan, Amman 11942, Jordan
* Correspondence: a.alhunaiti@ju.edu.jo

Abstract: Here, we report a visible light-induced-trimetallic catalyst (Cu$_{0.5}$Zn$_{0.5}$Fe$_2$O$_4$) prepared through green synthesis using *Tilia* plant extract. These nanomaterials were characterized for structural and morphological studies using powder x-ray diffraction (P-XRD), scanning electron microscopy (SEM) and thermogravimetric analysis (TGA). The spinel crystalline material was ~34 nm. In benign reaction conditions, the prepared photocatalyst oxidized various benzylic alcohols with excellent yield and selectivity toward aldehyde with 99% and 98%; respectively. Aromatic and aliphatic alcohols (such as furfuryl alcohol and 1-octanol) were photo-catalytically oxidized using Cu$_{0.5}$Zn$_{0.5}$Fe$_2$O$_4$, LED light, H$_2$O$_2$ as oxidant, 2 h reaction time and ambient temperature. The advantages of the catalyst were found in terms of reduced catalyst loading, activating catalyst using visible light in mild conditions, high conversion of the starting material and the recyclability up to 5 times without loss of the selectivity. Thus, our study offers a potential pathway for the photocatalytic nanomaterial, which will contribute to the advancement of photocatalysis studies.

Keywords: trimetallic; nanoparticles; iron oxide; photocatalytic; magnetic; alcohol; oxidation

Citation: Ghazzy, A.; Yousef, L.; Al-Hunaiti, A. Visible Light Induced Nano-Photocatalysis Trimetallic Cu$_{0.5}$Zn$_{0.5}$-Fe: Synthesis, Characterization and Application as Alcohols Oxidation Catalyst. *Catalysts* 2022, 12, 611. https://doi.org/10.3390/catal12060611

Academic Editors: Stéphanie Lambert and Julien Mahy

Received: 11 April 2022
Accepted: 27 May 2022
Published: 2 June 2022

Publisher's Note: MDPI stays neutral with regard to jurisdictional claims in published maps and institutional affiliations.

Copyright: © 2022 by the authors. Licensee MDPI, Basel, Switzerland. This article is an open access article distributed under the terms and conditions of the Creative Commons Attribution (CC BY) license (https://creativecommons.org/licenses/by/4.0/).

1. Introduction

The high demand to use renewable energy sources is required to reduce the dependency on fossil fuels and the possibility to reduce emissions in the atmosphere. Therefore, solar energy can be a good renewable alternative to fossil fuels. In recent years, nano semiconductor photocatalysis, as a "green" technology, has been widely used for conducting various applications. Outstanding stability, good photostability, nontoxicity and low price make spinel iron oxide the photocatalyst of choice for environmental remediation [1–4]. It is known that the UV region occupies only ~4% of the entire solar spectrum, while 45% of the energy belongs to visible light [5]. Therefore, developing efficient visible-light photocatalysts for environmental remediation has become an active research area in photocatalysis research [6–12]. In the photoelectric conversion process, the most important reaction involves hydroxyl ions on the semiconductor surface reacting with the holes, forming hydroxyl radicals (·OH), which is the main cause of the photocatalytic activity. The hydroxyl radical is a powerful, non-selective oxidant that can rapidly oxidize many organic compounds [13]. Many perovskite- or spinel-type complex oxides have been found to have selective visible-light-driven photoactivity using peroxides as oxidant. Among them, ferrites have also received considerable attention as magnetic nanoparticles (MNPs) with diverse applications metal oxides (e.g., Fe$_2$O$_3$, ZnFe$_2$O$_4$, NiCo$_2$O$_4$, MnFe$_2$O$_4$ and BiFeO$_3$), as it may show visible light photocatalytic alternative [14–19].

Trimetallic iron-based oxides possess attractive properties and widespread applications [20]. One of the essential applications of trimetallic NPs is catalysis [21–24]. The catalyst efficiency surges exponentially by increasing the number of coordination sites through a significant increase in the surface/volume ratio of NPs. Glucose oxidation produces gluconic acid under mild conditions catalyzed by Ag/Au/Pd NPs [24]. The cyclization reaction between diazodicarbonyl compounds and oxalyl chloride was carried out through green AuFeAg catalyst to afford α,β- or β,β-dichloroenones [25]. Moreover, large-scale sensitive and selective organic transformations achieved using trimetallic NPs in various forms such as core/shell [23], alloys [24] and layer-by-layer [25].

The catalytic activity can be dependent on preparation methods, as it affects the morphology and physiochemical properties of the material. Therefore, several methods have been applied in preparation of multi-metallic NPs. Among the known methods, electrochemical synthesis, chemical and hydrothermal synthesis, coprecipitation, sol-gel, mechanical alloying and solvothermal technologies are main techniques utilized [26,27]. Here we use coprecipitation technique using a plant extract as a reducing agent instead of unnecessary reagents. Generally, natural product extracts obtained from bio-renewable sources have a great potential as reducing, stabilizing material as it contains polyphenolic, glycoside and flavonoid compounds. Furthermore, the waxy material in plant extract can help in preventing the aggregation of powder nanoparticles, thus it can be applied for the greener production of mono- and multi-metallic NPs [28,29].

Oxidation of benzyl alcohol to produce the must-need industrial intermediate benzaldehyde is a crucial process in the industry. Within the aldehyde family, benzaldehyde is an essential building block for many industrial products [30]. The attractive interest of benzyl alcohol and benzaldehyde in the research area is noticeable as increase of preparation protocols for benzaldehydes production in the last decade, as well its applications. The heterogeneous catalytic oxidation of aryl or alkyl alcohol to supply various worthy chemicals such as aldehydes or carboxylic acids has been endorsed beneficial over homogenous catalytic techniques due to their recoverable fashion and the help of eco-friendly conditions like the use of H_2O_2 or O_2 as mild oxidants [31–34].

Herein, we report a trimetallic iron oxide-based NPs, as green and affordable catalyst using phyto mediated extract (*Telia*) in preparation of $Cu_{0.5}Zn_{0.5}Fe_2O_4$ that was applied as a photo-induced oxidation catalyst of benzyl alcohols and aliphatic alcohols. As shown in Scheme 1, benzyl alcohol oxidation is presented, exploiting mild reaction conditions. The characterization of the synthesized nanoparticles has been carried out in detail, employing the Rietveld refinement method using XRD data, SEM and TGA images.

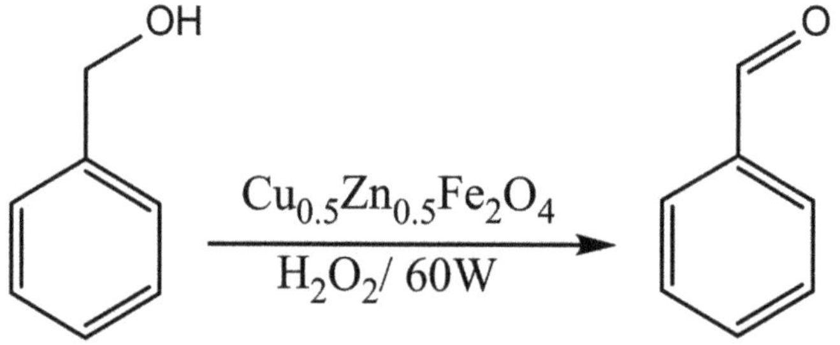

Scheme 1. Benzyl alcohol oxidation by using trimetallic Cu-Zn-Fe NPs in Acetonitrile.

To study the oxidation catalytic activity of Magnetic nanoparticles (MNP), benzyl alcohol was used as substrate model to optimize the reaction conditions. The present study has been extended to various alcohol substates to explore the possibilities of expanding the application of the catalyst, which shows a promising reactivity and applications. The novelty of the present work lies in the preparation of this magnetic trimetallic nanoparticles in a very simple, cheaper, eco-friendly coprecipitation synthesis process of nano particles, which can be used as a LED photo induced catalyst for oxidation of alcohols to obtain selectively aldehyde products.

2. Results and Discussion

2.1. Trimetallic Nanoparticles

The XRD patterns for the $Cu_{0.5}Zn_{0.5}FeO_4$ sample are shown in Figure 1. The XRD spectra for the samples with the diffraction peaks that appeared at 2θ values of 30.00, 35.40, 43.31, 50.41, 56.94 and 62.53 were assigned to the (220) (311) (400) (511) (440) and (550) planes, which was found to be in good agreement with $ZnFe_2O_4$ (PDF no. 01-089-1009) and Fe_3O_4 (PDF no. 01-088-0866). The signals consistent with the (220), (311), (400), (511), (440) and (533) reflect the spinel crystal structure of $CuFe_2O_4$ (PDF no.00-025-0283). These results show the successful loading of the cubic $Cu_{0.5}Zn_{0.5}Fe_2O_4$ spinel ferrites structures JCPDS: JCPDS: 00-051-0386. The lattice constant and crystallite size of the nanoparticles have been deliberated from the most conspicuous peak (311) using the Scherrer formula [35]. The X-ray diffraction nanoparticles reveal that $Cu_{0.5}Zn_{0.5}Fe_2O_4$ NPs possess a single-phase major cubic (fcc) spinel. The experimentally perceived d arrangement values and proportional strengths agree with those described in the records.

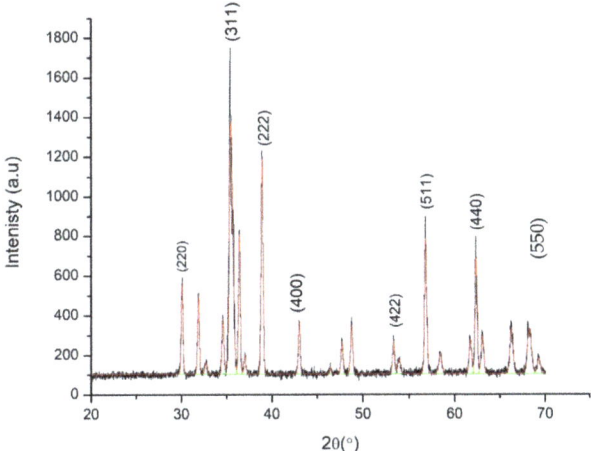

Figure 1. XRD patterns with Rietveld refinement the $Cu_{0.5}Zn_{0.5}Fe_2O_4$ sample.

The crystallite size (D) of each sample was calculated using the Stokes–Wilson Equation (1) [35,36]

$$D = \frac{\lambda}{\beta_c \cos \theta} \quad (1)$$

where λ is the X-ray wavelength, θ is the Bragg's angle of diffraction in degrees and β_c is the integral breadth of the diffraction peak corrected for the instrumental broadening. The diffraction peak used for the determination of the crystallite size was fitted with the relation:

$$\beta = \frac{A}{I_0} \quad (2)$$

Here A is the peak area and I_0 is the maximum intensity. The most intense peak (311) at $2\theta = 35.6°$ was fitted to provide information regarding crystallite size in the direction perpendicular to the corresponding crystallographic plane. In addition, the size of spinel phase was determined with different crystallographic directions using the (220) reflection at $2\theta = 30.0°$ and the (400) reflection at 43.04°; thus, the crystalline spinel $Cu_{0.5}Zn_{0.5}Fe_2O_4$ size was 34 nm.

The morphology of the prepared $CuZeFeO_4$ nanoparticles is explored by SEM image, as shown in Figure 2. The average particle size of the particles is ~31.2 nm.

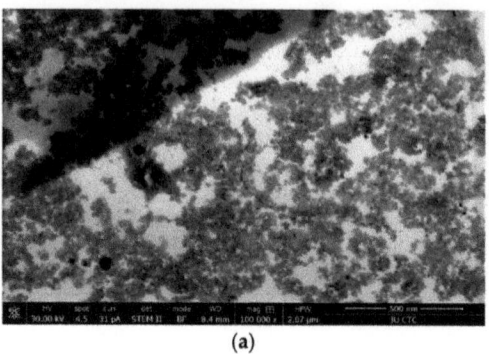 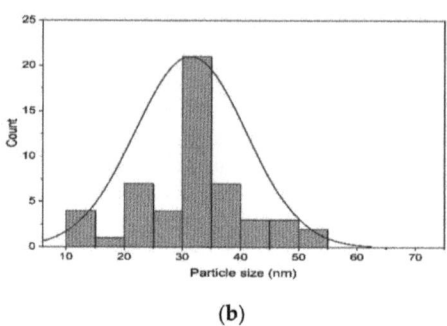

Figure 2. (a) FESEM images of the $Cu_{0.5}Zn_{0.5}Fe_2O_4$-NP; and (b) particle size distribution with a Gaussian fit average particle size (nm) = 31.2135 ± 0.23159.

2.2. TGA

Thermal stability of $Cu_{0.5}Zn_{0.5}Fe_2O_4$-NP was investigated by thermogravimetric analysis Figure 3. A gradual weight loss of 14 wt% was observed from 50–250 °C, owing to the reduction of water and CO_2 molecules [37]. In addition, the second drop of weight (12 wt%) at 250–450 °C can be attributed to the degradation of -OH ion from organic moieties of plant extract residues and decomposition of inorganic salts, which has two degradation mechanism involves both intermolecular and intramolecular transfer reactions [38]. The final loss of 11% occurred at 500–600 °C due to the formation of pure corresponding metal oxide. Notably, above 700 °C no weight loss was observed, which indicates the formation of $Cu_{0.5}Zn_{0.5}Fe_2O_4$-NP.

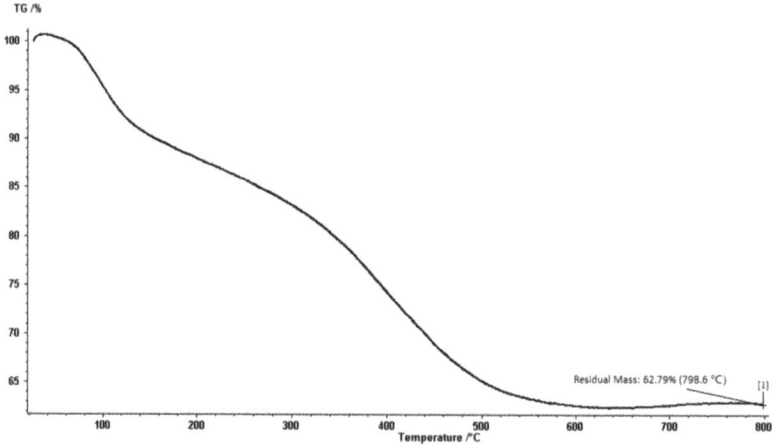

Figure 3. The thermogravimetric analysis (TGA) of the prepared nanoparticles.

2.3. Benzyl Alcohol Oxidation

The optimization of the catalytic reaction has been tested using benzyl alcohol as a model substrate, and the solvent effect is shown in Table 1. When using a different solvent, the reactivity changes drastically. To our delight, Acetonitrile has given the higher conversion for oxidizing benzyl alcohol (BA) into the anticipated aldehyde rather than isopropanol, while acetone and water have afforded a negligible amount of benzaldehyde. The high conversion of BA using acetonitrile can be explained via the formation of active radical species in the reaction media [39–41].

Table 1. Effect of solvents on the oxidation of benzyl alcohol.

Entry	Solvent	Yield * %
1	Acetonitrile	74
2	Isopropanol	67
3	Aceton	7
4	Water	2

* Benzaldehyde yield was analyzed using ^{1}H-NMR and UV-vis using internal standard. Reaction condition: benzyl alcohol = 0.1 mmol, catalyst = 15 mg, H_2O_2 = 250 mg, solvent 2 mL, light power 60 Watt, reaction time 4 h, temperature 25 °C.

In terms of optimizing reaction conditions, the influences of catalyst load amount, H_2O_2, on the BA substate have been shown in (Table 2). The blank experiments in entries 1 and 6 prove the need for catalyst and the oxidizing agent to produce an acceptable percentage of benzaldehyde. Excellent conversion (99%) of BA was obtained when the amount of catalyst load was 10–20 mg. Therefore, we used 10 mg of the catalyst to reduce the catalytic waste. Similarly, the H_2O_2 amount was optimized and 350 µL has provided the best targeted conversion into benzaldehyde. The role of the peroxide can be explained that the photo-induced electrons in the conduction band would dissolve oxygen and peroxide species which form peroxy and OH radicals intermediates. The reactive OH radicals species can oxidize alcohol into aldehyde.

Table 2. Quantitative optimization of Cu-Zn-Fe and oxidant amount.

Entry	Cu-Zn-Fe (mg)	H_2O_2 (µL)	Yield * %
1	0	350	1
2	7	350	95
3	10	350	99
4	15	350	99
5	20	350	89
6	10	0	9
7	10	250	54
8	10	350	99
9	10	500	99

* Benzaldehyde yield. Reaction condition: benzyl alcohol = 0.1 mmol, catalyst $Cu_{0.5}Zn_{0.5}Fe_2O_4$, ACN 2 mL, light power 60 Watt, reaction time 3 h, temperature 25 °C.

The effects of reaction time on the photocatalytic transformation of BA have been illustrated in Figure 4. The reaction was carried out up to 180 min, using the optimized conditions. In general, the oxidation rate of benzyl alcohol increased significantly when the reaction time raised for the first hour. The reaction ended up with a 99% yield in 120 min. This result shows that the catalyst is highly active in such conditions.

The catalytic performance is demonstrated by the four different illumination power from the LED source with an intensity of 7, 30, 60 and 120 W/cm^2, while the other conditions are set carefully at optimum conversion previous results. The reaction substrates included 10 mg of $Cu_{0.5}Zn_{0.5}Fe_2O_4$ catalyst, 1 mmol of benzyl alcohol, 350 mg of H_2O_2 and 2 mL of acetonitrile. With increasing illumination power by 23 W, the oxidation rate of benzyl alcohol surges dramatically to produce the highest yield 93%, while obviously the best conversion appears under 60 watt light power (Table 3).

Aliphatic and aromatic alcohol derivatives were chosen to explore the selective oxidation of various alcohol derivatives by $Cu_{0.5}Zn_{0.5}Fe_2O_4$ catalyst under the optimized conditions (Table 4). To our delight, the catalyst has shown a good to excellent reactivity toward both aliphatic and aromatic alcohols. Apparently, aromatic alcohols have shown high reactivity entries 1–4. The substituted benzylic alcohols have shown no effect on the obtained yield, while cinnamyl alcohol and the influence of allylic bond led to lower reactivity than the other aromatic substrates. This allows us to apply the catalyst to a lignin substrate model such as furfuryl alcohol, and indeed the catalyst has excellent yield and selectivity.

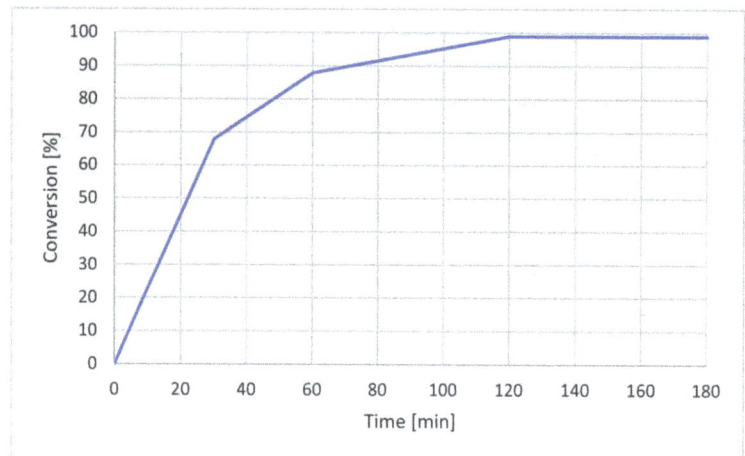

Figure 4. Effects of reaction time on the conversion of benzyl alcohol.

Table 3. Benzyl alcohol oxidation with different light power.

Entry	Light Power (W)	Yield * %
1	7	30.0
2	30	93.0
3	60	99
4	120	99

* Benzaldehyde yield. Reaction condition: (benzyl alcohol = 0.1 mmol, catalyst = 10 mg, Solvent ACN = 2 mL, H_2O_2 = 350 mg), reaction time 2 h, Temperature 25 °C.

Table 4. Alcohol derivatives on the selective oxidation of benzyl alcohol.

Entry	Substrate	Alcohol	Aldehyde	Yield * %
1	1-Octanol			29.2
2	2-Octanol			34.5
3	Furfuryl alcohol			99
4	Cinnamyl alcohol			61.4
5	3,4-dimethyl benzyl alcohol			99
6	Benzyl alcohol			99

* Benzaldehyde yield. Reaction condition: (alcohol = 0.1 mmol, catalyst = 10 mg, Solvent ACN = 2 mL, H_2O_2 = 350 µL, light power 60 Watt, reaction time 2 h, Temperature 25 °C.

When comparing our results to previously reported results, several studies conveyed to prepare adequate metallic catalysts to achieve this transformation, when most of them either use expensive metals like Pd, Pt and Rh [42–44] or electrical sources [45], while the reaction conditions depend on intense light power between 150 to 450 W, or high temperature 80 to 120 °C in both natural and synthetic sources. In addition, the percentage yield of benzaldehyde varies widely from around 28 to 99, shown in Table 5.

Table 5. A comparison studies using di- and trimetallic NPs as catalyst to oxidize alcohols.

Source	Metallic NPs	Reaction Conditions	Conversion %	Selectivity %	Ref.
Synthatic	Au-Pd/O-CNTs	2 h, 120 °C, Solvent free	28.3	96	[42]
Synthatic	Au@Ag/BiOCl–OV	1 h, Xenon 300 W, CH_3CN	92	99	[43]
Synthatic	Trimetallic PtPbBi	Electrical	-	-	[44]
Synthatic	2.5% Au + 2.5% Pd/TiO_2DP2	140 °C, 10 bar O_2	84	40	[45]
Synthatic	Pd/$H_2Ti_3O_7$ Nanowires	Halogen 150 W, 90 °C	49.5	76	[46]
Synthatic	TiO_2	LED 450 W, 36 h, O_2	99	99	[47]
Nature (Cacumen Platycladi)	Au–Pd/TiO_2	6 h, 90 °C, O_2	74.2	98	[48]
Nature (Oak fruit bark)	Pd NPs	12 h, 80 °C, K_2CO_3	95	-	[49]
This work (Tilia)	Trimetallic Cu-Zn-Fe	2 h, LED 60 W, H_2O_2	99	99	

2.4. Recyclability

The stability of the prepared materials when exposed to consecutive reaction cycles was tested. Therefore, we examine the recyclability by using benzyl alcohol as a substrate in the presence of 10 mg of the catalyst.

The catalyst was reused up to four successive runs without drastic loss in activity. The catalyst was washed with water/ethanol solution and oven dried pre-use for oxidation of benzyl alcohol (Figure 5). Interestingly, after the 5th run, the weight loss of the catalyst was 28 wt% using gravimetric assay. The drop in the reactivity can be due to the slight deactivation of the catalyst via leaching some of the metals in the nanoparticles, or poising the catalyst surface of unwanted reaction products. After the fifth cycle, the conversion dropped slightly without any loss of selectivity.

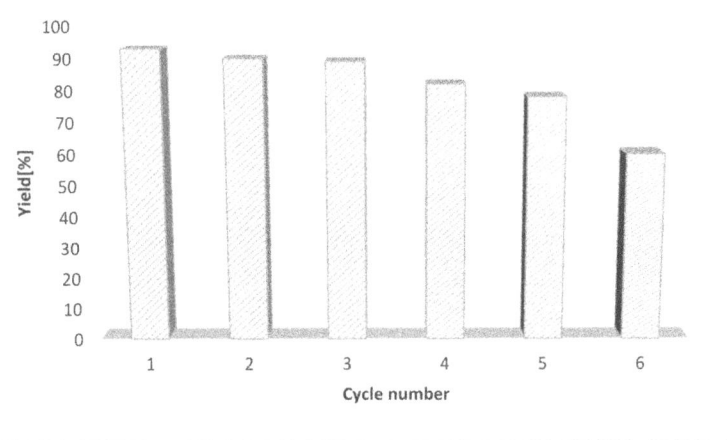

Figure 5. Recyclability test of $Cu_{0.5}Zn_{0.5}Fe_2O_4$-NP catalyst. The weight loss of catalyst after 5th runs was 28 wt%.

3. Experimental

3.1. Materials

The solvents used were purchased from sigma-Aldrich (Taufkirchen, Germany). $FeSO_4 \cdot 7H_2O$, $Zn(OAc)_2 \cdot 2H_2O$ $CuSO_4 \cdot 5H_2O$ and alcohols used were purchased from Sigma-Aldrich with 99.95% purity, and used as received. *Tilia* leaves extract was collected from a cultivated area in Amman, Jordan, during the flowering period in early spring of 2021. Sodium chloride (NaCl), acetic acid, sodium hydroxide and distilled water were used for all experiments. The crystallographic structure of MFe_2O_4 nano-ferrites was determined from the Powder X-ray diffraction (P-XRD, mnochromatic Cu-Kα radiation, nickel filter, 40 kV, 30 mA using Shimadzu XRD-7000, Japan) pattern, while the particle morphology and size distribution was determined with the aid of electron microscopy imaging (SEM, JOEL 6400 and FEI QUANTA 200, Japan).

3.2. Copper-Zinc Ferrite $Cu_{0.5}Zn_{0.5}Fe_2O_4$ NP Synthesis

The trimetallic ferrite nanoparticles were synthesized using an aqueous extract of *Telia*, as described previously [27]. Briefly, 1.12 g of $FeSO_4.7H_2O$ and 0.48 g of $Zn(OAC)_2.2H_2O$ and 0.48 g $CuSO_4 \cdot 5H_2O$ were dispersed in 20 mL distilled water for 1 h at room temperature under stirring, in order to obtain the homogeneous solution. A 50 mL of the aqueous plant extract was drop-wise added to the solution, and the mixture was stirred for 1 h and later the pH was controlled to obtain pH 10–12. The homogenous solution was then centrifuged, and the obtained solid products were collected and washed with distilled water and ethanol several times. Finally, the samples were dried in a vacuum oven at 60 °C for 6 h, and later calcinated at 1000 °C.

3.3. Characterization

The morphology and crystallinity of the synthesized $Cu_{0.5}Zn_{0.5}Fe_2O_4$ nanoparticles were determined via field emission scanning electron microscopy (SEM, FESEM JEOL JSM-6380) and (XRD, X'pert diffractometer using CuK α radiation), respectively. All measurements were performed at room temperature (25 ± 2 °C). Thermogravimetric analysis was accomplished using the System Setaram Setsys 12 TGA instrument (Setaram Instrumentation, Caluire, France), by heating the sample up to 800 °C at a rate of 10 °C·min^{-1}.

3.4. Photocatalytic Oxidation of Benzyl Alcohol to Benzaldehyde

To a stirred solution of 2 mL of acetonitrile containing 0.1 mmol benzyl alcohol, 15 mg of the prepared catalyst and 350 µL of H_2O_2 was added drop wise pre-irradiation. Subsequently, the suspension was irradiated by a 60 W LED, and the reaction was monitored by the absorbance using UV-vis for each sample.

3.5. Product Analysis

When the reaction ended, the magnetic catalyst was removed by applying an external magnet and the product was extracted by column chromatography and analyzed by ^{1}H-NMR and compared to the commercially available chemicals. The conversion percentage of benzyl alcohol was calculated by using the following Equation (3)

$$\text{Conversion (\%)} = [(C_0 - C_r)/C_0] \times 100\% \qquad (3)$$

where C_0 is the initial concentration of benzyl alcohol and C_r and C_t are the concentrations of benzyl alcohol and benzaldehyde, respectively.

3.6. Experimental Procedure for the Reuse of the Catalyst

The reaction was applied for benzyl alcohol oxidation in a 1.0 mmol scale. Keeping reaction conditions constant, except using the recycled $Cu_{0.5}Zn_{0.5}Fe_2O_4$ catalyst rather than fresh catalyst, the reaction was monitored. When the reaction ended, the catalyst was removed using an external magnet and the product was analyzed by either ^{1}H-NMR or

UV-Vis. The catalyst was collected and washed with water/ethanol (5 mL) three times and oven dried pre-reuse. The dried catalyst was applied for further catalytic cycle up to 5 times without any severe loss of reactivity.

4. Conclusions

In this study, we synthesized phyto-mediated trimetallic Np of $Cu_{0.5}Zn_{0.5}Fe_2O_4$ using a Telia extract. The NP catalyst was characterized by powder XRD, FESEM and TGA studies. The XRD indicated a spinel ferrite with average crystalline diameter ~34 nm and average particle size of 31.2 nm by the SEM. The trimetallic nanoparticles has great potential to act as a visible light photocatalyst using LED light for the oxidation of benzylic and aliphatic alcohols selectively to aldehyde, with high selectivity and reactivity under benign reaction conditions using 10 mg of catalyst, 3 equivalent H_2O_2, 60 W light and 2 h reaction time. Later, the catalyst was compared to previous works, and it shows a promising reactivity. The novelty of this study lies in the preparation of the magnetic trimetallic nanoparticles in a very simple, cheap and eco-friendly coprecipitation synthesis process of nano particles, which offers a potential pathway for photocatalytic oxidation reaction.

Author Contributions: Conceptualization, A.A.-H.; methodology, A.A.-H. and A.G.; software, L.Y. and A.A.-H.; validation, A.A.-H. and A.G.; formal analysis, A.A.-H. and L.Y.; investigation, A.G. and A.A.-H.; resources, A.A.-H.; writing—original draft preparation, A.G. and A.A.-H.; writing—review and editing, A.A.-H. and A.A.-H.; visualization, A.G. and A.A.-H.; supervision, A.A.-H.; project administration, A.A.-H.; and funding acquisition, A.A.-H. All authors have read and agreed to the published version of the manuscript.

Funding: This research was supported by Deanship of Academic Research at the University of Jordan (Grant No. 2364), and Al-Ahliyya Amman University.

Data Availability Statement: Not applicable.

Acknowledgments: The Jordanian Cell Therapy Center (CTC) is acknowledged for the SEM imaging. A.A.-H. acknowledges support by the Deanship of Scientific Research at the University of Jordan. A.G. would like to acknowledges support by Deanship of Scientific Research at Al-Ahliyya Amman University.

Conflicts of Interest: The authors declare no conflict of interest.

References

1. Ng, Y.H.; Lightcap, I.V.; Goodwin, K.; Matsumura, M.; Kamat, P.V. To What Extent Do Graphene Scaffolds Improve the Photovoltaic and Photocatalytic Response of TiO_2 Nanostructured Films? *J. Phys. Chem. Lett.* **2010**, *1*, 2222–2227. [CrossRef]
2. Zhang, H.; Lv, X.; Li, Y.; Wang, Y.; Li, J. P25-graphene Composite as a High Performance Photocatalyst. *ACS Nano* **2010**, *4*, 380–386. [CrossRef] [PubMed]
3. Woan, K.; Pyrgiotakis, G.; Sigmund, W. Photocatalytic Carbon Nanotube TiO_2 Composites. *Adv. Mater.* **2009**, *21*, 2233–2239. [CrossRef]
4. Kumaresan, L.; Mahalakshmi, M.; Palanichamy, M.; Murugesan, V. Synthesis, Characterization, and Photocatalytic Activity of Sr2þ Doped TiO_2 Nanoplates. *Ind. Eng. Chem. Res.* **2010**, *49*, 1480–1485. [CrossRef]
5. Zhang, Z.J.; Wang, W.Z.; Yin, W.Z.; Shang, M.; Wang, L.; Sun, S.M. Inducing Photocatalysis by Visible Light Beyond the Absorption Edge: Effect of Upconversion Agent on the Photocatalytic Activity of Bi_2WO_6. *Appl. Catal. B* **2010**, *101*, 68–73. [CrossRef]
6. Liang, Y.Y.; Wang, H.L.; Casalongue, H.S.; Chen, Z.; Dai, H.J. TiO_2 Nanocrystals Grown on Graphene as Advanced Photocatalytic Hybrid Materials. *Nano Res.* **2010**, *3*, 701–705. [CrossRef]
7. Vijayan, B.K.; Dimitrijevic, N.M.; Wu, J.S.; Gray, K.A. The Effects of Pt Doping on the Structure and Visible Light Photoactivity of Titania Nanotubes. *J. Phys. Chem. C* **2010**, *114*, 21262–21269. [CrossRef]
8. Romcevic, N.; Kostic, R.; Hazic, B.; Romcevic, M.; KuryliszynKudelska, I.; Dobrowolski, W.D.; Narkiewicz, U.; Sibera, D. Raman Scattering from ZnO Incorporating Fe Nanoparticles: Vibrational Modes and Low-Frequency Acoustic Modes. *J. Alloys Compd.* **2010**, *507*, 386–390. [CrossRef]
9. Cai, L.; Liao, X.; Shi, B. Using Collagen Fiber as a Template to Synthesize TiO_2 and Fe_x/TiO_2 Nanofibers and Their Catalytic Behaviors on the Visible Light-Assisted Degradation of Orange II. *Ind. Eng. Chem. Res.* **2010**, *49*, 3194–3199. [CrossRef]
10. Guo, R.Q.; Fang, L.A.; Dong, W.; Zheng, F.G.; Shen, M.R. Enhanced Photocatalytic Activity and Ferromagnetism in Gd Doped $BiFeO_3$ Nanoparticles. *J. Phys. Chem. C* **2010**, *114*, 21390–21396. [CrossRef]

11. Cai, W.D.; Chen, F.; Shen, X.X.; Chen, L.J.; Zhang, J.L. Enhanced Catalytic Degradation of AO$_7$ in the CeO$_2$-H$_2$O$_2$ System with Fe3þ Doping. *Appl. Catal. B* **2010**, *101*, 160–168. [CrossRef]
12. Shu, X.; He, J.; Chen, D. Visible-Light-Induced Photocatalyst Based on Nickel Titanate Nanoparticles. *Ind. Eng. Chem. Res.* **2008**, *47*, 4750–4753. [CrossRef]
13. Hirakawa, T.; Nosaka, Y. Properties of O$_2$ and OH Formed in TiO$_2$ Aqueous Suspensions by Photocatalytic Reaction and the Influence of H$_2$O$_2$ and Some Ions. *Langmuir* **2002**, *18*, 3247–3325. [CrossRef]
14. Xu, S.H.; Feng, D.L.; Shangguan, W.F. Preparations and Photocatalytic Properties of Visible-Light-Active Zinc Ferrite-Doped TiO$_2$ Photocatalyst. *J. Phys. Chem. C* **2009**, *113*, 2463–2467. [CrossRef]
15. Laokul, P.; Amornkitbamrung, V.; Seraphin, S.; Maensiri, S. Characterization and Magnetic Properties of Nanocrystalline CuFe$_2$O$_4$, NiFe$_2$O$_4$, ZnFe$_2$O$_4$ Powders Prepared by the Aloe Vera Extract Solution. *Curr. Appl. Phys.* **2011**, *11*, 101–108. [CrossRef]
16. Zhang, B.P.; Zhang, J.L.; Chen, F. Preparation and Characterization of Magnetic TiO$_2$/ZnFe$_2$O$_4$ Photocatalysts by a sol-gel Method. *Res. Chem. Intermed.* **2008**, *34*, 375–380. [CrossRef]
17. Gómez-Pastora, J.; Bringas, E.; Ortiz, I. Recent progress and future challenges on the use of high performance magnetic nano-adsorbents in environmental applications. *Chem. Eng. J.* **2014**, *256*, 187–204. [CrossRef]
18. Al-Hunaiti, A.; Ghazzy, A.; Sweidan, N.; Mohaidat, Q.; Bsoul, I.; Mahmood, S.; Hussein, T. Nano-Magnetic NiFe$_2$O$_4$ and Its Photocatalytic Oxidation of Vanillyl Alcohol—Synthesis, Characterization, and Application in the Valorization of Lignin. *Nanomaterials* **2021**, *11*, 1010. [CrossRef]
19. Al-Hunaiti, A.; Mohaidat, Q.; Bsoul, I.; Mahmood, S.; Taher, D.; Hussein, T. Synthesis and characterization of novel phyto-mediated catalyst, and its application for a selective oxidation of (VAL) into vanillin under visible light. *Catalysts* **2020**, *10*, 839. [CrossRef]
20. Sharma, G.; Kumar, D.; Kumar, A.; Ala'a, H.; Pathania, D.; Naushad, M.; Mola, G.T. Revolution from monometallic to trimetallic nanoparticle composites, various synthesis methods and their applications: A review. *Mater. Sci. Eng. C* **2017**, *71*, 1216–1230. [CrossRef]
21. Mondal, B.N.; Basumallick, A.; Chattopadhyay, P.P. Magnetic behavior of nanocrystalline Cu–Ni–Co alloys prepared by mechanical alloying and isothermal annealing. *J. Alloys Compd.* **2008**, *457*, 10–14. [CrossRef]
22. Roshanghias, A.; Bernardi, J.; Ipser, H. An attempt to synthesize Sn-Zn-Cu alloy nanoparticles. *Mater. Lett.* **2016**, *178*, 10–14. [CrossRef]
23. Lan, J.; Li, C.; Liu, T.; Yuan, Q. One-step synthesis of porous PtNiCu trimetallic nanoalloy with enhanced electrocatalytic performance toward methanol oxidation. *J. Saudi Chem. Soc.* **2019**, *23*, 43–51. [CrossRef]
24. Venkatesan, P.; Santhanalakshmi, J. Designed synthesis of Au/Ag/Pd trimetallic nanoparticle-based catalysts for Sonogashira coupling reactions. *Langmuir* **2010**, *26*, 12225–12229. [CrossRef]
25. Duan, H.; Wang, D.; Li, Y. Green chemistry for nanoparticle synthesis. *Chem. Soc. Rev.* **2015**, *44*, 5778–5792. [CrossRef]
26. Iravani, S.; Varma, R. Plant-derived Edible Nanoparticles and miRNAs: Emerging Frontier for Therapeutics and Targeted Drug-delivery. *ACS Sustain. Chem. Eng.* **2019**, *7*, 8055–8069. [CrossRef]
27. Imraish, A.; Al-Hunaiti, A.; Abu-Thiab, T.; Ibrahim, A.; Hwaitat, E.; Omar, A. Phyto-Facilitated Bimetallic ZnFe$_2$O$_4$ Nanoparticles via *Boswellia carteri*: Synthesis, Characterization, and Anti-Cancer Activity. *Anti-Cancer Agents Med. Chem.* **2021**, *21*, 1767–1772. [CrossRef]
28. Iravani, S.; Varma, R.S. Plants and plant-based polymers as scaffolds for tissue engineering. *Green Chem.* **2019**, *21*, 4839–4867. [CrossRef]
29. Pugh, S.; McKenna, R.; Halloum, I.; Nielsen, D.R. Engineering *Escherichia coli* for Renewable Benzyl Alcohol Production. *Metab. Eng. Commun.* **2015**, *2*, 39–45. [CrossRef]
30. Pillai, U.R.; Sahle-Demessie, E. Oxidation of alcohols over Fe^{3+}/montmorillonite-K10 using hydrogen peroxide. *Appl. Catal. A-Gen.* **2003**, *245*, 103–109. [CrossRef]
31. O'Brien, P.; Lopez-Tejedor, D.; Benavente, R.; Palomo, J.M. Pd Nanoparticles-Polyethylenemine-Lipase Bionanohybrids as Heterogeneous Catalysts for Selective Oxidation of Aromatic Alcohols. *ChemCatChem* **2018**, *10*, 4992–4999. [CrossRef]
32. Alshammari, H.M.; Alshammari, A.S.; Humaidi, J.R.; Alzahrani, S.A.; Alhuaimess, M.S.; Aldosari, O.F.; Hassan, H. Au-Pd Bimetallic Nanocatalysts Incorporated into Carbon Nanotubes (CNTs) for Selective Oxidation of Alkenes and Alcohol. *Processes* **2020**, *8*, 1380. [CrossRef]
33. An, H.; Deng, C.; Sun, Y.; Lv, Z.; Cao, L.; Xiao, S.; Zhao, L.; Yin, Z. Design of Au@Ag/BiOCl–OV photocatalyst and its application in selective alcohol oxidation driven by plasmonic carriers using O$_2$ as the oxidant. *CrystEngComm* **2020**, *22*, 6603–6611. [CrossRef]
34. Smit, J.; Wijn, H.P.J.; Ferrites; Mahmood, S.H. Properties and Synthesis of Hexaferrites. In *Hexaferrite Permanent Magnetic Material*; Wiley: New York, NY, USA, 1959; p. 54.
35. Mahmood, S.H. Magnetic anisotropy in fine magnetic particles. *J. Magn. Magn. Mater.* **1993**, *118*, 359–364. [CrossRef]
36. Narayanasamy, A.; Jeyadevan, B.; Chinnasamy, C.N.; Ponpandian, N.; Greneche, J.M. Structural, magnetic and electrical properties of spinel ferrite nanoparticles. In Proceedings of the Ninth International Conference on Ferrites, San Francisco, CA, USA, 3 January 2005; pp. 867–875.
37. Rincón-Granados, K.L.; Vázquez-Olmos, A.R.; Vega-Jiménez, A.; Ruiz, F.; Garibay-Febles, V.; Ximénez-Fyvie, L. Preparation, characterization and photocatalytic activity of NiO, Fe$_2$O$_3$ and NiFe$_2$O$_4$. *Materialia* **2021**, *15*, 177–182.

38. Marmisollé, W.A.; Azzaroni, O. Recent developments in the layer-by-layer assembly of polyaniline and carbon nanomaterials for energy storage and sensing applications. From synthetic aspects to structural and functional characterization. *Nanoscale* **2016**, *8*, 9890–9918. [CrossRef]
39. Jiang, C.; Markutsya, S.; Tsukruk, V.V. Compliant, robust, and truly nanoscale free-standing multilayer films fabricated using spin-assisted layer-by-layer assembly. *Adv. Mater.* **2004**, *16*, 157–161. [CrossRef]
40. Kroschwitz, J.I.; Howe-Grant, M. *Kirk-Othmer Encyclopedia of Chemical Technology*; John Wiley and Sons: New York, NY, USA, 1991; p. 127.
41. Al-Hunaiti, A.; Al-Said, N.; Halawani, L.; Haija, M.A.; Baqaien, R.; Taher, D. Synthesis of magnetic $CuFe_2O_4$ nanoparticles as green catalyst for toluene oxidation under solvent-free conditions. *Arab. J. Chem.* **2020**, *13*, 4945–4953. [CrossRef]
42. Karimi, B.; Abedi, S.; Clark, J.; Budarin, V. Highly efficient aerobic oxidation of alcohols using a recoverable catalyst: The role of mesoporous channels of SBA-15 in stabilizing palladium nanoparticles. *Angew. Chem. Int. Ed.* **2006**, *17*, 4776–4779. [CrossRef]
43. Wang, C.; Chen, W.; Chang, H.T. Enzyme Mimics of Au/Ag Nanoparticles for Fluorescent Detection of Acetylcholine. *Anal. Chem.* **2012**, *84*, 9706–9712. [CrossRef]
44. Zhu, Z.; Liu, F.; Fan, J.; Li, Q.; Min, Y.; Xu, Q. C_2 Alcohol Oxidation Boosted by Trimetallic PtPbBi Hexagonal Nanoplates. *ACS Appl. Mater. Interfaces* **2020**, *12*, 52731–52740. [CrossRef] [PubMed]
45. Miedziak, P.; Sankar, M.; Dimitratos, N.; Lopez-Sanchez, J.A.; Carley, A.F.; Knight, D.W.; Brian, T.; Christopher, J.K.; Hutchings, G.J. Oxidation of benzyl alcohol using supported gold–palladium nanoparticles. *Catal. Today* **2011**, *164*, 315–319. [CrossRef]
46. Higashimoto, S.; Kitao, N.; Yoshida, N.; Sakura, T.; Azuma, M.; Ohue, H.; Sakata, Y. Selective photocatalytic oxidation of benzyl alcohol and its derivatives into corresponding aldehydes by molecular oxygen on titanium dioxide under visible light irradiation. *J. Catal.* **2009**, *266*, 279–285. [CrossRef]
47. Du, M.; Zeng, G.; Huang, J.; Sun, D.; Li, Q.; Wang, G.; Li, X. Green photocatalytic oxidation of benzyl alcohol over noble-metal-modified $H_2Ti_3O_7$ nanowires. *ACS Sustain. Chem. Eng.* **2019**, *7*, 9717–9726. [CrossRef]
48. Hong, Y.; Jing, X.; Huang, J.; Sun, D.; Odoom-Wubah, T.; Yang, F.; Li, Q. Biosynthesized bimetallic Au–Pd nanoparticles supported TiO_2 for solvent-free oxidation of benzyl alcohol. *ACS Sustain. Chem. Eng.* **2014**, *2*, 1752–1759. [CrossRef]
49. Veisi, H.; Hemmati, S.; Qomi, M. Aerobic oxidation of benzyl alcohols through biosynthesized palladium nanoparticles mediated by Oak fruit bark extract as an efficient heterogeneous nanocatalyst. *Tetrahedron Lett.* **2017**, *58*, 4191–4496. [CrossRef]

Article

Natural Clay Modified with ZnO/TiO$_2$ to Enhance Pollutant Removal from Water

Julien G. Mahy [1,*], Marlène Huguette Tsaffo Mbognou [1,2,3], Clara Léonard [1], Nathalie Fagel [4], Emmanuel Djoufac Woumfo [2] and Stéphanie D. Lambert [1]

[1] Department of Chemical Engineering—Nanomaterials, Catalysis & Electrochemistry, University of Liège, B6a, Quartier Agora, Allée du Six Août 11, 4000 Liège, Belgium; tsaffombognou@yahoo.fr (M.H.T.M.); leonardclara1@gmail.com (C.L.); stephanie.lambert@uliege.be (S.D.L.)
[2] Laboratoire de Physico-Chimie des Matériaux Minéraux, University of Yaounde I, Yaounde 337, Cameroon; edjoufac2000@yahoo.fr
[3] Institute of Geological and Mining Research (IRGM), Ministère de la Recherche Scientifique et de L'innovation du Cameroun, Yaoundé 4410, Cameroon
[4] Laboratoire Argiles, Géochimie et Environnements Sédimentaires (AGEs), Department of Geology, Faculty of Sciences, University of Liège, 4000 Liège, Belgium; nathalie.fagel@uliege.be
* Correspondence: julien.mahy@uliege.be; Tel.: +32-366-3563

Abstract: Raw clays, extracted from Bana, west Cameroon, were modified with semiconductors (TiO$_2$ and ZnO) in order to improve their depollution properties with the addition of photocatalytic properties. Cu^{2+} ions were also added to the clay by ionic exchange to increase the specific surface area. This insertion of Cu was confirmed by ICP-AES. The presence of TiO$_2$ and ZnO was confirmed by the detection of anatase and wurzite, respectively, using X-ray diffraction. The composite clays showed increased specific surface areas. The adsorption property of the raw clays was evaluated on two pollutants, i.e., fluorescein (FL) and p-nitrophenol (PNP). The experiments showed that the raw clays can adsorb FL but are not efficient for PNP. To demonstrate the photocatalytic property given by the added semiconductors, photocatalytic experiments were performed under UVA light on PNP. These experiments showed degradation up to 90% after 8 h of exposure with the best ZnO-modified clay. The proposed treatment of raw clays seems promising to treat pollutants, especially in developing countries.

Keywords: smectite; adsorption; photocatalysis; pollutant removal; environment remediation

1. Introduction

The population growth, intensive industrialization, and agricultural practices that occurred in recent decades have led to an increase in environmental pollution, which is now considered a global crisis [1]. This scourge has its origins in the constant improvement in the standard of living and the strong demands of consumers. In Cameroon, for example, many cotton, pharmaceutical, fertilizer, tanning, and pesticide manufacturing industries release pollutants such as dyes, pesticides, or bacteria into the environment, leading to disturbances of aquatic fauna and constituting a risk for human health [2]. Faced with this alarming situation, the global demand for water, the most vital natural resource, is increasing [3] and at the same time, the quality of freshwater sources is declining due to the presence of emerging contaminants. Most of these contaminants escape conventional wastewater treatment offered by wastewater treatment plants. The presence of these emerging pollutants in the environment is a matter of concern for most environmental agencies in developing countries [4]. This water should be treated as part of the recycling of wastewater that can be used by low-income populations for watering vegetable crops and washing cars and clothes in order to allow these populations to have a profitable and healthy economic activity.

In order to limit the arrival of these various types of refractory contaminants into the environment, effective and ecological treatment strategies have been developed, such as the use of local clays widely available in Cameroon from kaolinites, andosols, illites, and smectites [5], and globally, the use of adsorption as an efficient process to remove pollutants [6]. Clays have been the subject of different characterizations and applications [7]. For nearly three decades, many research works have been carried out on clay materials from Cameroon and their applications [8]. The search for new deposits and the characterization and valuation of clay materials are still relevant today.

Advanced oxidation processes (AOPs) have been applied in several sectors for the treatment of surface and groundwater [9,10] and for the elimination of odors and volatile organic compounds [11], as well as for water discoloration, the degradation of phytosanitary and pharmaceutical products [12], the production of molecules such as H_2 [13], and water disinfection [14]. AOPs can be used either as an oxidative pretreatment leading to easily biodegradable compounds, or as a tertiary treatment method for the removal or complete mineralization of residual pollutants [15]. This process is based on the generation of radical species able to degrade organic pollutants thanks to the use of a photocatalyst material activated by UV radiation [16]. The most-used UV-sensitive photocatalysts are TiO_2 and ZnO [17–19]. Different composites of photocatalysts have already been developed for pollutant removal [20–24].

In this work, a combination of adsorption and photocatalysis through the synthesis of mixed materials based on smectite-TiO_2 or smectite-ZnO is presented. Two types of pollutants are explored, one dye and one pesticide-type pollutant: fluorescein (FL) and p-nitrophenol (PNP), respectively. The physico-chemical properties of the pure and mixed materials are determined as well as their adsorption and photocatalytic activities. The production of mixed materials allows the use of a material already present in Cameroon and the addition of small fraction (<30%) of photocatalysts to increase the pollutant removal efficiency of the clay. The efficiency of the process and the cost can be studied and compared to other known methods.

The advantages of using semiconductor-modified clay materials for pollutant removal in water in developing countries are numerous: (i) the materials are composed primarily of natural material (the clay) directly located in the country where the pollution will be treated; (ii) the semiconductor material loading stays low (<30 wt %), reducing the cost of production; (iii) ZnO and TiO_2 are the most common semiconductor materials and can be produced with green synthesis with low use of organic reagents; (iv) the composite material presents both high adsorption capacity and photocatalytic properties, increasing its depollution properties compared to bare materials; and (v) the process for the production of the composite materials is simple.

2. Results and Discussion

2.1. Composition

Macroscopically, the raw clays, the Cu^{2+}-modified clays, and the TiO_2-modified clays are pale yellow. The ZnO-modified clays are slightly gray. The main compositions of the six different clay samples, determined by ICP-AES, are presented in Table 1.

The clays contain 9–21% of Si, 5–11% of Al, and 1–4% of Fe with an atomic Si/Al ratio equal to 2, consistent with a smectic composition [25]. The amount of copper increases up to 0.8% in the Cu^{2+}-modified samples (Table 1). The percentage of ZnO reaches 28.1% and 30.3% in the Clay/ZnO and Clay/ZnO/Cu^{2+} samples, respectively. The percentage of TiO_2 is 28.8% and 27.6% in the Clay/TiO_2 and Clay/TiO_2/Cu^{2+} samples, respectively.

The XRD patterns of the eight samples (Figure 1) allow us to estimate the crystallinity of the samples.

Table 1. Sample compositions by ICP-AES.

	Al	Si	Fe	Cu	TiO_2	ZnO
	wt %	wt %	wt %	wt %	wt %	wt %
Bare Clay	10.1	20.9	3.7	<0.1	<0.1	<0.1
Clay/Cu^{2+}	11.5	21.2	4.2	0.8	<0.1	<0.1
Clay/ZnO	5.2	9.2	1.6	<0.1	<0.1	28.1
Clay/ZnO/Cu^{2+}	6.4	11.6	1.8	0.4	<0.1	30.3
Clay/TiO_2	5.7	10.2	1.6	<0.01	28.8	<0.1
Clay/TiO_2/Cu^{2+}	5.9	11.6	1.2	0.3	27.6	<0.1

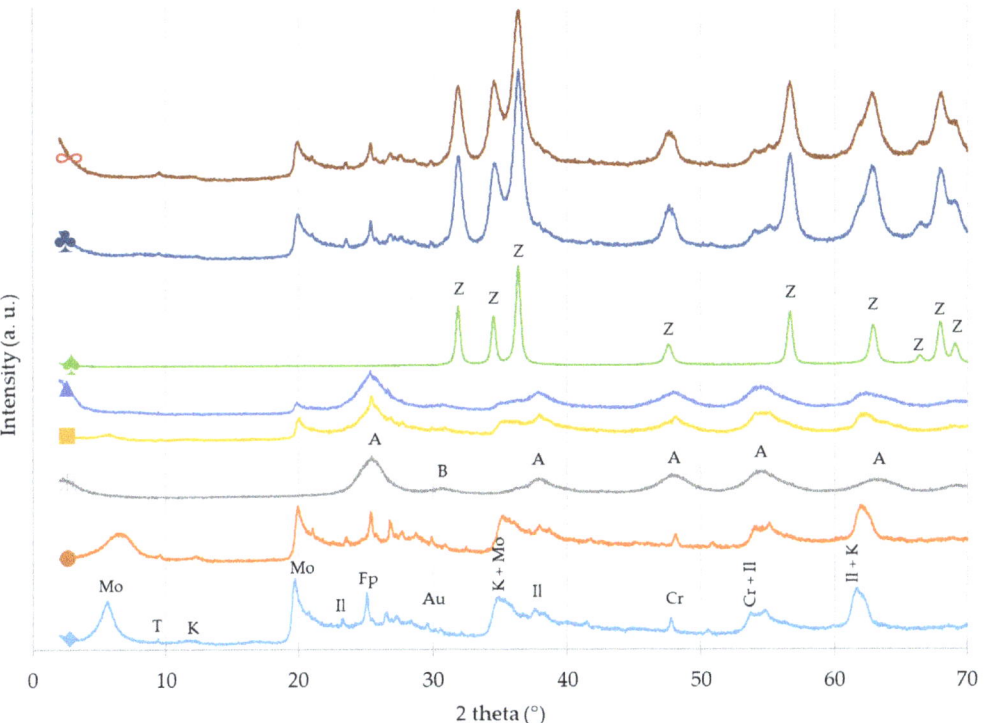

Figure 1. XRD patterns of samples: (♦) Bare Clay, (●) Clay/Cu^{2+}, (*) pure TiO_2, (■) Clay/TiO_2, (▲) Clay/TiO_2/Cu^{2+}, (♠) pure ZnO, (♣) Clay/ZnO, (∞) Clay/ZnO/Cu^{2+}. The positions of the reference peaks are indicated on the three pure materials (Bare Clay, TiO_2, and ZnO) by the following letters: (A) anatase, (B) brookite, (Z) wurzite, (Mo) montmorillonite, (T) talc, (K) kaolinite, (Il) illite, (Fp) feldspar, (Au) augite, and (Cr) cristobalite. The positions are not indicated on the composites materials to not overload the figure.

The Bare Clay (♦) is mainly composed of smectite, which is a family of different clay minerals observed in Figure 1 (all the following phases are observed: augite, cristobalite, montmorillonite, illite, kaolinite, feldspar, and talc). Smectite forms an important group of the phyllosilicate family of minerals, which are distinguished by layered structures composed of polymeric sheets of SiO_4 tetrahedra linked to sheets of (Al, Mg, Fe) $(O,OH)_6$ octahedra (Figure 2) [26–28].

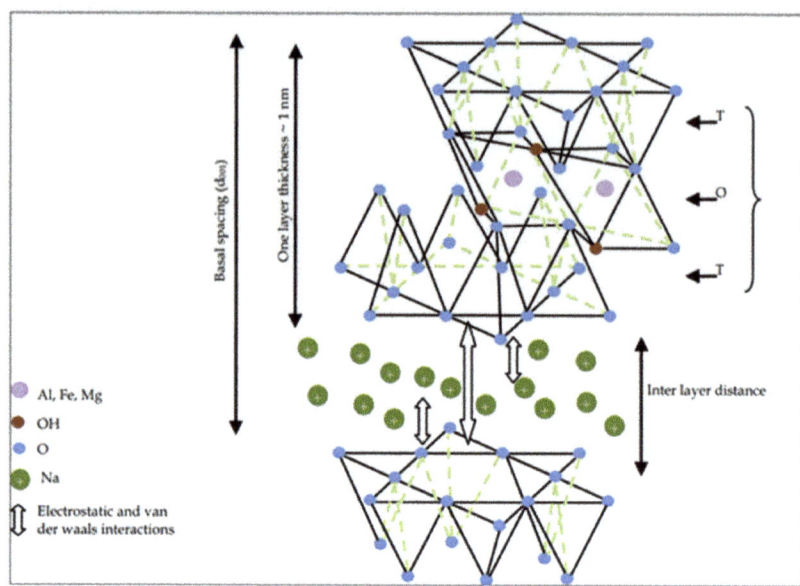

Figure 2. Smectite structure scheme from [26].

When Cu^{2+} ions are introduced to the network (Clay/Cu^{2+} sample, in orange (●) in Figure 1), a similar XRD pattern was recorded; however, the peak around 5–6° was spread due to the Cu^{2+} insertion.

The pure TiO_2 sample (Figure 1, pattern in gray (*)) is composed of anatase with a small amount of brookite (denoted A and B in Figure 1). These mixed phases were previously observed in aqueous sol-gel synthesis [29]. The pure ZnO sample (Figure 1, pattern in green (♠)) is made of wurtzite phase, as expected with this synthesis method [30].

The XRD results (Figure 1) confirm the successful production of hybrid clay–photocatalytic materials. Indeed, when the clay is modified with TiO_2, the corresponding XRD patterns (patterns in yellow (■) and mid blue (▲) in Figure 1) present the characteristic TiO_2 and clay peaks for both Clay/TiO_2 and Clay/TiO_2/Cu^{2+} samples if the peak positions are compared to the bare samples. The XRD patterns of the ZnO-modified clays (patterns in red (∞) and dark blue (♣) in Figure 1) likely present characteristic peaks of both wurtzite and clays.

2.2. Texture and Morphology

Table 2 presents the specific surface areas (S_{BET}) of the different samples, ranging from 30 to 325 m^2/g. The Bare Clay sample has a relatively low specific surface area (45 m^2/g), which increases slightly when Cu^{2+} ions are intercalated (55 m^2/g). This increase comes from the insertion of the cations in the smectite network [28]; indeed, this insertion is observed in the XRD patterns (Figure 1) with the spread of the peak around 5°. The pure TiO_2 sample presents an S_{BET} value equal to 180 m^2/g, in agreement with literature data [29]. When the clay is modified with TiO_2, S_{BET} increases to 325 and 240 m^2/g for Clay/TiO_2 and Clay/TiO_2/Cu^{2+} samples, respectively. This is logical, as these composite materials are produced with nanospheres of TiO_2, which have high specific surface area. They can also enter the clay network to expand the material and thus increase its specific surface area. The pure ZnO sample presents a low S_{BET} value (30 m^2/g). When clay is modified with ZnO, the specific surface area increases for Clay/ZnO sample (125 m^2/g), but it stays relatively low for Clay/ZnO/Cu^{2+} (50 m^2/g). The increased surface area of the Clay/ZnO sample could come from an insertion of some ZnO particles inside the clay network.

Table 2. Specific surface areas of samples.

Sample	Specific Surface Area (m²/g) ± 5
Bare Clay	45
Clay/Cu^{2+}	55
Pure TiO_2	180
Clay/TiO_2	325
Clay/Cu^{2+}/TiO_2	240
Pure ZnO	30
Clay/ZnO calciné à 300 °C	125
Clay/Cu^{2+}/ZnO	50

Concerning the nitrogen adsorption–desorption isotherms, two different types are observed between all samples: (i) type I isotherm, with a sharp increase at low pressure followed by a plateau corresponding to microporous solid; and (ii) type IV isotherm, characterized by a broad hysteresis at high pressure (mesoporous solid). Samples containing TiO_2 have type I isotherms, and the other samples have type IV isotherms. As an example, the isotherms of Bare Clay and Clay/TiO_2 samples are plotted in Figure 3. The other isotherms are represented in Figures S1 and S2 in the Supplementary Materials.

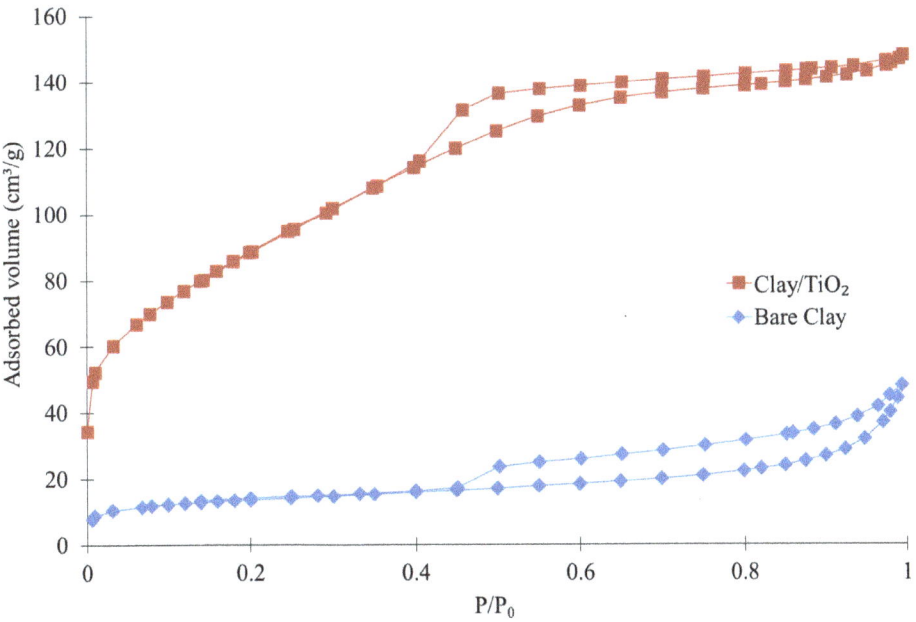

Figure 3. Nitrogen adsorption–desorption isotherms for (♦) Bare Clay and (■) Clay/TiO_2 samples.

SEM pictures of several samples are presented in Figure 4 for Bare Clay, Clay/TiO_2, and Clay/ZnO at two different magnifications. Bare Clay and Clay/ZnO samples have a similar aspect (Figure 4a,c), with large, granular powder, while the Clay/TiO_2 powder is finely dispersed (Figure 4b). These observations are in agreement with the higher specific surface area of Clay/TiO_2 and Clay/TiO_2/Cu^{2+} samples, characteristic of smaller hybrid particles and resulting in smaller voids between particles. This finely dispersed aspect comes from the TiO_2 nanoparticles, which are very small (5–10 nm), as observed in the TEM pictures of pure TiO_2 samples (Figure 5a). Contrarily, the pure ZnO sample has larger particles (Figure 5b), indicating that the composite material with clay is more similar to the Bare Clay.

Figure 4. SEM pictures of (**a**) Bare Clay, (**b**) Clay/TiO$_2$, and (**c**) Clay/ZnO samples at 1000× magnification; (**d**) Bare Clay, (**e**) Clay/TiO$_2$, and (**f**) Clay/ZnO at 2500× magnification.

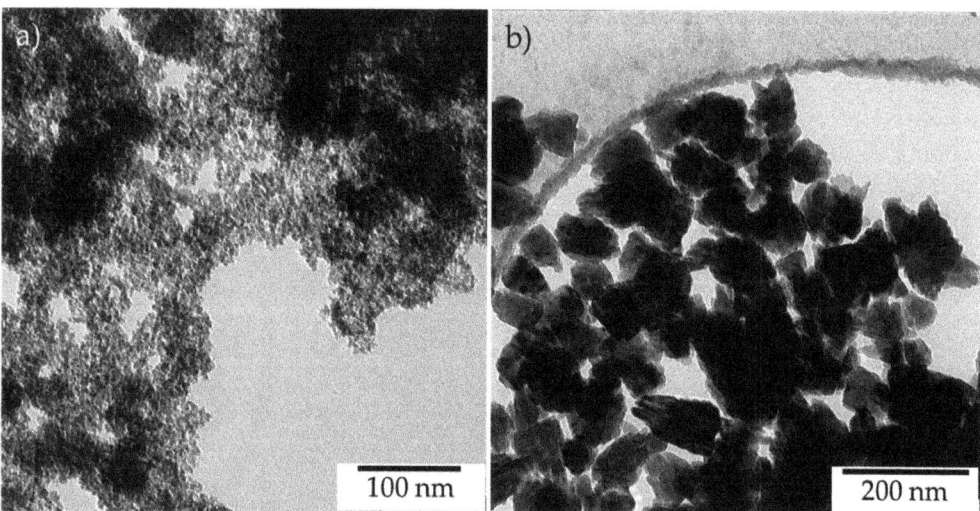

Figure 5. TEM pictures of (**a**) pure TiO$_2$ and (**b**) pure ZnO.

The samples with Cu^{2+} (Clay/Cu^{2+}, Clay/TiO$_2$/Cu^{2+} and Clay/ZnO/Cu^{2+}) have similar aspects and are represented in Supplementary Figure S3.

ICP results (Table 1) confirmed the presence of the semiconductor materials in the composite materials.

2.3. Adsorption Study

The experimental results of fluorescein adsorption are transformed with the following equation to determine the amount of FL adsorbed per g of clay (q_e):

$$q_e = \frac{(C_0 - C_e) * V}{W} \quad (1)$$

where C_0 and C_e are the initial and equilibrium liquid-phase concentrations of FL (mg$_{FL}$ L^{-1}), respectively; V is the volume of the FL solution (L); and W is the mass of clay used (g$_C$).

q_e in function of C_e is represented in Figure 6a after 6 h of adsorption for Bare Clay and Clay/Cu^{2+} samples. Similar curves are obtained for both samples; indeed, they have similar specific surface areas (Table 2).

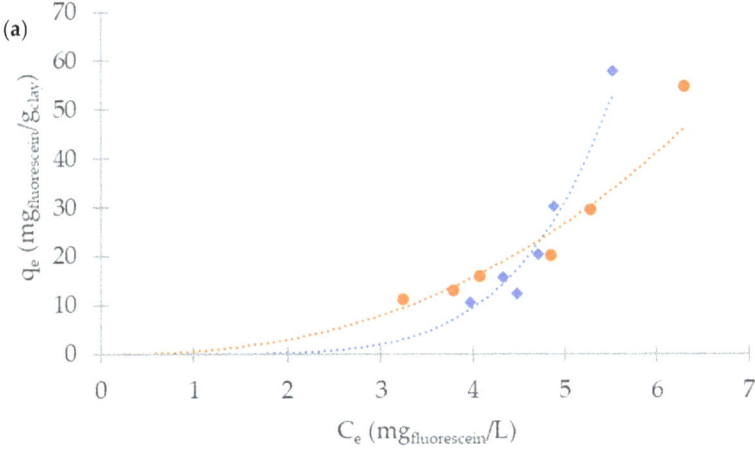

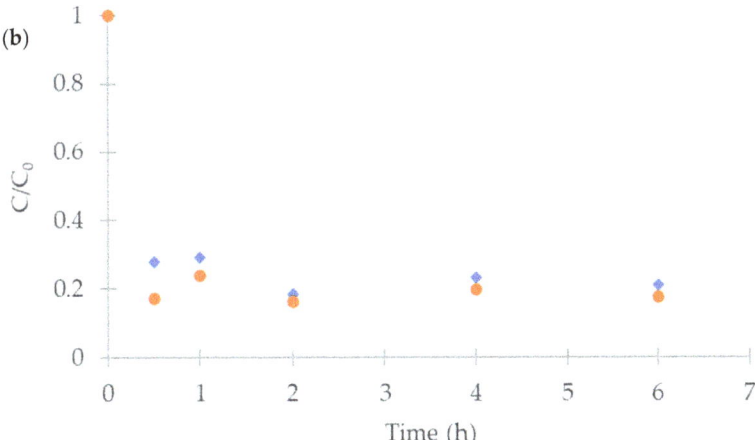

Figure 6. (a) Experimental fluorescein adsorption experiment representing the amount of FL adsorbed per g of clay in function of the equilibrium liquid-phase concentrations of FL after 6 h adsorption tests for the 6 different concentrations of powder samples for (♦) Bare Clay and (•) Clay/Cu^{2+} samples. (b) C/C_0 evolution with time for 30 mg concentrations of powder samples for (♦) Bare Clay and (•) Clay/Cu^{2+}.

In Figure 6b, the evolution of the FL concentration (C/C_0) with time is represented for Bare Clay and Clay/Cu^{2+} with 30 mg concentrations of powder samples. After 0.5 h of the experiment, more than 75% of FL were adsorbed for both samples (Figure 6b) and the concentration did not decrease much after 6 h; thus, the equilibrium was reached.

An example of the FL UV-visible spectrum is given in Figure S4 in the Supplementary Materials.

The PNP adsorption study shows that PNP is not adsorbed on the clay. Indeed, the concentration in solution remains constant with time. The removal of this kind of pollutant requires photocatalytic properties.

2.4. Photocatalytic Activity

As observed in the previous section, the clay can adsorb some pollutants such as dye, but it is not efficient to adsorb, for instance, PNP. Therefore, the clay was modified with photocatalysts to degrade molecules that cannot be efficiently adsorbed. Adsorption experiments in the dark (to avoid an interaction with room light which can activate the photocatalytic materials) were performed on all eight samples in contact with PNP for 8 h. No change in the PNP concentration was observed, showing that none of the samples adsorb the PNP molecule.

The photocatalytic property was evaluated on the PNP degradation under UVA illumination after 8 h of exposure (Figure 7a). An example of the PNP UV-visible spectrum is given in Figure S5 in the Supplementary Materials.

The Bare Clay and Clay/Cu^{2+} samples originally had no photocatalytic properties. However, such properties were attained after treatment with either TiO_2 or ZnO. PNP was degraded from 45% to 92% according to the samples. Clay/ZnO/Cu^{2+} is the most efficient material, with PNP degradation of 92%. The pure TiO_2 and ZnO materials reached 100% PNP degradation, as observed in previous studies [18,19].

For the same amount of semiconductor material (TiO_2 or ZnO), the ZnO-modified clay is more efficient than the TiO_2-modified one. As previously observed [18,31], ZnO materials have better activity than TiO_2, due to fewer recombinations of photogenerated species. The addition of Cu^{2+} ions increases the photoactivity due to an additional photo-Fenton effect that improves the PNP degradation [32]. Indeed, Cu^{2+} ions can react with water when exposed to UV radiation to produce additional $^\bullet OH$ radicals [32]. These radicals can degrade the organic molecules and thus enhance the photoactivity [32]. The equation of Cu^{2+} photo-Fenton effects is the following [32]:

$$Cu^{2+} + H_2O + h\nu \rightarrow Cu^+ + {}^\bullet OH + H^+$$

where h is the Planck constant (6.63×10^{-34} J.s) and ν is the light frequency (Hz).

For the two best composite materials (Clay/TiO_2/Cu^{2+} and (•) Clay/ZnO/Cu^{2+}), the evolution of the PNP degradation over time is presented in Figure 7b. The evolution of the degradation is linear; then, the degradation is first order.

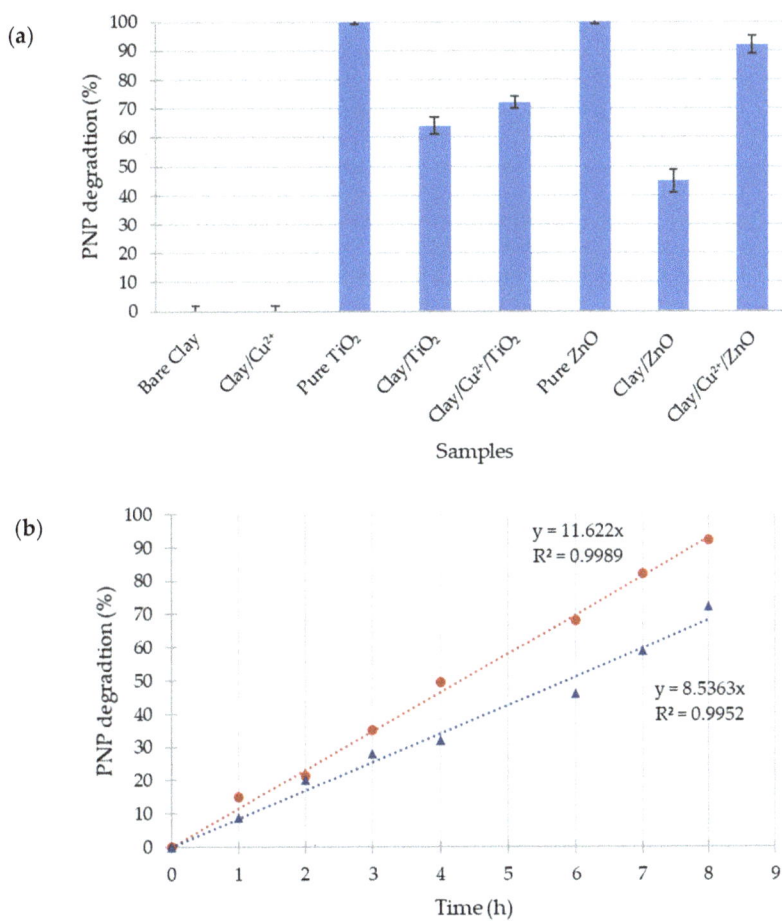

Figure 7. (a) PNP degradation (%) under UVA illumination for 8 h with all samples and (b) PNP degradation evolution over 8 h for the two best composite samples, (▲) Clay/TiO$_2$/Cu^{2+} and (•) Clay/ZnO/Cu^{2+}.

3. Materials and Methods

3.1. Description of the Clay and Modification with Cu^{2+} Ions

3.1.1. Presentation of the Clay

The clay material was whitish in color, sampled at 20 cm depth. The UTM coordinates of the sampling were north 5°06′08.3″ and east 10°17′28.0″, at an altitude of 1423 m. These coordinates corresponded to Mont Batcha, commonly called Bakotcha, in the district of Bana (West Cameroon). This region has an equatorial climate, characterized by average annual rainfall of 1300–2500 mm and a mean annual temperature of 21.23 °C [33]. The vegetation is highly anthropogenized post-forestry savannah with remains of a persisting semi-deciduous forest in areas of difficult accessibility [34]. The dominant soil types are red ferralitic soils, associated with brunified and hydromorphic soils [35]. The sampled clay was air-dried in the laboratory to a constant weight before grinding and sieving in a 160 µm diameter sieve.

3.1.2. Modification of Clay with Ions Cu^{2+} or Interfoliar Cation Exchange

This treatment does not destroy the structure of the clay material and it allows the insertion of ions (as shown in Figure 2). We used the following reagents: copper (II) sulfate pentahydrate (>98.0% from LabChem, Gauteng, South Africa), barium sulfate (99%, pure, from Laboratoriumdiscounter), clay powder (>160 µm), and distilled water.

In order to produce a homogeneous cation exchange, 50 g of clay was mixed under stirring in 0.1 M of $CuSO_4$ solution for 4 h. After 2 h rest, the supernatant was poured, and the agitation was repeated with a new solution of 0.1 M of $CuSO_4$. This operation was repeated twice, and excess Cu^{2+} and SO_4^{2-} ions were washed with distilled water until the Baryum test (precipitation test) became negative. The homoionic Cu^{2+} clay material was then oven-dried at 110 °C overnight.

3.2. Synthesis of Pure TiO_2 and ZnO Photocatalysts

3.2.1. ZnO Sample

Pure zinc oxide powders were synthesized by the sol-gel method following Benhebal et al. [30,36]. The reagents were zinc acetate dihydrate (≥98%), oxalic acid dihydrate (≥99%), and absolute ethanol (ACS grade). They were obtained from BIOCHEM, Chemopharma (Cosne-Cours-sur-Loire, France), of analytical grade, and used directly as purchased.

Zinc acetate dihydrate (10.98 g) was treated with ethanol (300 mL) at 60 °C. The salt was completely dissolved in about 30 min. Oxalic acid dihydrate (12.6 g) was dissolved in ethanol (200 mL) at 60 °C for 30 min. The oxalic acid solution was added slowly, with stirring, to the hot ethanolic zinc solution, and the mixture was stirred for 90 min at 50 °C. The resulting gel was placed in an oven at 80 °C for 24 h. The product was calcined at 400 °C for 4 h. The color of the pure ZnO powder was white.

3.2.2. TiO_2 Sample

Pure titanium oxide powders were synthesized by the sol-gel method of Mahy et al. [32]. The reagents used were titanium (IV) tetraisopropoxide (TTIP > 97%, Sigma-Aldrich, St. Louis, MO, USA), nitric acid (HNO_3, 65%, Merck, Darmstadt, Germany), isopropanol (IsoP, 99.5%, Acros, Hull, Belgium), and distilled water.

Nitric acid HNO_3 (65%, Merck) was used to acidify 250 mL of distilled water to pH 1. Then, 15 mL of TTIP was added to 15 mL of isopropanol (IsoP), and the mixture was stirred for 30 min at room temperature. The resulting solution of TTIP + IsoP mixture was added to acidified water under controlled stirring. The liquid was left under stirring for 4 h at 80 °C. The obtained sol had a clear blue color. Then, the sol was dried for 10 h under an ambient air flow to obtain a xerogel. The powders were dried at 100 °C for 1 h and a pure TiO_2 powder of yellowish-white color was obtained [32].

3.3. Synthesis of Hybrid Clay/Photocatalyst Materials

3.3.1. Clay/ZnO Materials

For the preparation of modified clays with ZnO, the procedure was similar as for pure ZnO material. However, when the oxalic acid solution was added slowly with stirring to the hot ethanolic zinc solution, 10 g of clay materials was added, and the mixture was left under stirring for 90 min at 50 °C. The resulting gel was placed in an oven at 80 °C for 24 h. The product was calcined at 400 °C for 4 h. The ZnO-modified clay powders were light gray in color.

3.3.2. Clay/TiO_2 Materials

For the preparation of hybrid clay/TiO_2 powders, the same protocol of preparation of pure TiO_2 powder was used with the addition of 10 g of clay material. When the mixture TTIP + IsoP was obtained, it was added to acidified water under controlled stirring and the liquid was left under stirring for 4 h at 80 °C. To the obtained sol, clear blue in color, 10 g of clayey material was added and left under stirring for 2 h. The soil was dried for 24 h under an ambient air flow. The powders were dried at 100 °C for 1 h and hybrid clay/TiO_2 powders were obtained.

3.4. Characterization of Samples

The actual composition of the bare and modified clays was determined by inductively coupled plasma–atomic emission spectroscopy (ICP–AES), equipped with an ICAP 6500 THERMO Scientific device (Waltham, MA, USA). The mineralization is fully described in [32]; however, we used HF instead of HNO_3.

The crystallographic properties were observed through the X-ray diffraction (XRD) patterns recorded with a Bruker D8 Twin-Twin powder diffractometer (Bruker, Billerica, MA, USA) using Cu-Kα radiation.

The specific surface area of samples was determined by nitrogen adsorption–desorption isotherms in an ASAP 2420 multi-sampler volumetric device from Micromeritics (Norcross, GA, USA) at 77 °K.

SEM micrographs were obtained using a Jeol-JSM-6360LV microscope (Tokyo, Japan) under high vacuum at an acceleration voltage of 20 kV.

Transmission electron microscopy was performed on the LEO 922 OMEGA Energy Filter Transmission Electron Microscope (Zeiss, Oberkochen, Germany) operating at 120 kV. Sample preparation consisted of dispersing a few milligrams of each sample in water, using sonication. Then, a few drops of the supernatant were placed on a holed carbon film deposited on a copper grid (CF-1.2/1.3-2 Cu-50, C-flat™, Protochips, Morrisville, NC, USA).

3.5. Adsorption Experiments

Concerning the adsorption experiments, only the Bare Clay and Clay/Cu^{2+} samples were assessed. The adsorption of two types of model pollutants, fluorescein (FL) and p-nitrophenol (PNP), was tested. For an adsorption experiment, 6 vials were prepared containing 5, 10, 15, 20, 25, and 30 mg of powder clay and 20 mL of pollutant solution in water. The samples were under continuous stirring. The remaining concentration in solution was evaluated every hour for 6 h with a Genesys 10S UV-Vis spectrophotometer (Thermo Scientific) after filtration with a syringe filter. The main absorption peaks were located at 317 and 485 nm for PNP and FL, respectively, as shown in Figures S4 and S5 in the Supplementary Materials. The initial concentration of FL was 6×10^{-5} M and 10^{-4} M for PNP.

3.6. Photocatalytic Experiments

The degradation of p-nitrophenol (PNP) was studied under UVA light (λ = 365 nm) to determine the photocatalytic activity of the synthesized material. The lamp was an Osram Sylvania, Blacklight-Bleu Lamp, F 18W/BLB-T8, considered as monochromatic at 365 nm.

Each sample was placed in a Petri dish with 20 mL of 10^{-4} M of PNP solution in water. The degradation of PNP was evaluated from absorbance measurements with a Genesys 10S UV-Vis spectrophotometer (Thermo Scientific) at λ = 317 nm. Previously, adsorption tests were performed in the dark (dark tests) to show whether PNP was adsorbed on the surface of samples. A blank test, consisting of irradiating the pollutant solution for 24 h in a Petri dish without any catalyst, showed that PNP concentration under UVA illumination remained constant. The Petri dishes with catalysts and pollutants were stirred on orbital shakers and illuminated for 8 h. Aliquots of PNP were sampled at 0, 4, and 8 h. The photocatalytic degradation was equal to the total degradation of PNP, taking the catalyst adsorption (dark test) into account. Each photocatalytic measurement was triplicated to assess the reproducibility of the data. In each box, the catalyst concentration was 1 g/L.

4. Conclusions

In this work, natural clays were used to remove pollutants from water by adsorption and photocatalysis processes. The approach was applied on smectite-rich Cameroon clays.

The clays were preliminarily treated with Cu^{2+} and then with semiconductors TiO_2 and ZnO to produce hybrid clays. The aim was to increase the depollution efficiency of these modified clayey materials by their photocatalytic properties. The protocol was controlled by XRD and ICP-AES measurements. The modified clays displayed an increase in their specific surface areas in comparison with natural clay properties. XRD confirmed the presence of crystalline TiO_2 and ZnO.

The adsorption experiments confirmed the bare clays can adsorb fluorescein, but they were not efficient on other pollutants, namely p-nitrophenol. The addition of semiconductor materials improved the degradation of the pollutants when exposed to UVA light. Photocatalytic experiments on PNP gave degradation levels of 70% to 90% after 8 h of exposition with the TiO_2- and ZnO-modified clays, respectively.

This study emphasizes the importance of composite clays to remove pollutants via adsorption and photocatalysis processes. Such approaches offer an opportunity, especially in developing countries, to use natural clay materials with slight modifications for water purification.

Supplementary Materials: The following are available online at https://www.mdpi.com/article/10.3390/catal12020148/s1, Figure S1: Nitrogen adsorption desorption isotherms for (♦) pure TiO_2 and (■) pure ZnO samples, Figure S2: Nitrogen adsorption desorption isotherms for (♦) Clay/Cu^{2+}, (▲) Clay/TiO_2/Cu^{2+}, (×) Clay/ZnO/Cu^{2+} and (■) Clay/ZnO samples, Figure S3: SEM pictures of (a) Clay/Cu^{2+}, (b) Clay/TiO_2/Cu^{2+} and (c) Clay/ZnO/Cu^{2+} samples at a 1000x magnification, Figure S4: FL UV/visible spectrum for (●) initial FL solution and (▲) after 6 h in adsorption experiment with bare Clay sample; Figure S5: PNP UV/visible spectrum for (●) initial PNP solution and (■) after 8 h in photocatalytic experiment with Clay/TiO_2 sample.

Author Contributions: Conceptualization, methodology, investigation, analysis, and writing, J.G.M., M.H.T.M. and S.D.L.; writing—original draft preparation, J.G.M., M.H.T.M. and S.D.L.; XRD characterizations analysis, M.H.T.M. and N.F.; adsorption experiments, J.G.M. and C.L.; supervision, funding acquisition, and project administration, E.D.W. and S.D.L. All the authors corrected the paper before submission and during the revision process. All authors have read and agreed to the published version of the manuscript.

Funding: This research was funded by PACODEL/University of Liège, bourse de mobilité doctorale.

Data Availability Statement: The raw/processed data required to reproduce these findings cannot be shared at this time as these data are part of an ongoing study.

Acknowledgments: J.G.M. and S.D.L. thank the Belgian National Funds for Scientific Research (F.R.S.-FNRS) for his postdoctoral fellowship and her senior associate researcher position, respectively. The authors thank the CARPOR platform of the University of Liège and its manager, Alexandre Léonard, for the nitrogen adsorption–desorption measurements.

Conflicts of Interest: The authors declare no conflict of interest.

References

1. Kemgang Lekomo, Y.; Mwebi Ekengoue, C.; Douola, A.; Fotie Lele, R.; Christian Suh, G.; Obiri, S.; Kagou Dongmo, A. Assessing Impacts of Sand Mining on Water Quality in Toutsang Locality and Design of Waste Water Purification System. *Clean. Eng. Technol.* **2021**, *2*, 100045. [CrossRef]
2. Auriol, M.; Filali-Meknassi, Y.; Dayal Tyagi, R. Présence et Devenir Des Hormones Stéroïdiennes Dans Les Stations de Traitement Des Eaux Usées. Occurrence and Fate of Steroid Hormones in Wastewater Treatment Plants. *Rev. Des Sci. L'eau* **2007**, *20*, 89–108.
3. Ekengoue, C.M.; Lele, R.F.; Dongmo, A.K. Influence De L'exploitation Artisanale Du Sable Sur La Santé Et La Sécurité Des Artisans Et L'environnement: Cas De La Carrière De Nkol'Ossananga, Région Du Centre Cameroun. *Eur. Sci. J. ESJ* **2018**, *14*, 246. [CrossRef]
4. Available online: Https://Minepded.Gov.Cm/Fr/ (accessed on 18 November 2021).
5. Nkoumbou, C.; Njopwouo, D.; Villiéras, F.; Njoya, A.; Yonta Ngouné, C.; Ngo Ndjock, L.; Tchoua, F.M.; Yvon, J. Talc Indices from Boumnyebel (Central Cameroon), Physico-Chemical Characteristics and Geochemistry. *J. Afr. Earth Sci.* **2006**, *45*, 61–73. [CrossRef]

6. Filice, S.; Bongiorno, C.; Libertino, S.; Compagnini, G.; Gradon, L.; Iannazzo, D.; la Magna, A.; Scalese, S. Structural Characterization and Adsorption Properties of Dunino Raw Halloysite Mineral for Dye Removal from Water. *Materials* **2021**, *14*, 3676. [CrossRef]
7. Jacques Richard, M. *Mineralogie et Proprietes Physico-Chimiques des Smectites de Bana et Sabga (Cameroun). Utilisation Dans La Décoloration d' Une Huile Végétale Alimentaire*; Université de Liège: Liège, Belgique, 2013.
8. Djoufac Woumfo, E.; Elimbi, A.; Panczer, G.; Nyada Nyada, R.; Njopwouo, D. Physico-Chemical and Mineralogical Characterization of Garoua Vertisols (North Cameroon). *Ann. Chim.* **2006**, *31*, 75–90.
9. Léonard, G.L.-M.; Malengreaux, C.M.; Mélotte, Q.; Lambert, S.D.; Bruneel, E.; van Driessche, I.; Heinrichs, B. Doped Sol–Gel Films vs. Powders TiO_2: On the Positive Effect Induced by the Presence of a Substrate. *J. Environ. Chem. Eng.* **2016**, *4*, 449–459. [CrossRef]
10. Parsons, S. *Advanced Oxidation Processes for Water and Wastewater Treatment*; IWA Publishing: London, UK, 2004; ISBN 1843390175.
11. Bhowmick, M.; Semmens, M.J. Ultraviolet Photooxidation for the Destruction f VOCs in Air. *Wat. Res.* **1994**, *28*, 2407–2415. [CrossRef]
12. Ikehata, K.; El-Din, M.G. Aqueous Pesticide Degradation by Hydrogen Peroxide/Ultraviolet Irradiation and Fenton-Type Advanced Oxidation Processes: A Review. *J. Environ. Eng. Sci.* **2006**, *5*, 81–135. [CrossRef]
13. Filice, S.; Fiorenza, R.; Reitano, R.; Scalese, S.; Sciré, S.; Fisicaro, G.; Deretzis, I.; la Magna, A.; Bongiorno, C.; Compagnini, G. TiO_2 Colloids Laser-Treated in Ethanol for Photocatalytic H2 Production. *ACS Appl. Nano Mater.* **2020**, *3*, 9127–9140. [CrossRef]
14. Goncharuk, V.V.; Potapchenko, N.G.; Savluk, O.S.; Kosinova, V.N.; Sova, A.N. Study of Various Conditions for O3/UV Disinfection of Water. *Khimiya Tecknol. Vody* **2003**, *25*, 487–496.
15. Drogui, P.; Blais, J.-F.; Mercier, G. Review of Electrochemical Technologies for Environmental Applications. *Recent Pat. Eng.* **2007**, *1*, 257–272. [CrossRef]
16. Douven, S.; Mahy, J.G.; Wolfs, C.; Reyserhove, C.; Poelman, D.; Devred, F.; Gaigneaux, E.M.; Lambert, S.D. Efficient N, Fe Co-Doped TiO_2 Active under Cost-Effective Visible LED Light: From Powders to Films. *Catalysts* **2020**, *10*, 547. [CrossRef]
17. Mahy, J.G.; Wolfs, C.; Vreuls, C.; Drot, S.; Dircks, S.; Boergers, A.; Tuerk, J.; Hermans, S.; Lambert, S.D. Advanced Oxidation Processes for Waste Water Treatment: From Lab-Scale Model Water to on-Site Real Waste Water. *Environ. Technol.* **2021**, *42*, 3974–3986. [CrossRef] [PubMed]
18. Mahy, J.G.; Lejeune, L.; Haynes, T.; Body, N.; de Kreijger, S.; Elias, B.; Marcilli, R.H.M.; Fustin, C.A.; Hermans, S. Crystalline ZnO Photocatalysts Prepared at Ambient Temperature: Influence of Morphology on p-Nitrophenol Degradation in Water. *Catalysts* **2021**, *11*, 1182. [CrossRef]
19. Bodson, C.J.; Heinrichs, B.; Tasseroul, L.; Bied, C.; Mahy, J.G.; Man, M.W.C.; Lambert, S.D. Efficient P- and Ag-Doped Titania for the Photocatalytic Degradation of Waste Water Organic Pollutants. *J. Alloys Compd.* **2016**, *682*, 144–153. [CrossRef]
20. Cheng, T.; Gao, H.; Liu, G.; Pu, Z.; Wang, S.; Yi, Z.; Wu, X.; Yang, H. Preparation of Core-Shell Heterojunction Photocatalysts by Coating CdS Nanoparticles onto Bi4Ti3O12 Hierarchical Microspheres and Their Photocatalytic Removal of Organic Pollutants and Cr(VI) Ions. *Colloids Surf. A Physicochem. Eng. Asp.* **2022**, *633*, 127918. [CrossRef]
21. Xiong, S.; Yin, Z.; Zhou, Y.; Peng, X.; Yan, W.; Liu, Z.; Zhang, X. The Dual-Frequency (20/40 KHz) Ultrasound Assisted Photocatalysis with the Active Carbon Fiber-Loaded Fe3+-$TiO2$ as Photocatalyst for Degradation of Organic Dye. *Bull. Korean Chem. Soc.* **2013**, *34*, 3039–3045. [CrossRef]
22. Li, Y.; Li, M.; Xu, P.; Tang, S.; Liu, C. Efficient Photocatalytic Degradation of Acid Orange 7 over N-Doped Ordered Mesoporous Titania on Carbon Fibers under Visible-Light Irradiation Based on Three Synergistic Effects. *Appl. Catal. A Gen.* **2016**, *524*, 163–172. [CrossRef]
23. Tang, N.; Li, Y.; Chen, F.; Han, Z. In Situ Fabrication of a Direct Z-Scheme Photocatalyst by Immobilizing CdS Quantum Dots in the Channels of Graphene-Hybridized and Supported Mesoporous Titanium Nanocrystals for High Photocatalytic Performance under Visible Light. *RSC Adv.* **2018**, *8*, 42233–42245. [CrossRef]
24. Lin, X.; Li, M.; Li, Y.; Chen, W. Enhancement of the Catalytic Activity of Ordered Mesoporous TiO_2 by Using Carbon Fiber Support and Appropriate Evaluation of Synergy between Surface Adsorption and Photocatalysis by Langmuir-Hinshelwood (L-H) Integration Equation. *RSC Adv.* **2015**, *5*, 105227–105238. [CrossRef]
25. Ndé, H.S.; Tamfuh, P.A.; Clet, G.; Vieillard, J.; Mbognou, M.T.; Woumfo, E.D. Comparison of HCl and H_2SO_4 for the Acid Activation of a Cameroonian Smectite Soil Clay: Palm Oil Discolouration and Landfill Leachate Treatment. *Heliyon* **2019**, *5*, e02926. [CrossRef] [PubMed]
26. Olad, A. 7 Polymer/Clay Nanocomposites. In *Advances in Diverse Industrial Applications of Nanocomposites*; Reddy, B., Ed.; InTechOpen: London, UK, 2011.
27. Yeop Lee, S.; Jin Kim, S. Expansion of smectite by hexadecyltrimethylammonium. *Clays Clay Miner.* **2002**, *50*, 435–445.
28. Theo Kloprogge, J.; Komarnenl, S.; Amonetie, J.E. Synthesis of smectite clay minerals: A critical review. *Clays Clay Miner.* **1999**, *47*, 529–554. [CrossRef]
29. Mahy, J.G.; Léonard, G.L.-M.; Pirard, S.; Wicky, D.; Daniel, A.; Archambeau, C.; Liquet, D.; Heinrichs, B. Aqueous Sol-Gel Synthesis and Film Deposition Methods for the Large-Scale Manufacture of Coated Steel with Self-Cleaning Properties. *J. Sol-Gel Sci. Technol.* **2017**, *81*, 27–35. [CrossRef]
30. Benhebal, H.; Chaib, M.; Leonard, A.; Lambert, S.D.; Crine, M. Photodegradation of Phenol and Benzoic Acid by Sol-Gel-Synthesized Alkali Metal-Doped ZnO. *Mater. Sci. Semicond. Process.* **2012**, *15*, 264–269. [CrossRef]

31. Léonard, G.L.-M.; Pàez, C.A.; Ramírez, A.E.; Mahy, J.G.; Heinrichs, B. Interactions between Zn2+ or ZnO with TiO_2 to Produce an Efficient Photocatalytic, Superhydrophilic and Aesthetic Glass. *J. Photochem. Photobiol. A Chem.* **2018**, *350*, 32–43. [CrossRef]
32. Mahy, J.G.; Lambert, S.D.; Léonard, G.L.M.; Zubiaur, A.; Olu, P.Y.; Mahmoud, A.; Boschini, F.; Heinrichs, B. Towards a Large Scale Aqueous Sol-Gel Synthesis of Doped TiO_2: Study of Various Metallic Dopings for the Photocatalytic Degradation of p-Nitrophenol. *J. Photochem. Photobiol. A Chem.* **2016**, *329*, 189–202. [CrossRef]
33. Geology, K.G. Geology, Petrology and Geochemistry of the Tertiary Bana Volcano-Plutonic Complex, West Cameroon, Central Africa. Ph.D. Thesis, Kobe University, Kobe, Japan, 2004.
34. Aboubakar, Y. *Etude Pédologique Du Terroir de Bana*; ORSTOM: Yaounde, Cameroon, 1974.
35. Bi Tra, T. *Etude Pédologique et Cartographique à L'échelle 1/50000 d'un Secteur de L'ouest-Cameroun (Région de Bafang)*; ORSTOM: Yaounde, Cameroon, 1980.
36. Benhebal, H.; Chaib, M.; Crine, M.; Leonard, A.; Lambert, S.D. Photocatalytic Decolorization of Gentian Violet with Na-Doped (SnO_2 and ZnO). *Chiang Mai J. Sci.* **2016**, *43*, 584–589.

Article

Synergistic Effect on Photocatalytic Activity of Co-Doped NiTiO₃/g-C₃N₄ Composites under Visible Light Irradiation

Duc Quang Dao [1], Thi Kim Anh Nguyen [1], Thanh-Truc Pham [2] and Eun Woo Shin [1,*]

1. School of Chemical Engineering, University of Ulsan, Daehakro 93, Nam-gu, Ulsan 44610, Korea; quangdao.ys@gmail.com (D.Q.D.); nguyenthikimanhtb@gmail.com (T.K.A.N.)
2. Material Technology Department, Faculty of Applied Science, HCMC University of Technology and Education (HCMUTE), No. 1 Vo Van Ngan Street, Linh Chieu Ward, Thu Duc District, Ho Chi Minh City 700000, Vietnam; trucpt@hcmute.edu.vn
* Correspondence: ewshin@ulsan.ac.kr; Tel.: +82-52-259-2253

Received: 3 November 2020; Accepted: 14 November 2020; Published: 16 November 2020

Abstract: Co-doped NiTiO₃/g-C₃N₄ composite photocatalysts were prepared by a modified Pechini method to improve their photocatalytic activity toward methylene blue photodegradation under visible light irradiation. The combination of Co-doped NiTiO₃ and g-C₃N₄ and Co-doping into the NiTiO₃ lattice synergistically enhanced the photocatalytic performance of the composite photocatalysts. X-ray photoelectron spectroscopy results for the Co-doped NiTiO₃/g-C₃N₄ composite photocatalysts confirmed Ti-N linkages between the Co-doped NiTiO₃ and g-C₃N₄. In addition, characteristic X-ray diffraction peaks for the NiTiO₃ lattice structure clearly indicated substitution of Co into the NiTiO₃ lattice structure. The composite structure and Co-doping of the C-x composite photocatalysts (x wt % Co-doped NiTiO₃/g-C₃N₄) not only decreased the emission intensity of the photoluminescence spectra but also the semicircle radius of the Nyquist plot in electrochemical impedance spectroscopy, giving the highest k_{app} value (7.15×10^{-3} min^{-1}) for the C-1 composite photocatalyst.

Keywords: NiTiO₃; g-C₃N₄; composite photocatalyst; recombination; photodegradation; charge separation efficiency

1. Introduction

Graphitic carbon nitride (g-C₃N₄) is a metal-free polymeric semiconductor having high thermal and chemical stability with good electronic and optical properties [1,2]. Its bandgap of ~2.7 eV is appropriate for absorption of visible light. Due to these features, g-C₃N₄ has been widely applied as a potential photocatalyst for environmental remediation and artificial photosynthesis [3]. However, g-C₃N₄ also has some drawbacks as a photocatalyst, including fast recombination of photo-induced electron-hole pairs and a low specific surface area [3–5]. To overcome these weaknesses, many researchers have tested a range of solutions.

In recent years, composite materials prepared from g-C₃N₄ have been widely used for photocatalysis under visible light irradiation owing to their efficient charge separation of photo-excited electron-hole pairs and a narrow band gap [4,6–14]. Metal oxide photocatalysts, such as TiO₂ and ZnO, have been individually studied and as inorganic components in g-C₃N₄ composite photocatalysts [4,6–9]. For example, when ZnO, a UV-responsive photocatalyst, was combined with g-C₃N₄ as a composite photocatalyst, photocatalytic activity was improved by enhanced electron-hole separation through the Z-scheme mechanism [8,9]. Inorganic materials used in composite photocatalysts have been extended to double metal oxides, such as NiTiO₃, which can be prepared by wet methods and has a narrow bandgap responsive to visible light irradiation [10–15]. In comparison with the performance of individual photocatalytic components, the composite photocatalyst ($NiTiO_3$/g-C_3N_4) had a much higher photocatalytic activity toward photodegradation of dye pollutants [12,14], H_2 production [10], and nitrobenzene removal [15]. Further modification of the $NiTiO_3$/g-C_3N_4 composite photocatalyst was achieved by doping heteroatoms into the $NiTiO_3$ lattice structure [13]. Molybdenum (Mo) doping was used to modify the structural, electrical, and optical properties of $NiTiO_3$ [13,16], resulting in a high photocatalytic activity [13]. In a previous study, cobalt (Co) has been used as a heteroatom dopant, substituting Ni sites of the $NiTiO_3$ lattice, suggesting potential for use as a dopant for $NiTiO_3$/g-C_3N_4 composite photocatalysts [16].

In this study, we prepared Co-doped $NiTiO_3$/g-C_3N_4 composite photocatalyst by a modified Pechini method to reveal the effects of Co-doping into the $NiTiO_3$ lattice and compositing with g-C_3N_4 on photocatalytic activity for methylene blue (MB) photodegradation. The composite photocatalysts were characterized with the use of X-ray diffraction (XRD), Fourier transform infrared (FT-IR) spectroscopy, field-emission scanning electron microscopy (FE-SEM), X-ray photoelectron spectroscopy (XPS), ultra violet (UV)/visible absorption spectroscopy, photoluminescence (PL) spectroscopy, and electrochemical impedance spectroscopy (EIS). The heteroatom, Co, was doped into the $NiTiO_3$ lattice structure to modify the optical and electrochemical properties of the $NiTiO_3$/g-C_3N_4 composite photocatalyst. Furthermore, the Co-doped $NiTiO_3$/g-C_3N_4 composite photocatalysts had a higher photocatalytic activity than $NiTiO_3$/g-C_3N_4 composite or Co-doped $NiTiO_3$ photocatalysts, which we attribute to the high charge separation efficiency of Co-doped $NiTiO_3$/g-C_3N_4 composite photocatalyst.

2. Results and Discussion

2.1. Morphological and Structural Properties

A detailed description of the prepared materials in this study is listed in Table 1. The samples denoted as N-x, C, and C-x in Table 1 represent pure or Co-doped $NiTiO_3$, pure g-C_3N_4, and composites of g-C_3N_4 and N-x, respectively.

Table 1. Physicochemical properties and photodegradation rate constants of the prepared photocatalysts.

Sample	Description	d (nm) [a]	E_g (eV) [b]	$k_{app} \times 10^3$ (min^{-1}) [c]	R^2
N-0	Pure $NiTiO_3$	32.39	2.97	0.34	0.989
N-1	Co-doped $NiTiO_3$ (Co = 1%)	35.06	2.76	3.17	0.990
N-3	Co-doped $NiTiO_3$ (Co = 3%)	36.54	2.91	1.95	0.995
C	Pure g-C_3N_4	-	2.65	0.44	0.983
C-0	g-C_3N_4/N-0 composite	21.62	2.67	0.79	0.984
C-1	g-C_3N_4/N-1 composite	27.32	2.53	7.15	0.992
C-3	g-C_3N_4/N-3 composite	33.83	2.45	5.66	0.987

[a] Average crystallite sizes estimated using the Scherrer equation on the basis of the (1 0 4) diffraction; [b] Band gap calculated from UV–Vis spectra; [c] Apparent rate constant of MB photodegradation reactions.

The phase structure of as-prepared g-C_3N_4 (C), Co-doped $NiTiO_3$/g-C_3N_4 composites (C-x), and Co-doped $NiTiO_3$ oxides (N-x) were characterized by XRD. All composites and oxides had typical diffraction peaks of the $NiTiO_3$ ilmenite phase with the space group $R\bar{3}$ (JCPDS 33-0960), (Figure 1). Peaks at 24.02°, 32.98°, 35.56°, 40.74°, 49.32°, 53.86°, 62.34°, 64.00°, and 71.46°, correspond to (0 1 2), (1 0 4), (1 1 0), (1 1 $\bar{3}$), (0 2 4), (1 1 $\bar{6}$), (0 1 8), (1 2 $\bar{4}$), (3 3 0), and (1 0 10) planes, respectively. No diffraction peaks attributed to Co-containing phases were detected, indicating that cations in the lattice were substituted by Co without affecting the crystal structure. However, doping of Co ions induced a 0.1° shift of the characteristic peak positions to higher diffraction angles and the peak width narrowed owing to a change in the $NiTiO_3$ crystallite size [13,17,18]. The Scherrer equation was employed to estimate the average crystallite sizes ($d_{(104)}$) from the (1 0 4) diffraction plane (Table 1). For both the oxide and composites families, the crystallite size increased as the Co content was increased.

This result is consistent with FE-SEM observations of the photocatalysts (Figure S1, see Supplementary Materials). In theory, doping of Co cations is regulated by Pauling's rule, which means that Co can enter the $NiTiO_3$ lattice by replacing Ni octahedral sites [19]. In addition, the characteristic XRD peaks of g-C_3N_4 at 13.1° and 27.2°, which are indexed as the (1 0 0), and (0 0 2) planes, respectively, were difficult to identify in the x-ray diffraction (XRD) patterns of the composite photocatalysts because of their relatively low intensity, suggesting that the compositing led to a high dispersion of C_3N_4 in the composite photocatalysts.

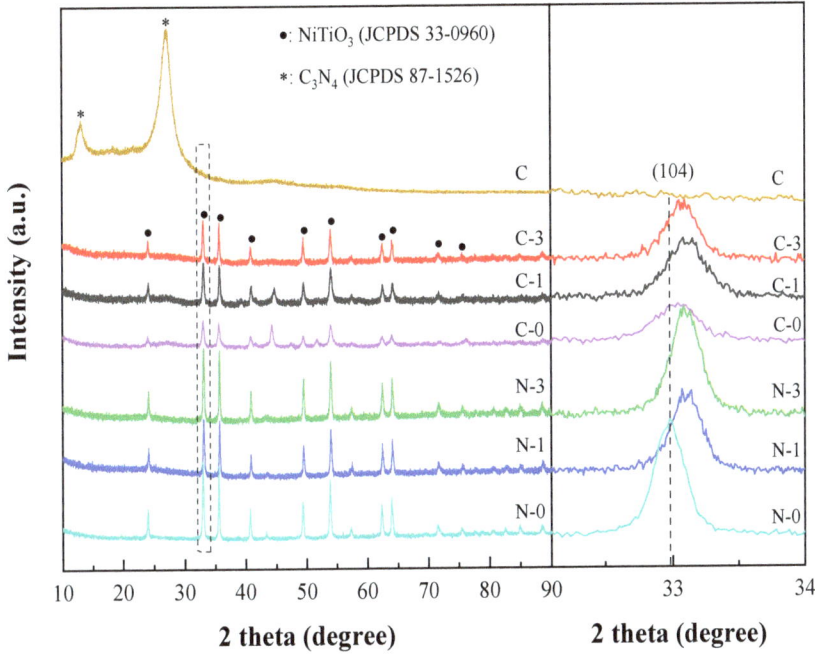

Figure 1. XRD patterns of the prepared photocatalysts.

FT-IR spectra of the prepared materials are shown in Figure 2. Characteristic vibrational peaks attributed to g-C_3N_4 were observed in all three C-x composites, consistent with previous reports [12]. A sharp peak at 807 cm^{-1} is assigned to bending vibration of heptazine rings on g-C_3N_4 [20]. A series of peaks ranging from 1225 to 1636 cm^{-1} are assigned to typical stretching modes of aromatic C-N and C=N in tri-s-triazine rings. Furthermore, there is a broad band in the range of 3000–3600 cm^{-1} assigned to the stretching vibrations of N-H bonds, associated with primary and secondary amino groups, and of O-H groups from adsorbed water [12]. The major phase of $NiTiO_3$ oxides (N-x) was identified in all N-x samples based on characteristic vibrational bands corresponding to oxygen-metal bonds at 451 cm^{-1} (Ti-O-Ni stretch), 551 cm^{-1} (Ni-O stretch), 655 cm^{-1} (Ti-O stretch), and 731 cm^{-1} (O-Ti-O bend) [17]. The Co doping induced no change in the FT-IR spectra, indicating that no CoO_x phases formed in either the oxides or the composites.

2.2. Electrical and Optical Properties

The interfacial electronic states of the prepared catalysts were further investigated by XPS analysis. We examined the C 1s and N 1s core-level spectra for g-C_3N_4 and the composites (Figure 3a,b) to identify the electrochemical state of g-C_3N_4. For pure g-C_3N_4, the N 1s spectra in Figure 3a should be fitted by three peaks [21,22]. The peaks at 397.6 and 398.4 eV derive from sp^2-hybridized nitrogen atoms (C=N-C), and tertiary nitrogen (N-C$_3$), respectively. The peak at 399.7 eV is assigned to amino groups (C-N-H). In the composite, another small peak appeared at 396.5 eV, in addition to above-mentioned peaks, corresponding to Ti-N linkages between NiTiO$_3$ and g-C$_3$N$_4$ [13]. The C-1 composite photocatalyst had a higher amount of nitrogen involved in Ti-N bonding than C-0 and C-3, based on the area ratio of the Ti-N peak and the total N 1s peaks. Thus, the C-1 composite had stronger interactions between the inorganic (NiTiO$_3$) and organic (C$_3$N$_4$) components than the remaining composites. In the C 1s spectra (Figure 3b), peak fitting revealed features at 284.1, 285.5, 287.4, and 288.8 eV, which we assign to external carbon contamination, C-NH$_2$, C-(N)$_3$ coordination, and N-C=N in the aromatic rings of g-C$_3$N$_4$ in the four samples, respectively [23].

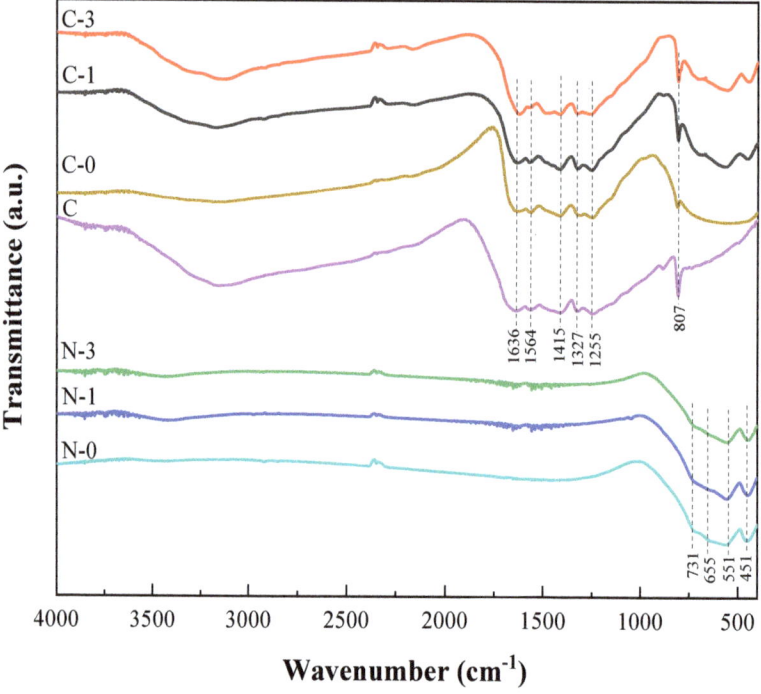

Figure 2. FT-IR spectra of the prepared photocatalysts.

The XPS data for metal components in the oxides and the composites are shown in Figure 3c,d. The XPS data of Ni 2p (Figure 3c) had doublets at 854.9 and 872.5 eV corresponding to the Ni 2p$_{3/2}$ and Ni 2p$_{1/2}$ states, respectively. Broad peaks at ca. 861 and 878 eV represent Ni^{2+} satellites [24,25]. In the Ti 2p XPS data (Figure 3d), peaks of Ti 2p$_{3/2}$ and Ti 2p$_{1/2}$ at 458.6 and 464.4 eV correspond to the Ti^{4+} oxidation state, and those at 457.5 and 463.1 eV are assigned to Ti^{3+} [13]. In addition, the appearance of Ti-N features in the composites reconfirmed the presence of Ti-N bonding and a relatively high Ti-N atomic percentage in the C-1 composite photocatalyst compared with other composite photocatalysts. The formation of the Ti-N linkages between the NiTiO$_3$ lattice and g-C$_3$N$_4$ can decrease the recombination rate of the photo-generated charges in the composite photocatalysts owing to high charge separation efficiency. As a result, the Co-doped NiTiO$_3$/g-C$_3$N$_4$ composite photocatalysts can have a high photocatalytic activity. Notably, at higher Co contents, the binding energies of the Ni 2p$_{1/2}$ and Ti 2p peaks slightly shifted, suggesting electronic interactions between the structured oxides and the doped Co component [17].

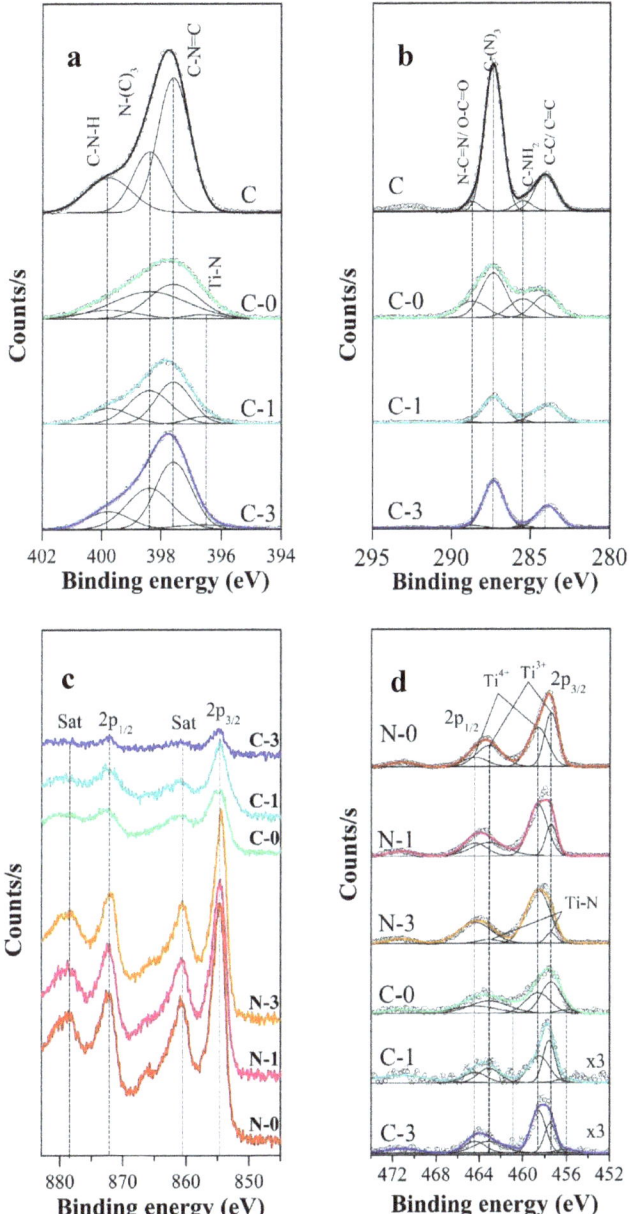

Figure 3. XPS data of (**a**) N 1s, (**b**) C 1s, (**c**) Ni 2p, and (**d**) Ti 2p for the prepared photocatalysts.

Optical properties of photocatalysts are important to their performance; hence, the light absorption of the prepared materials was measured by UV-visible diffuse reflectance spectra (Figure 4a). The band gap—an important factor in evaluating the photoactivity of semiconductors [26]—was estimated from the following formula,

$$(\alpha h\upsilon)^n = k(h\upsilon - E_g), \tag{1}$$

where α, k, hυ, and E_g are the absorption coefficient, a constant related to the effective masses associated with the conduction and valence bands, the absorption energy, and band gap, respectively (Table 1 and Figure S2, see Supplementary Materials). Each of the prepared catalysts had a visible light response owing to band gaps within the range of 2.45–2.97 eV. The band gaps of the composites were lower than those of the oxides (Figure S2), suggesting interactions between the individual components. This same phenomenon has been previously reported [13].

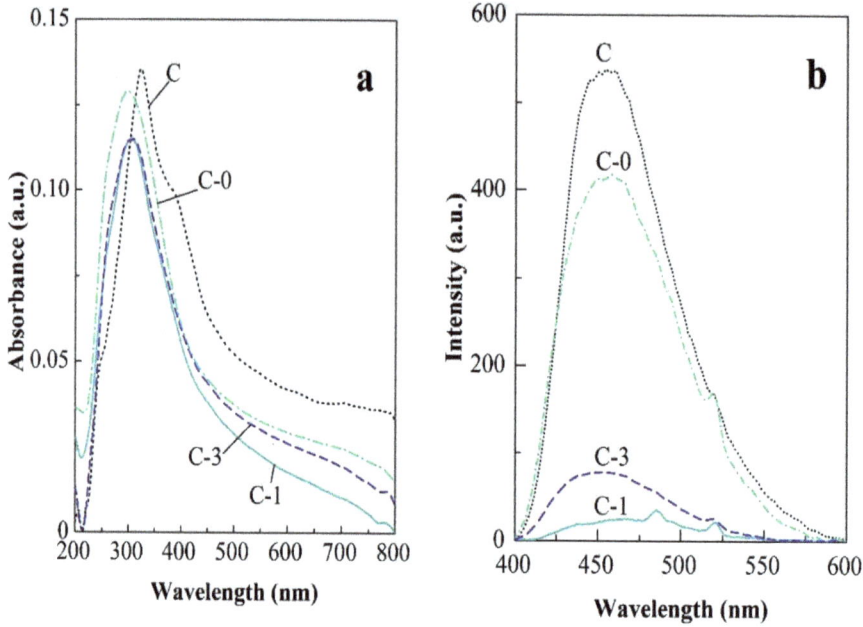

Figure 4. (a) UV–Vis absorption spectra, and (b) PL spectra of g-C_3N_4 and C-x composites.

The interfacial charge transfer process has an important role in photocatalytic performance. Therefore, we measured the PL emission spectra to investigate the recombination process of photo-induced charge pairs. Similar to pure g-C_3N_4, all the composites had a broad peak ca. 460 nm with a tail extending to 600 nm in Figure 4b. However, the PL emission intensity for the composites was lower than that of pure g-C_3N_4, indicating faster photoelectron transfer within the composites, when Co was doped into the lattice structure. The higher PL intensity of C-3 compared with C-1 might be explained by aggregation of Co and weak interfacial effects, which accelerated the recombination rate of electrons and holes.

To further examine the charge transfer separation and photogenerated exciton separation efficiency of the prepared catalysts, we performed EIS (Figure 5). In general, the smaller the semicircle radius in the EIS Nyquist plot, the lower the recombination rate. The semicircular radius (Figure 5) increased in the order, C-x < N-x < C, indicating that charge transfer processes in the composite photocatalysts were promoted by Co doping. The Co dopant in the composite photocatalysts acted as an impurity state, which accepted electrons excited from the conduction band, to decrease the recombination rate. Therefore, on the basis of the above PL and EIS results, we confirm that Co-doping of the composite photocatalysts improved the efficiency of the charge separation process of photo-induced electrons and holes.

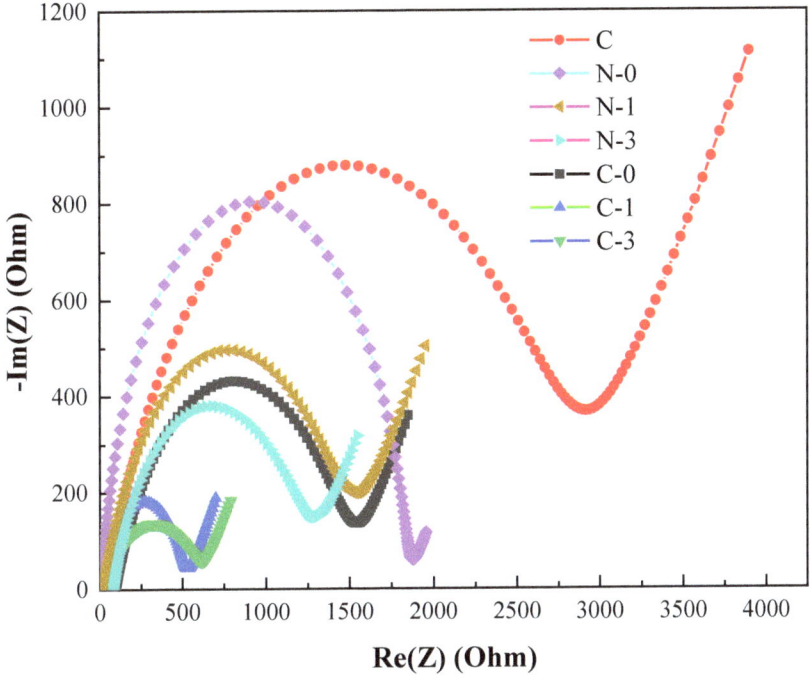

Figure 5. EIS Nyquist plot of the prepared photocatalysts.

2.3. Photocatalytic Performance

In this study, the performance of the prepared photocatalysts was evaluated by the MB photocatalytic degradation tests under visible light irradiation, and corresponding results are shown in Figure 6 and Table 1. A pseudo-first-order model was used to better understand the reaction kinetics of the MB degradation [27], where the apparent rate constant, k_{app}, was calculated from the slope of the plot t vs $\ln(C/C_0)$ by the equation,

$$\ln(C/C_0) = -k_{app}t \qquad (2)$$

where C_0 and C are the concentrations of MB initially and at a given time, respectively. The concentrations were measured by UV-Vis spectra for MB (Figure S3, see Supplementary Materials). The determination coefficients (R^2 values) for the linear regression were higher than 0.97, indicating that the photodegradation reaction rate constants of all reactions followed an apparent-first-order reaction model [23]. A blank test confirmed that MB was stable under visible light in the absence of the photocatalysts over the same period of time with rate constant $k_{app} = 0.01 \times 10^{-3}$ min^{-1}. The apparent rate constants of pure g-C_3N_4 (C) and NiTiO$_3$ (N-0), and their composite (C-0) were relatively low with the values of 0.44×10^{-3}, 0.34×10^{-3}, and 0.79×10^{-3} min^{-1}, respectively, which were similar to results of a previous study [13]. The rates for the Co-doped NiTiO$_3$ oxide (N-x) were higher at 3.17×10^{-3} min^{-1} for N-1 and 1.95×10^{-3} min^{-1} for N-3. Furthermore, the Co-doped composites (C-1 and C-3) had outstanding photocatalytic activity with k_{app} values of 7.15×10^{-3} min^{-1} for the former and 5.66×10^{-3} min^{-1} for the latter. This performance was approximately 3 times as high as that of the oxide catalysts at the same Co-doping content, indicating a synergistic effect between the Co doping and the g-C_3N_4 coupling.

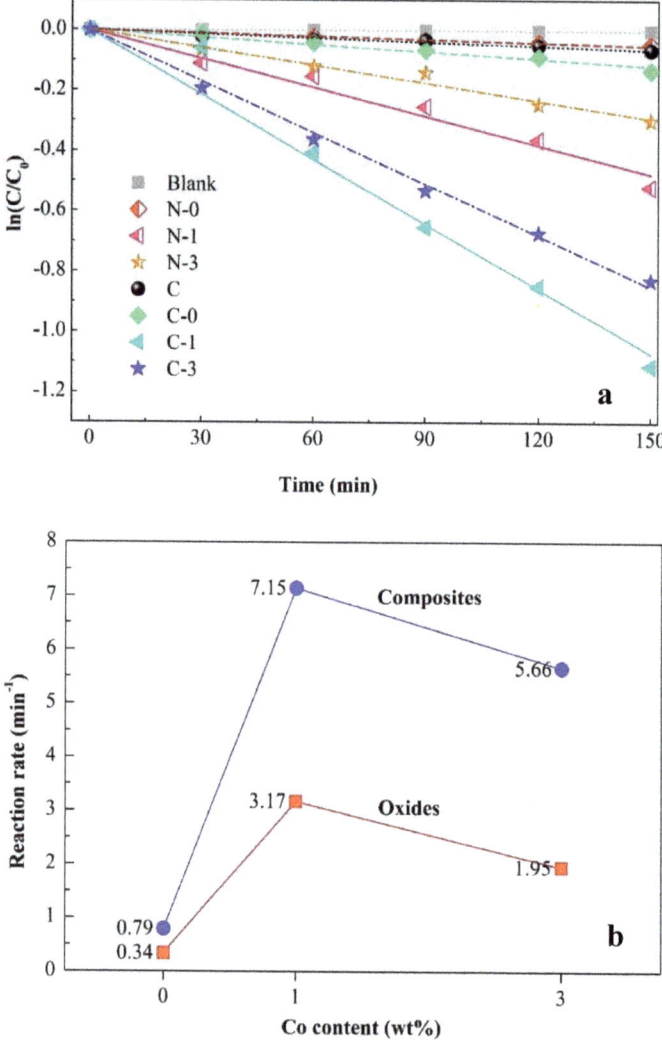

Figure 6. (a) Photocatalytic degradation of MB (initial concentration = 10 ppm) fitted to the apparent-first-order kinetics model. (b) Trend of the photodegradation rate constant as a function of the Co content.

3. Materials and Methods

3.1. Materials

In this study, dicyandiamide (DCDA 99%), titanium *n*-butoxide (Ti(OC$_4$H$_9$)$_4$ ≥ 97%), cobalt nitrate hexahydrate (Co(NO$_3$)$_2$·6H$_2$O 98%), and nickel nitrate hexahydrate (NiNO$_3$·6H$_2$O 98%) were obtained from Sigma-Aldrich Korea (Gyounggi, South Korea). Anhydrous ethyl alcohol (C$_2$H$_5$OH 99.9%) was supplied from Samchun Pure Chemical Company Ltd. (PyoungTaek, Korea). Citric acid monohydrate (C$_6$H$_8$O$_7$·H$_2$O 99.5%) was supplied from OCI Company Ltd. (Seoul, Korea). Distilled water was used for all of solution preparation.

3.2. Synthesis of Photocatalysts

Bulk g-C_3N_4 was synthesized from DCDA. A 5-mg portion of DCDA was ground for 10 min, and the obtained fine powder was transferred to a crucible boat before being placed in the middle of a tube furnace. The tube was purged with N_2 gas to remove air and humidity for 1 h, and then heated to 550 °C for 4 h (ramping rate = 10 °C/min) to initiate the thermal polymerization. Next, the yellow solid sample was washed and centrifuged several times with an ethanol/distilled water mixture. The resultant sample was dried at 80 °C for 15 h and finally ground to a fine powder in a mortar. Bulk g-C_3N_4 is denoted as C.

Pure $NiTiO_3$ and Co-doped $NiTiO_3$ were prepared by modified Pechini method. A mixture of 0.75 g citric acid and 100 mL ethanol was magnetically stirred for 15 min at room temperature. Cobalt nitrate hexahydrate was added and the mixture continuously stirred to achieve a homogeneous mixture for 30 min. Afterwards, titanium n-butoxide and nickel nitrate hexahydrate were dropped into the mixture, which was then stirred for 1 h to obtain a transparent green solution. The amounts of precursors were calculated on the basis of the nickel and titanium molar ratio of 1:1. The obtained solution was then transferred to a Teflon-lined autoclave for further solvothermal treatment at 160 °C for 6 h (ramp rate = 2 °C/min). The resultant solid was collected and washed several times with ethanol by centrifugation, then completely dried at 80 °C in air and calcinated at 600 °C for 5 h (ramp rate = 2 °C/min). Pure $NiTiO_3$ is denoted as N-0. The Co-doped $NiTiO_3$ is denoted N-x, where x is the Co weight percent (wt %) over the total weight of Co-doped $NiTiO_3$ (x = 0, 1, 3).

The composite photocatalysts were synthesized by mixing N-x with DCDA in a mortar for 10 min at a weight ratio of 1:1 and then the mixtures were transferred to a crucible boat covered with aluminum foil. The samples were calcinated in the N_2 atmosphere of the tube furnace at 500 °C for 4 h (ramping rate = 10 °C/min). The obtained dark yellow solids were washed and centrifuged, dried at 80 °C for 15 h, and finally ground again into powder. These composite photocatalysts are denoted as C-0, C-1, and C-3.

3.3. Characterization Techniques

The morphologies of all photocatalyst samples were analyzed by field-emission scanning electron microscopy (FE-SEM; JSM-600F JEOL, Tokyo, Japan). XRD, Rigaku D/MAZX 2500 V/PC high-power diffractometer, Rigaku Corp., Tokyo, Japan) with a Cu Kα X-ray source operating at a wavelength of λ = 1.5415 Å was used to determine the crystalline structures of the obtained samples in the range of 10°–90° at a scan rate of 2° (2θ)/min. Functional groups of the prepared photocatalysts were characterized with a Fourier transform infrared (FT-IR, Nicolet 380 spectrometer, thermo Scientific Nicolet iS5 with an iD1 transmission accessory, Waltham, MA, USA). Elemental compositions and electronic states of the elements were examined by XPS, Thermo Scientific K-Alpha system, Waltham, MA, USA). The optical properties of the photocatalysts were analyzed by ultraviolet–visible diffuse reflectance (UV–Vis, Analytik Jena SPECORD 210 Plus 190 spectroscope, Jena, Germany) and photoluminescence measurements (PL, Agilent Cary Eclipse fluorescence spectrophotometer, Santa Clara, CA, USA) at room temperature with a 473-nm diode laser. Electrochemical impedance spectroscopy measurements (EIS, BioLogic Science Instruments VSP, Seyssinet–Pariset, France) were conducted with the use of an electrochemical analyzer in a three-electrode quartz cell with 1 M NaOH electrolyte solution to study the recombination of the photogenerated charged carriers. A 3-W visible light bulb was used as the visible light source. A Ag/AgCl electrode and platinum wire were used as the reference electrode, and counter electrode, respectively.

3.4. Photocatalytic Tests

Photocatalytic activities were studied in the degradation of methylene blue in an aqueous solution (MB, Riedel-de Haen, Germany; initial concentration C_o = 10 ppm). A 50 mL portion of MB solution containing 10 mg of the prepared photocatalyst was magnetically stirred for 30 min at room temperature in a dark chamber to obtain an equilibrium adsorption state. The solution was irradiated for 180 min by four surrounding visible-light bulbs (model GB22100(B)EX-D, Eltime, 100 W). The concentration of MB was measured with a UV–Vis spectrometer (SPECORD 210 Plus spectroscope, Analytik Jena, Germany) at λ_{max} = 664 nm. For each analysis point, a 1 mL portion of MB solution was collected by syringe with polytetrafluoroethylene membrane filter 0.2 µm (Whatman GmbH, Dassel, Germany).

4. Conclusions

In this study, Co-doped $NiTiO_3$ oxides (N-x) and Co-doped $NiTiO_3$/g-C_3N_4 composites (C-x) were successfully synthesized by the modified Pechini method. The results of various characterization methods and photocatalytic MB degradation reactions under visible light irradiation indicate that Co doping in composite photocatalysts enhanced the interaction between inorganic ($NiTiO_3$) and organic (g-C_3N_4) components in the composites through the formation of Ti-N linkages and an impurity state for better charge transfer efficiency. The photocatalytic activity and separation efficiency of the photogenerated electrons in the photocatalysts followed the order of C-x > N-x > C.

Supplementary Materials: The following are available online at http://www.mdpi.com/2073-4344/10/11/1332/s1, Figure S1: FE-SEM images of N-1, N-3, C-1, and C-3; Figure S2: Calculated band gap of photocatalysts as a function of Co wt %; Figure S3: UV-Vis spectra of MB photodegraded with C-1 photocatalyst as a function of reaction time.

Author Contributions: D.Q.D. performed the experiments and wrote a draft of the paper; T.K.A.N. performed the experiments and contributed to data analysis; T.-T.P. contributed to data collections; E.W.S. supervised the work and polished the paper. All authors have read and agreed to the published version of the manuscript.

Funding: This research was supported by the National Research Foundation of Korea (NRF) grant funded by the Korea government (MSIT) (No. 2018R1A2B6004219 and No. 2020R1A4A4079954).

Acknowledgments: In this section you can acknowledge any support given which is not covered by the author contribution or funding sections. This may include administrative and technical support, or donations in kind (e.g., materials used for experiments).

Conflicts of Interest: The authors declare no conflict of interest.

References

1. Bandyopadhyay, A.; Ghosh, D.; Kaley, N.M.; Pati, S.K. Photocatalytic activity of g-C_3N_4 quantum dots in visible light: Effect of physicochemical modifications. *J. Phys. Chem. C* **2017**, *121*, 1982–1989. [CrossRef]
2. Wang, X.; Maeda, K.; Thomas, A.; Takanabe, K.; Xin, G.; Carlsson, J.M.; Domen, K.; Antonietti, M. A metal-free polymeric photocatalyst for hydrogen production from water under visible light. *Nat. Mater.* **2009**, *8*, 76–80. [CrossRef]
3. Ong, W.-J.; Tan, L.-L.; Ng, Y.H.; Yong, S.-T.; Chai, S.-P. Graphitic carbon nitride (g-C_3N_4)-based photocatalysts for artificial photosynthesis and environmental remediation: Are we a step closer to achieving sustainability? *Chem. Rev.* **2016**, *116*, 7159–7329. [CrossRef] [PubMed]
4. Li, J.; Zhang, M.; Li, X.; Li, Q.; Yang, J. Effect of the calcination temperature on the visible light photocatalytic activity of direct contact Z-scheme g-C_3N_4-TiO_2 heterojunction. *Appl. Catal. B Environ.* **2017**, *212*, 106–114. [CrossRef]
5. Kumar, S.; Kumar, B.; Baruah, A.; Shanker, V. Synthesis of magnetically separable and recyclable g-C_3N_4-Fe_3O_4 hybrid nanocomposites with enhanced photocatalytic performance under visible-light irradiation. *J. Phys. Chem. C* **2013**, *117*, 26135–26143. [CrossRef]
6. Wang, J.; Huang, J.; Xie, H.; Qu, A. Synthesis of g-C_3N_4/TiO_2 with enhanced photocatalytic activity for H_2 evolution by a simple method. *Int. J. Hydrogen Energy* **2014**, *39*, 6354–6363. [CrossRef]
7. Qu, A.; Xu, X.; Xie, H.; Zhang, Y.; Li, Y.; Wang, J. Effects of calcining temperature on photocatalysis of g-C_3N_4/TiO_2 composites for hydrogen evolution from water. *Mater. Res. Bull.* **2016**, *80*, 167–176. [CrossRef]
8. Jung, H.; Pham, T.-T.; Shin, E.W. Interactions between ZnO nanoparticles and amorphous g-C_3N_4 nanosheets in thermal formation of g-C_3N_4/ZnO composite materials: The annealing temperature effect. *Appl. Surf. Sci.* **2018**, *458*, 369–381. [CrossRef]

9. Jung, H.; Pham, T.-T.; Shin, E.W. Effect of g-C_3N_4 precursors on the morphological structures of g-C_3N_4/ZnO composite photocatalysts. *J. Alloys Compd.* **2019**, *788*, 1084–1092. [CrossRef]
10. Zeng, Y.; Wang, Y.; Chen, J.; Jiang, Y.; Kiani, M.; Li, B.; Wang, R. Fabrication of high-activity hybrid $NiTiO_3$/g-C_3N_4 heterostructured photocatalysts for water splitting to enhanced hydrogen production. *Ceram. Int.* **2016**, *42*, 12297–12305. [CrossRef]
11. Qu, Y.; Zhou, W.; Ren, Z.; Du, S.; Meng, X.; Tian, G.; Pan, K.; Wang, G.; Fu, H. Facile preparation of porous $NiTiO_3$ nanorods with enhanced visible-light-driven photocatalytic performance. *J. Mater. Chem.* **2012**, *22*, 16471–16476. [CrossRef]
12. Pham, T.-T.; Shin, E.W. Thermal formation effect of g-C_3N_4 structure on the visible light driven photocatalysis of g-C_3N_4/$NiTiO_3$ Z-scheme composite photocatalysts. *Appl. Surf. Sci.* **2018**, *447*, 757–766. [CrossRef]
13. Pham, T.-T.; Shin, E.W. Inhibition of charge recombination of $NiTiO_3$ photocatalyst by the combination of Mo-doped impurity state and Z-scheme charge transfer. *Appl. Surf. Sci.* **2020**, *501*, 143992. [CrossRef]
14. Pham, T.-T.; Shin, E.W. Influence of g-C_3N_4 Precursors in g-C_3N_4/$NiTiO_3$ Composites on Photocatalytic Behavior and the Interconnection between g-C_3N_4 and $NiTiO_3$. *Langmuir* **2018**, *34*, 13144–13154. [CrossRef] [PubMed]
15. Wang, H.; Yuan, X.; Wang, H.; Chen, X.; Wu, Z.; Jiang, L.; Xiong, W.; Zhang, Y.; Zeng, G. One-step calcination method for synthesis of mesoporous gC_3N_4/$NiTiO_3$ heterostructure photocatalyst with improved visible light photoactivity. *RSC Adv.* **2015**, *5*, 95643–95648. [CrossRef]
16. Jiang, K.; Pham, T.-T.; Kang, S.G.; Men, Y.; Shin, E.W. Modification of the structural properties of $NiTiO_3$ materials by transition metal dopants: The dopant size effect. *J. Alloys Compd.* **2018**, *739*, 393–400. [CrossRef]
17. Pham, T.-T.; Kang, S.G.; Shin, E.W. Optical and structural properties of Mo-doped $NiTiO_3$ materials synthesized via modified Pechini methods. *Appl. Surf. Sci.* **2017**, *411*, 18–26. [CrossRef]
18. Lenin, N.; Karthik, A.; Sridharpanday, M.; Selvam, M.; Srither, S.R.; Arunmetha, S.; Paramasivam, P.; Rajendran, V. Electrical and magnetic behavior of iron doped nickel titanate (Fe^{3+}/$NiTiO_3$) magnetic nanoparticles. *J. Magn. Magn. Mater.* **2016**, *397*, 281–286. [CrossRef]
19. Carter, C.B.; Norton, M.G. *Ceramic Materials: Science and Engineering*; Springer: New York, NY, USA, 2007.
20. Hu, S.; Ma, L.; You, J.; Li, F.; Fan, Z.; Lu, G.; Liu, D.; Gui, J. Enhanced visible light photocatalytic performance of g-C_3N_4 photocatalysts co-doped with iron and phosphorus. *Appl. Surf. Sci.* **2014**, *311*, 164–171. [CrossRef]
21. Gao, D.; Xu, Q.; Zhang, J.; Yang, Z.; Si, M.; Yan, Z.; Xue, D. Defect-related ferromagnetism in ultrathin metal-free g-C_3N_4 nanosheets. *Nanoscale* **2014**, *6*, 2577–2581. [CrossRef]
22. Tian, N.; Huang, H.; Liu, C.; Dong, F.; Zhang, T.; Du, X.; Yu, S.; Zhang, Y. In situ co-pyrolysis fabrication of CeO_2/g-C_3N_4 n–n type heterojunction for synchronously promoting photo-induced oxidation and reduction properties. *J. Mater. Chem. A* **2015**, *3*, 17120–17129. [CrossRef]
23. Nguyen, T.K.A.; Pham, T.-T.; Nguyen-Phu, H.; Shin, E.W. The effect of graphitic carbon nitride precursors on the photocatalytic dye degradation of water-dispersible graphitic carbon nitride photocatalysts. *Appl. Surf. Sci.* **2020**, *537*, 148027. [CrossRef]
24. Inceesungvorn, B.; Teeranunpong, T.; Nunkaew, J.; Suntalelat, S.; Tantraviwat, D. Novel $NiTiO_3$/Ag_3VO_4 composite with enhanced photocatalytic performance under visible light. *Catal. Commun.* **2014**, *54*, 35–38. [CrossRef]
25. Bellam, J.B.; Ruiz-Preciado, M.A.; Edely, M.; Szade, J.; Jouanneaux, A.; Kassiba, A.H. Visible-light photocatalytic activity of nitrogen-doped $NiTiO_3$ thin films prepared by a co-sputtering process. *RSC Adv.* **2015**, *5*, 10551–10559. [CrossRef]
26. Radecka, M.; Rekas, M.; Trenczek-Zajac, A.; Zakrzewska, K. Importance of the band gap energy and flat band potential for application of modified TiO_2 photoanodes in water photolysis. *J. Power Sources* **2008**, *181*, 46–55. [CrossRef]
27. Phan, T.T.N.; Nikoloski, A.N.; Bahri, P.A.; Li, D. Heterogeneous photo-Fenton degradation of organics using highly efficient Cu-doped $LaFeO_3$ under visible light. *J. Ind. Eng. Chem.* **2018**, *61*, 53–64. [CrossRef]

Publisher's Note: MDPI stays neutral with regard to jurisdictional claims in published maps and institutional affiliations.

© 2020 by the authors. Licensee MDPI, Basel, Switzerland. This article is an open access article distributed under the terms and conditions of the Creative Commons Attribution (CC BY) license (http://creativecommons.org/licenses/by/4.0/).

Article

In-Depth Structural and Optical Analysis of Ce-modified ZnO Nanopowders with Enhanced Photocatalytic Activity Prepared by Microwave-Assisted Hydrothermal Method

Otman Bazta [1,2,*], Ana Urbieta [3], Susana Trasobares [1], Javier Piqueras [3], Paloma Fernández [3], Mohammed Addou [2], Jose Juan Calvino [1] and Ana Belén Hungría [1]

[1] Department of Materials Science and Metallurgical Engineering and Inorganic Chemistry, University of Cadiz, 11003 Cadiz, Spain; susana.trasobares@uca.es (S.T.); jose.calvino@uca.es (J.J.C.); ana.hungria@uca.es (A.B.H.)

[2] Department of Physics, Abdelmalek Essaadi University Faculty of Science and Technology of Tangier FST, Tangier 93000, Morocco; mohammed_addou@yahoo.com

[3] Department of Materials Physics, Complutense University of Madrid, 28040 Madrid, Spain; anaur@fis.ucm.es (A.U.); piqueras@fis.ucm.es (J.P.); arana@ucm.es (P.F.)

* Correspondence: otman.bazta@alum.uca.es

Received: 29 March 2020; Accepted: 12 May 2020; Published: 15 May 2020

Abstract: Pure and Ce-modified ZnO nanosheet-like polycrystalline samples were successfully synthesized by a simple and fast microwave-based process and tested as photocatalytic materials in environmental remediation processes. In an attempt to clarify the actual relationships between functionality and atomic scale structure, an in-depth characterization study of these materials using a battery of complementary techniques was performed. X-ray diffraction (XRD), field emission-scanning electron microscopy (FE-SEM), high-resolution transmission electron microscopy (HRTEM), high-angle annular dark field-scanning transmission electron microscopy (HAADF-STEM), energy-dispersive X-Ray spectroscopy-scanning transmission electron microscopy (STEM-XEDS), photoluminescence spectroscopy (PL) and UV–Visible absorption spectroscopy were used to evaluate the effect of Ce ions on the structural, morphological, optical and photocatalytic properties of the prepared ZnO nanostructures. The XRD results showed that the obtained photocatalysts were composed of hexagonal, wurtzite type crystallites in the 34–44 nm size range. The SEM and TEM showed nanosheet-shaped crystallites, a significant fraction of them in contact with bundles of randomly oriented and much smaller nanoparticles of a mixed cerium–zinc phase with a composition close to $Ce_{0.68}Zn_{0.32}O_x$. Importantly, in clear contrast to the prevailing proposals regarding this type of materials, the STEM-XEDS characterization of the photocatalyst samples revealed that Ce did not incorporate into the ZnO crystal lattice as a dopant but that a heterojunction formed between the ZnO nanosheets and the Ce–Zn mixed oxide phase nanoparticles instead. These two relevant compositional features could in fact be established thanks to the particular morphology obtained by the use of the microwave-assisted hydrothermal synthesis. The optical study revealed that in the ZnO:Ce samples optical band gap was found to decrease to 3.17 eV in the samples with the highest Ce content. It was also found that the ZnO:Ce (2 at.%) sample exhibited the highest photocatalytic activity for the degradation of methylene blue (MB), when compared to both the pure ZnO and commercial TiO_2-P25 under simulated sunlight irradiation. The kinetics of MB photodegradation in the presence of the different photocatalysts could be properly described using a Langmuir–Hinshelwood (LH) model, for which the ZnO:Ce (2 at.%) sample exhibited the highest value of effective kinetic constant.

Keywords: Ce-modified ZnO; electron microscopy; methylene blue (MB) degradation; photoluminescence; photocatalysis

1. Introduction

Nowadays, extensive efforts and various strategies have been developed to control the increasing amounts and variety of water pollutants linked to the disposal of industrial waste. In this context, semiconductor photocatalysis technology has attracted substantial interest as an effective approach for the degradation of different hazardous compounds present in polluted waters [1,2]. Among the various semiconductor photocatalysts, ZnO has drawn widespread interest owing to its exceptional features, as its wide optical band gap at room temperature (3.3 eV), large exciton binding energy (60 meV) and low cost coupled with its non-toxicity and availability [3–7]. Moreover, ZnO has proved its high potential to be exploited in a wide range of technological applications such as sensing [8], optoelectronic devices [9,10], transparent electronics [11] and in particular, as a photocatalyst to completely eliminate a variety of hazardous organic molecules under UV/solar irradiation [12].

However, regarding the photocatalytic applications, ZnO presents some drawbacks. These include the high recombination probability of the photogenerated electron–hole (e^-/h^+) pairs. This process inhibits the diffusion of the separated charge carriers to reach the surface of ZnO, thus deteriorating its photocatalytic efficiency [13]. Moreover, early studies have reported the activity deterioration of ZnO photocatalysts, ascribed to the photo-dissolution of ZnO. Neppolian et al. [14] noticed a decrease in the activity of ZnO photocatalysts and it was found to be due to photo-dissolution by self-oxidation. Kong et al. [15] investigated the photocatalytic performance of Ta-doped ZnO. Their finding revealed that the photocatalytic activity was negatively affected at a low pH which was due to the photo-dissolution of ZnO.

The use of rare earth as dopants (such as Ce, Tb, La, Nd and Eu) [16–19] has raised great interest as a route to enhance the electrical and luminescent properties of ZnO in order to overcome the mentioned drawbacks. In particular, Ce has attracted significant attention because of its proved capacity to trap the photoinduced charge carriers due to its unique 4f-electronic configuration and the defects introduced in the ZnO matrix as a consequence of the doping process [20–22]. For instance, Ismail et al. [23] reported the successful preparation of Ce-doped ZnO via a hydrothermal route and its high photocatalytic performance towards methylene blue (MB) oxidation under UV irradiation. Similarly, Shi et al. [24] reported the important role of Ce doping on improving the luminescence properties of Ce-doped ZnO obtained by a sol–gel process. On the other hand, Subash et al. [25] synthesized a Ce-doped Ag–ZnO material by a solvothermal route, which showed a photocatalytic activity in the degradation of Naphthol blue black dye under visible excitation, higher than that of the undoped and commercial ZnO, TiO_2–P25 and Ag–ZnO. As in the previous works, Liang et al [26] attributed the enhanced visible light absorption ability of ZnO/Ce to the formation of defects which can act as photoelectron trap centers due to the incorporation of Ce into the ZnO lattice. In all of these studies the photocatalytic performance of ZnO/Ce was found to be ameliorated but a detailed study about the presence of Ce as a dopant into the ZnO host structure was not provided. In this respect, Cerrato et al. [27] reported on a visible light active CeO_2–ZnO photocatalyst prepared via precipitation, where it was shown that Ce did not enter as dopant in ZnO but rather formed isolated CeO_2 nanoparticles deposited on the surface of the zinc oxide crystallites. Moreover, a specific approach was applied by Cerrato et al. to monitor and quantify the charge separation of the Ce–ZnO materials. However, neither detailed optical studies nor photocatalysis tests were performed in that contribution. An analytical study of the Ce-containing phase deposited onto the surface of ZnO was not performed either. In fact, the identification of the CeO_2 phase was done based on measurements of lattice spacing values on high-resolution transmission electron microscopy (HRTEM) images. In a more recent paper [28], the formation of nanosized CeO_2/ZnO heterojunctions have also been proposed to rationalize the optical and photocatalytic performance of Ce-modified ZnO materials with Ce contents below 10 at.%. In this case, the composition of the Ce-containing phase decorating the surface of the ZnO aggregates was investigated by means of energy-dispersive X-Ray spectroscopy-scanning transmission electron microscopy (STEM-XEDS). Nevertheless, the complex structure of the material, formed by the assembly of a large number of ZnO sheets into micron-sized spheres decorated on the surface by Ce-containing nanosized crystallites, made it impossible to isolate the energy-dispersive X-Ray spectroscopy (XEDS) signal coming from the ZnO support from that of the supported Ce phase. Once again, in this paper the assignment of the Ce-containing phase was done on the basis of d-spacing measurements on HRTEM images.

Summarizing the previously reported data, there is still some controversy about the incorporation of Ce into the host lattice of ZnO. Likewise, the actual chemical nature of the Ce-containing phase, which some authors detect as giving rise to a heterojunction type system, is to date not fully clear.

To clarify these two major questions, an in-depth characterization by both structural and analytical techniques at the sub-nanometer and atomic scale is compulsory. It is anyhow clear that the synthesis procedures may have a quite important effect on the final structure of these Ce-modified nanomaterials. Concerning the synthesis techniques, several routes have been tested to enhance the properties of pure and rare earth (RE)-doped ZnO-based materials with various morphologies, such as pyrolysis [29], hydrothermal synthesis [30], magnetron co-sputtering deposition [16], sonochemical synthesis [31] or pulsed laser deposition [32]. In fact, most published surveys involve expensive, time consuming procedures which usually require high-temperature conditions. However, as far as we know, the preparation of the Ce-modified ZnO photocatalysts via a microwave-assisted hydrothermal process and the characterization of their photocatalytic performance under simulated sunlight irradiation have to date not been reported, despite the very interesting features which characterize this route, such as low cost, short synthesis times and mild temperature conditions. Therefore, in this work we used this approach in the synthesis of Ce/ZnO photocatalysts. In addition, a detailed study, by means of scanning and conventional transmission electron microscopy techniques, was carried out to exactly determine the actual distribution of the lanthanide within the materials prepared by this alternative route.

In particular, we focused on the optimization of the synthesis conditions of Ce-modified ZnO photocatalysts. The structural characterization of the obtained products, by X-ray diffraction (XRD), scanning electron microscopy (SEM) and by scanning transmission electron microscopy, (S)TEM, as well as by optical techniques, like photoluminescence spectroscopy (PL) at room temperature, were also carried out. Additionally, the photocatalytic performances of TiO_2–P25, ZnO and the microwave-assisted hydrothermal Ce-modified ZnO photocatalysts under simulated sunlight irradiation were compared.

2. Results and Discussion

2.1. Structural Properties

Figure 1a shows the XRD patterns of the pure and Ce-modified ZnO photocatalysts. All the identified peaks in the different samples correspond to the hexagonal wurtzite lattice structure of ZnO (JCPDS No. 36-1451). In samples with low cerium concentration (1% and 2%), no diffraction maxima corresponding to the phases related to cerium are observed. This fact does not rule out the possibility that this type of phases was present, since they could have formed at a concentration and size too low to be detectable in the X-ray diffraction experiment. In the case of ZnO:Ce 3%, a weak peak was detected at $2\theta = 28.81°$ (see Figure 1b) corresponding to the (111) plane of the cubic fluorite CeO_2 structure [33], indicating that at least part of the cerium content was not incorporated into the structure of ZnO.

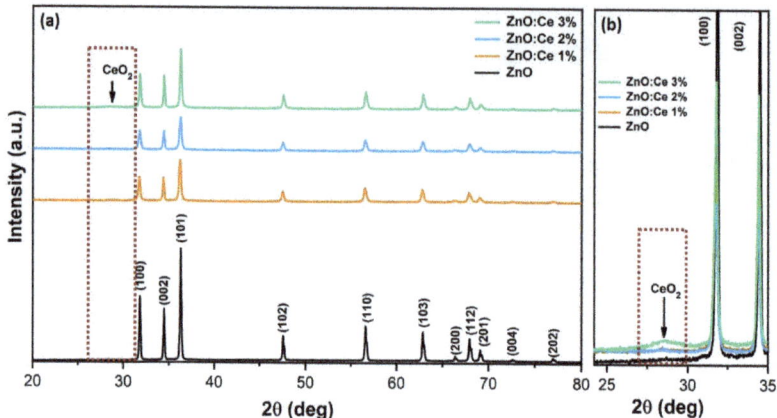

Figure 1. (a) XRD patterns of the ZnO and Ce/ZnO photocatalysts at different Ce concentrations. (b) Zoom on the region in the XRD illustrating the presence of the (111) peak of the cubic fluorite CeO$_2$ structure.

In general, the samples show intense wurtzite structure diffraction maxima, indicating a crystalline nature in all the samples. A slight broadening of the XRD peaks was noticed in the modified samples suggesting the formation of crystals with smaller sizes with respect to the pure ZnO. The Scherrer formula was applied to the XRD patterns to quantify the influence of Ce on the ZnO crystallite size.

The crystallite size of the different samples was calcluated through a complete analysis of the XRD diagram by refining using PowderCell. The following crystalline domain (Table 1) size values were obtained including in the analysis the whole set of diffraction peaks:

Table 1. The dependence of the crystallite size on the Ce content.

Sample	d (nm)–ZnO
ZnO	65
ZnO-1%Ce	47
ZnO-2%Ce	40
ZnO-3%Ce	69

The morphology of the ZnO and Ce-modified ZnO photocatalysts was investigated by FE-SEM and the obtained results are displayed in Figure 2a–d. The SEM images clearly reveal the growth of smooth, randomly distributed nanosheets in the 500 nm^{-1} µm size range with thicknesses around 50 nm. The incorporation of different amounts of Ce leads to the appearance of small agglomerates of a porous structure on the smooth ZnO nanosheets (Ce-containing aggregates are highlighted within Figure 2).

In order to determine the nature of these new structures and their interaction with the ZnO sheets, it is essential to perform a study by means of transmission electron microscopy to obtain both the structural information using the HRTEM and the compositional information using the STEM-XEDS. The observations via STEM are fully consistent with the SEM analysis in terms of morphology and the dimensionality of the different components of the samples. Thus, it is clear from the STEM images that the pure ZnO photocatalyst presents a nanosheet morphology, as shown in Figure 3a, whereas in the case of the Ce-modified ZnO photocatalysts, shown in Figure 3b–d, very small heterogeneously-distributed aggregates are observed on the surface of the ZnO nanosheets with regions showing a high concentration of these nanoparticulated aggregates. White dashed rectangles have been added in Figure 3b–d to show some of these aggregates, which are mostly in contact with the ZnO sheets, although there is also a minority of them isolated.

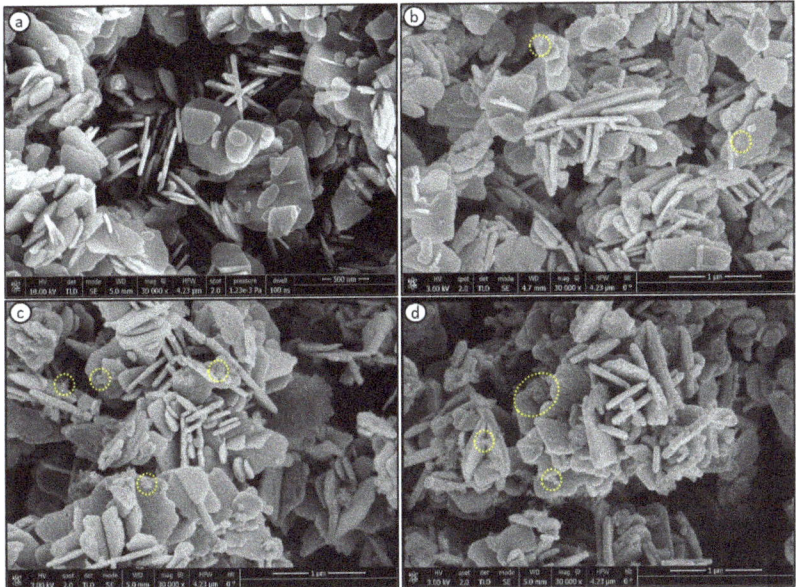

Figure 2. SEM images of (**a**) ZnO, (**b**) ZnO:Ce 1%, (**c**) ZnO:Ce 2% and (**d**) ZnO:Ce 3%.

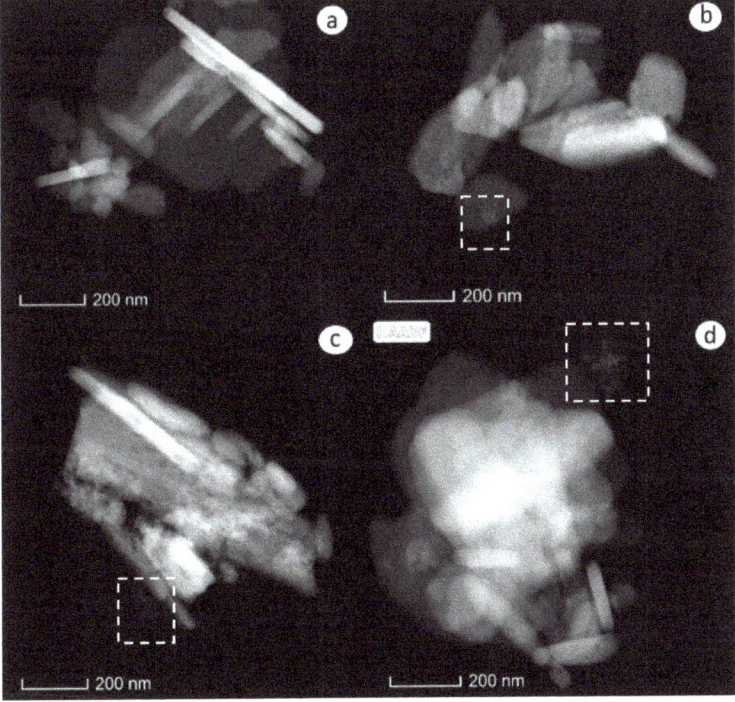

Figure 3. High-angle annular dark field-scanning transmission electron microscopy (HAADF-STEM) images of (**a**) ZnO, (**b**) ZnO:Ce 1%, (**c**) ZnO:Ce 2% and (**d**) ZnO:Ce 3%.

In order to characterize the composition of the samples, STEM-XEDS maps were acquired at the different sample locations in each catalyst. In the ZnO sample, as expected, when a compositional map was recorded, like that of the area shown in Figure 4a, the only cation that appears is Zn, both in the image, shown in Figure 4b, and in the spectrum, shown in Figure 4c. In the latter, the only additional signals are those due to oxygen and copper from the grid. The structural characterization of these ZnO oxide sheets by HRTEM is complicated due to their large thickness, unless they are correctly oriented. Figure 4d shows a high-resolution image of the ZnO sample where the typical spacing of the (002) wurtzite plane can be measured.

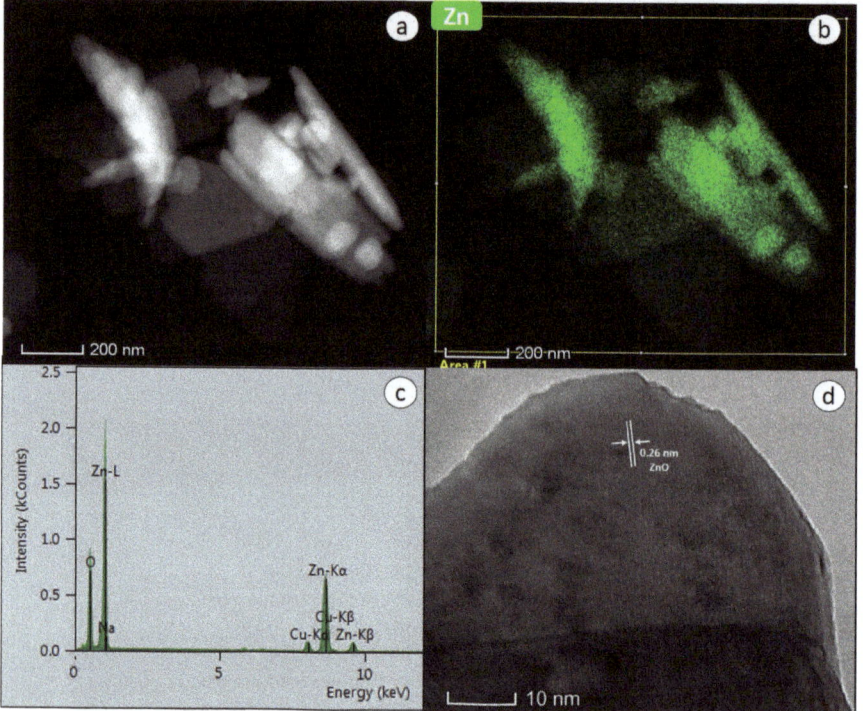

Figure 4. (**a**) Sample ZnO HAADF-STEM image. (**b**) energy-dispersive X-Ray spectroscopy (XEDS) elemental map showing the spatial distribution of Zn corresponding to the area displayed in **a**. (**c**) energy-dispersive X-ray (XED) spectra of the area marked in **b**. (**d**) HRTEM image.

The analysis of the cerium-modified samples by STEM-XEDS shows analogous results for the specimens with 1%, 2% and 3% cerium so that they are described together for simplicity. When acquiring a map of an area in which both sheets and nanoparticulate aggregates are observed in the high-angle annular dark field-scanning transmission electron microscopy (HAADF-STEM) image, such as those shown in Figures 5a, 6a and 7a, the corresponding maps show that the sheets are rich in Zn (displayed in green on the maps) and the nanoparticulate aggregates are rich in cerium (displayed in red on the maps). Figures 5b–d, 6b–d and 7b–d show the maps of Ce, Zn and the combination of both elements for the samples with 1%, 2% and 3% cerium, respectively. Once the maps were acquired it was possible to extract the sum spectrum of all the spectra acquired in each of the pixels contained in a selected area and thus determine and quantify which elements were present in that area. In the three samples containing cerium, when an area was chosen where there was apparently only Zn, i.e., an area where green predominated, such as the areas selected with a yellow square in Figures 5d, 6d and 7d—the spectra show only the presence of O, Zn and Cu, as shown in Figures 5e, 6e and 7e for the ZnO:Ce 1%,

2% and 3% samples, respectively. When these spectra are quantified, small amounts of cerium appear, always below 0.1 at.%. Therefore, although this may at first sight suggest that a tiny amount of cerium (<0.1%) has entered the structure of zinc oxide as a dopant, the absence of peaks corresponding to cerium in the XEDS spectra seems to indicate that these amounts of cerium correspond in all cases to the background of the XEDS spectrum. In fact, the quantification of the amount of cerium present in the sample that contained only ZnO, resulted in similar lanthanide amounts (lower than 0.1%), which confirmed that these signals corresponded to the background of the XEDS spectrum. The results of these analyses, as well as those performed on some of the spectra showing the absence of peaks corresponding to cerium, are shown in Figure 2 of the supporting information (Figure S2). Accordingly, STEM-XEDS indicates that the wurtzite phase was made up of pure ZnO.

However, the most interesting part of the XEDS maps relates to the analysis of the cerium-rich areas, displayed in red on the images. The quantification of the areas marked with blue squares and their corresponding spectra, shown in Figures 5f, 6f and 7f for the ZnO:Ce 1%, 2% and 3% samples, respectively, reveals in all cases that these areas contained both cerium and zinc. Figure 3 of the supporting information shows that the analysis of 25 of these areas belonging to the three samples with different cerium contents provides an average Zn:Ce ratio of 0.47, i.e., very close to 1:2, corresponding to an oxide with the composition $Ce_{0.68}Zn_{0.32}O_x$. Therefore, in the preparations described in the present article when the precursors of zinc and cerium were added simultaneously into the microwave-assisted hydrothermal synthesis, the nanoparticles of a mixed phase of cerium and zinc with a Zn:Ce atomic ratio of 1:2 formed in parallel with the growth of pure ZnO nanosheets. The former was heterogeneously distributed on the surface of the ZnO crystallite, giving rise to a network of $ZnO-Ce_{0.68}Zn_{0.32}O_x$ contacts or nano-heterojunctions. A sample with the composition $Ce_{0.68}Zn_{0.32}O_x$ was prepared under similar conditions. The obtained sample was analyzed by electron microscopy in HRTEM, HAADF-STEM and STEM-XEDS modes. It was observed that the sample was made up of the small crystals (3–5 nm) of a mixed oxide phase (see Figure S7), with a composition of approximately $Ce_{0.85}Zn_{0.15}O_x$, and nanoflakes containing only ZnO and the ZnO nanoflakes being a minority in this sample. In addition, the XEDS maps of one of the areas with mixed cerium–zinc oxide nanoparticles are shown (see Figure S8). It can therefore be seen that the synthesis method and conditions used did not lead to a single-phase product comprising only the Ce/Zn mixed oxide.

It is important to recall at this point that the synthesis of Zn–Ce mixed oxides has been described in the literature [34], showing a solubility limit for Zn in CeO_2 between 20 and 30 mol % in particles of about 5 nm in diameter. In the samples prepared in this work, the size of the nanoparticles of the Ce–Zn phase was even smaller, around 3 nm, as can be clearly seen in Figure 9b, which may have modified the solubility limit of Zn to slightly higher values, around 32 mol %. The formation of nanocomposites in which small crystals of a CeO_2 rich phase are deposited on ZnO larger size crystals has been reported in the literature after a one-pot hydrothermal synthesis [35], though in this case a compositional analysis of the cerium-rich phase was not performed, so the exact composition of these nanoparticles was not determined but was assumed to correspond to pure CeO_2. Additional XEDS maps of the three samples with different cerium loads were included in the supporting information.

An additional study was carried out using electron energy loss spectroscopy (EELS) to verify, with a more sensitive technique capable of detecting even single atoms, the absence of cerium in zinc oxide nanosheets. The EELS experiment was performed working in the spectrum imaging (SI) mode [36], which allows the correlation of the analytical and structural information on selected regions of the material under study. In particular, an area of 56.5 × 54.4 nm was analysed simultaneously acquiring the HAADF (Figure 8b) and EELS signal every 0.6 nm while the electron beam was scanned across the selected area of the sample (green square, Figure 8a). In the case of EELS-SI, the experiment was acquired in DUAL mode, acquiring the zero loss peak and the Ce-M$_{4,5}$ signal simultaneously using an energy dispersion of 0.025 eV. Figure 8c displays a chemical map, the area in red being where Ce is present. The EELS spectra extracted from area 1 and 2 are displayed in Figure 8d, where the absence of a Ce signal in the green region of Figure 8c is evident, as previously suggested by the XEDS measurements.

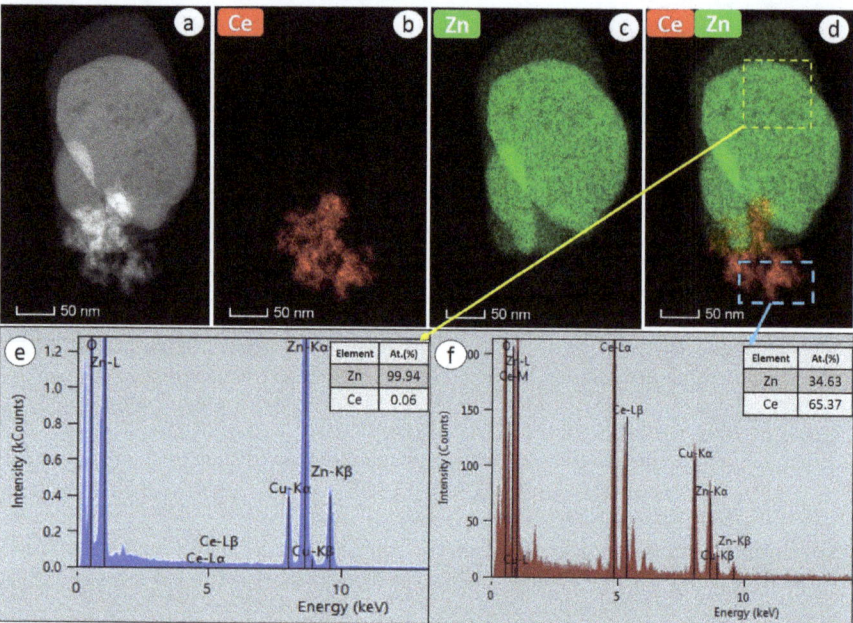

Figure 5. Sample ZnO:Ce 1%. (**a**) HAADF-STEM image and the XEDS elemental maps showing the spatial distribution of (**b**) Ce, (**c**) Zn and (**d**) both elements together corresponding to the area displayed in (**a**), (**e,f**) and the XED spectra of the areas marked in (**d**).

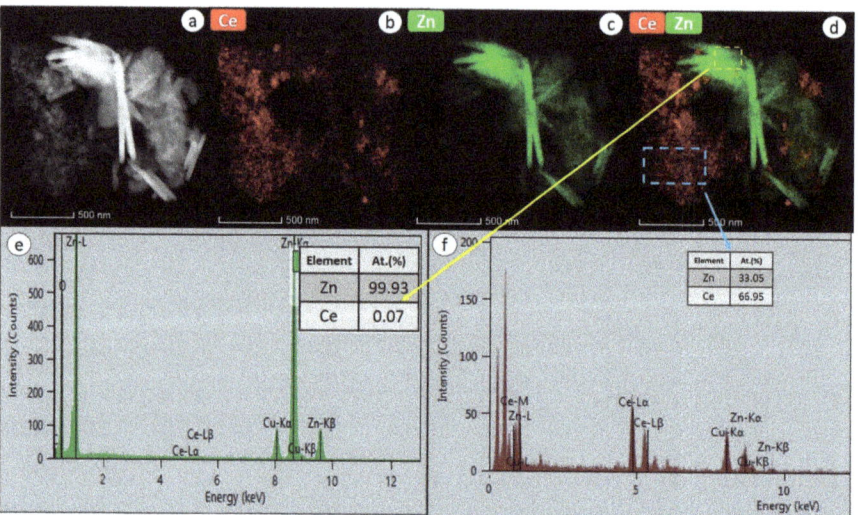

Figure 6. Sample ZnO–Ce 2% (**a**) HAADF-STEM image and the XEDS elemental maps showing the spatial distribution of (**b**) Ce, (**c**) Zn and (**d**) both elements together corresponding to the area displayed in (**a**), (**e,f**) and the XED spectra of the areas marked in (**d**).

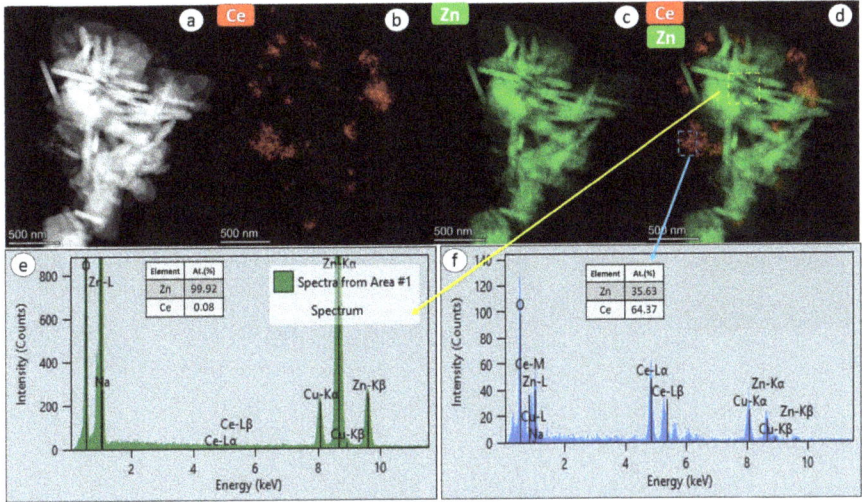

Figure 7. Sample ZnO:Ce 3% (**a**) HAADF-STEM image and the XEDS elemental maps showing the spatial distribution of (**b**) Ce, (**c**) Zn and (**d**) both elements together corresponding to the area displayed in (**a**), (**e**,**f**) and the XED spectra of the areas marked in (**d**).

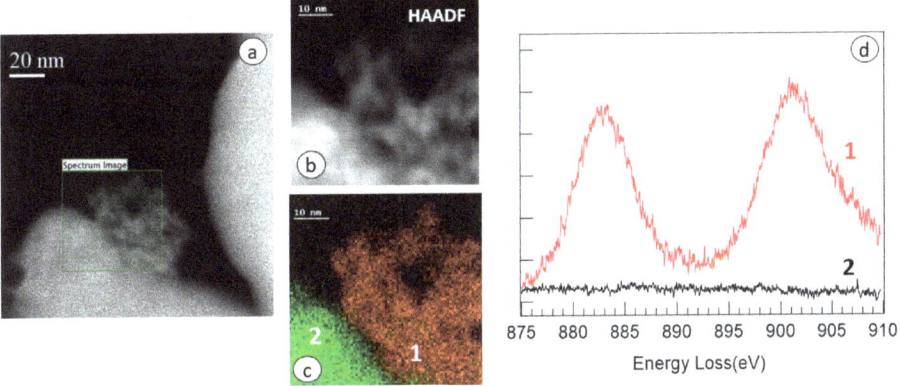

Figure 8. (**a**) High-angle annular dark field (HAADF) image of the ZnO:Ce 2% sample showing the spectrum imaging (SI) area. (**b**) HAADF image simultaneously recorded to the electron energy loss spectroscopy (EELS) acquisition. (**c**) Chemical map, the area in red being where Ce is present. (**d**) EELS spectra extracted from area 1 and 2.

In order to determine, in addition to the composition, the structure of the different phases present in the nanocomposite, a HRTEM analysis of the samples was carried out. Figure 9 shows the images corresponding to the sample with an intermediate Ce content, ZnO:Ce 2%. The analysis of the other two specimens provided similar results. Figure 9 shows two areas of the ZnO:Ce 2% sample in which both ZnO sheets and small particles of the mixed Ce–Zn phase appear. In the digital diffraction patterns (DDP) from the larger particle, inset in Figure 9a, the spacing and angle values of the ZnO wurtzite structure can be identified, in this particular case corresponding to a crystallite of this phase along the [010] zone axis. The analysis of the particle placed on the surface of the ZnO crystallite in Figure 9b shows that the spacing values and angles correspond to a crystal with a fluorite structure, just as that of pure CeO2 does, oriented along the [001] zone axis, as indicated in the DDP, shown as inset. The incorporation of Zn into the fluorite-type structure of cerium oxide without distorting the lattice parameters can be possibly accounted for taking into consideration the size of the Zn^{2+} and Ce^{4+} cations. Thus, Zn^{2+} cations, with a radius of 0.74 Å, could incorporate into the positions of the Ce^{4+} ions, of 0.97 Å. Therefore, when the cerium and zinc precursors are brought into contact, part of the Zn^{2+} cations could substitute Ce^{4+} ions or, alternatively, locate at interstitial positions in the CeO_2 lattice to form a $Ce_{1-x}Zn_xO_2$- solid solution. On the contrary, because of its larger size, it is possibly difficult for Ce^{4+} to enter into the smaller crystalline lattice of ZnO to substitute Zn^{2+}.

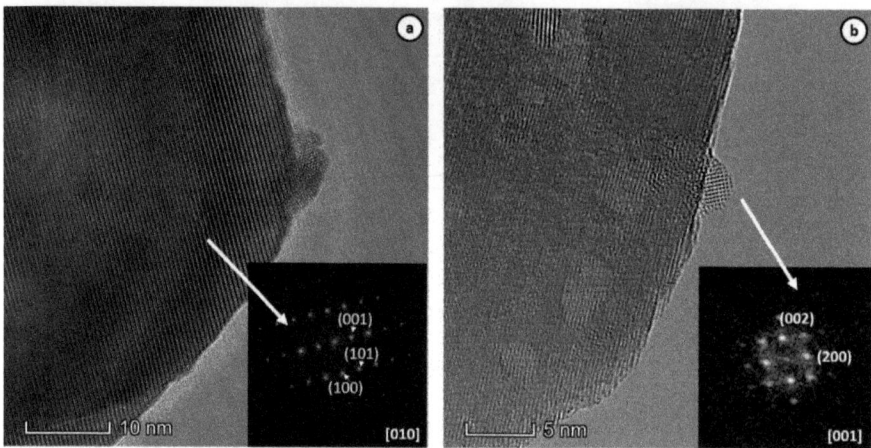

Figure 9. HRTEM images and the corresponding digital diffraction patterns (DDP) of the ZnO:Ce 2% sample where the wurtzite structure of ZnO (**a**) and the fluorite structure present in a supported $Ce_xZn_{1-x}O_y$ nanoparticle (**b**) can be identified.

2.2. Optical Absorption

UV–Visible absorption spectroscopy studies were performed to examine the influence of the Ce addition on the optical properties of ZnO. Pure and Ce-modified ZnO photocatalysts with increasing Ce contents were dispersed in absolute ethanol and the recorded spectra are shown in Figure 10a. A strong excitonic absorption peak (λ_{max}) was detected at 380 nm, corresponding to E_{ex} = 3.26 eV, mainly due to the electron transition from O 2p to Zn 3d, corresponding to the band-to-band transition of the ZnO energy band-structure [37], and was calculated using the following formula $E_{ex} = hc/\lambda_{max}$, where h is Planck's constant and c defines the speed of light. The obtained values are summarized in Table 2. The optical band gap was determined by the first derivative of the absorbance with respect to the photon energy model [38], Figure 10b–e. In this method, the optical bandgap value corresponds to the maximum in the derivative spectra at the lower energy sides.

Note that the optical bandgap decreased from 3.46 eV for ZnO to 3.17eV (see Figure 10b–e) for the maximum Ce content. Since the actual nanostructure of the different Ce-modified samples involves varying amounts of two different phases (ZnO and the $Ce_{0.68}Zn_{0.32}O_x$ mixed oxide), the values determined for the band gap at the macroscopic level must be intermediate between the band-gap values of these two components. The value for the pure ZnO sample is 3.46 eV. Though we do not have a value for the pure $Ce_{0.68}Zn_{0.32}O_x$ mixed oxide, as we previously commented it was not possible to prepare it under the same experimental conditions as the rest of the samples, we can take as a reference that of the CeO_2 with the fluorite structure as the mixed oxide, which amounts to 3.19 eV. The values we are measuring are in fact in the 3.5–3.2 eV range, which is in good agreement with a mixture of the two components. Moreover, our data indicate a monotonous decrease in the band-gap of the Ce-modified samples with the lanthanide content, which is in good agreement with an increasing content of the $Ce_{0.68}Zn_{0.32}O_x$ mixed oxide phase. This behavior has been already observed for other modified ZnO systems, and has been tentatively attributed to the competition between Moss–Burstein and band renormalization effects [39].

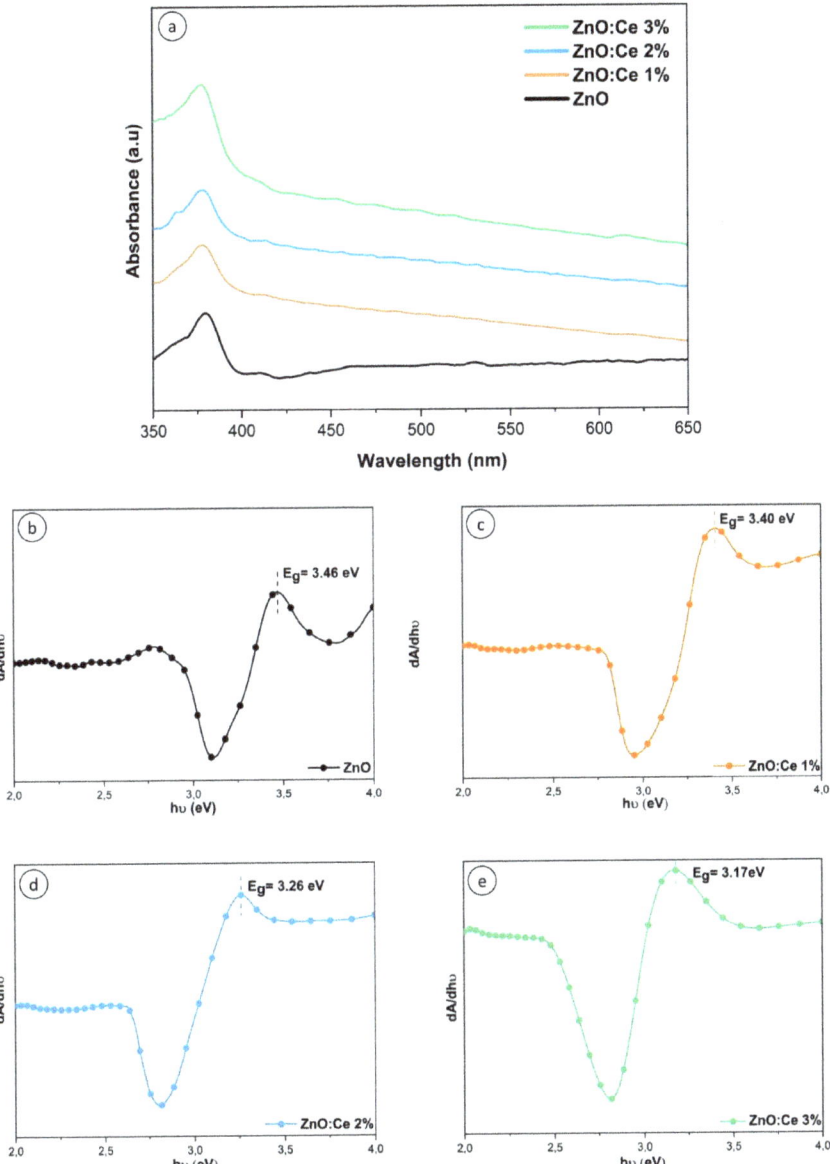

Figure 10. (a) UV–Vis absorption spectra of the ZnO and ZnO:Ce photocatalysts. (**b**–**e**) Plots of the first derivative of the optical absorbance with respect to the photon energy of the Ce-modified ZnO photocatalysts.

Table 2. The dependence of the band gap and exciton energy on the Ce content.

Sample	E_g (eV)	E_{ex} (eV)
ZnO	3.46	3.26
ZnO:Ce 1 at.%	3.40	3.28
ZnO:Ce 2 at.%	3.26	3.28
ZnO:Ce 3 at.%	3.17	3.29

2.3. PL Spectroscopy

Figure 11 displays the PL emission spectra of the pure and Ce-modified ZnO photocatalysts at different Ce contents. The emission peak intensity of the Ce-modified photocatalysts was noticeably decreased when compared to the pure ZnO. This observation could very likely be attributed to the fact that Ce^{4+} ions may act as electron trapping centers, which could contribute, for example, to inhibit the recombination of free carriers on the surface of ZnO. It is important to recall at this respect the reducibility of Ce^{4+} ions in CeO_2, a key property in most of the applications of this oxide. Likewise, it may also result as relevant the fact that the incorporation of aliovalent (+3 and +2) ions into the host fluorite structure resulted in a large increase in the reducibility of ceria-based mixed oxides with respect to the pure CeO_2 [40]. Thus, ZnO:Ce 2 at.% exhibited the lowest emission peak intensity, which can be considered as an optimal Ce concentration for the lowest electron–hole recombination rate. This low PL emission could contribute to improve the photocatalytic performance, as described below.

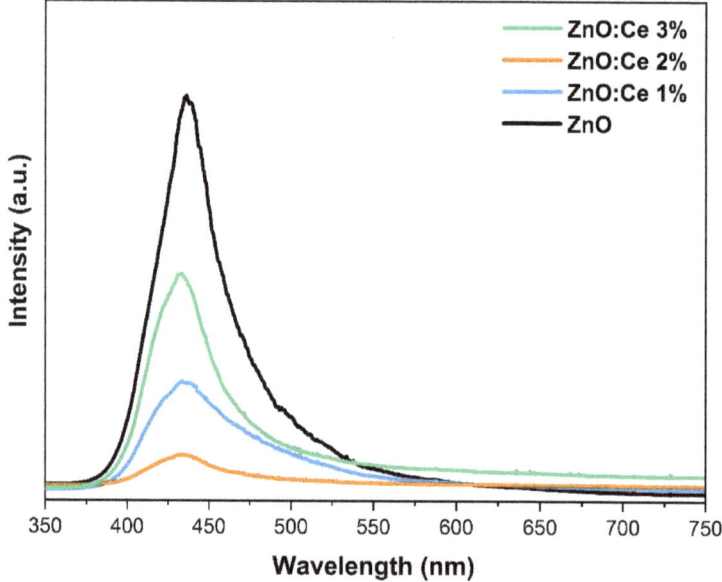

Figure 11. Photoluminescence spectroscopy (PL) spectra of the pure and Ce-modified ZnO photocatalysts.

2.4. Photocatalytic Degradation of Methylene Blue by ZnO:Ce Photocatalysts

The photocatalytic activity of the different synthesized photocatalysts was tested for the decolorization of MB under simulated sunlight illumination and subsequently compared with that of the reference commercial TiO_2 P-25 at 25 °C. Generally, the degradation of MB leads to the formation of water, sulfate, carbon dioxide and nitrate [41].

Figure S9 a depicts the variation with time of the main absorption band of the target (MB) in the presence of the ZnO:Ce 2 at.% photocatalyst under sunlight illumination. The results show that the intensity of the intense absorption band centered at λ_{max} = 663 nm decreased noticeably during the irradiation process, confirming the potential of the material to decompose MB.

Figure 12 displays the degradation profile of MB in the presence of TiO_2 (P25), pure ZnO, Ce-modified ZnO photocatalysts at different Ce contents and $Ce_{0.68}Zn_{0.32}O_x$ under the same experimental conditions. In this plot, C_0 and C_t are the concentrations at the adsorption–desorption equilibrium and at time t, respectively. It is clearly observed that all the Ce-modified ZnO photocatalysts exhibited a higher performance towards the decomposition of MB compared to the pure ZnO and TiO_2 photocatalysts. The photodegradation percentages of the dye reached about 49%, 63%, 80%, 95% and 65%, for TiO_2, ZnO, ZnO:Ce 1%, ZnO:Ce 2% and ZnO:Ce 3%, respectively, after 120 min of irradiation (see Figure S9b). The sample with stoichiometry $Ce_{0.68}Zn_{0.32}O_x$, but actually showing the $Ce_{0.85}Zn_{0.15}O_x$ phase with some contacts between $Ce_{0.85}Zn_{0.15}O_x$ and ZnO, shows the worst activity of all the samples studied, reaching a photodegradation percentage of 7% after 120 min of irradiation. Interestingly, the ZnO:Ce 2 at.% photocatalyst showed the best decolorization performance. In fact, a 2 at.%. Ce addition almost guarantees a complete removal of the MB concentrations essayed, in good agreement with the expectations posed by the PL results and being probably favored by the larger surface area of this sample compared to the rest. The negligible loss of MB under self-decolorization conditions, i.e., in the absence of any photocatalysts after 120 min irradiation is presented in Figure 12, which clearly shows that the decomposition of MB is due to the photo-activation by ZnO:Ce. Our results also evidence a superior performance of the materials prepared in this work compared to those of other previously reported reference materials [42–44].

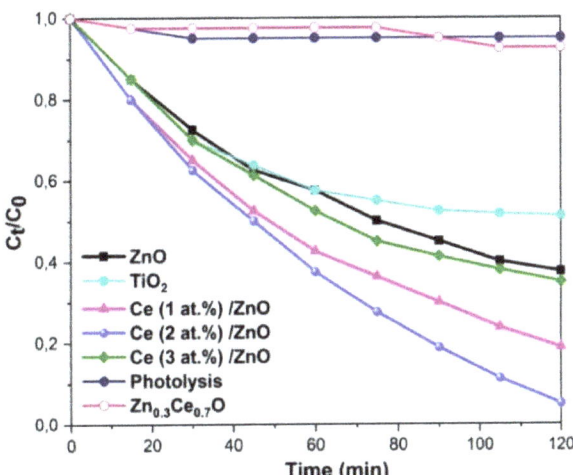

Figure 12. Photodegradation of methylene blue (MB) in the presence of the different ZnO:Ce photocatalysts, TiO_2, $Ce_{0.7}Zn_{0.3}O_x$ and without any photocatalyst.

The Langmuir–Hinshelwood (LH) model was successfully utilized to describe the kinetics of heterogeneous catalytic processes in which an intermediate adsorption of the reactants over the catalytically active phase was required prior to reaction, the latter being the rate-limiting step of the process. The overall reaction rate of the LH model can be expressed by the following equation [45]:

$$r = -\frac{dC}{dt} = \frac{k_1 K_2 \, C}{1 + C K_2} \quad (1)$$

where r stands for the rate of reaction that changes with time; k_1 and K_2 represent the rate (kinetic) and adsorption (thermodynamic) constants, respectively. C is the concentration of the organic pollutant.

The integration of the left-hand side of the Equation (1) from t = 0 to t = t and of the right-hand side from the starting concentration C_0 to the final concentration C leads to the following equation:

$$\frac{\ln(C_0/C)}{(C_0 - C)} = k_1 K_2 \cdot \frac{t}{(C_0 - C)} - K_2 \qquad (2)$$

A linear fit of the LH plots ($\ln(C_0/C)/(C_0-C)$ vs. $t/(C_0-C)$) allows one to determine the value of both K_2 and k_1 from the intercept and slope. Figure S10 shows the LH plots for the MB photodegradation process by the TiO_2, ZnO and ZnO:Ce photocatalysts. From the figure it is evident that all the photodegradation tests can be fitted to LH model. The values of the effective rate constant (i.e., the product $k_1 \cdot K_2$) obtained after fitting are summarized in Table 3. From the table, the sample ZnO:Ce 2 at.% is the one that exhibits the highest value of this parameter. As seen from the photocatalytic activity tests, a further increase in Ce content (>2 at.%) leads to a deterioration of the photocatalytic performance of the samples. The results obtained from the fitting to the LH model are fully in line with the presence of a larger surface area (Table 4) and the absorption and photoluminescence findings. In fact, the minimum PL emission in the ZnO:Ce 2 at.% system matches perfectly well with the optimum photocatalytic activity observed for this sample.

Table 3. Numerical values of the effective rate constant estimated using the LH model.

Sample	k (M. min^{-1})
ZnO	5.2×10^{-3}
P25	1.6×10^{-3}
ZnO:Ce 1 at.%	1.3×10^{-2}
ZnO:Ce 2 at.%	3.9×10^{-2}
ZnO:Ce 3 at.%	5.8×10^{-3}

Table 4. Numerical values of the effective rate constant estimated using the LH model.

Sample	Surface Area m^2.g^{-1})
ZnO	10
ZnO:Ce 1 at.%	10
ZnO:Ce 2 at.%	22
ZnO:Ce 3 at.%	15

With the available information is not possible to suggest a precise mechanism for the photocatalytic process, but from the present results the following, quite relevant facts, to date not clarified in the previously available literature, become clear: (1) the RE dopant does not incorporate into the host lattice of ZnO as reported in a high number of papers; (2) a secondary $Ce_{0.68}Zn_{0.32}O_x$ phase is formed in which the two metals are mixed at the atomic scale but maintaining the fluorite structure of pure CeO_2. The formation of this phase has not been detected in any previous study; (3) the Ce-modified materials are formed by agglomerates of the $Ce_{0.68}Zn_{0.32}O_x$ fluorite phase in contact with the surfaces of the ZnO nanosheets, which give rise to $Ce_{0.68}Zn_{0.32}O_x$/ZnO nano-heterojunctions. Such structure, which has been facilitated by the particular morphology of the ZnO component in the form of isolated nanoplates, is of interest with regards to any further explanation based on DFT calculations or proposals about electronic transfer phenomena across the observed heterojunctions. The precise nature of these heterojunctions is of great importance, as is shown by the fact that the sample with a low number of contacts between $Ce_{0.85}Zn_{0.15}O_x$ and ZnO showed almost negligible activity. The present results show that the improvements in the photocatalytic activities in the ZnO:Ce systems were not related to changes in the band structure of ZnO due to RE doping or to an interaction between the band structures at the CeO_2/ZnO hererojunctions. Similarly, previous DFT calculations to explain changes in band-gap or electron-transfer phenomena through CeO_2/ZnO heterojunctions do not apply when rationalizing the observed photocatalytic changes after the modification of ZnO with Ce in our samples.

New calculations should be performed that consider a composite ZnO/Ce$_{0.68}$Zn$_{0.32}$O$_x$ system in which, at least partially, nanosized heterojunctions between these two phases are formed. This is part of the on-going research in our lab concerning these systems.

At a preliminary level, it could be tentatively hypothesized that the decrease in the band gap in ZnO would allow us to improve the absorption in the UV–Vis range and promote the formation of electron–hole pairs. The contact with the reducible Ce$_{0.68}$Zn$_{0.32}$O$_x$ phase would allow us to decrease the recombination of the pairs, through e.g., the trapping of electron by promotion of Ce^{4+} to Ce^{3+} transitions, which are well known to be enhanced in mixed ceria-oxides. This separation of the charge carriers would explain the decrease in PL and would also allow the promotion of both the anodic process at the hole trapping sites and the cathodic process, which is very likely at sites neighbored by reduced Ce^{3+} species, since these can be easily reoxidised to Ce^{4+} and close the photocatalytic process.

3. Experimental

3.1. Materials

Zinc acetate dihydrate (Zn(O$_2$CCH$_3$)$_2$(H$_2$O)$_2$), cerium nitrate (CeN$_3$O$_9$·6H$_2$O), sodium hydroxide (NaOH), and hexamethylenetetramine (HMT) (C$_6$H$_{12}$N$_4$) were purchased from Sigma-Aldrich (St.Louis, MO, USA). Methylene blue (MB) was selected as a target organic pollutant and was purchased from Sigma-Aldrich (St.Louis, MO, USA). Milli-Q water was used in all the preparations. All the reagents were used as received without further purification.

3.2. Preparation of Pure and Ce-Modified ZnO Photocatalysts

ZnO and Ce-modified ZnO photocatalysts were prepared via a microwave-assisted hydrothermal process. The preparation procedure of the pure ZnO was as follows: 6 g of zinc acetate and 4 g of HMT were dissolved in 100 mL of MQ water. Then, 50 mL of 0.3 M aq. NaOH solution was added dropwise into the above solution and stirred continuously for 1 h at room temperature. The uniform mixture was ultrasonicated for 1 h. The resultant milky mixture was transferred and sealed in two 100 mL Teflon-lined autoclave reactors and then heated to 180 °C in a microwave oven with a controlled power of 400 W for 18 min. The white precipitates formed were centrifuged and thoroughly washed by repeated centrifugation–redispersion with MQ water and then absolute ethanol. The white precipitates collected were dried in an air oven maintained at 110 °C for 12 h. A similar procedure was applied to prepare Ce-modified photocatalysts with a cationic percentage of cerium of 1%, 2% and 3%, the corresponding samples being designated ZnO:Ce 1%, ZnO:Ce 2% and ZnO:Ce 3%. In this case, the initial zinc acetate and HMT solution also contained the required amounts of the Ce(NO$_3$)$_3$ precursor.

3.3. Characterization

The photocatalysts were characterized by X-ray diffraction (XRD) using a Bruker D8 advance X-ray diffractometer (λ = 0.154056 nm) (Bruker, Billerica, MA, USA). The surface morphology was investigated via scanning electron microscopy (FE-SEM, Nova NanoSEM 450) (Thermo Fisher Scientific, Waltham, MA, USA). Transmission electron microscopy (TEM) was performed with a TEM/STEM FEI Talos F200X G2 microscope, equipped with 4 Super-X SDDs. HAADF-STEM images and XEDS maps were acquired using a high brightness electron probe in combination with a highly stable stage which minimized sample drift. Elemental maps were acquired with a beam current of 100 pA and a dwell time of 50 µs which resulted in a total acquisition time of approximately 12 min (Thermo Fisher Scientific, Waltham, MA, USA). Electron energy loss spectroscopy (EELS) experiments were performed using a double aberration-corrected FEI Titan^3 Themis 60–300 microscope operated at 80Kv (Thermo Fisher Scientific, Waltham, MA, USA).

Optical absorption spectra of all the photocatalysts were obtained by dispersing the powders in absolute ethanol and then recorded using a Shimadzu UV-1603 spectrophotometer (Shimadzu corporation, Kyoto, Japan), with absolute ethanol as the reference medium. Photoluminescence (PL) spectra were recorded using a confocal Horiba Jobin Yvon LabRAM HR800 (Horiba, Kyoto, Japan) with a UV He-Cd laser operating at 325 nm as the excitation source. For the determination of the specific area of the different photocatalysts, the nitrogen adsorption–desorption isotherms at 77 K were recorded (QuantaChrome-Anton Paar, Graz, Austria). The samples were degassed at 250 °C in vacuum prior to the measurements.

3.4. Photodegradation

The photocatalytic activity of the Ce/ZnO photocatalysts was tested by the decolorization of methylene blue (MB) dye under simulated sunlight irradiation. Typically, 25 mg of the photocatalyst was dispersed in 300 mL of a 10 mg/L MB solution. Before irradiation, and to attain the adsorption–desorption equilibrium between the surface of the photocatalyst and the MB, the resulting mixture was stirred for 60 min in the dark and aerated by a pump to provide oxygen and complete mixing of the solution. Subsequently, the solution was continuously irradiated with a Xenon lamp placed approximately 40 cm above the reaction solution (the lamp spectrum is included in Figure 1 of the supporting information section). Then, 5 mL of the suspension were extracted every 15 min and centrifuged at 5000 rpm for 10 min, which was repeated four times to remove the photocatalysts from the suspension. The residual concentration of the MB was evaluated using a UV–Vis spectrophotometer.

The photocatalytic efficiency was estimated through the following formula:

$$\text{Degradation } (\%) = \frac{C_0 - C_t}{C_0} \times 100 \tag{3}$$

where C_0 corresponds to the initial concentration of the MB dye, and C_t represents the concentration of the MB at different times.

4. Conclusions

In summary, we succeeded in synthesizing a series of Ce-modified ZnO nanosheets which depicted a highly efficient photocatalytic activity using a simple and fast microwave-assisted hydrothermal route. The crystal phase, morphology, elemental composition, optical absorption, photoluminescence and photocatalytic activity of the prepared materials were investigated. The intense diffraction peaks evidenced that the obtained photocatalysts were highly crystalline. A red shift of the optical band gap was revealed by UV–Vis measurements. The PL data confirmed a high charge separation efficiency in these materials. The as-synthesized products exhibited excellent sunlight driven MB photodegradation activity, higher than those of both the ZnO and commercially available TiO_2. This improved photocatalytic performance confirms that the ZnO:Ce system offers potential applications for environmental remediation.

The in-depth characterization performed by the STEM techniques revealed some new, structural features which have not been considered in previous works. The present study indicates the absence of ZnO doping by Ce, which forms a $Ce_{0.68}Zn_{0.32}O_x$ mixed oxide phase instead. It is suggested that the unique redox properties of this mixed oxide phase together with the modification in the ZnO band gap and the presence of the $ZnO/Ce_{0.68}Zn_{0.32}O_x$ nano-heterojunctions may explain the improvement in the photocatalytic performance. Finally, this work highlights the key role of advanced STEM techniques in the analysis of this type of photocatalysts, particularly to rationalize their behavior on the basis of structurally meaningful correct data.

Supplementary Materials: The following are available online at http://www.mdpi.com/2073-4344/10/5/551/s1, Figure S1: Emission spectrum of Xenon lamp utilized within the present study. Figure S2: (a)The graph shows the atomic percentages of cerium present in apparently cerium-free zones of samples ZnO, ZnO:Ce1%, ZnO:Ce2%, ZnO:Ce3%, and a sample prepared as a reference consisting in a physical mixture of CeO_2 and ZnO (with 2% atomic loading of cerium). An analysis performed in a holey carbon area of the grids free of sample is also considered. All loads, including those acquired in an unmodified ZnO sample, are less than 0.1%. As can be seen in the expanded spectrum shown in image (b), acquired in an area of the ZnO:Ce 2% sample, which shows an atomic cerium content of 0.06%, there is only noise at the position at which the cerium peaks should appear. Figure S3: The graph shows 25 analyses of the atomic Zn/Ce ratio in cerium-rich areas of the three samples modified with different amounts of cerium. The dotted line corresponds to the mean value of 0.47, which corresponds to a composition of the mixed oxide $Ce_{0.68}Zn_{0.32}O_x$. Figure S4: Sample ZnO–Ce 1% (a) HAADF-STEM image and the XEDS elemental maps showing the spatial distribution of (b) Ce, (c) Zn and (d) both elements together corresponding to the area displayed in (a). Figure S5: Sample ZnO–Ce 2% (a) HAADF-STEM image and the XEDS elemental maps showing the spatial distribution of (b) Ce, (c) Zn and (d) both elements together corresponding to the area displayed in (a). Figure S6: Sample ZnO–Ce 3% (a) HAADF-STEM image and the XEDS elemental maps showing the spatial distribution of (b) Ce, (c) Zn and (d) both elements together corresponding to the area displayed in (a). Figure S7: HAADF-STEM images of the sample at low magnification showing the major polycrystalline aggregates together with the ZnO nanoflakes. Figure S8: Sample ZnO–Ce0.68Zn0.32Ox XEDS elemental maps showing the spatial distribution of (b) Ce, (c) Zn and (d) both elements together corresponding to the area displayed in (a). Figure S9: (a) Absorbance spectra of the MB aqueous solution in the presence of the ZnO:Ce 2% photocatalyst at increasing irradiation times. (b) Photodegradation vs. the irradiation time of different samples. Figure S10: LH plots of the MB degradation by the Ce/ZnO and P25 photocatalysts.

Author Contributions: Conceptualization, O.B., J.J.C. and A.B.H.; methodology, O.B.; investigation, O.B. and S.T.; resources, J.J.C. and A.B.H.; writing—original draft preparation, O.B.; writing—review and editing, A.U., S.T., J.P., P.F., M.A., J.J.C. and A.B.H.; supervision, A.U., P.F. and A.B.H.; project administration, J.J.C. and A.B.H.; All authors have read and agreed to the published version of the manuscript.

Funding: This work was supported by MINECO/FEDER (MAT 2016-81118-P, MAT 2017- 87579-R and RED2018-102609-T. O.B. thanks Aula del Estrecho fellowship.

Conflicts of Interest: The authors declare no conflict of interest.

References

1. Kostedt IV, W.L.; Ismail, A.A.; Mazyc, D.W. Impact of heat treatment and composition of ZnO-TiO2 nanoparticles for photocatalytic oxidation of an azo dye. *Ind. Eng. Chem. Res.* **2008**, *47*, 1483–1487. [CrossRef]
2. Abu Tariq, M.; Faisal, M.; Muneer, M. Semiconductor-mediated photocatalysed degradation of two selected azo dye derivatives, amaranth and bismarck brown in aqueous suspension. *J. Hazard. Mater.* **2005**, *127*, 172–179. [CrossRef]
3. Faisal, M.; Khan, S.B.; Rahman, M.M.; Jamal, A.; Asiri, A.M.; Abdullah, M.M. Synthesis, characterizations, photocatalytic and sensing studies of ZnO nanocapsules. *Appl. Surf. Sci.* **2011**, *258*, 672–677. [CrossRef]
4. Zhang, H.H.; Pan, X.H.; Li, Y.; Ye, Z.Z.; Lu, B.; Chen, W.; Huang, J.Y.; Ding, P.; Chen, S.S.; He, H.P.; et al. The role of band alignment in p-type conductivity of Na-doped ZnMgO: Polar versus non-polar. *Appl. Phys. Lett.* **2014**, *104*. [CrossRef]
5. Masuda, Y.; Kato, K. Aqueous synthesis of ZnO rod arrays for molecular sensor. *Cryst. Growth Des.* **2009**, *9*, 3083–3088. [CrossRef]
6. Ong, C.B.; Ng, L.Y.; Mohammad, A.W. A review of ZnO nanoparticles as solar photocatalysts: Synthesis, mechanisms and applications. *Renew. Sustain. Energy Rev.* **2018**, *81*, 536–551. [CrossRef]
7. Bazta, O.; Urbieta, A.; Piqueras, J.; Fernández, P.; Addou, M.; Calvino, J.J.; Hungría, A.B. Enhanced UV emission of Li–Y co-doped ZnO thin films via spray pyrolysis. *J. Alloy. Compd.* **2019**, *808*, 151710. [CrossRef]
8. Xing, X.; Chen, T.; Li, Y.; Deng, D.; Xiao, X.; Wang, Y. Flash synthesis of Al-doping macro-/nanoporous ZnO from self-sustained decomposition of Zn-based complex for superior gas-sensing application to n-butanol. *Sens. Actuators B Chem.* **2016**, *237*, 90–98. [CrossRef]
9. Minami, T.; Miyata, T.; Ihara, K.; Minamino, Y.; Tsukada, S. Effect of ZnO film deposition methods on the photovoltaic properties of ZnO-Cu2O heterojunction devices. *Thin Solid Film.* **2006**, *494*, 47–52. [CrossRef]
10. Bazta, O.; Urbieta, A.; Piqueras, J.; Fernández, P.; Addou, M.; Calvino, J.J.; Hungría, A.B. Influence of yttrium doping on the structural, morphological and optical properties of nanostructured ZnO thin films grown by spray pyrolysis. *Ceram. Int.* **2018**. [CrossRef]
11. Heo, Y.W.; Norton, D.P.; Tien, L.C.; Kwon, Y.; Kang, B.S.; Ren, F.; Pearton, S.J.; Laroche, J.R. ZnO nanowire growth and devices. *Mater. Sci. Eng. R Rep.* **2004**, *47*, 1–47. [CrossRef]
12. Wahab, R.; Tripathy, S.K.; Shin, H.S.; Mohapatra, M.; Musarrat, J.; Al-Khedhairy, A.A.; Kumar Kaushik, N. Photocatalytic oxidation of acetaldehyde with ZnO-quantum dots. *Chem. Eng. J.* **2013**, *226*, 154–160. [CrossRef]
13. Lu, Y.; Lin, Y.; Wang, D.; Wang, L.; Xie, T.; Jiang, T. A high performance cobalt-doped ZnO visible light photocatalyst and its photogenerated charge transfer properties. *Nano Res.* **2011**, *4*, 1144–1152. [CrossRef]

14. Neppolian, B. Solar/UV-induced photocatalytic degradation of three commercial textile dyes. *J. Hazard. Mater.* **2002**, *89*, 303–317. [CrossRef]
15. Kong, J.-Z.; Li, A.-D.; Li, X.-Y.; Zhai, H.-F.; Zhang, W.-Q.; Gong, Y.-P.; Li, H.; Wu, D. Photo-degradation of methylene blue using Ta-doped ZnO nanoparticle. *J. Solid State Chem.* **2010**, *183*, 1359–1364. [CrossRef]
16. Teng, X.M.; Fan, H.T.; Pan, S.S.; Ye, C.; Li, G.H. Influence of annealing on the structural and optical properties of ZnO: Tb thin films. *J. Appl. Phys.* **2006**, *100*, 053507. [CrossRef]
17. Zong, Y.; Li, Z.; Wang, X.; Ma, J.; Men, Y. Synthesis and high photocatalytic activity of Eu-doped ZnO nanoparticles. *Ceram. Int.* **2014**, *40*, 10375–10382. [CrossRef]
18. Khatamian, M.; Khandar, A.A.; Divband, B.; Haghighi, M.; Ebrahimiasl, S. Heterogeneous photocatalytic degradation of 4-nitrophenol in aqueous suspension by Ln (La 3+, Nd 3+ or Sm 3+) doped ZnO nanoparticles. *J. Mol. Catal. A Chem.* **2012**, *365*, 120–127. [CrossRef]
19. Sin, J.C.; Lam, S.M.; Lee, K.T.; Mohamed, A.R. Preparation and photocatalytic properties of visible light-driven samarium-doped ZnO nanorods. *Ceram. Int.* **2013**, *39*, 5833–5843. [CrossRef]
20. Chouchene, B.; Chaabane, T.B.; Balan, L.; Girot, E.; Mozet, K.; Medjahdi, G.; Schneider, R. High performance Ce-doped ZnO nanorods for sunlight-driven photocatalysis. *Beilstein J. Nanotechnol.* **2016**, *7*, 1338–1349. [CrossRef]
21. Tan, W.K.; Abdul Razak, K.; Lockman, Z.; Kawamura, G.; Muto, H.; Matsuda, A. Photoluminescence properties of rod-like Ce-doped ZnO nanostructured films formed by hot-water treatment of sol-gel derived coating. *Opt. Mater.* **2013**, *35*, 1902–1907. [CrossRef]
22. Ahmad, M.; Ahmed, E.; Zafar, F.; Khalid, N.R.; Niaz, N.A.; Hafeez, A.; Ikram, M.; Khan, M.A.; Hong, Z. Enhanced photocatalytic activity of Ce-doped ZnO nanopowders synthesized by combustion method. *J. Rare Earths* **2015**, *33*, 255–262. [CrossRef]
23. Faisal, M.; Ismail, A.A.; Ibrahim, A.A.; Bouzid, H.; Al-Sayari, S.A. Highly efficient photocatalyst based on Ce doped ZnO nanorods: Controllable synthesis and enhanced photocatalytic activity. *Chem. Eng. J.* **2013**, *229*, 225–233. [CrossRef]
24. Shi, Q.; Wang, C.; Li, S.; Wang, Q.; Zhang, B.; Wang, W.; Zhang, J.; Zhu, H. Enhancing blue luminescence from Ce-doped ZnO nanophosphor by Li doping. *Nanoscale Res. Lett.* **2014**, *9*, 1–7. [CrossRef] [PubMed]
25. Subash, B.; Krishnakumar, B.; Velmurugan, R.; Swaminathan, M.; Shanthi, M. Synthesis of Ce co-doped Ag-ZnO photocatalyst with excellent performance for NBB dye degradation under natural sunlight illumination. *Catal. Sci. Technol.* **2012**, *2*, 2319–2326. [CrossRef]
26. Liang, Y.; Guo, N.; Li, L.; Li, R.; Ji, G.; Gan, S. Preparation of porous 3D Ce-doped ZnO microflowers with enhanced photocatalytic performance. *RSC Adv.* **2015**, *5*, 59887–59894. [CrossRef]
27. Cerrato, E.; Gionco, C.; Paganini, M.C.; Giamello, E.; Albanese, E.; Pacchioni, G. Origin of Visible Light Photoactivity of the CeO 2 /ZnO Heterojunction. *ACS Appl. Energy Mater.* **2018**, *1*, 4247–4260. [CrossRef]
28. Zhu, L.; Li, H.; Xia, P.; Liu, Z.; Xiong, D. Hierarchical ZnO Decorated with CeO2 Nanoparticles as the Direct Z-Scheme Heterojunction for Enhanced Photocatalytic Activity. *ACS Appl. Mater. Interfaces* **2018**, *10*, 39679–39687. [CrossRef]
29. St John, J.; Coffer, J.L.; Chen, Y.; Pinizzotto, R.F. Synthesis and characterization of discrete luminescent erbium-doped silicon nanocrystals. *J. Am. Chem. Soc.* **1999**, *121*, 1888–1892. [CrossRef]
30. Sarro, M.; Gule, N.P.; Laurenti, E.; Gamberini, R.; Paganini, M.C.; Mallon, P.E.; Calza, P. ZnO-based materials and enzymes hybrid systems as highly efficient catalysts for recalcitrant pollutants abatement. *Chem. Eng. J.* **2018**, *334*, 2530–2538. [CrossRef]
31. Karunakaran, C.; Gomathisankar, P.; Manikandan, G. Preparation and characterization of antimicrobial Ce-doped ZnO nanoparticles for photocatalytic detoxification of cyanide. *Mater. Chem. Phys.* **2010**, *123*, 585–594. [CrossRef]
32. Ducll're, J.R.; Doggett, B.; Henry, M.O.; McGlynn, E.; Rajendra Kumar, R.T.; Mosnier, J.P.; Perrin, A.; Guilloux-Viry, M. (20–23) ZnO thin films grown by pulsed laser deposition on Ce O2 -buffered r -sapphire substrate. *J. Appl. Phys.* **2007**, *101*. [CrossRef]
33. Yousefi, M.; Amiri, M.; Azimirad, R.; Moshfegh, A.Z. Enhanced photoelectrochemical activity of Ce doped ZnO nanocomposite thin films under visible light. *J. Electroanal. Chem.* **2011**, *661*, 106–112. [CrossRef]

34. Kellici, S.; Gong, K.; Lin, T.; Brown, S.; Clark, R.J.H.; Vickers, M.; Cockcroft, J.K.; Middelkoop, V.; Barnes, P.; Perkins, J.M.; et al. High-throughput continuous hydrothermal flow synthesis of Zn-Ce oxides: Unprecedented solubility of Zn in the nanoparticle fluorite lattice. *Philos. Trans. R. Soc. A Math. Phys. Eng. Sci.* **2010**, *368*, 4331–4349. [CrossRef] [PubMed]
35. He, G.; Fan, H.; Wang, Z. Enhanced optical properties of heterostructured ZnO/CeO2 nanocomposite fabricated by one-pot hydrothermal method: Fluorescence and ultraviolet absorption and visible light transparency. *Opt. Mater.* **2014**, *38*, 145–153. [CrossRef]
36. Jeanguillaume, C.; Colliex, C. Spectrum-image: The next step in EELS digital acquisition and processing. *Ultramicroscopy* **1989**, *28*, 252–257. [CrossRef]
37. Qin, H.; Li, W.; Xia, Y.; He, T. Photocatalytic activity of heterostructures based on ZnO and N-doped ZnO. *ACS Appl. Mater. Interfaces* **2011**, *3*, 3152–3156. [CrossRef]
38. Othman, A.A.; Ali, M.A.; Ibrahim, E.M.M.; Osman, M.A. Influence of Cu doping on structural, morphological, photoluminescence, and electrical properties of ZnO nanostructures synthesized by ice-bath assisted sonochemical method. *J. Alloy. Compd.* **2016**, *683*, 399–411. [CrossRef]
39. Kaneva, N.; Bojinova, A.; Papazova, K.; Dimitrov, D. Photocatalytic purification of dye contaminated sea water by lanthanide (La 3+, Ce 3+, Eu 3+) modified ZnO. *Catal. Today* **2015**, *252*, 113–119. [CrossRef]
40. Hiley, C.I.; Fisher, J.M.; Thompsett, D.; Kashtiban, R.J.; Sloan, J.; Walton, R.I. Incorporation of square-planar Pd2+ in fluorite CeO2: Hydrothermal preparation, local structure, redox properties and stability. *J. Mater. Chem. A* **2015**, *3*, 13072–13079. [CrossRef]
41. Houas, A.; Lachheb, H.; Ksibi, M.; Elaloui, E.; Guillard, C.; Herrmann, J.-M. Photocatalytic degradation pathway of methylene blue in water. *Appl. Catal. B Environ.* **2001**, *31*, 145–157. [CrossRef]
42. Sin, J.C.; Lam, S.M.; Lee, K.T.; Mohamed, A.R. Preparation of cerium-doped ZnO hierarchical micro/nanospheres with enhanced photocatalytic performance for phenol degradation under visible light. *J. Mol. Catal. A Chem.* **2015**, *409*, 1–10. [CrossRef]
43. Yayapao, O.; Thongtem, S.; Phuruangrat, A.; Thongtem, T. Sonochemical synthesis, photocatalysis and photonic properties of 3% Ce-doped ZnO nanoneedles. *Ceram. Int.* **2013**, *39*, 563–568. [CrossRef]
44. Lang, J.; Wang, J.; Zhang, Q.; Li, X.; Han, Q.; Wei, M.; Sui, Y.; Wang, D.; Yang, J. Chemical precipitation synthesis and significant enhancement in photocatalytic activity of Ce-doped ZnO nanoparticles. *Ceram. Int.* **2016**, *42*, 14175–14181. [CrossRef]
45. Kumar, K.V. Langmuir—Hinshelwood kinetics—A theoretical study. *Catal. Commun.* **2008**, *9*, 82–84. [CrossRef]

© 2020 by the authors. Licensee MDPI, Basel, Switzerland. This article is an open access article distributed under the terms and conditions of the Creative Commons Attribution (CC BY) license (http://creativecommons.org/licenses/by/4.0/).

Article

Ceramized Fabrics and Their Integration in a Semi-Pilot Plant for the Photodegradation of Water Pollutants

Lara Faccani, Simona Ortelli *, Magda Blosi and Anna Luisa Costa

Institute of Science and Technology for Ceramics—Italian National Research Council, Via Granarolo 64, I-48018 Faenza, RA, Italy; lara.faccani@istec.cnr.it (L.F.); magda.blosi@istec.cnr.it (M.B.); anna.costa@istec.cnr.it (A.L.C.)
* Correspondence: simona.ortelli@istec.cnr.it; Tel.: +39-0546-69-9729

Abstract: The use of nano-photocatalysts for the water/wastewater purifications, particularly in developing regions, offers promising advantages over conventional technologies. TiO_2-based photocatalysts deposited on fabrics represent an efficient solution for obtaining heterogeneous photocatalysts, which are easily adaptable in the already installed water treatment plants or air purification systems. Despite the huge effort spent to develop and characterize novel nano-photocatalysts, which are especially active under solar light, knowledge gaps still persist for their full-scale application, starting from the reactor design and scale-up and the evaluation of the photocatalytic efficiency in pre-pilot scenarios. In this study, we offered easily scalable solutions for adapting TiO_2-based photocatalysts, which are deposited on different kinds of fabrics and implemented in a 6 L semi-pilot plant, using the photodegradation of Rhodamine B (RhB) as a model of water pollution. We took advantage of a multi-variable optimization approach to identify the best design options in terms of photodegradation efficiency and turnover frequency (TOF). Surprisingly, in the condition of use, the irradiation with a light-emitting diode (LED) visible lamp appeared as a valid alternative to the use of UV LED. The identification of the best design options in the semi-pilot plant allowed scaling up the technology in a 100 L pilot plant suitable for the treatment of industrial wastewater.

Keywords: photodegradation; nanoparticles; semi-pilot plant; fabric

1. Introduction

Clean water is essential for life, and ensuring the availability and sustainable management of water and sanitation became in 2015 one of 17 goals world leaders agreed upon in order to achieve a better world in 2030 [1]. Organic compounds, toxic pesticides, and manure emissions from each industry are polluting drinking water and rivers, which is becoming a worldwide contamination with increased severity. The wide area of water pollution, diversification, and non-biodegradable problems has become a problem that cannot be solved by the natural cleansing cycle [2]. New technologies, as the nanotechnology, are developing to improve the current methods of remediation of water [3–6]. Metal oxide semiconductor nanoparticles (NPs), in particular TiO_2, have attracted particular attention due to their photocatalytic properties [7]. The photocatalytic activity of TiO_2 is induced by UV light excitation with the consequent formation of electron$^-$/hole$^+$ pairs. The released electrons can react with water and oxygen molecules on the surface to form free radicals [8]. These species are very reactive and able to degrade most of the organic compounds as well as biological contaminants to its mineral components, i.e., carbon dioxide and water [9].

The use of nano-photocatalysts for the water/wastewater purifications, particularly in developing regions, offers promising advantages over conventional technologies, such as low cost, simplicity, environmental friendliness, wide-ranging efficiency, and the capacity to break down traces of a wide variety of organic molecules, including viruses and chlorine-resistant organisms [10]. Despite the huge effort spent to develop and characterize novel nano-photocatalysts, which are especially active under solar light, knowledge gaps still

Citation: Faccani, L.; Ortelli, S.; Blosi, M.; Costa, A.L. Ceramized Fabrics and Their Integration in a Semi-Pilot Plant for the Photodegradation of Water Pollutants. *Catalysts* **2021**, *11*, 1418. https://doi.org/10.3390/catal11111418

Academic Editors: Stéphanie Lambert and Julien Mahy

Received: 20 October 2021
Accepted: 19 November 2021
Published: 22 November 2021

Publisher's Note: MDPI stays neutral with regard to jurisdictional claims in published maps and institutional affiliations.

Copyright: © 2021 by the authors. Licensee MDPI, Basel, Switzerland. This article is an open access article distributed under the terms and conditions of the Creative Commons Attribution (CC BY) license (https://creativecommons.org/licenses/by/4.0/).

persist for their full-scale application, starting from the reactor design and scale-up and the evaluation of the photocatalytic efficiency in pre-pilot scenarios [11,12]. One of the main objectives addressed by more recent studies is to extend the use of TiO_2-based photocatalysts to solar (visible) light for the application in areas without electricity or as a sustainable solution to avoid the use of bio-hazardous and costly UV light [13–16]. To respond to this very urgent and necessary need, different pollutant treatments with visible light-responsive photocatalysts have been developed, which are mainly based on TiO_2 modification [17]; nevertheless, the majority of works are carried out on a laboratory scale, and the introduction of other materials into TiO_2 dramatically increases the complexity of the photocatalyst preparation and cost, and the modification with heavy metals or harmful organics could even improve the degree of environment pollution [18,19].

Moreover, the scale-up of nano-TiO_2-based photocatalytic technology in water depuration systems is strongly influenced by the need to immobilize the photocatalysts in large available, low weight, high mechanical flexibility supports, allowing an easy implementation in many different geometry water treatment reactors and the easy recovery and regeneration of photocatalysts. So, thanks to their affinity for TiO_2 NPs, hydrophilic fabrics are particularly suitable supports for hosting nanostructured coatings with high washing fastness [13,17].

In this work, we focused on the optimization of TiO_2-based photocatalytic fabrics, which were implemented in a semi-pilot plant scale (6 L capacity) reactor, investigating the quantum efficiency of TiO_2, when irradiated by both UV and visible light-emitting diode (LED) lights, using Rhodamine B (RhB) as a reaction model and identified the best design options, comparing catalyst and fabric properties, process parameters, and type of irradiation.

2. Results and Discussion

2.1. Characterization of TiO_2-Based Nanosuspension

The hydrodynamic diameter, Z-potential, and $pH_{i.e.,p.}$ values of pristine materials and TiO_2-based nanosuspensions are reported in Table 1. A slight increase in the hydrodynamic diameter and zeta potential is observed in samples obtained after resin treatment (TACR and SiO_2-R). This is due to a decrease in the colloidal stability coupled with the pH change. On the other hand, a significant increase in the hydrodynamic diameter in the TiO_2:SiO_2 sample is caused by both the steric hindrance of SiO_2 heterocoagulated on the TiO_2 surface [20] and the consequent electrostatic destabilization due to the progressive neutralization of the TiO_2 surface charge with the increase in negatively charged SiO_2 content. This was further demonstrated by the shift of the $pH_{i.e.,p.}$ toward acidic pH (Table 1). The TiO_2:SiO_2 sample was obtained by the self-assembled heterocoagulation process between TAC and SiO_2-R, which exhibit, at the working pH = 4, potentials opposite in sign and high enough to preserve colloidal stabilization (Figure 1). Therefore, TAC and SiO_2-R were able to promote the heterocoagulation between positive TiO_2 and negative SiO_2 nanosurfaces.

Table 1. Hydrodynamic diameter (nm), Z-potential (mV), and $pH_{i.e.,p.}$ of pristine materials and TiO_2-based nanosuspension.

Sample	d_{DLS} (nm)	Z Potential (mV)	pH	$pH_{i.e.,p.}$
TAC	27	+36	1.5	7.7
TACR	29	+45	4	5.2
SiO_2	30	−45	7	<1.5
SiO_2-R	37	−34	4	<1.5
TiO_2:SiO_2	317	+38	4	5.2

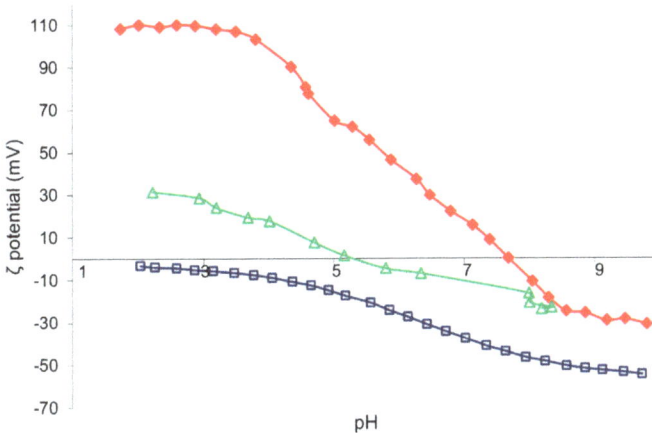

Figure 1. Z-potential vs. pH curves of TAC (red), SiO_2 (blue), and $TiO_2:SiO_2$ (green) nanosuspension.

2.2. Characterization of Fabrics

A basic morphological characterization of fabrics, used as support for obtaining nano-TiO_2-coated photocatalysts, was carried out by optical microscope (Figure 2).

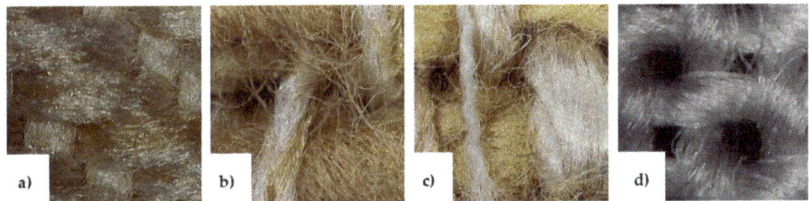

Figure 2. Optical microscopy images of untreated fabrics: (**a**) SP; (**b**) SC; (**c**) SM; (**d**) C.

The images (Figure 2) show the differences of color, warp, and weft of the target fabrics. SP, SC, and SM fabrics are characterized by very intertwined fibers, whilst the weave of the C fabric is more regular and expanded than other fabrics. This can explain the significant differences in the absorption of the TiO_2-based nanosuspensions (TACR and $TiO_2:SiO_2$), as demonstrated by the add-on percentage (AO%) reported in Table 2. In fact, the fabric C being characterized by a large weave and low weight (Table 3) adsorbs half the amount adsorbed by the other fabrics.

Table 2. AO% parameters calculated.

Fabric	AO% (TACR)	AO% ($TiO_2:SiO_2$)
SP	5.9	8
SC	8.4	n.a.
SM	6.2	n.a.
C	3.8	3

n.a. not available.

Table 3. Fabrics used as supports for nano-TiO$_2$-based coatings.

Code	Images	Composition	g/m^2
SP		65% cotton 35% polyester	450
SC		65% cotton 35% polyester	500
SM		Not available	640
C		100% cotton	100

In order to evaluate the hydrophilicity of fabrics and estimate the adsorption capacity of the textile supports, before and after the TiO$_2$ treatment (TACR), contact angle measurements were performed. The results are reported in Table 4.

Table 4. Contact angle measurements on untreated and TACR coated fabrics.

Fabric	Untreated	Coated
SP	121 ± 1	121 ± 3
SC	113 ± 1	n.d.
SM	n.d.	n.d.
C	n.d.	122 ± 4

n.d. not determined.

In general, all samples show hydrophilic properties, and no significant differences between untreated and treated fabrics are found. Specifically, the SM fabric exhibits very high hydrophilic behavior both on untreated and treated samples. In fact, the water drop is absorbed so quickly that during the contact angle measurements, the values are not detectable. This also occurs for untreated C and treated SC fabrics. In the case of SC fabric, an increase in hydrophilicity induced by TiO$_2$ treatment is found. Otherwise, a decrease in hydrophilicity induced by TiO$_2$ treatment was found in the C sample fabric. In any case, no clear correlation between the variation of hydrophilicity between different samples and the photocatalytic efficiency was found, as the results discussed in the following paragraphs show.

2.3. Optimization of Photocatalytic Process

2.3.1. Effect of TiO$_2$-Based Coatings Composition

Using SP as the target fabric, we evaluated the photocatalytic efficiency of TACR and TiO$_2$:SiO$_2$-based coatings under both UV and visible light irradiation. Very low differences were observed between TiO$_2$ and TiO$_2$:SiO$_2$ compositions (Table 5). This result confirms the different mechanism occurring when the photodegradation of TiO$_2$ is tested at a liquid and gas state. In fact, in a previous study [21], using NO$_x$ abatement (DeNO$_x$) as the experimental model, we found that the presence of SiO$_2$ significantly improved the efficiency of the photocatalyst.

Table 5. Comparison between different TiO$_2$-based photocatalysts. Tests carried out with SP fabrics under UV and visible light.

Irradiation Light	Coating	Photocatalytic Efficiency %
Visible	TACR	49
	TiO$_2$:SiO$_2$	51
UV	TACR	64
	TiO$_2$:SiO$_2$	60

The RhB photodegradation of the two photocatalysts over time is represented in Figure 3, showing a progressive decrease in the absorbance of the RhB peak at 554 nm. Moreover, for both photocatalysts, a blue-shift of the λ_{max} was detected. This is associated to a de-ethylation of RhB molecules, which is in agreement with the hypothesized RhB degradation mechanism, in the presence of supported photocatalysts. The assessment of photocatalytic efficiency was done by considering the maximum of the absorbance peaks, allowing an estimation of the overall reactivity, because we referred to the capacity of the catalyst to photodegrade the dye and its by-products. The higher shift detected in the case of the TiO$_2$:SiO$_2$ photocatalyst can be attributed to a complete conversion of the N,N,N',N'-tetraethylated rhodamine molecule (λ_{max} 554 nm) to de-ethylated rhodamine (λ_{max} 498 nm), as a consequence of the attack of oxidative radicals against N-ethyl group [22], which was not achieved in the case of TACR. This was further confirmed by Chen et al. [23] that showed different absorption mechanisms and consequently different degradation mechanisms using TiO$_2$ or the TiO$_2$:SiO$_2$ composite. In fact, they declare that in the case of RhB absorption on TiO$_2$:SiO$_2$, the chromophore is absorbed by the photocatalyst through the diethylamino groups while in the case of TiO$_2$, it is absorbed through the carboxyl groups. This difference results in an attack of the chromophore ring and its cleavage in the RhB-TiO$_2$ case, while in the RhB-TiO$_2$:SiO$_2$ case, the auxochromic groups are attacked and produce the de-ethylation of the alkylamine group. Moreover, they found that the blue-shift phenomenon due to the RhB de-ethylation is more evident under visible light than UV, which is because the UV radiation directly excites the TiO$_2$, while under visible light, it is the RhB absorbed on the surface of the photocatalyst to subsequently produce the active oxygen species.

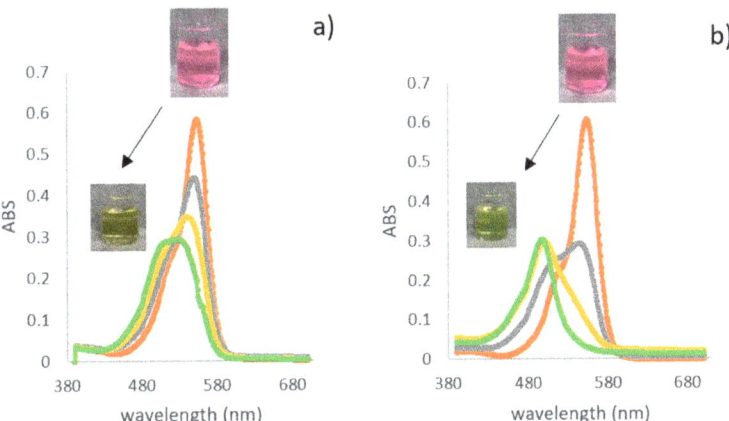

Figure 3. Absorbance of RhB solutions after irradiation under visible light at 0 min (orange), 40 min (gray), 80 min (yellow), and 100 min (green). Time 0: λ_{max} 554 nm; time 100: λ_{max} 527 nm for TACR (**a**) and 498 nm for TiO$_2$:SiO$_2$ (**b**).

2.3.2. Effect of Temperature

The dependence of photocatalytic performance on temperature has been widely investigated in the literature, and it is still under debate, with increasing temperature promoting phenomena such as the desorption of adsorbed reactants and the rate of recombination of photogenerated electron/hole pairs that are detrimental for the photocatalysis [24,25]. Otherwise, it is well known that the temperature influences the reaction kinetics, enhancing the activation energy and so speeding the photodegradation process [26,27]. Photodegradation tests using SP fabrics coated with TiO_2:SiO_2, under both UV and visible light sources, were compared at three working temperatures: 15 °C, 25 °C (room temperature), and 38 °C. The results are shown in Table 6. Under UV LED irradiation, the photocatalytic activity decreases by increasing the temperature, whilst in the case of visible LED, no significant trend was observed. The dependence of photodegradation efficiency by temperature is the result of synergetic (increase in activation energy, increase in charge transfer kinetic) and detrimental effects (the recombination of charge carriers and the altered adsorption equilibrium of reactants such as dye molecules, water, and oxygen) [24,26–28]. Therefore, in this case, the best compromise is working at room temperature, matching environmental and economic requirements [29].

Table 6. Effect of temperature. Tests carried out with TiO_2:SiO_2-coated SP samples, under both UV and visible LEDs.

Irradiation	Temperature °C	Photocatalytic Efficiency %
Visible	15	55
	25	51
	38	57
UV	15	63
	25	60
	38	59

2.3.3. Effect of Fabric Substrate

In order to investigate the effect of fabric substrates vs. type of irradiation (UV and visible light sources), we carried out photocatalytic tests with TACR-coated fabrics, and the results are reported in Table 7.

Table 7. Effect of fabric substrates. Tests carried out with TACR at T° = 25 °C, under both UV and visible light sources.

Fabric	Photocatalytic Efficiency %	
	UV LED	Visible LED
SP	49	64
SC	64	54
SM	65	61
C	67	57

Overall samples showed a comparable photodegradation efficiency despite the type of fabrics used and the type of irradiation source. Nevertheless, the fabric composition and structure seem to affect the efficiency; the SP sample is even more reactive under a visible source. The reactivity shown by the samples irradiated by visible LED was surprising, considering that the TACR crystalline phase, corresponding to anatase with 16% of brookite [30], has a band-gap, previously measured of 3.26 eV, that restricts its use only to the ultra-violet range of light. Considering the UV light fraction intensity measured on the fabric surface—UV LED (48.5 W/m^2) and visible LED (4.3 × 10^{-3} W/m^2), it is evident that in the case of a visible lamp, the few photons achieving the fabric surface have enough energy to activate the catalyst, and they are responsible for the photodegradation reactivity, which is comparable with that obtained irradiating the samples with UV LED, with a UV irradiation intensity that is five orders of magnitude higher. This result is quite unexpected because even if it is reported that a few photons of energy (i.e., as low as 1×10^{-2} W/m^2) can induce the photo-generation of electron–hole pairs [31], high intensities (400–1000 W/m^2) are needed to achieve a high photocatalytic reaction rate, particularly in water disinfection treatment [32]. The really low intensity of ultraviolet radiation (UVR) needed to activate our photocatalysts encourages their use and activation under solar irradiation; considering that the UV light portion of the yearly average solar irradiance at sea level, on a clear day, is about a few units W/m^2, we can reasonably estimate that the UVR intensity of the sun is in large excess of the amount needed to activate our photocatalysts. From this perspective, the proposed TiO$_2$-based photocatalysts do not require costly and time-consuming doping treatments to be activated under visible LED or solar light, with consequent benefits from safety, environmental, and economic points of view [31].

In order to better compare the photocatalytic ability of coated fabrics and select the most suitable fabric support, we calculated the TOF parameter (Table 8. The results show that the C fabric presents the highest photoactivity, both under UV and visible LEDs. Considering the natural source of cotton (100% biodegradable), its high availability at low cost and the shown photo-induced reactivity under visible LED, which is one order of magnitude higher than the other photocatalyst, also irradiated by UV light, it is evident that the cotton photocatalyst under visible LED becomes the best choice, matching the criteria of sustainability and safety.

Table 8. TOF parameter calculated at time 100 min. Tests carried out with TACR at T° = 25 °C, under both UV and visible light sources.

Fabric	Photocatalytic Efficiency %	
	UV LED	Visible LED
SP	7.5×10^{-5}	9.8×10^{-5}
SC	8.9×10^{-5}	7.5×10^{-5}
SM	8.6×10^{-5}	8.8×10^{-5}
C	1.02×10^{-3}	8.7×10^{-4}

2.4. Process Scalability

The treated fabrics were integrated and tested in the LED-based semi-pilot photocatalytic reactor of Figure 4a. Figure 4b shows the pilot reactor based upon the best design options identified performing tests with the semi-pilot reactor. The performances obtained with the pilot plant and the evaluation about the scalability of the semi-pilot plant are the objective of a future study.

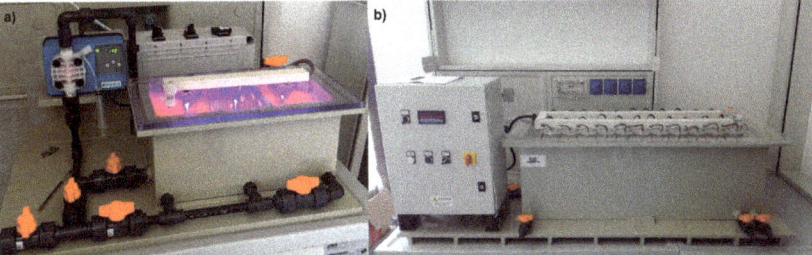

Figure 4. (a) LED-based semi-pilot photocatalytic reactor tested (6 L), used in this study; (b) Up-scaled reactor (100 L) built upon the best design options in the present study.

3. Materials and Methods

3.1. Materials

TiO_2 nanosol (NAMA41, 6 wt %), called TAC and SiO_2 nanosol 40 wt % (Ludox® HS-40) were purchased from Colorobbia (Sovigliana, Vinci (FI), Italy) and Grace Davison (USA), respectively. Rhodamine B (dye content ≈ 95%) target dye, ion excharger Dowex® 1 X 8 basic anion excharger resin and ion excharger Dowex® 50 W X 4 acidic cation exchanger resin were purchased from Sigma Aldrich (Milano, Italy).

3.2. TiO_2-Based Nanosuspensions

Acid TiO_2 nanosol (TAC, pH 1.5) was used to prepare two TiO_2-based nanosuspensions: TACR and TiO_2:SiO_2 suspensions. TACR was obtained diluting TAC at the 3 wt % concentration with distilled water (DI) water and treated with ion exchanger Dowex® 1 X 8 basic anion exchanger resin in order to increase the pH from 1.5 to 4. This increase in pH is necessary in order to avoid fabric damage caused by acidity, and the residual by the synthesis of original TiO_2 can reduce the photocatalytic activity [30]. TiO_2:SiO_2 3 wt % was prepared by the heterocoagulation method. SiO_2 nanosol 40 wt % (Ludox® HS-40) was diluted at the concentration 3 wt % with DI water and treated with ion exchange Dowex® 50 W X 4 acidic cationic exchanger resin in order to decrease the pH from 10 to 4 (SiO_2-R). TiO_2 suspension (TACR) was dropped into SiO_2-R suspension. The TiO_2:SiO_2 sample was obtained through an electrostatic interaction between negatively charged silica nanoparticles and positively charged titania nanoparticles. The electrostatic interactions between SiO_2 and TiO_2 surfaces are promoted by mixing the sols in well-defined ratios and by ball milling for 24 h with zirconia spheres (diameter 5 mm) as grinding media.

3.3. Ceramized Fabric

The coated fabrics was obtained via the dip-pad-curing method. The fabric was washed in an ultrasonic bath for 15 min in DI water and dried in an oven at 100 °C. Then, the washed fabric was dipped in a TiO_2-based suspension for 5 min, squeezed in two rolls to eliminate the excess of suspension (pad stage), dried in a stove at 100 °C, and finally cured at 130 °C for 10 min in order to well fix the NPs to the fabric. A single impregnation was carried out achieving the final dry add-on value (AO%), which is defined as the percent amount of the finishing agent added to the fabric with respect to the initial weight of the latter, i.e.,

$$AO\% = \frac{W_f - W_i}{W_i} \times 100 \qquad (1)$$

where w_i and w_f are the weights of the fabric before and after the dip-pad-curing process.

We tested four fabrics different in color, weight, and structure described in Table 3.

3.4. Semi-Pilot Plant and Irradiation Source

The photocatalytic tests were carried out in a 6 L semi-pilot plant [33], as schematized in Figure 5. The semi-pilot plant was designed and built by RAFT s.r.l., (Montelupo Fiorentino (FI) Italy). It hosts two plastic windows for supporting ceramized fabrics (14.8 × 11.4 cm; 100 cm² fabric exposed area for each support), and on the top, there are three holes for UV or visible light lamps. The homogeneity of water flow was ensured through a peristaltic pump, and the feed bath was thermostated by a chiller (Julabo, F12). The LED strip lights were provided by the Wiva Group (Firenze, Italy). The visible LED light is characterized by a wide emission spectrum (Figure 6a) with a main peak at wavelength = 452 nm and a second peak at wavelength = 569 nm. The UV LED light has a very narrow emission spectrum with λ_{max} = 384 nm (Figure 6b). The irradiance was calculated placing a radiometric UV probe (UV-A 315–400 nm) on the fabric surface in order to evaluate the UV component irradiance of the two light-emitting sources that resulted in 4.3 × 10^{-3} W/m² for visible light and 150 W/m² for UV light.

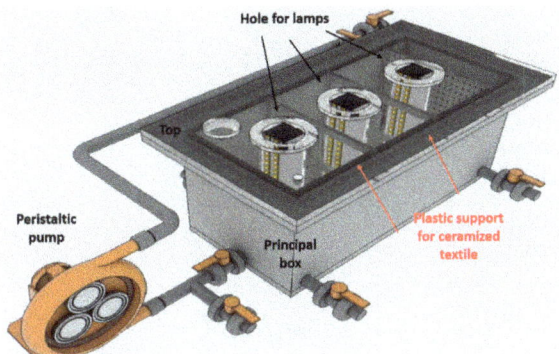

Figure 5. Schematized representation of a 6 L semi-pilot plant.

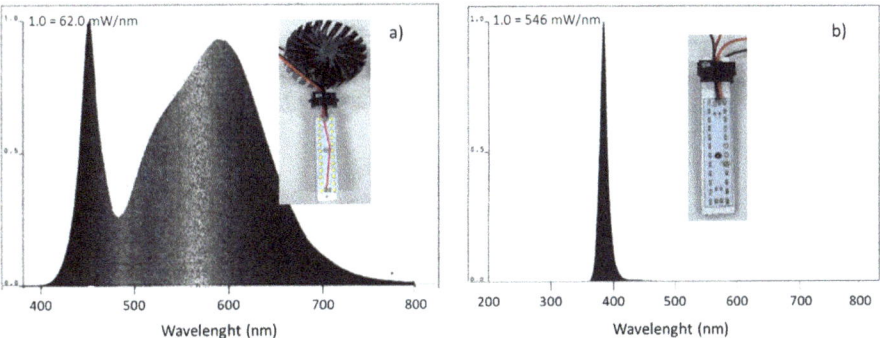

Figure 6. Emission spectrum of (**a**) visible LED light and (**b**) UV LED light, including photographs of the corresponding LED strips, mounting an air-cooling fan.

3.5. Characterization

3.5.1. Dynamic Light Scattering/Electrophoretic Light Scattering

The hydrodynamic diameters and zeta potential distribution of TiO_2-based suspensions, dispersed at 0.3 wt % in DI water, were evaluated by using a Zetasizer instrument (Malvern Instruments, Zetasizer Nano-ZS, Malvern, UK) based on the dynamic light scattering (DLS) and electrophoretic light scattering (ELS) techniques. For particle size distribution evaluation, about 1 mL of the sample was measured consecutively three times at 25 °C. The size distribution (nm) is reported as the intensity-weighted mean diameter derived from the cumulant analysis (Z-average) and is the average of three independent measurements. The reliability of the measurements was controlled by using the automatic attenuator (kept between 6 and 8) and the intercept autocorrelation function (<0.9) as quality criteria [4].

The zeta potential was measured on 700 µL of the sample at 25 °C, setting the measurement time, the attenuator position, and the applied voltage to automatic. After a 2 min temperature equilibration step, the samples were measured three times, and the data were obtained by averaging the three measurements. The data of zeta potential (mV) are derived by electrophoretic mobility using Smoluchowski's formula. The reliability of the measurements was controlled by check the phase plot graph.

The same instrument coupled with an automating titrating system was used to create zeta potential vs. pH curve to identify the pH at which the zeta potential sets to zero, namely the isoelectric point ($pH_{i.e.,p.}$). The titrants used were 0.1M KOH solution and 0.1M HCl solution.

3.5.2. Characterization on Fabrics

The untreated fabrics were morphologically observed by optical microscope using a Hirox 3D digital microscope, RH200 with a magnitude of lens X35 and X50. Specifically, we observed the fabric weave, thickness, and color of a single fiber.

The hydrophilicity of untreated and TACR-treated fabrics was evaluated by contact angle analyses. The measurements were carried out using a KRUSS DSA 30 S instrument, the sessile drop as the drop deposition method at room temperature (25 °C), 20 µL drop volume, and a tangent-fitting method.

3.6. Rhodamine B Degradation Tests

The photodegradation tests were carried out using Rhodamine B (RhB) as a model of organic trace pollutant. RhB is a synthetic dye that is commonly use in water remediation due to the easy detection of small concentrations by UV-Vis absorption analysis using a single beam spectrophotometer Hach Lange, DR3900. RhB imparts a deep magenta hue to its water solutions and displays a well-defined absorbance peak at 554 nm. In our experiment, we used 3.5 mg/L RhB concentration. The lamp was switched on outside the plant 30 min before starting the test in order to stabilize the power of emission; simultaneously, the ceramized textile was put into RhB solution to reach the adsorption–desorption equilibrium. In order to evaluate the degradation kinetics, an aliquot of 3 mL was withdrawn and analyzed every 20 min (A_x) through UV-Vis analysis in the range of 350–700 nm to a final reaction time of 100 min. The A_x was measured in correspondence to the maximum of the absorbance peak detected, taking into account the shift of the initial absorbance peak of RhB. Before starting the degradation tests, after 30 min of absorption, the initial absorbance (A_0) was determined. The order of photocatalytic degradation reactions was ascertained from the pseudo first-order kinetic model:

$$\ln\left(\frac{A_0}{A_x}\right) = k * t. \quad (2)$$

The photocatalytic efficiency was calculated at t = 100 min. It indicates the ratio between the amount of reagent consumed and the amount of reagent initially present in the reaction environment, and it was determined by the following formula:

$$\text{Photocatalytic efficiency } (\%) = \left(1 - \frac{A_x}{A_0}\right) * 100 \quad (3)$$

where A_x is the peak value at time t and A_0 is the peak value at time 0. In order to facilitate the comparison between the different photocatalysts, normalizing for the amount of catalyst and the time of exposure, the turnover frequency (TOF) parameter was calculated. The TOF parameter was calculated by following equation:

$$TOF = \frac{\left(\frac{\text{mol of product}}{t\,(s)}\right)}{\text{mol of catalyst}} \quad (4)$$

where
mol of product is calculated as the initial concentration of reagent per efficiency reached at the time s
mol of catalyst are the moles of the catalyst deposited on the exposure area of fabric calculated by the AO% parameter.

4. Conclusions

In response to the still persisting knowledge gaps for the full-scale application of nano-photocatalysts in water/wastewater purification systems, we immobilized TiO_2-based nanoparticles as the coating of fabrics, obtaining large available, low cost, highly flexible photocatalysts that allow an easy implementation in many different geometry water treatment reactors and the easy recovery and regeneration of photocatalysts. We implemented the obtained photocatalytic fabrics in a 6 L capacity semi-pilot plant and evaluated the degradation of RhB dye, which was used as a probe molecule, simulating the water pollution. We adopted a multi-variables optimization approach to look for the photoreactor operating parameters that mostly affected the photocatalytic performance and identified the best design option also in response to safety and sustainable criteria. We found that the 100% biodegradable cotton fabric irradiated by visible LED is the best candidate, because it showed a TOF higher than all the other samples, which was irradiated by both UV and visible light sources. The really low intensity of UV radiation-activating fabrics under visible LED expands the applicability of the technology to solar light, since the measured intensity of the UV radiation component, in visible LED, is much lower than the solar yearly average one. The findings from the multi-variable optimization study were translated into updated recommendations for the design and the technical application of these efficient and low-cost TiO_2-based photocatalysts, which are suitable for developing a range of technologies aimed at environmental protection. The good results obtained encouraged the scale-up of the 6 L semi-pilot plant up to the 100 L pilot plant that has been built, even if the evaluation of photocatalytic performances is still under investigation.

Author Contributions: A.L.C., S.O., M.B. and L.F. conceived and designed the experiments; L.F. performed the experiments and analyzed the data; L.F., S.O. and A.L.C. wrote the paper. All authors have read and agreed to the published version of the manuscript.

Funding: This work was supported by the "ASINA" (Anticipating Safety Issues at the Design Stage of NAno Product Development) European project (H2020—GA 862444).

Acknowledgments: The authors acknowledge RAFT s.r.l, (Montelupo Fiorentino (FI), Italy) and Wiva Group SpA, (Firenze, Italy) for the support provided to the design and production of the (semi) pilot and pilot plants and the LED-lamps, respectively, used in this work.

Conflicts of Interest: The authors declare no conflict of interest.

References

1. Organizzazione delle Nazioni Unite. Trasformare il Nostro Mondo: L'Agenda 2030 per lo Sviluppo Sostenibile (Agenda2030). In *Risoluzione Adottata Dall'assemblea Gen. 25 Settembre 2015*; 2015. Available online: https://unric.org/it/wp-content/uploads/sites/3/2019/11/Agenda-2030-Onu-italia.pdf.1 (accessed on 1 November 2021).
2. Lee, S.Y.; Park, S.J. TiO_2 photocatalyst for water treatment applications. *J. Ind. Eng. Chem.* **2013**, *19*, 1761–1769. [CrossRef]
3. Qu, X.; Alvarez, P.J.J.; Li, Q. Applications of nanotechnology in water and wastewater treatment. *Water Res.* **2013**, *47*, 3931–3946. [CrossRef]
4. Varenne, F.; Hillaireau, H.; Bataille, J.; Smadja, C.; Barratt, G.; Vauthier, C. Application of validated protocols to characterize size and zeta potential of dispersed materials using light scattering methods. *Colloids Surf. A Physicochem. Eng. Asp.* **2019**, *560*, 418–425. [CrossRef]
5. Gehrke, I.; Geiser, A.; Somborn-Schulz, A. Innovations in nanotechnology for water treatment. *Nanotechnol. Sci. Appl.* **2015**, *8*, 1–17. [CrossRef]
6. Qu, X.; Brame, J.; Li, Q.; Alvarez, P.J.J. Nanotechnology for a safe and sustainable water supply: Enabling integrated water treatment and reuse. *Acc. Chem. Res.* **2013**, *46*, 834–843. [CrossRef]
7. Liu, H.; Ma, H.T.; Li, X.Z.; Li, W.Z.; Wu, M.; Bao, X.H. The enhancement of TiO_2 photocatalytic activity by hydrogen thermal treatment. *Chemosphere* **2003**, *50*, 39–46. [CrossRef]
8. Yuranova, T.; Laub, D.; Kiwi, J. Synthesis, activity and characterization of textiles showing self-cleaning activity under daylight irradiation. *Catal. Today* **2007**, *122*, 109–117. [CrossRef]
9. Bellardita, M.; Di Paola, A.; Palmisano, L.; Parrino, F.; Buscarino, G.; Amadelli, R. Preparation and photoactivity of samarium loaded anatase, brookite and rutile catalysts. *Appl. Catal. B Environ.* **2011**, *104*, 291–299. [CrossRef]
10. Adly, M.S.; El-Dafrawy, S.M.; El-Hakam, S.A. Application of nanostructured graphene oxide/titanium dioxide composites for photocatalytic degradation of rhodamine B and acid green 25 dyes. *J. Mater. Res. Technol.* **2019**, *8*, 5610–5622. [CrossRef]

11. Khodadadian, F.; de Boer, M.W.; Poursaeidesfahani, A.; van Ommen, J.R.; Stankiewicz, A.I.; Lakerveld, R. Design, characterization and model validation of a LED-based photocatalytic reactor for gas phase applications. *Chem. Eng. J.* **2018**, *333*, 456–466. [CrossRef]
12. Li, R.; Li, T.; Zhou, Q. Impact of titanium dioxide (TiO_2) modification on its application to pollution treatment—A review. *Catalysts* **2020**, *10*, 804. [CrossRef]
13. Costa, A.L.; Ortelli, S.; Blosi, M.; Albonetti, S.; Vaccari, A.; Dondi, M. TiO_2 based photocatalytic coatings: From nanostructure to functional properties. *Chem. Eng. J.* **2013**, *225*, 880–886. [CrossRef]
14. Mahanta, U.; Khandelwal, M.; Suresh Deshpande, A. $TiO_2@SiO_2$ nanoparticles for methylene blue removal and photocatalytic degradation under natural sunlight and low-power UV light. *Appl. Surf. Sci.* **2021**, *576*, 151745. [CrossRef]
15. Basavarajappa, P.S.; Patil, S.B.; Ganganagappa, N.; Reddy, K.R.; Raghu, A.V.; Reddy, C.V. Recent progress in metal-doped TiO_2, non-metal doped/codoped TiO_2 and TiO_2 nanostructured hybrids for enhanced photocatalysis. *Int. J. Hydrogen Energy* **2020**, *45*, 7764–7778. [CrossRef]
16. Dong, X.; Xu, J.; Kong, C.; Zeng, X.; Wang, J.; Zhao, Y.; Zhang, W. Synthesis of β-$FeOOH/TiO_2/SiO_2$ by melting phase separation-hydrothermal method to improve photocatalytic performance. *Ceram. Int.* **2021**, *47*, 32303–32309. [CrossRef]
17. Ortelli, S.; Malucelli, G.; Blosi, M.; Zanoni, I.; Costa, A.L. NanoTiO_2 @DNA complex: A novel eco, durable, fire retardant design strategy for cotton textiles. *J. Colloid Interface Sci.* **2019**, *546*, 174–183. [CrossRef] [PubMed]
18. Factories, H.; Study, A.C.; Plant, P.; Joseane, J.; Mesa, M.; Sebasti, J.; Gonz, W.; Rojas, H.; Murcia Mesa, J.J.; Hernández Niño, J.S.; et al. Photocatalytic Treatment of Stained Wastewater Coming from Handicraft Factories. A Case Study at the Pilot Plant Level. *Water* **2021**, *13*, 2705.
19. Sciscenko, I.; Mestre, S.; Climent, J.; Valero, F.; Escudero-Oñate, C.; Oller, I.; Arques, A. Magnetic photocatalyst for wastewater tertiary treatment at pilot plant scale: Disinfection and enrofloxacin abatement. *Water* **2021**, *13*, 329. [CrossRef]
20. Qian, W.; Zhaoqun, W.; Xuanfeng, K.; Xiaodan, G.; Gi, X. A facile strategy for controlling the self-assembly of nanocomposite particles based on colloidal steric stabilization theory. *Langmuir* **2008**, *24*, 7778–7784. [CrossRef]
21. Ortelli, S.; Poland, C.A.; Baldi, G.; Costa, A.L. Silica matrix encapsulation as a strategy to control ROS production while preserving photoreactivity in nano-TiO_2. *Environ. Sci. Nano* **2016**, *3*, 602–610. [CrossRef]
22. Ortelli, S.; Blosi, M.; Albonetti, S.; Vaccari, A.; Dondi, M.; Costa, A.L. TiO_2 based nano-photocatalysis immobilized on cellulose substrates. *J. Photochem. Photobiol. A Chem.* **2014**, *276*, 58–64. [CrossRef]
23. Chen, F.; Zhao, J.; Hidaka, H. Highly selective deethylation of Rhodamine B: Adsorption and photooxidation pathways of the dye on the TiO_2/SiO_2 composite photocatalyst. *Int. J. Photoenergy* **2003**, *5*, 209–217. [CrossRef]
24. Meng, Y.; Xia, S.; Pan, G.; Xue, J.; Jiang, J.; Ni, Z. Preparation and photocatalytic activity of composite metal oxides derived from Salen-Cu(II) intercalated layered double hydroxides. *Korean J. Chem. Eng.* **2017**, *34*, 2331–2341. [CrossRef]
25. Meng, F.; Liu, Y.; Wang, J.; Tan, X.; Sun, H.; Liu, S.; Wang, S. Temperature dependent photocatalysis of g-C_3N_4, TiO_2 and ZnO: Differences in photoactive mechanism. *J. Colloid Interface Sci.* **2018**, *532*, 321–330. [CrossRef]
26. Barakat, N.A.M.; Kanjwal, M.A.; Chronakis, I.S.; Kim, H.Y. Influence of temperature on the photodegradation process using Ag-doped TiO_2 nanostructures: Negative impact with the nanofibers. *J. Mol. Catal. A Chem.* **2013**, *366*, 333–340. [CrossRef]
27. Chen, Y.W.; Hsu, Y.H. Effects of reaction temperature on the photocatalytic activity of TiO_2 with Pd and Cu cocatalysts. *Catalysts* **2021**, *11*, 966. [CrossRef]
28. Liu, B.; Zhao, X.; Parkin, I.P.; Nakata, K. *Charge Carrier Transfer in Photocatalysis*; Elsevier: Amsterdam, The Netherlands, 2020; Volume 31, ISBN 9780081028902.
29. Alisawi, H.A.O. Performance of wastewater treatment during variable temperature. *Appl. Water Sci.* **2020**, *10*, 89. [CrossRef]
30. Ortelli, S.; Costa, A.L.; Dondi, M. TiO_2 nanosols applied directly on textiles using different purification treatments. *Materials* **2015**, *8*, 7988–7996. [CrossRef]
31. Chong, M.N.; Jin, B.; Chow, C.W.K.; Saint, C. Recent developments in photocatalytic water treatment technology: A review. *Water Res.* **2010**, *44*, 2997–3027. [CrossRef]
32. Rincón, A.G.; Pulgarin, C. Photocatalytical inactivation of E. coli: Effect of (continuous-intermittent) light intensity and of (suspended-fixed) TiO_2 concentration. *Appl. Catal. B Environ.* **2003**, *44*, 263–284. [CrossRef]
33. Ortelli, S.; Costa, A.L.; Torri, C.; Samorì, C.; Galletti, P.; Vinais, C.; Varesano, A.; Bonura, L.; Bianchi, G. Innovative and sustainable production of Biopolymers. In *Factories of the Future: The Italian Flagship Initiative*; Tolio, T., Copani, G., Terkaj, W., Eds.; Springer International Publishing: Cham, Swizerland, 2019; Chapter 6, ISBN 9783319943589.

Article

Cytostatic Drug 6-Mercaptopurine Degradation on Pilot Scale Reactors by Advanced Oxidation Processes: UV-C/H_2O_2 and UV-C/TiO_2/H_2O_2 Kinetics

Luis A. González-Burciaga [1], Juan C. García-Prieto [2], Manuel García-Roig [2], Ismael Lares-Asef [1], Cynthia M. Núñez-Núñez [3,*] and José B. Proal-Nájera [1,*]

[1] Instituto Politécnico Nacional, CIIDIR-Unidad Durango, Calle Sigma 119, Fracc. 20 de Noviembre II, Durango 34220, Mexico; luis.gonzalez.iq@gmail.com (L.A.G.-B.); ismaelares@yahoo.com (I.L.-A.)
[2] Universidad de Salamanca, Centro de Investigación y Desarrollo Tecnológico del Agua, Campo Charro s/n, 37080 Salamanca, Spain; jcgarcia@usal.es (J.C.G.-P.); mgr@usal.es (M.G.-R.)
[3] Universidad Politécnica de Durango, Carretera Durango-México km 9.5, Col. Dolores Hidalgo, Durango 34300, Mexico
* Correspondence: cynthia.nunez@unipolidgo.edu.mx (C.M.N.-N.); jproal@ipn.mx (J.B.P.-N.)

Abstract: 6-Mercaptopurine (6-MP) is a commonly used cytostatic agent, which represents a particular hazard for the environment because of its low biodegradability. In order to degrade 6-MP, four processes were applied: Photolysis (UV-C), photocatalysis (UV-C/TiO_2), and their combination with H_2O_2, by adding 3 mM H_2O_2/L (UV-C/H_2O_2 and UV-C/TiO_2/H_2O_2 processes). Each process was performed with variable initial pH (3.5, 7.0, and 9.5). Pilot scale reactors were used, using UV-C lamps as radiation source. Kinetic calculations for the first 20 min of reaction show that H_2O_2 addition is of great importance: in UV-C experiments, highest k was reached under pH 3.5, k = 0.0094 min^{-1}, while under UV-C/H_2O_2, k = 0.1071 min^{-1} was reached under the same initial pH; similar behavior was observed for photocatalysis, as k values of 0.0335 and 0.1387 min^{-1} were calculated for UV-C/TiO_2 and UV-C/TiO_2/H_2O_2 processes, respectively, also under acidic conditions. Degradation percentages here reported for UV-C/H_2O_2 and UV-C/TiO_2/H_2O_2 processes are above 90% for all tested pH values. Ecotoxicity analysis of samples taken at 60 min in the photolysis and photocatalysis processes, suggests that contaminant degradation by-products present higher toxicity than the original compound.

Keywords: 6-mercaptopurine photodegradation; wastewater; cytostatic; emerging pollutants; photocatalysis

1. Introduction

Emerging pollutants (EPs) could be defined as a new class of products, synthetic or natural, that have found their way into humans and animals due to their recent discovered sources [1–3]. Nowadays, there are no guidelines or legislative background for EPs management; for that reason, these pollutants are not commonly monitored in the environment or included in routine monitoring programs and their fate, occurrence, and environmental effects are not well understood [1,2]. Drugs are a major group of EPs that have been detected in wastewaters, surface waters, groundwater and drinking water. In the last decade their presence has been reported in the aquatic cycle in the order of ng/L and in some cases a few µg/L, having a negative impact on the environment. [4–7].

Cytostatic drugs, also called anticancer or antineoplastic agents, are used in hospitals for the treatment of cancer via chemotherapy. These drugs represent a high risk regarding human health and a particular hazard for the environment because of an increasing demand for the chemotherapy treatment and their carcinogenic, genotoxic, mutagenic, and cytotoxic properties, even at low concentrations, added to a low biodegradability. Therefore, it is crucial to monitor their presence in the environment to extend the research in order to classify their by-products, as they have biotoxic potential [4–6,8,9]. Different organisms have been used to assess the toxicity of certain compounds in the environment, including aquatic bacteria *Vibrio fischeri* [6,10,11], which emits light under normal conditions. When the enzymatic activity of this organism is affected by toxins, bioluminescence decreases [12]. Advances in analytical instruments have allowed the detection of cytostatic drugs and the further transformation by-products in environmental samples, even at very low concentration (ng/L). Recognizing the presence of these compounds has driven the research on potential methods for their removal from water bodies [4,5,7].

Many studies have been carried out with different processes for the removal of cytostatic drugs. Zhang et al. 2013 [13], conducted a review article on the removal of cytostatic drugs present in aquatic environments; in the cited works, 21 compounds were found in hospital effluents, influents and effluents from wastewater treatment plants, and surface waters by different detection methods like SPE GC-MS, LC-MS/MS, SPE HPLC-MS/MS, LLE GC-MS, SPE UPLC-MS/MS, ICP-MS, among others. The authors gathered information from studies between the years 2005–2012, where cytostatic drugs were removed or degraded by advanced technologies, such as membrane bioreactors, membrane filtration methods, electrochemical oxidation, catalytic oxidation, UV/H_2O_2, and ozonation, with results ranging from 20% to 100%.

In 2017, a research carried out by Franquet-Griell et al. [14] explored the degradation of 16 cytostatics commonly detected in hospital effluents, through hydrolysis, biodegradation in a sequential batch reactor, UV-C process, UV-H_2O_2, and simulated sunlight reactor. The best results for all pollutants were achieved with the UV/H_2O_2 process.

Antineoplastic drugs play an essential role in the development of chemotherapy treatment [15]. 6-mercaptopurine (6-MP) is an anticancer agent that has the additional application as immuno-suppressive and anti-inflammatory agent [15,16]. 6-MP has been widely used for the treatment of acute lymphoblastic leukemia, chronic myeloid leukemia, choriocarcinoma, Crohn's disease, and psoriatic arthritis [15–18]. When it is administrated in chemotherapy, other drugs must be used to extend the duration of remission achieved [19], which can last from two to three years after diagnosis [20]. The initial dose in Mexico for adults varies from 75 mg/m^2/day [21] to 100 mg/m^2/day [22–24]. Whereas the starting dose in the United Kingdom, the Nordic countries and the United States is 75 mg/m^2/day, in most of Europe, the initial dose is 50 mg/m^2/day [20]. The percentage of drug excreted unchanged through the urine is 50%, so half of the high doses of 6-MP administered for cancer treatment end up in sewage [13]. The doses of 6-MP for the treatment of diseases such as intestinal arterial insufficiency, Crohn's disease, ulcerative colitis and inflammatory bowel disease, range from 1 to 2.5 mg/kg/day [25,26]. Due to their characteristics, high doses and administration over large periods of time, presence and removal of cytostatic drugs as 6-MP from wastewater, have become a challenge.

Heterogenous photocatalysis (HPC) has become the most distinctive, popular, effective and promising treatment technique for the removal of recalcitrant contaminants, such as pharmaceutical wastewater. The mechanism of HPC consists in the capability of the semiconductors to generate hydroxyl radicals (•OH) in situ, which are extremely reactive and strong oxidizing agents (E° = 2.8 eV), and can lead to further reactions generating harmless products as CO_2 and H_2O [27–30]. The photocatalyst is the core of the photocatalytic process, most photocatalysts are metal oxides such as: titanium dioxide (TiO_2), zinc oxide (ZnO), zinc sulfide (ZnS), tungsten trioxide (WO_3), cadmium sulfide (CdS), tin dioxide (SnO_2), and gallium phosphide (GaP), among others [31,32]. TiO_2 is widely used in environmental applications due to its low cost and properties, such as chemical stability, light absorption, it is biologically inert and resistant to chemical corrosion, and can be used at ambient temperature and pressure [27–29,32].

HPC involves a series of simultaneous oxidative and reductive reactions on the surface of the semiconductor, initiated by the irradiation of TiO_2 with UV light to excite the catalyst (Equation (1)). When the energy becomes equal to or greater than the band gap, the electrons in the valance band (e^-_{VB}) are promoted to the conduction band (CB) leaving behind a hole (h^+_{VB}) in the surface of the photocatalyst. While the h^+_{VB} is positive enough to generate •OH in the surface of the photocatalyst that contains absorbed water (Equations (2) and (3)), the CB with a negative charge, reduces the oxygen in the solution in order to produce another series of •OH (Equations (4) and (5)) [27–30]:

$$TiO_2 + \hbar\upsilon \rightarrow e^-_{CB} + h^+_{VB} \quad (1)$$

$$H_2O + h^+ \rightarrow \bullet OH + H^+ \quad (2)$$

$$OH^- + h^+ \rightarrow \bullet OH \quad (3)$$

$$O_2 + e^- \rightarrow O_2^- \quad (4)$$

$$O_2 + 2e^- \rightarrow O_2^{2-} \quad (5)$$

•OH could also be generated through reactions between some intermediate products:

$$2O_2^- + 2H^+ \rightarrow H_2O_2 + O_2 \quad (6)$$

$$H_2O_2 + e^- \rightarrow \bullet OH + OH^- \quad (7)$$

Formed •OH reacts with organic compounds R (either chemical compounds or microorganisms) present in the sample, generating a mineralization reaction:

$$\bullet OH + R \rightarrow CO_2 + H_2O \quad (8)$$

The objective of this work was to apply photolytic and photocatalytic processes on pilot scale reactors, using UV-C lamps, for the degradation of 6-MP at different conditions (acidic, basic and neutral media), and comparing the results when hydrogen peroxide (H_2O_2) is added as an oxidant in the processes. It also analyses the toxicity of 6-MP and possible transformation by-products, generated during experimentation, using marine bioluminescent bacteria *V. fischeri*.

2. Results and Discussion

2.1. UV-C and UV-C/H_2O_2 6-MP Degradation Experiments

Experiments performed in the absence in radiation (control experiments), showed a degradation of 20% after 80 min of reaction, so the effect of H_2O_2 addition alone is not enough to reach the degradation data here presented for experiments with radiation.

The 6-MP degradation reached through 2 h UV irradiation only, was lower than 70% at the three tested pH values. Direct photolysis of molecules present in wastewater is possible when irradiating at 254 nm wavelength, but the process is not effective to achieve total mineralization of contaminants [33]. According to Smaranda et al. [34], an energy of 3.13 eV can excite electrons present in 6-MP; given this, radiation under 396 nm wavelengths will excite the molecule. As radiation used in this research was 254 nm, breaking of the molecule was expected. In the past, UV light was reported to be responsible for photodegradation of azathioprine, a parent compound to 6-MP [34]. In fact, 6-MP presents maximum absorbance at approximately 340 nm [35] and in aqueous solution, generates reactive oxygen species (ROS) when irradiated with UV radiation in the presence of molecular oxygen [35,36]. Thiopurines are known to absorb UV-A radiation and are converted to an unstable excited triplet state that can interact with oxygen to form a thiopurine radical and superoxide [35]. Superoxide radical ($O_2^{\bullet-}$) is able to degrade substituted aromatic compounds with high absorption in the UV range but has a low oxidizing power [37].

Figure 1 shows 6-MP degradation by UV-C and UV-C/H_2O_2 experiments. The best results (6-MP degradation > 99%), were achieved at initial pH of 3.5 and 7, when H_2O_2 was added.

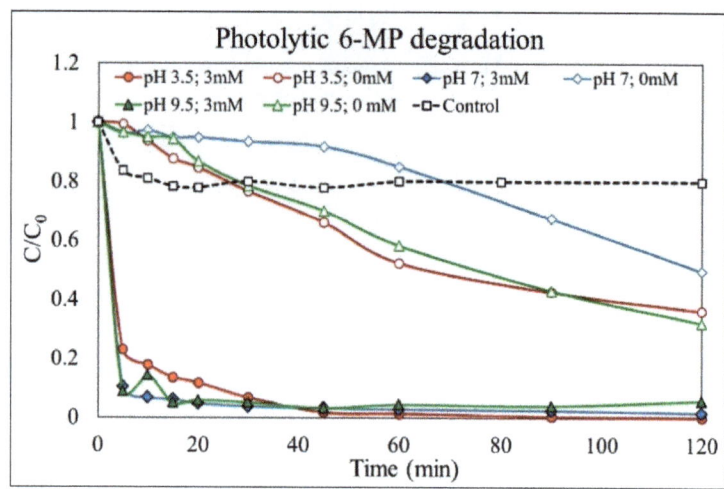

Figure 1. 6-MP degradation in the UV-C (empty markers) and UV-C/H$_2$O$_2$ (filled markers) processes with and without H$_2$O$_2$ addition and three different initial pH: 3.5 (circles), 7 (rhombuses) and 9.5 (triangles). Control experiment is represented by the segmented line.

Irradiation below 280 nm wavelength, breaks O-O bonds of H$_2$O$_2$ into HO• [38], thus degradation enhancement is to be expected. Best degradation percentage after only 5 min treatment is in fact achieved in alkaline conditions; however, the best results at the end of treatment (120 min) were achieved at acidic and neutral pH, being a possible explanation the formation of less reactive HO$_2^•$ radicals in HO• excess, through (Equation (9)), this is similar to what other authors observed [34,38–40]:

$$HO• + H_2O_2 \rightarrow H_2O + HO_2^• \tag{9}$$

HO$_2^•$ radicals can later react as follows [40]:

$$HO_2^• + H_2O_2 \rightarrow HO• + H_2O + O_2 \tag{10}$$

$$2HO_2^• \rightarrow H_2O_2 + O_2 \tag{11}$$

$$HO_2^• + HO• \rightarrow H_2O + O_2 \tag{12}$$

2.2. UV-C/TiO$_2$ and UV-C/TiO$_2$/H$_2$O$_2$ 6-MP Degradation Experiments

Experiments performed in the absence in radiation (control experiments), showed a degradation of 16% after 80 min of reaction, so the effect of H$_2$O$_2$ addition alone is not enough to reach the degradation reached by the experiments in the presence of radiation.

Photocatalysis showed a similar behavior than UV-C and UV-C/H$_2$O$_2$ 6-MP degradation experiments (Section 2.1): addition of H$_2$O$_2$ improved considerably 6-MP degradation. When no H$_2$O$_2$ is added, a fast concentration drop is presented under acidic condition, but after 120 min of treatment, alkaline conditions present better results (Figure 2).

In the presence of TiO$_2$, photocatalyst isoelectric point confers extra importance to initial pH. As TiO$_2$ isoelectric point is around 6.5 [41], at lower solution pH its surface would be positively charged, whereas at higher pH the surface would be negatively charged [42]. The isoelectric point of 6-MP is 9.47, calculated from its pKa values of 7.77 and 11.17 [43]. Nevertheless, the performance of UV-C/TiO$_2$/H$_2$O$_2$ is more efficient that obtained with UV-C/TiO$_2$ and no remarkable differences were observed at all tested pH when H$_2$O$_2$ was added (92.48–100% degradation), authors infer that the repulsive forces between particles do not play an essential role in these experimental conditions and the importance of adding H_2O_2 is evident.

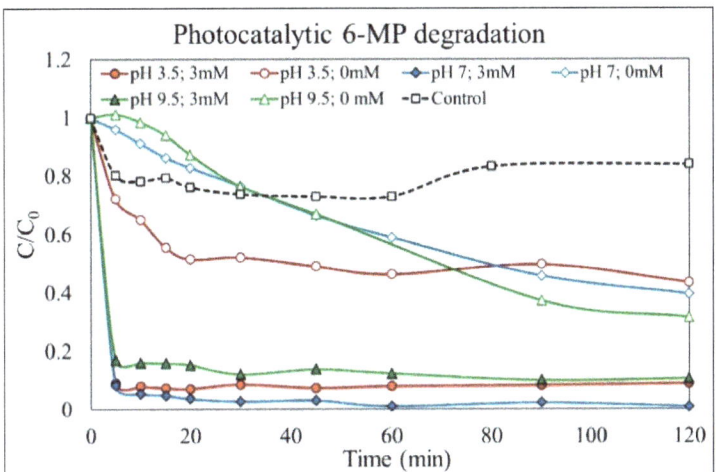

Figure 2. 6-MP degradation in the UV-C/TiO_2 (empty markers) and UV-C/TiO_2/H_2O_2 (filled markers) processes with and without H_2O_2 addition and three different initial pH: 3.5 (circles), 7 (rhombuses) and 9.5 (triangles). Control experiment is represented by the segmented line.

2.3. Kinetic Analysis

Table 1 presents calculated photolytic constants (k_{ph}), operational photocatalytic constants (k_{op}) and half-life times ($t_{1/2}$) for the tested 6-MP degradation processes, which follow a first order reaction model for the first 20 min of reaction. In processes where no H_2O_2 was added (UV-C and UV-C/TiO_2), higher constants and shorter half-life times are reached under acidic conditions (Table 1). It is known that NO_3^- and NO_2^- present in water, when excited under light, result in the formation of HO• [33]; as HNO_3 was used to lower the initial pH of the solution, nitric acid dissociation is to be blamed by the presence of such ions, which provides an additional pathway for radical formation, explaining the better results found under a 3.5 pH.

Even though 6-MP maximum degradation (100%) achieved with UV-C/H_2O_2 processes is higher under acidic and neutral pH (Table 1), higher k_{ph} and shorter $t_{1/2}$ were found at alkaline conditions. Best results should appear at alkaline conditions experiments given that the conjugate anion of H_2O_2 increases with pH and favors HO• production [37].

When H_2O_2 is added to the photolysis UV_{254} process, better results are expected due to its photolytic dissociation, which yields hydroxyl radicals and provides an additional pathway to oxidation [44]. Analyzing the processes where H_2O_2 was added against the ones without it, such improvement proves to be truth.

When comparing K_{ph} values of UV-C/H_2O_2 against K_{op} values of UV-C/TiO_2/H_2O_2 processes, it becomes evident that photocatalytic processes are better for 6-MP degradation, because photocatalytic constant values are higher than the photolytic, except in the case of UV-C/H_2O_2 process under an initial pH of 9.5, which yielded the highest K. On the other hand, when comparing the K_{ph} of the processes with H_2O_2 against the K_{ph} of the processes without it, the importance of adding the oxidant to the reaction becomes clear. The same behavior can be observed in the photocatalysis processes.

Table 1. Rate photolytic constant (k_{ph}), operational photocatalytic constants (k_{op}) and half-life time ($t_{1/2}$) for 6-MP degradation with and without H_2O_2 added, at three pH.

Process	$pH_{initial}$	k_{ph} (min^{-1})	R^2	$t_{1/2}$ (min)	Degradation % at 120 min
UV-C	3.5	0.0094	0.9562	73.74	65.60
	7	0.0025	0.7494	277.26	52.24
	9.5	0.0063	0.8622	110.02	69.90
UV-C/H_2O_2	3.5	0.1071	0.8076	6.47	100
	7	0.1616	0.7693	4.29	100
	9.5	0.1975	0.7174	3.51	98.34

Process	$pH_{initial}$	k_{op} (min^{-1})	R^2	$t_{1/2}$ (min)	Degradation % at 120 min
UV-C/TiO_2	3.5	0.0335	0.9340	20.69	58.60
	7	0.0099	0.9970	70.01	61.70
	9.5	0.0072	0.8198	96.27	71
UV-C/TiO_2/H_2O_2	3.5	0.1387	0.6183	4.99	94.40
	7	0.1725	0.7614	4.02	100
	9.5	0.0857	0.5526	8.09	92.48

As K values were calculated using 20 min of reaction data, R^2 values are greatly affected by the fast degradation in the first 5 min of the processes where H_2O_2 is added. As can be seen in Figures 1 and 2, a rapid descent in concentration was measured in the first 5 min of reaction, so the line drawn between first and last experimental points (0 and 20 min for kinetic calculations) leaves out the points in between, resulting in low R^2 values.

When H_2O_2 is added, hydroxyl radicals can be generated through decomposition from hydroxide peroxide when irradiated (Equation (13)), but also by the contact of hydrogen peroxide and the electron in the conduction band of the titanium dioxide (Equation (14)), reinforcing pollutant degradation by increasing OH• production [33,34]:

$$H_2O_2 + h\upsilon \rightarrow 2HO\bullet \tag{13}$$

$$H_2O_2 + e_{CB^-} \rightarrow \bullet OH + OH^- \tag{14}$$

2.4. Ecotoxicity Dynamics

According to Cramer's rules, 6-MP has a toxicity level III (Toxtree database version 3.1.0), which means that initial samples of 6-MP caused more than 20% diminution of *Vibrio fischeri* bioluminescence.

Ecotoxicity analysis of samples taken at 60 min in the photolysis and photocatalysis processes, suggests that contaminant degradation generated by-products with higher toxicity than the original compound. Ecotoxicity of samples taken at the end of each experiment shows that even if the toxicity decreases in the second half of the process, it is still higher than the original one, presumably by the combination of original contaminant and its byproducts. A similar behavior was observed in the past by Chatzimpaloglou et al. 2021 [45], for photocatalytic degradation of the antineoplastic drug irinotecan, and concluded that the toxicity increase is due to transformation products formation at the first minutes of the process.

The increase in toxicity at 60 min of experimentation is due to two by-products, purine-6-sulfinate and purine-6-sulfonate. According to Moore (1998), these compounds are the main oxidation products of 6-MP and generate strongly oxidizing SO_2^- and SO_3^- radicals with extensive reactivity, capable of breaking DNA. The presence of these compounds, in addition to traces of 6-MP, increases the toxicity towards the end of the process. The toxicity decreases with the subsequent degradation of the remnant of the parent compound and the by-products, but as mineralization is not reached, the solution remains toxic [46].

2.5. Residual H_2O_2

Remnant peroxide in photocatalysis experiments averaged 36.94% whereas photolysis consumed around 67.02% of applied oxidant. The results presented in Figure 3 suggest that H_2O_2 dosage can be lowered up to around 2 mM for experiments of 120 min, and the rapid decrease in concentration in the first five minutes of experimentation seems to demonstrate that a rapid formation of hydroxyl radicals under 254 nm radiation is the main source of degradation. Rapid degradation of H_2O_2 in the first 5 min of reaction, matches fast degradation of pollutant measured in the same time period.

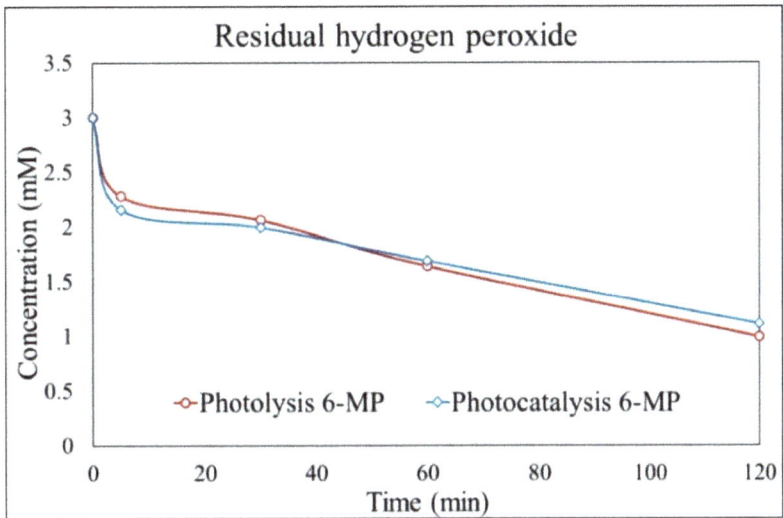

Figure 3. Hydrogen peroxide consumption during 6-MP photolysis (circles) and photocatalysis (rhombuses) degradation.

2.6. 6-MP by-Products Formation

Analysis by HPLC-MS of samples from UV-C/H_2O_2 and UV-C/TiO_2/H_2O_2 6-MP degradation processes were performed in normal phase and reverse phase, and showed the existence of various degradation by-products. Starting from 6-mercaptopurine in the presence of dissolved oxygen and favored in basic pH, anionic salts purine-6-sulfinate and purine-6-sulfonate were found in the analysis of the samples taken. Other possible product can be hypoxanthine, caused by the decrement of oxygen in the solution. These results match the suggested by Hemmens and Moore [36]. The increase of free radicals produced by the photocatalytic processes and the acidic pH of the solution, favors the creation of 7H-purine-6-sulfonic acid, also found in the HPLC-MS analysis. The photodegradation pathway may occur as shown in the Figure 4.

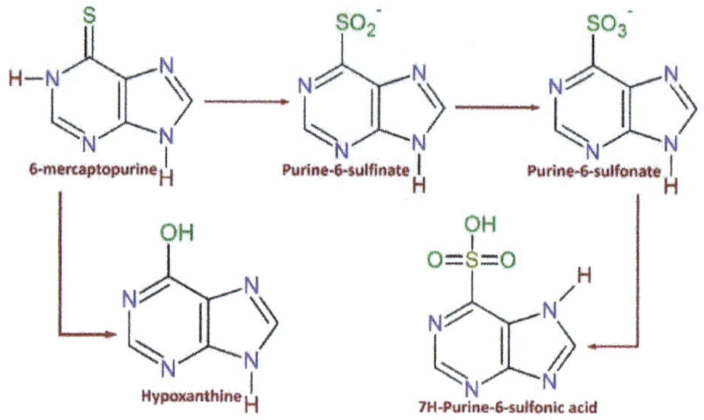

Figure 4. Possible 6-MP degradation pathways.

3. Materials and Methods

3.1. Sample Preparation and Reagents

Samples were produced by dissolving the commercial drug 6-MP (Purinethol®, Aspen Pharmacare, La Lucia, Durban, South Africa) in distilled water to an initial concentration of 6 mg/L, based on the works of degradation of cytostatics by heterogeneous photocatalysis carried out by Molinari et al. 2008 [47], and Evgenidou et al. [48], where the initial concentrations of contaminant were 8 mg/L and 1 mg/L, respectively. Due to reactors characteristics, in UV-C experiments, a sample volume of 50 L was used, meanwhile UV-C/TiO$_2$ required volume was 25 L. H$_2$O$_2$ was obtained from Labbox Labware (CAS: 7722-84-1, Barcelona, Cataluña, Spain). NaCl used for the ecotoxicity method was purchased from Panreac Química (CAS: 7647-14-5, Barcelona, Cataluña, Spain) and the bacteria *Vibrio fischeri* was part of the commercial kit WATERTOX™ (Environmental Bio-detection Products Inc., Burlington, ON, Canada). Titanium oxysulfate was purchased from Sigma-Aldrich (CAS: 13825-74-6, St. Louis, MO, USA).

3.2. UV-C and UV-C/H$_2$O$_2$ 6-MP Degradation Processes

UV reactor used for UV-C and UV-C/H$_2$O$_2$ experiments is shown in Figure 5, modified from Núñez-Núñez et al. [41]. The reactor system consists of a 200 L tank for the water sample, a 1 hp pump for sample recirculation through the system, a 50 µm filter to prevent large particles from entering the main reactor body, and a rotameter in order to measure the water flow going into the reactor. The reactor, main part of the system, is a compartment of stainless steel with the inlet on the bottom part and the outlet on the top.

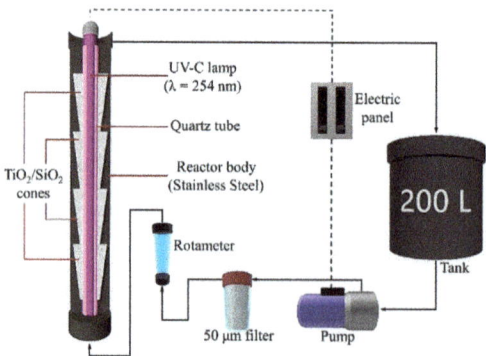

Figure 5. Reactor system used for UV-C and UV-C/H$_2$O$_2$ 6-MP degradation processes.

A low-pressure mercury lamp (254 nm radiation peak; T5 Philips, Amsterdam, The Netherlands) was used as irradiation source. Such lamp was placed inside of a transparent quartz tube to prevent it from entering in contact with the sample. The tube is introduced on the top end of the reactor and occupies its center, so the radiation hits the polished reactor internal surface and reflects back to the sample [49,50].

UV-C and UV-C/H_2O_2 experiments were performed at 650 L/h flow rate. Once sample recirculation started in the system, pH was adjusted with 65% (v/v) HNO_3 or 0.1 M NaOH solutions, to tested pH 3.5, 7 and 9.5. pH values were selected given the TiO_2 isoelectric point of 6.5: an acidic pH, were TiO_2 surface is positively charged; basic pH, were TiO_2 surface is negatively charged; and neutral pH, were TiO_2 superficial charge is low.

The effect of H_2O_2 on the degradation of 6-MP was also tested, the experiments were carried out with the addition of 3 mM H_2O_2/L and the absence of it at each pH. After pH adjustment and H_2O_2 addition (when required), the first water sample was taken (time 0) and lamp was turned on in order to start the process. Aliquot samples were taken at 5, 10, 15, 20, 30, 45, 60, 90, and 120 min of reaction in order to perform the analysis and quantify 6-MP degradation.

3.3. UV-C/TiO$_2$ and UV-C/TiO$_2$/H$_2$O$_2$ 6-MP Degradation Processes

For photocatalysis degradation processes, the commercial AOP1 model reactor (Bright-Water Environmental, Harleston, Norfolk, UK) was used. This reactor is composed by a titanium cylinder covered in the internal wall by a TiO_2 layer. Radiation is provided by a lamp emitting radiation at 254 nm and sample recirculation is achieved by a 0.85 kW water pump (Jet Inox 850, super-ego tools, Abadiano, Vizcaya, Spain). Figure 6 shows the reactor, which is 75 mm in diameter and 475 mm long. A flow rate of 500 L/h was set. Excepting flow rate, photocatalysis experiments were conducted in the same conditions and procedure than UV-C and UV-C/H_2O_2 experiments: same tested pH values (3.5, 7, and 9.5), and in the presence or absence of H_2O_2 (0 and 3 mM).

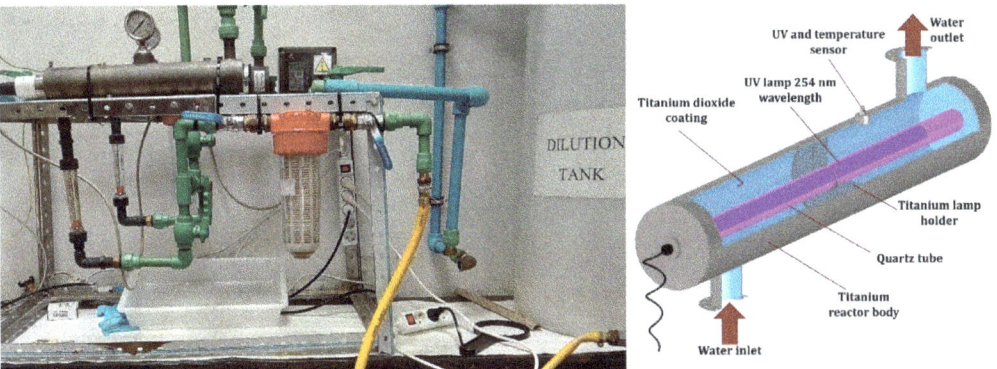

Figure 6. Commercial photocatalytic UV reactor used in UV-C/TiO$_2$ and UV-C/TiO$_2$/H$_2$O$_2$ 6-MP degradation processes (**left**); Titanium AOP1 reactor (**right**).

3.4. Control Experiments

In order to measure the effect of hydrogen peroxide on samples, control experiments in absence of radiation were performed. Sample was prepared as described above and 3 mM of H_2O_2/L of sample was added. The recirculation was started, but the lamp was kept off. Degradation was measured at different times. Control experiments were performed for both photocatalysis and photolysis processes, under an initial sample pH of 7.

3.5. Chemical Analysis

3.5.1. 6-MP Analysis

Water samples at initial time (0 min) were taken directly from the solution tank; subsequent samples were taken at the end of the system, right where it connects to the solution tank. All samples were analyzed by UV–VIS spectrophotometry (T80+ UV–VIS spectrophotometer, PG Instruments Ltd., Alma Park, Leicestershire, UK) at 324 nm wavelength to measure 6-MP concentration. Calibration curve (R^2 = 0.9996) was built using 6-MP monohydrated purchased from Sigma-Aldrich (CAS: 6112-76-1, St. Louis, MO, USA), dissolved in deionized water.

3.5.2. Ecotoxicity Measurement

Water samples taken at 0, 60, and 120 min reaction times from selected experiments were tested for ecotoxicity in order to measure and compare potential environmental damage by 6-MP and its photodegradation by-products. Such analyses were performed by the UNE-EN ISO 11348-3 method [51], which measures the effect of the sample on *V. fischeri* luminescence. Luminescence was measured with a luminometer Bacterial Systems BG-1 (GEM Biomedical Inc., Lewisville, TX, USA). Experiments selected for the analysis were photolysis, under initial pH of 3.5 and photocatalysis, under initial pH of 7, both with H_2O_2 addition.

3.5.3. Residual H_2O_2 Analysis

Residual H_2O_2 concentration was measured in samples taken at 5-, 30-, 60-, and 120-min reaction times. Quantitative analysis was performed following the method previously used by Klamerth [52], which is based in the formation of a yellow solution when H_2O_2 and titanium oxysulfate are mixed, absorbance of the solution is then measured at 410 nm wavelength. Calibration line was built using H_2O_2 solutions with concentrations ranging from 0.5 up to 3 mM.

3.5.4. 6-MP by-Products Formation

Water samples from experiments were analyzed by mass spectrometry in order to determine the possible chemical structures of 6-MP degradation by-products. For comparison purposes, samples from times 0 and 120 min reaction times were analyzed by an Agilent 1100 HPLC coupled to an UV detector and a mass spectrometer Agilent Trap XCT and HPLC Surveyor MS with a LTQ spectrometer.

3.6. Kinetic Analysis

Analysis of 6-MP concentration at different times of the reaction were used to determine reaction operational photocatalytic constants (k_{op}) and photolytic constants (k_{ph}) of the first order reaction by plotting ln[6-MP] against time in the first 20 min of reaction [41,53]. Once k_{op} and k_{ph} were determined, half-life time ($t_{1/2}$ = ln 2/k_{op}; $t_{1/2}$ = ln 2/k_{ph}) of 6-MP degradation of each process was calculated.

4. Conclusions

Experiments show that 6-MP is degraded by both UV-C and UV-C/TiO_2 processes, and degradation higher than 50% can be achieved, under acidic, neutral, or basic pH conditions. Hydrogen peroxide addition to photolysis and photocatalysis experiments greatly improves 6-MP degradation, as degradation percentages here reported for UV-C/H_2O_2 and UV-C/TiO_2/H_2O_2 processes are above 90% for different initial pH values.

pH lowering through HNO_3 addition, improves degradation as kinetic results signal, but such effect can be observed only in absence of H_2O_2; when the oxidant is present, its effect is the most important factor in the experiments, which points to HO• formation by H_2O_2 molecule rupture under UV-C radiation provided. Nevertheless, special care should be put into the hydrogen peroxide addition, as an excess could result in a detrimental effect over degradation. In this research, H_2O_2 addition (3 mM/L) ended up being excessive, so a dose of 2 mM, instead of 3, should be tested for 2 h duration experiments.

Even though, when H_2O_2 is added to the processes, 6-MP degradation over 80% is reached in the first 10 min of reaction, toxicity actually increased, demonstrating that byproducts are formed, and such byproducts, or their combination with remaining 6-MP, possess a greater toxicity than the original compound.

Author Contributions: Methodology: L.A.G.-B. and J.B.P.-N.; validation: L.A.G.-B., I.L.-A., and J.B.P.-N.; formal analysis: L.A.G.-B. and C.M.N.-N.; investigation: L.A.G.-B. and C.M.N.-N.; resources: J.C.G.-P., M.G.-R., and J.B.P.-N.; data curation: L.A.G.-B., M.G.-R., and J.C.G.-P.; writing—original draft preparation: L.A.G.-B.; writing—review and editing: L.A.G.-B., C.M.N.-N., and J.B.P.-N.; visualization: L.A.G.-B.; supervision: I.L.-A. and J.B.P.-N.; project administration: J.B.P.-N.; funding acquisition: J.C.G.-P., M.G.-R., and J.B.P.-N. All authors have read and agreed to the published version of the manuscript.

Funding: First author thanks the Consejo Nacional de Ciencia y Tecnología (CONACyT), who provided funding through the doctorate scholarship granted. Support was received also from Instituto Politécnico Nacional (IPN/SIP project 20190247and 20200670). The content does not necessarily reflect the views and policies of the funding organizations.

Data Availability Statement: Data is contained within the article or supplementary material.

Acknowledgments: The authors would like acknowledge the Centro de Investigación y Desarrollo Tecnológico del Agua of Salamanca University, Spain, for facilitating the use of its facilities to carry out experimentation, and Servicio General de Espectrometría de Masas of Salamanca University (SGEM-USAL) for sample analysis.

Conflicts of Interest: The authors declare no conflict of interest.

References

1. Geissen, V.; Mol, H.; Klumpp, E.; Umlauf, G.; Nadal, M.; van der Ploeg, M.; van de Zee, S.E.A.T.M.; Ritsema, C.J. Emerging pollutants in the environment: A challenge for water resource management. *Int. Soil Water Conserv. Res.* **2015**, *3*, 57–65. [CrossRef]
2. Dimpe, K.M.; Nomngongo, P.N. Current sample preparation methodologies for analysis of emerging pollutants in different environmental matrices. *Trends Anal. Chem.* **2016**, *82*, 199–207. [CrossRef]
3. Belver, C.; Bedia, J.; Rodriguez, J.J. Zr-doped TiO_2 supported on delaminated clay materials for solar photocatalytic treatment of emerging pollutants. *J. Hazard. Mater.* **2017**, *332*, 233–242. [CrossRef] [PubMed]
4. Calza, P.; Medana, C.; Sarro, M.; Rosato, V.; Aigotti, R.; Baiocchi, C.; Minero, C. Photocatalytic degradation of selected anticancer drugs and identification of their transformation products in water by liquid chromatography–high resolution mass spectrometry. *J. Chromatogr. A* **2014**, *1362*, 135–144. [CrossRef] [PubMed]
5. Roig, B.; Marquenet, B.; Delpla, I.; Bessonneau, V.; Sellier, A.; Leder, C.; Thomas, O.; Bolek, R.; Kummerer, K. Monitoring of methotrexate chlorination in water. *Water Res.* **2014**, *57*, 67–75. [CrossRef]
6. Lutterbeck, C.A.; Baginska, E.; Machado, Ê.L.; Kümmerer, K. Removal of the anti-cancer drug methotrexate from water by advanced oxidation processes: Aerobic biodegradation and toxicity studies after treatment. *Chemosphere* **2015**, *141*, 290–296. [CrossRef] [PubMed]
7. Kanakaraju, D.; Glass, B.D.; Oelgemöller, M. Advanced oxidation process-mediated removal of pharmaceuticals from water: A review. *J. Environ. Manag.* **2018**, *219*, 189–207. [CrossRef]
8. Kosjek, T.; Negreira, N.; López de Alda, M.; Barceló, D. Aerobic activated sludge transformation of methotrexate: Identification of biotransformation products. *Chemosphere* **2015**, *119*, S42–S50. [CrossRef]
9. Lai, W.W.; Hsu, M.H.; Lin, A.Y. The role of bicarbonate anions in methotrexate degradation via UV/TiO_2: Mechanisms, reactivity and increased toxicity. *Water Res.* **2017**, *112*, 157–166. [CrossRef]
10. Białk-Bielińska, A.; Mulkiewicz, E.; Stokowski, M.; Stolte, S.; Stepnowski, P. Acute aquatic toxicity assessment of six anti-cancer drugs and one metabolite using biotest battery e Biological effects and stability under test conditions. *Chemosphere* **2017**, *189*, 689–698. [CrossRef] [PubMed]
11. Chen, S.; Blaney, L.; Chen, P.; Deng, S.; Hopanna, M.; Bao, Y.; Yu, G. Ozonation of the 5-fluorouracil anticancer drug and its prodrug capecitabine: Reaction kinetics, oxidation mechanisms, and residual toxicity. *Front. Environ. Sci. Eng.* **2019**, *13*, 59. [CrossRef]
12. Lin, A.Y.; Hsueh, J.H.; Hong, P.K. Removal of antineoplastic drugs cyclophosphamide, ifosfamide, and 5-fluorouracil and a vasodilator drug pentoxifylline from wastewaters by ozonation. *Environ. Sci. Pollut. Res.* **2015**, *22*, 508–515. [CrossRef]
13. Zhang, J.; Chang, V.W.C.; Giannis, A.; Wang, J.-Y. Removal of cytostatic drugs from aquatic environment: A review. *Sci. Total Environ.* **2013**, *445–446*, 281–298. [CrossRef] [PubMed]

14. Franquet-Griell, H.; Medina, A.; Sans, C.; Lacorte, S. Biological and photochemical degradation of cytostatic drugs under laboratory conditions. *J. Hazard. Mater.* **2017**, *323*, 319–328. [CrossRef] [PubMed]
15. Suresh, S.; Athimoolan, S.; Sridhar, B. XRD, vibrational spectra and quantum chemical studies of an anticancer drug: 6-Mercaptopurine. *Spectrochim. Acta A Mol. Biomol. Spectrosc.* **2015**, *146*, 204–213. [CrossRef] [PubMed]
16. Li, A.P.; Peng, J.D.; Zhou, M.; Zhang, J. Resonance light scattering determination of 6-mercaptopurine coupled with HPLC technique. *Spectrochim. Acta A Mol. Biomol. Spectrosc.* **2016**, *154*, 1–7. [CrossRef]
17. Jin, M.; Mou, Z.L.; Zhang, R.L.; Liang, S.S.; Zhang, Z.Q. An efficient ratiometric fluorescence sensor based on metal-organic frameworks and quantum dots for highly selective detection of 6-mercaptopurine. *Biosens. Bioelectron.* **2017**, *91*, 162–168. [CrossRef]
18. Zerra, P.; Bergsagel, J.; Keller, F.G.; Lew, G.; Pauly, M. Maintenance Treatment with Low-Dose Mercaptopurine in Combination with Allopurinol in Children with Acute Lymphoblastic Leukemia and Mercaptopurine-Induced Pancreatitis. *Pediatr. Blood Cancer* **2015**, *63*, 712–715. [CrossRef]
19. Lennard, L.; Rees, C.A.; Lilleyman, J.S.; Maddocks, J.L. Childhood leukaemia: A relationship between intracellular 6- mercaptopurine metabolites and neutropenia. *Br. J. Clin. Pharmacol.* **1983**, *16*, 359–363. [CrossRef] [PubMed]
20. Schmiegelow, K.; Nielsen, S.N.; Frandsen, T.L.; Nersting, J. Mercaptopurine/Methotrexate maintenance therapy of childhood acute lymphoblastic leukemia: Clinical facts and fiction. *Pediatr. Hematol. Oncol. J.* **2014**, *36*, 503–517. [CrossRef]
21. Gutiérrez, O. Asociación de los Polimorfismos de la TPMT*1,*2,*3A,*3B,*3C y la MTHFR (677C>T y 1298A>C) con la Farmacocinética de la 6-Mercaptopurina, las Reacciones Adversas y la Susceptibilidad a Leucemia Linfoblástica Aguda en Pacientes Pediátricos. Ph.D. Dissertation, Centro Interdisciplinario de Investigación para el Desarrollo Integral Regional, Durango, México, June 2016.
22. Consejo de Salubridad General. *Cuadro Básico y Catálogo de Medicamentos*; Comisión Interinstitucional del Cuadro Básico y Catálogo de Insumos del Sector Salud: Ciudad de México, México, 2017.
23. Mejía, R. *Cuadro Básico y Catálogo Institucional, Oncología*; Secretaría de Salud de la Ciudad de México: Ciudad de México, México, 2018.
24. Instituto Mexicano del Seguro Social. *Cuadro Básico de Medicamentos Instituto Mexicano del Seguro Social*; Dirección de Prestaciones Médicas, Unidad de Atención Médica: Ciudad de México, México, 2019.
25. Medscape.com. Available online: https://reference.medscape.com/drug/purinethol-purixan-mercaptopurine-342094 (accessed on 15 February 2020).
26. Drugs.com. Available online: https://www.drugs.com/dosage/mercaptopurine.html (accessed on 2 February 2020).
27. Sarkar, S.; Das, R.; Choi, H.; Bhattacharjee, C. Involvement of process parameters and various modes of application of TiO_2 nanoparticles in heterogeneous photocatalysis of pharmaceutical wastes—A short review. *RSC Adv.* **2014**, *4*, 57250–57266. [CrossRef]
28. Chekir, N.; Tassalit, D.; Benhabiles, O.; Merzouk, N.K.; Ghenna, M.; Abdessemed, A.; Issaadi, R. A comparative study of tartrazine degradation using UV and solar fixed bed reactors. *Int. J. Hydrog. Energy* **2017**, *42*, 8948–8954. [CrossRef]
29. Lee, C.M.; Palaniandy, P.; Dahlan, I. Pharmaceutical residues in aquatic environment and water remediation by TiO_2 heterogeneous photocatalysis: A review. *Environ. Earth Sci.* **2017**, *76*, 611. [CrossRef]
30. Ahmed, S.N.; Haider, W. Heterogeneous photocatalysis and its potential applications in water and wastewater treatment: A review. *Nanotechnology* **2018**, *29*, 1–30. [CrossRef]
31. Tokode, O.; Prabhu, R.; Lawton, L.A.; Robertson, P.K.J. UV LED Sources for Heterogeneous Photocatalysis. *Environ. Photochem. Part III* **2014**, *35*, 159–179. [CrossRef]
32. Yasmina, M.; Mourad, K.; Mohammed, S.H.; Khaoula, C. Treatment Heterogeneous Photocatalysis; Factors Influencing the Photocatalytic Degradation by TiO_2. *Energy Procedia* **2014**, *50*, 559–566. [CrossRef]
33. Wang, J.L.; Xu, L.J. Advanced Oxidation Processes for Wastewater Treatment: Formation of Hydroxyl Radical and Application. *Crit. Rev. Environ. Sci. Technol.* **2012**, *42*, 251–325. [CrossRef]
34. Smaranda, I.; Nila, A.; Manta, C.-M.; Samohvalov, D.; Gherca, D.; Baibarac, M. The influence of UV light on the azathioprine photodegradation: New evidences by photoluminescence. *Results Phys.* **2019**, *14*, 102443. [CrossRef]
35. Karran, P.; Attard, N. Thiopurines in current medical practice: Molecular mechanisms and contributions to therapy-related cancer. *Nat. Rev. Cancer* **2008**, *8*, 24–36. [CrossRef] [PubMed]
36. Hemmens, V.J.; Moore, D.E. Photo-oxidation of 6-Mercaptopurine in Aqueous Solution. *J. Chem. Soc.* **1984**, *2*, 209–211. [CrossRef]
37. Litter, M.I. Introduction to Photochemical Advanced Oxidation Processes for Water Treatment. *Environ. Photochem. Part II* **2005**, *2*, 325–366. [CrossRef]
38. Legrini, O.; Oliveros, E.; Braun, A.M. Photochemical processes for water treatment. *Chem. Rev.* **1993**, *93*, 671–698. [CrossRef]
39. Kusic, H.; Koprivanac, N.; Bozic, A.L. Minimization of organic pollutant content in aqueous solution by means of AOPs: UV- and ozone-based technologies. *Chem. Eng. J.* **2006**, *123*, 127–137. [CrossRef]
40. Baxendale, J.H.; Wilson, J.A. The photolysis of hydrogen peroxide at high light intensities. *Trans. Faraday Soc.* **1956**, *53*, 344–356. [CrossRef]
41. Núñez-Núñez, C.M.; Cháirez-Hernández, I.; García-Roig, M.; García-Prieto, J.C.; Melgoza-Alemán, R.M.; Proal-Nájera, J.B. UV-C/H_2O_2 heterogeneous photocatalytic inactivation of coliforms in municipal wastewater in a TiO_2/SiO_2 fixed bed reactor: A kinetic and statistical approach. *React. Kinet. Mech. Catal.* **2018**, *125*, 1159–1177. [CrossRef]
42. Parks, G.A. The isoelectric points of solid oxides, solid hydroxides and aqueous hydroxo complex systems. *Chem. Rev.* **1965**, *65*, 177–198. [CrossRef]

43. Connors, K.A.; Amidon, G.L.; Stella, V.J. *Chemical Stability of Pharmaceuticals: A Handbook for Pharmacists*, 2nd ed.; Wiley: Hoboken, NJ, USA, 1986; p. 544.
44. Pablos, C.; Marugán, J.; van Grieken, R.; Serrano, E. Emerging micropollutant oxidation during disinfection processes using UV-C, UV-C/H_2O_2, UV-A/TiO_2 and UV-A/TiO_2/H_2O_2. *Water Res.* **2013**, *47*, 1237–1245. [CrossRef] [PubMed]
45. Chatzimpaloglou, A.; Christophoridis, C.; Fountoulakis, I.; Antonopoulou, M.; Vlastos, D.; Bais, A.; Fytianos, K. Photolytic and photocatalytic of antineoplastic drug irinotecan. Kinetic study, identification of transformation products and toxicity evaluation. *Chem. Eng. J.* **2021**, *405*, 1–17. [CrossRef]
46. Moore, D.E. Mechanism of photosensitization by phototoxic drugs. *Mutat. Res.* **1998**, *422*, 165–173. [CrossRef]
47. Molinari, R.; Caruso, A.; Argurio, P.; Poerio, T. Degradation of the drugs Gemfibrozil and Tamoxifen in pressurized and de-pressurized membrane photoreactors using suspended polycrystalline TiO_2 as catalyst. *J. Membr. Sci.* **2008**, *319*, 54–63. [CrossRef]
48. Evgenidou, E.; Ofrydopoulou, A.; Malesic-Eleftheriadou, N.; Nannou, C.; Ainali, N.M.; Christodoulou, E.; Bikiaris, D.N.; Kyzas, G.Z.; Lambropoulou, D.A. New insights into transformation pathways of a mixture of cytostatic drugs using Polyester-TiO_2 films: Identification of intermediates and toxicity assessment. *Sci. Total Environ.* **2020**, *741*, 140394. [CrossRef] [PubMed]
49. Garcés-Giraldo, L.F.; Mejía-Franco, E.A.; Santamaría-Arango, J.J. La fotocatálisis como alternativa para el tratamiento de aguas residuales. *Rev. Lasallista Investig.* **2004**, *1*, 83–92.
50. Pantoja-Espinoza, J.C.; Proal-Nájera, J.B.; García-Roig, M.; Cháirez-Hernández, I.; Osorio-Revilla, G.I. Comparative efficiencies of coliform bacteria inactivation in municipal wastewater by photolysis (UV) and photocatalysis (UV/TiO_2/SiO_2). Case: Treatment wastewater plant of Salamanca, Spain. *Rev. Mex. Ing. Quim.* **2015**, *14*, 119–135.
51. AENOR. *Determinación del Efecto Inhibidor de Muestras de Agua Sobre la Luminiscencia de Vibrio Fischeri (Ensayo de Bacterias Luminiscentes)*; UNE-EN ISO 11348-3; AENOR: Madrid, Spain, 2009.
52. Klamerth, N. Application of a Solar Photo-Fenton for the Treatment of Contaminants in Municipal Wastewater Effluents. Ph.D. Dissertation, University of Almería, Almería, Spain, June 2011.
53. Kuhn, H.; Försterling, H.D. *Principles of Physical Chemistry*, 2nd ed.; Wiley: West Sussex, UK, 2000; pp. 750–751.

Article

Comparison of Advanced Oxidation Processes for the Degradation of Maprotiline in Water—Kinetics, Degradation Products and Potential Ecotoxicity

Nuno P. F Gonçalves [1,*], Zsuzsanna Varga [2], Edith Nicol [2], Paola Calza [1] and Stéphane Bouchonnet [2]

1 Department of Chemistry, University of Turin, 10124 Torino, Italy; Paola.calza@unito.it
2 Laboratoire de Chimie Moléculaire, CNRS/Ecole Polytechnique, Institut Polytechnique de Paris, 91128 Palaiseau, France; zsuzsanna.varga@polytechnique.edu (Z.V.); edith.nicol@polytechnique.edu (E.N.); stephane.bouchonnet@polytechnique.edu (S.B.)
* Correspondence: nunopaulo.ferreiragoncalves@unito.it

Abstract: The impact of different oxidation processes on the maprotiline degradation pathways was investigated by liquid chromatography-high resolution mass spectrometry (LC/HRMS) experiments. The in-house SPIX software was used to process HRMS data allowing to ensure the potential singular species formed. Semiconductors photocatalysts, namely Fe-ZnO, Ce-ZnO and TiO_2, proved to be more efficient than heterogeneous photo-Fenton processes in the presence of hydrogen peroxide and persulfate. No significant differences were observed in the degradation pathways in the presence of photocatalysis, while the SO_4^- mediated process promote the formation of different transformation products (TPs). Species resulting from ring-openings were observed with higher persistence in the presence of SO_4^-. In-silico tests on mutagenicity, developmental/reproductive toxicity, Fathead minnow LC_{50}, D. magna LC_{50}, fish acute LC_{50} were carried out to estimate the toxicity of the identified transformation products. Low toxicant properties were estimated for TPs resulting from hydroxylation onto bridge rather than onto aromatic rings, as well as those resulting from the ring-opening.

Citation: Gonçalves, N.P.F; Varga, Z.; Nicol, E.; Calza, P.; Bouchonnet, S. Comparison of Advanced Oxidation Processes for the Degradation of Maprotiline in Water—Kinetics, Degradation Products and Potential Ecotoxicity. *Catalysts* **2021**, *11*, 240. https://doi.org/10.3390/catal11020240

Keywords: maprotiline; advanced oxidation processes; LC/HRMS coupling; structural elucidation; kinetics; ecotoxicity estimations

Academic Editors: Stephanie Lambert and Julien Mahy

Received: 13 January 2021
Accepted: 8 February 2021
Published: 11 February 2021

Publisher's Note: MDPI stays neutral with regard to jurisdictional claims in published maps and institutional affiliations.

Copyright: © 2021 by the authors. Licensee MDPI, Basel, Switzerland. This article is an open access article distributed under the terms and conditions of the Creative Commons Attribution (CC BY) license (https://creativecommons.org/licenses/by/4.0/).

1. Introduction

A consequence of human activities is the arising of so-called contaminants of emerging concern (CECs) and their subsequent detection in surface-, ground-, and drinking water worldwide [1–3]. Particular attention should be paid to compounds that are scarcely biodegradable, such as pharmaceuticals, personal care products, hormones, food additives, artificial sweeteners, pesticides, and dyes among others [4–6]. Due to their distinct chemical properties (polarity, functional groups, solubility, ...) and potential persistence, CECs are poorly removed by conventional wastewater treatment methodologies and are thus frequently detected in the wastewater treatment plant (WWTP) effluents [7]. These effluents act as a constant low-level source of such substances to the environment. For example, antidepressants and psychotic drugs, as a result of their fast increasing consumption [8,9] and high stability, have been recently observed in surface water and wastewater [10]. Their environmental hazardousness is not related to their (usually low) concentrations, but to their continuous discharge, persistence, and high biological activity [11]. Like many other pharmaceuticals, antidepressants may be excreted in their native form or as metabolites and enter the water bodies via different pathways. This class of drugs is of particular concern as it affects not only the central nervous system but also the reproduction, growth, and immune functions [12,13]. Moreover, tricyclic and tetracyclic antidepressants are known to have several side-effects when consumed by humans and their exposure to fish increased mortality, developmental retardation, and morphological anomalies [12]. Maprotiline is a tetracyclic antidepressant drug with fast-growing consumption that has been detected in surface water and wastewater at µg L^{-1} concentrations [14,15]. A study monitoring pharmaceuticals in the Baltic Sea region [16] detected maprotiline in around 50% of influents and effluents wastewater treatment plants with only 44% of removal by the conventional wastewater treatment methodologies. Alarmingly, the study also reports significant bioaccumulation with a concentration of more than 170 µg/kg of maprotiline in blue mussels in the same region.

Advanced oxidation processes (AOPs) have emerged as a promising approach to efficiently remove a wide range of recalcitrant pollutants by reaction with generated strong radical oxidants (mostly hydroxyl radical •OH and sulfate radical $SO_4^{•-}$) [17]. Among AOPs, heterogeneous photocatalysis processes demonstrated to be efficient in achieving the degradation and in most cases the mineralization of a wide range of pollutants [18–20]. TiO_2 and ZnO materials have been extensively investigated for water treatment because they are non-toxic, chemically, and biologically inert, inexpensive and they have a high capability to generate reactive species [21–24]. The introduction of doping agents in the reticular structure of zinc oxide materials is described to reduce the charge recombination increasing in this way the photocatalytic efficiency [23,25]. In previous studies, we demonstrated that the addition of iron [26] and cerium [27] in proper amounts promote interesting photocatalytic properties, increasing the removal efficiency of water pollutants.

Other AOPs, widely investigated due to their high effectiveness, are based on (photo)-Fenton-like processes, characterized by the formation of reactive species mediated by iron. The use of an iron-based heterogeneous catalyst allows the catalyst recovery and reusability with a low amount of iron released in solution, thus avoiding sludge formation and simplifying the process in comparison with the conventional homogeneous process [18]. In previous studies, we demonstrated that the stabilization of Fe(II) ions in the magnetite structure, essentials to keep the system active, can be achieved by coating them with humic acid substances [28]. The use of magnetite particles coated with humic acid (Fe_3O_4/HA) as catalysts for heterogeneous (photo)-Fenton processes, proved to effectively activate hydrogen peroxide and persulfate, for the degradation of pollutants in water [28,29].

Despite the efficiency of some oxidation technologies, many questions remain about pollutant degradation pathways. The generated transformation products (TPs) can be even more toxic and persistent than the parent molecule itself. For example, relatively safe pharmaceuticals, such as ibuprofen and naproxen are described to form TPs that are many times more toxic than the parent compound [30,31]. Sinclair et al., analyzing the toxicity of hundreds of TPs formed from the degradation of several synthetic chemicals, reported that even if the majority were less or presented similar toxicity of the parent molecule, 20% were >3 times more toxic and 9% were >10 times more toxic than the parent compound [32]. Thus, it is important not only to degrade the parent molecule, but also to study the degradation products and ideally achieve their mineralization [33–35].

In a previous study, maprotiline degradation in river water proved to occur through the formation of 12 intermediates resulting from hydroxylation/oxidation and ring-opening [36]. Conversely, photocatalytic experiments proved to lead to more complex degradation pathways with the formation of a significantly higher number of TPs, resulting mainly from the drug multi-hydroxylation. Moreover, by retrospective analysis of wastewater effluents samples, we ascertained that several of these species are also formed during the wastewater treatment processes, thus being released into the environment. This fact raises concerns about how the wastewater oxidation processes can impact the degradation pathways and consequently the toxicity of the released intermediates and products.

In the present work, we studied the impact of different oxidation technologies on the removal of maprotiline and their effect on the degradation pathways by monitoring the disappearance kinetics and the formation of transformation products by LC/HRMS; in-silico toxicity, estimations have been carried out on the TPs as well. For this purpose, different AOPs were considered: (i) semiconductors photocatalysis: Iron-doped zinc oxide

(Fe-ZnO), cerium doped zinc oxide (Ce-ZnO) and TiO$_2$; (ii) heterogeneous photo-Fenton: magnetite coated with humic acid (Fe$_3$O$_4$/HA) for the activation of H$_2$O$_2$ and S$_2$O$_8^{2-}$).

2. Results

2.1. Maprotiline Photoinduced Degradation

For each advanced oxidation technology, maprotiline removal was followed by LC/HRMS in ESI positive ion mode, monitoring the protonated molecule at m/z 278.1911. As reported in Figure 1, no significant maprotiline removal was observed using direct photolysis under UVA irradiation, due to its low radiation absorption above 280 nm [37]. These results are in agreement with the long half-life time previously observed in river water under irradiation [36,38]. The experiments performed by means of different oxidation approaches showed that all the materials were able to remove maprotiline completely in less than 40 min under irradiation. The semiconductor photocatalysts proved to be the most efficient, removing the drug with kinetics faster than those of the photo-Fenton systems.

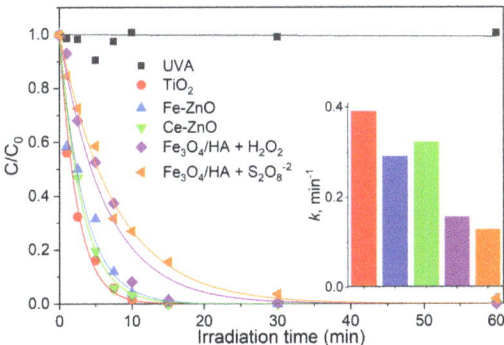

Figure 1. Maprotiline removal in the presence of different oxidation processes under UVA irradiation: photocatalysis using TiO$_2$, Fe-ZnO or Ce-ZnO, and heterogeneous photo-Fenton using Fe$_3$O$_4$/HA for the activation of H$_2$O$_2$ or S$_2$O$_8^{2-}$. The inset compares the kinetic constant values.

The photo-Fenton-like processes, Fe$_3$O$_4$/HA exhibited slightly faster degradation kinetics in the presence of hydrogen peroxide comparing with the same system in the presence of persulfate. In all cases, a monoexponential decay was observed for maprotiline with the degradation time: C/C$_0$ = exp(−k × t), where C$_0$ is the maprotiline initial concentration, C that at the reaction time t, and the pseudo-first-order kinetic constant k.

2.2. Maprotiline Transformation Products

Considering the aim of this study and how the different photocatalytic systems may affect the generation of TPs, samples were firstly analyzed by direct infusion in ESI positive mode, without any sample treatment, and the SPIX software was used to process HRMS data, to select statistically relevant ions from complex spectra. It should be noted that because of direct infusion analysis isomeric species are not distinguishable. Among the different degradation systems, SPIX allowed to ascertain maprotiline degradation and to extract a total of 10 m/z ratios appearing and disappearing over the photocatalytic degradation time, recognized as potential transformation products. The output data for the extracted ions and the kinetic models best fitting the ion abundance evolution over time are presented in Table S1 (Supplementary Materials). SPIX software allowed to ensure that potential singular species formed depending on the degradation system were not neglected due to operator-related subjectivity. For more information about SPIX functioning, please refer to Nicol et al. [39].

LC/HRMS experiments allowed identifying thirty-six species, interestingly all matching with those previously identified using the SPIX software in the form of several isomers, as presented in Table 1. The species identified were named based on their M+H$^+$, e.g., 294-A corresponds to a isomer at m/z 294. mass In agreement with our previous studies [36], maprotiline degradation derives in several isomeric forms resulting from hydroxyl group addition(s) in several positions, namely mono- (m/z 294), di- (m/z 310), tri- (m/z 326) and tetra-hydroxylated (m/z 342) derivatives. From those, the corresponding dehydrogenated mono- (m/z 292), di- (m/z 308), and tri- hydroxylated (m/z 324) species were also identified. Species resulting from the ring-opening were also observed (m/z 284, 258, and 234).

Table 1. Summary of TPs resulting from maprotiline degradation under different AOPs.

Compound	[M+H]$^+$	Formula	Δ (ppm)	r.t. (min)	TiO$_2$	Fe-ZnO	Ce-ZnO	Fe$_3$O$_4$/HA	
								H$_2$O$_2$	S$_2$O$_8^{2-}$
Maprotiline	278.1911	C$_{20}$H$_{24}$N	−2.81	9.7					
294-A [a]	294.1858	C$_{20}$H$_{24}$NO	−1.98	7.3	+	+	+	+	−
294-B [a]				7.8	+	+	+	+	+
294-C [a]				8.0	+	+	+	+	+
294-D				8.3	+	+	+	+	+
294-E				8.9	+	+	+	+	+
310-A	310.1807	C$_{20}$H$_{24}$NO$_2$	−2.00	6.9	+	+	+	+	−
310-B				7.2	+	+	+	+	−
310-C				7.4	+	+	+	+	−
310-D				8.0	+	+	+	+	+
310-E				8.2	+	+	+	+	−
310-F				8.5	+	+	+	+	+
326-A	326.1757	C$_{20}$H$_{24}$NO$_3$	−2.13	6.1	+	+	+	+	+
326-B [a]				6.9	+	+	+	+	+
326-C [a]				7.6	+	+	+	+	+
342-A [a]	342.1707	C$_{20}$H$_{24}$NO$_4$	−2.09	6.0	+	+	+	+	+
342-B [a]				6.4	+	+	+	+	+
342-C [a]				7.1	+	+	+	+	+
342-D				7.6	+	+	+	+	+
292-A	292.1701	C$_{20}$H$_{22}$NO	−2.06	6.9	+	+	+	+	−
292-B [a]				7.9	+	+	+	+	−
292-C [a]				8.5	+	+	+	+	−
292-D				8.6	+	+	+	+	−
292-E				9.1	+	+	+	+	+
308-A	308.1651	C$_{20}$H$_{22}$NO$_2$	−1.96	6.4	+	+	+	+	−
308-B [a]				7.3	+	+	+	+	−
308-C [a]				7.7	+	+	+	+	+
324-A	324.1599	C$_{20}$H$_{22}$NO$_3$	−1.70	5.8	+	+	+	+	+
324-B				6.4	−	−	−	−	+
324-C				7.0	−	−	−	−	+
324-D				7.9	−	−	−	−	+
284-A	284.1650	C$_{18}$H$_{22}$NO$_5$	−1.75	6.7	+	+	+	+	+
284-B				7.3	+	+	+	+	+
284-C [a]				7.7	+	−	−	+	+
284-D				8.4	−	−	−	−	+
234 [a]	234.1492	C$_{14}$H$_{20}$NO$_2$	−1.70	4.4	+	+	+	+	+
258	258.1493	C$_{16}$H$_{20}$NO$_2$	−1.77	6.3	+	+	−	+	+

(+) detected; (−) not observed. [a] observed in wastewater treatment effluents by retrospective analysis [36].

The formed TPs evolution over time depending on the oxidation process are shown in Figure S1. This evidenced the formation of all TPs within a few minutes, exhibiting maximum intensity after 5 to 10 min of treatment and their complete disappearance after 30 min in presence of semiconductors photocatalysts and H$_2$O$_2$ activated by Fe$_3$O$_4$/HA. However, most of those species have a slower degradation profile when mediated by the persulfate, with maximum intensity observed at longer irradiation times even if also completely removed within 1 h, except the ion m/z 234 that is still present at the end of the experiment (see Figure S1). In a general way, most of the identified TPs are common to all the oxidation processes with the exception of that using persulfate, which induces a specific behavior regarding some TPs. For instance, the persulfate system leads to the formation of four isomers with the formula $C_{20}H_{22}NO_3$ (MH^+ at m/z 324) against only one (324-A) for other AOPs. In the same way, the species 284-D and 308-C are also specifically formed with persulfate when other 284 and 308 isomers are common to all oxidation processes. Contrariwise, only one isomer (292-E) was observed for the ion at m/z 292 in the degradation promoted by the persulfate system, as shown in Table 1 and better evidenced in Figure S1. Similarly, only two out of the six species resulting from the dihydroxylation (MH^+ at m/z 310) were formed in the persulfate oxidation process. The collision induced dissociation data presented in Table S2 (Supplementary Materials) allowed making the correspondence between the TPs detected in the present study and those reported in a study devoted to UV-Vis maprotiline degradation, in which CID experiments led to propose TPs' chemical structures [36].

2.3. In Silico Bioassays

The in-silico toxicity was evaluated on maprotiline and on the previously elucidated TPs [36]; the results as summarized in Table 2. The mutagenicity prediction was performed using a consensus-based on four different models. Maprotiline was found to be non-mutagenic, in agreement with the available experimental data [40], as well as for all the TPs. The developmental/reproductive toxicity was estimated by the library model implementing a virtual library of toxicant compounds with endpoint indicators based on toxicant/non-toxicant [41]. Maprotiline, as well as 292-E, 284-B, 234, and 258 were estimated as non-toxicant, while, the remaining TPs were predicted as a toxicant, with the potential to impact the developmental/reproductive abilities.

Table 2. Toxicity values estimated by the VEGA software for maprotiline and the main degradation products.

Compound		Mutagenicity (AMES test)	Developmental/ Reproductive Toxicity	F. Minnow LC50 (96 h) (mg L^{-1})	D. Magna LC50 (48 h) (mg L^{-1})	Fish Acute LC50 (48 h) (mg L^{-1})
MPT		Negative	Non-toxicant	2.61	0.08	1.29
294-A		Negative	Toxicant	3.13	0.20	1.68
294-B		Negative	Toxicant	7.15	0.31	2.43
294-C		Negative	Toxicant	3.10	0.21	1.61

Table 2. Cont.

Compound		Mutagenicity (AMES test)	Developmental/ Reproductive Toxicity	F. Minnow LC50 (96 h) (mg L^{-1})	D. Magna LC50 (48 h) (mg L^{-1})	Fish Acute LC50 (48 h) (mg L^{-1})
294-D		Negative	Toxicant	2.00	0.22	1.58
294-E		Negative	Toxicant	2.52	0.23	1.57
292-A		Negative	Toxicant	3.21	0.16	2.06
292-B		Negative	Toxicant	2.59	0.18	2.17
292-C		Negative	Toxicant	3.19	0.16	1.59
292-D		Negative	Toxicant	2.06	0.18	1.86
292-E		Negative	Non-toxicant	8.82	0.28	3.39
310-A		Negative	Toxicant	8.02	5.25	3.85

Table 2. Cont.

Compound	Mutagenicity (AMES test)	Developmental/ Reproductive Toxicity	F. Minnow LC50 (96 h) (mg L^{-1})	D. Magna LC50 (48 h) (mg L^{-1})	Fish Acute LC50 (48 h) (mg L^{-1})
310-B	Negative	Toxicant	26.12	6.29	5.70
310-E	Negative	Toxicant	2.37	4.44	3.14
310-F	Negative	Toxicant	13.88	5.47	5.54
284-A	Negative	Toxicant	0.47	0.14	5.58
284-B	Negative	Non-toxicant	8.81	0.26	8.00
284-C	Negative	Toxicant	1.92	0.12	5.11
284-D	Negative	Toxicant	2.09	0.12	4.35
234	Negative	Non-toxicant	48.19	36.35	5.89

Table 2. Cont.

Compound		Mutagenicity (AMES test)	Developmental/ Reproductive Toxicity	F. Minnow LC50 (96 h) (mg L^{-1})	D. Magna LC50 (48 h) (mg L^{-1})	Fish Acute LC50 (48 h) (mg L^{-1})
258	[structure]	Negative	Non-Toxic	4.97	0.73	5.89

Fathead minnow LC50 (96 h) test represents the concentration that results in the mortality of half of the fish population (*Pimephales promelas*) in 96 h. The software-based on EPA model estimates similar toxicity for the drug and TPs resulting from the drug mono-hydroxylation and the corresponding dehydrogenated species, except for the species resulting from the addition of a hydroxyl group onto the ethylene bridge (294-B and 292-E), which were estimated 3/4 times less toxic. Among the di-hydroxylated species, those resulting from addition onto the ethylene bridge were estimated as significantly less toxic, comparing with the one resulting from hydroxylation onto the phenyl ring (310-E). Among all TPs, only the specie 284-A was estimated as significantly toxicant, with an LC50 value much lower (0.47 mg L^{-1}) than that of maprotiline (2.61 mg L^{-1}).

The predicted lethal concentration dose for half of the population of Daphnia Magna within 48 h (EPA model) is lower for maprotiline (0.08 mg L^{-1}) than for all the tested TPs (ranging from 0.12 to 36.35 mg L^{-1}). Even if less toxic, the species with m/z 294, 292, 284, and 258 were considered as toxicants while those resulting from dihydroxylation were estimated as slightly toxic. Likewise, the species with m/z 234 was estimated as having a significantly lower toxic effect. A similar trend was estimated for fish acute lethal dose concentration where the TPs are predicted as less toxicant (LC50 between 1.58 and 8.00 mg L^{-1}) than the parent compound (LC50 estimated at 1.29 mg L^{-1}). As observed before, TPS resulting from the ring-opening and addition of multiple hydroxyl groups were estimated as less toxicant, i.e., m/z 310 isomers, particularly those resulting from the bridge hydroxylation.

3. Discussion

As evidenced in Figure 1, semiconductor catalysts proved to be the most efficient for maprotiline removal. No significant difference was observed among the investigated semiconductors, with a slightly faster removal efficiency observed for TiO$_2$. These results contrast with the substantially higher efficiency of Ce-ZnO materials compared to TiO$_2$ reported for the degradation of several x-ray contrast agents [42]. Additionally, it is worth noting that no considerable differences in the TPs evolution over time were observed due to the semiconductor applied, as evidenced in Figure S1 (Supplementary Materials). This can be justified by the fact that all those photocatalysts have a similar mechanism for the generation of reactive species. The photoexcitation mechanisms of TiO$_2$ and ZnO based photocatalysts are characterized by the formation of HO$^\bullet$ radicals, that can then react with organic pollutants [43].

As shown in Figure 1, the heterogeneous photo-Fenton catalyst in the presence of persulfate showed a slightly lower efficiency towards maprotiline removal than hydrogen peroxide. In previous studies, Fe$_3$O$_4$/HA demonstrated high efficiency for water pollutants removal through the catalytic activation of hydrogen peroxide and persulfate [29]. By the addition of isopropanol, a selective HO$^\bullet$ scavenger, to the hydrogen peroxide system the strong inhibition in the degradation rate pointed out the HO$^\bullet$ radical as the main responsible during the oxidation process. When in the presence of persulfate, the addition of radical scavengers allowed to confirm the SO$_4^{\bullet-}$ as responsible for the degradation process.

The lower reduction potential of persulfate radical ($E^o_{SO_4^{\bullet-}/SO_4^{2-}} = 2.4$ V) [44,45] to react with organic pollutants compared to that hydroxyl radical ($E^o_{HO^\bullet + H^+/H_2O} = 2.74$ V), [45] could be the explanation for the slower degradation rate. Moreover, the reaction of HO$^\bullet$ with organic pollutants mainly involves: (i) Hydrogen atom abstraction, reaction possible for the oxidation of saturated hydrocarbons and (ii) radical addition to double bonds or aromatic rings, usually reacting very fast, sometimes controlled by the diffusion. For sulfate radical, one-electron-reactions are favored—one electron is transferred from the pollutant to SO$_4^{\bullet-}$ generating the molecular radical and sulfate ion (Equation (1), followed by pollutants hydroxylation with water or OH$^-$ [46,47]:

$$R + SO_4^{\bullet-} \rightarrow R^{\bullet+} + SO_4^{2-} \tag{1}$$

The electron transfer reaction is easier with pollutants that have electron-donating groups and less likely for those with electron-withdrawing, such as aromatic compounds [48]. This point, together with the H abstraction reaction and the addition reaction may explain the faster maprotiline degradation for HO$^\bullet$ mediated processes in comparison with those using SO$_4^{\bullet-}$.

Yang et al. reported a slower ibuprofen degradation kinetics using SO$_4^{\bullet-}$ instead of HO$^\bullet$; a theoretical study attributed this effect to the steric hindrance of SO$_4^{\bullet-}$, for which the reaction process presents energy barriers significantly higher than that involving HO$^\bullet$ [49]. These intrinsic differences between SO$_4^{\bullet-}$ and HO$^\bullet$ could also explain the differences in TPs formation. In the case of the persulfate system, the formation of only one m/z 292 isomer out of five, issued from H$_2$ elimination from m/z 294 photoproducts, could be justified by the steric hindrance of SO$_4^{\bullet-}$. The fact that only the specie 292-E, which specifically results from H$_2$ elimination from the alkyl chain is in good agreement with this hypothesis.

As a consequence of the lower efficiency of the persulfate system, one could expect fewer TPs resulting from multi-hydroxylation but this was not the case. In fact, the SO$_4^{\bullet-}$ seems to promote the formation of more multi-hydroxylated isomers at m/z 324 rather than the HO$^\bullet$ mediated processes.

From the in-silico toxicity predictions shown in Table 2 and better evidenced by the frequency as a function of Log (LC$_{50}$) presented in Figure 2, all the TPs were estimated with similar or lower toxicant properties of the antidepressant drug for both *Daphnia magna* and *Fish acute*. In the case of *Fathead minnow*, the species 284-A, which carries two aldehyde functions, was predicted as significantly more toxicant than all other TPs. Based on the predicted LC$_{50}$ values, TPs can be dispatched into the 4 categories of concern defined by the EPA (Environmental Protection Agency): category 1 (LC$_{50}$ < 0.1 mg L^{-1}), category 2 (0.1 < LC$_{50}$ < 1 mg L^{-1}), category 3 (1 < LC$_{50}$ < 10 mg L^{-1}) and category 4 (10 < LC$_{50}$ < 100 mg L^{-1}), [50]. The frequency distribution of TPs, shown in Figure 2, made it clear that potentially higher toxicant properties of several intermediates comparing with the parent compound. These results are in agreement with those previously observed by the *vibrio fischeri* toxicity test for the mixture obtained during the maprotiline degradation in the presence of TiO$_2$ photocatalyst [36]. It was reported an increase in the bioluminescence inhibition of the target organism during the first degradation stage attributed to the slightly more toxic TPs.

A correlation was found between the number of hydroxyl groups and the predicted toxicity, e.g., dehydroxylated isomers of TPs with m/z at 310 are estimated with higher lethal dose concentrations than the mono-hydroxylated species (m/z 294). Lower toxicant properties were estimated for TPs resulting from hydroxylation onto the bridge rather than onto aromatic rings, e.g., 294-B seems less toxic than the others m/z 294 isomers. As evidenced in Table 2, this species has been previously identified by retrospective analysis of wastewater treatment plant effluents resulting from current remediation processes [36]. Additionally, the species resulting from ring-opening (m/z 284, 234, and 258), observed with higher persistence for degradation mediated by SO$_4^{\bullet-}$, showed significantly lower toxicant effect than maprotiline for all the different endpoint indicators, except for 284-A. It should be pointed out that even if the m/z 234 was still present at the end of the irradiation time, considering its lower toxicant properties, no significant harmful impacts can be anticipated by using persulfate rather than other oxidative processes. Notably, this species was also reported in the wastewater treatment plant effluents (Table 2).

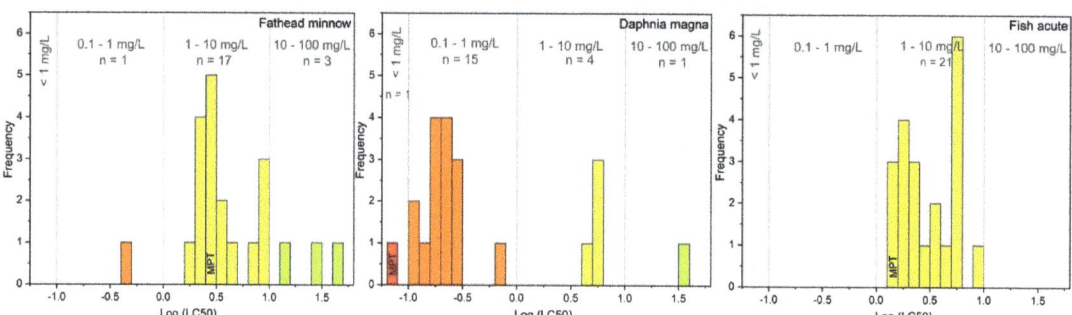

Figure 2. Distribution of TPs estimated Log (LC_{50}) thresholds (mg L^{-1}), based on EPA acute aquatic concerning categories.

Even if the estimated lethal dose toxicity for maprotiline and respective TPs was in the mg L^{-1} concentration range—considerably higher than what is expected to be released in the environment—studies with environmentally relevant concentrations are strongly needed to understand the potentially harmful effects of those substances on aquatic biota. For example, fluoxetine (another antidepressant drug) was estimated with acute toxicity LC_{50} = 5.91 mg L^{-1} on *Daphnia magna* [51]. However, it has been reported to alter the swimming behavior and camouflage efficiency of invertebrates even at very low concentrations (1–100 ng L^{-1}) [52,53]. Additionally, Weinberger et al., at the same concentration range observed a significant impact on mating, defensive, aggressive and isolation behaviors of fish fathead minnow [54]. These issues raise the importance of the wastewater treatment plants to efficiently removing the unregulated contaminants of emerging concern but also their degradation products. Also, it is of high interest to extend the studies on mixture effects as different contaminants and TPs can have a synergistic effect in terms of toxicity.

4. Materials and Methods

4.1. Chemicals

Maprotiline hydrochloride (CAS: 10347-81-6), was purchased from Sigma-Aldrich (Milan, Italy). All precursors and reagents used for the synthesis of materials and subsequent analyzes were purchased from Sigma-Aldrich (Milan, Italy) too. $FeCl_3 \cdot 6H_2O$ and $FeSO_4 \cdot 7H_2O$ were purchased from Carlo Erba Reagents (Milan, Italy); humic acid sodium salts (technical, 50–60% as HA) from Aldrich-Chemie. Acetonitrile and formic acid were purchased from Sigma Aldrich (Saint-Quentin-Fallavier, France). TiO_2 (Evonik P25, Frankfurt, Germany), to avoid possible interference from ions adsorbed on the photocatalyst, was irradiated and washed with distilled water, until there were no detectable signals due to chloride, sulfate, and sodium ions.

4.2. Preparation of Catalysts

4.2.1. Fe(0.5%)ZnO and Ce(1%)ZnO

Zinc oxide materials doped with iron and cerium were prepared by the hydrothermal method previously described [25,42]. In detail, the doping agent, $FeCl_3$ or $CeCl_3 \cdot 7H_2O$, was added in the desired molar concentration to a 1 M solution of $Zn(NO_3)_2 \cdot 6H_2O$ (20 mL) under continuous stirring. A solution of NaOH 4 M was added dropwise until reaching a pH value of 10–11. The solution was transferred into a 100 mL Teflon-lined autoclave, completed with water up to 50 mL, and kept at 175 °C for 15 h. The resulting precipitate was washed with water and recovered by centrifugation (6000 rpm for 10 min), repeating the procedure 2 times. After drying at 70 °C overnight, the powders were homogenized using a mortar and pestle.

4.2.2. Fe_3O_4/HA

Humic coated magnetite particles were prepared by co-precipitation under nitrogen atmosphere as previously reported [28]. Briefly, 35 mL of a solution of $FeCl_3 \cdot 6H_2O$ 0.68 M and $FeSO_4 \cdot 7H_2O$ 0.43 M (molar ratio Fe(III)/Fe(II) = 1.5) were added to 65 mL of deoxygenated water at 90 °C, under vigorous mechanic stirring and N_2 continuous flow. 10 mL of ammonium hydroxide (25%) and 50 mL of a 0.5 wt. % HA solution were added rapidly and sequentially. The reaction was kept for 30 min at 90 °C, then cooled down to room temperature under continuous nitrogen flow. The obtained magnetite particles (Fe_3O_4/0.5 HA) were centrifuged and washed four times with 40 mL of water; they were then dried in a tube furnace under nitrogen flow at 80 °C for 15 h and manually crumbled.

4.3. Methods

4.3.1. Photodegradation Experiments under UVA Irradiation

The photocatalytic experiments were carried out in Pyrex glass cells kept under magnetic stirring and containing 5 mL of maprotiline (5 mg L^{-1}) and catalyst (100 mg L^{-1}) solution at room temperature. Samples were irradiated for different times up to 60 min using a Cleo 6x15 W TL-D Actinic BL (290–400 nm range, 90 ± 2 W m^{-2}) lamp (Philips, Milan, Italy) with maximum emission at 365 nm, measured with a CO.FO.ME.GRA (Milan, Italy) power meter.

The photo-Fenton-like experiments were performed in the previously optimized conditions [28]. In detail, the hydrogen peroxide (1.0 mM) was added to the suspension of Fe_3O_4/HA (100 mg L^{-1}) adjusted at pH 3, while sodium persulfate (1.0 mM) was added to the suspension at pH 6. The pH was adjusted using H_2SO_4. After irradiation, 0.33 mL of methanol was added to the 5 mL of the suspension to quench the thermal Fenton reaction. All the samples were filtered using a 0.45 µm hydrophilic PTFE filter (Minisart® SRP15 Syringe Filter 17574, Sartorius, Göettingen, Germany). An equilibration time of 30 min was always applied to follow possible adsorption and in-dark degradation.

4.3.2. Analysis

LC-MS and LC-MS^n measurements were performed on an Acquity HPLC system (Water Technologies, Guyancourt, France) coupled with a Bruker SolarixXR FT-ICR 9.4 T MS instrument (Bruker Daltonics, Bremen, Germany). An Agilent Pursuit XRs^{ULTRA} C18 (length 50 mm, diameter 2 mm, particle size 2.8 µm) column was used with an Agilent HPLC MetaGuard (Pursuit XRs C18, 3 µm, 2 mm) guard column (Agilent Technologies, Les Ulis, France). The flow was set to 0.2 mL/min, the total run time was 22 min with an acquisition time of 15 min. Gradient flow was used with solvent A being water with 0.1% formic acid and solvent B being acetonitrile with 0.1% formic acid. The LC method consisted of holding 95% of solvent A and 5% of solvent B for 3 min, after which a gradient of 9 min was applied until the ratio of the two solvents reached 50–50%. The ratio of solvent A was promptly decreased to 5% and kept at this ratio until 17.1 min total run time, after it was reset to the initial 95%. After every measurement, 5 min were left for equilibration and washing at this latter solvent ratio.

The injection volume was 2 µL and 10 µL for LC-MS and LC-MS^n experiments, respectively. The sample manager was kept at 4 °C for better preservation. Each sample was prepared for the analysis by adding 10% volume of a mixture of acetonitrile/formic acid (0.1%). Electrospray ionization was used as the ion source in positive mode with a sample flow of 0.2 mL/min. The capillary voltage was 4000 V and the spray shield was set at −500 V. Nitrogen was used as nebulizer gas (1 bar) and drying gas (8 L/min, 250 °C). The detection range was 57.7–1000 m/z, in broadband mode, with a data acquisition size of 4 Mpts and a data reduction of noise of 97%. For MS^2 experiments, isolation was carried out with a 1 m/z window and collision-induced dissociation experiments were performed with energies of 5, 10, 15, and 20 V. Preliminary in-cell experiments were performed to assess the optimal parameters (isolation window, quadrupole voltage, and collision energy) for the MS^3 experiments. Measurements were also completed on the FT-ICR using direct infusion. In this case, the flow rate was 120 µL/h, with an acquisition of 4 Mpts, using the same m/z range as for LC-MS measurements. The ion accumulation time was 0.1 s and 100 scans were recorded for each spectrum. The Bruker DataAnalysis software was used for data processing.

Dealing with HRMS complex data sets is generally tricky, mainly because of (i) the experimental errors and uncertainties induced during sample preparation and analysis, (ii) the intrinsic variability of samples, (iii) the operator's subjectivity when facing mass spectra, including hundreds or thousands of ions. That is why the SPIX software has been recently developed, which is capable of analyzing a set of mass spectra at once, taking into account the intrinsic variability of the samples and that induced by HRMS measurements [39]. One feature of SPIX is to extract from a set of complex spectra all the ions that undergo significant change over time during a given process; it carries out statistical calculations aiming at fitting the evolution of an ion abundance to one of the exponential kinetic models contained in a library or a polynomial model which fits best the evolution of the ion. This permits to reveal the decrease of a parent compound, as well as the appearance and eventual disappearance of intermediates and products. The software attributes an R^2 value for the goodness of the fitting, as well as a statistical significance value (p-value).

In the present study, 5 measurements were performed over time, for each catalytic system; SPIX was run with the mass spectra data set of each system, in order to monitor maprotiline photodegradation and reveal intermediates and products that are statistically relevant considering experimental uncertainties. Experiments were performed using positive direct infusion MS, with minimal a priori sample preparation.

4.3.3. In-Silico Bioassays

Preliminary ecotoxicity assessments were performed using the VEGA QSAR software (website: https://www.vegahub.eu/, version 1.1.5 beta 22, accessed on 25 November 2020) developed by Politecnico di Milano, Italy [55]. VEGA is based on quantitative structure-activity relationship model platforms, which provide detailed information and analysis to support a toxicity prediction from different models. The models on VEGA platform were built following the OECD principles for acceptance of QSAR models for regulatory use.

The endpoint for each individual model is assessed based on a scale of reliability. Mutagenicity (AMES test) was predicted by CONSENSUS model supported by four different models (CAESAR—version 2.1.13; SarPy—version 1.0.7; ISS—version 1.0.2 and KNN—version 1.0.0). Developmental/Reproductive Toxicity models were estimated by the library (PG) (version 1.1.0). The tests for *Fathead minnow* LC_{50} (96 h) and *Daphnia magna* LC_{50} (48 h) were predicted by EPA model (version 1.0.7) based on T.E.S.T. software developed by US EPA [52]. The quantitative model for Fish acute toxicity (LC_{50}) was performed by the toxicity IRFMN model (version 1.0.0).

5. Conclusions

Semiconductor oxides proved to be more efficient for maprotiline removal than photo-Fenton-like processes. The degradation trends followed the order: TiO_2 > Ce-ZnO > Fe-ZnO > Fe_3O_4/HA + H_2O_2 > Fe_3O_4/HA + $S_2O_4^{2-}$. The degradation process mediated by $SO_4^{\bullet-}$ led to slower degradation and slower intermediates disappearance in comparison with $OH^\bullet$ mediated processes. The in-house developed SPIX software tackled the subjectivity, finding those with statistically significant change throughout the experimental time. A total of thirty-six TPs were identified using LC-HRMS; they result from drug multi-hydroxylation, oxidation, and ring-opening. The application of different AOPs demonstrated to have an impact on the formation of diverse TPs attributed mainly to the oxidation species involved. In-silico bioassays estimated similar or lower toxicity for the majority of the investigated TPs compared to the parent compound, finding no correlation between the applied oxidation process and the toxicity.

Supplementary Materials: The following are available online at https://www.mdpi.com/2073-4344/11/2/240/s1, Figure S1: Evolution over time of maprotiline degradation products formed in the presence of different catalysts under UVA irradiation., Table S1: Summary of SPIX software output data from maprotiline and potential TPs, Table S2: Summary of [M+H]$^+$ and their main fragments from CDI experiments, together with the empirical formula.

Author Contributions: Conceptualization, P.C., N.P.F.G. and Z.V.; software, Z.V.; validation, Z.V., E.N. and S.B.; investigation, N.P.F.G. and Z.V.; data curation, N.P.F.G. and Z.V.; writing—original draft preparation, N.P.F.G.; writing—review and editing, N.P.F.G., P.C., Z.V. and S.B.; supervision, P.C. and S.B.; funding acquisition, P.C. and S.B. All authors have read and agreed to the published version of the manuscript.

Funding: This work is part of a project that has received funding from the European Union's Horizon 2020 research and innovation program under the Marie Skłodowska-Curie grant agreement No 765860. Authors also acknowledge the financial support for experiments provided by the Region Council of Auvergne, from the "Féderation des Recherches en Environnement" through the CPER "Environment" founded by the Region Auvergne, the French government, FEDER from the European Community and from CAP 20e25 I-site project. Financial support from the National FT-ICR network (FR 3624 CNRS) for conducting the research is gratefully acknowledged.

Data Availability Statement: Not applicable.

Conflicts of Interest: The authors declare no conflict of interest.

References

1. Taheran, M.; Naghdi, M.; Brar, S.K.; Verma, M.; Surampalli, R.Y. Emerging contaminants: Here today, there tomorrow! *Environ. Nanotechnol. Monit. Manag.* **2018**, *10*, 122–126. [CrossRef]
2. Petrie, B.; Barden, R.; Kasprzyk-Hordern, B. A review on emerging contaminants in wastewaters and the environment: Current knowledge, understudied areas and recommendations for future monitoring. *Water Res.* **2014**, *72*, 3–27. [CrossRef] [PubMed]
3. Starling, M.C.V.M.; Amorim, C.C.; Leão, M.M.D. Occurrence, control and fate of contaminants of emerging concern in environmental compartments in Brazil. *J. Hazard. Mater.* **2019**, *372*, 17–36. [CrossRef]
4. Ebele, A.J.; Abdallah, M.E.; Harrad, S. Pharmaceuticals and personal care products (PPCPs) in the freshwater aquatic environment. *Emerg. Contam.* **2017**, *3*, 1–16. [CrossRef]
5. Das, S.; Ray, N.M.; Wan, J.; Khan, A.; Chakraborty, T.; Ray, M.B. Micropollutants in Wastewater: Fate and Removal Processes. In *Physico-Chemical Wastewater Treatment and Resource Recovery*; In Tech: Rijeka, Croatia, 2017; pp. 75–107.
6. Lindberg, R.H.; Östman, M.; Olofsson, U.; Grabic, R.; Fick, J. Occurrence and behaviour of 105 active pharmaceutical ingredients in sewage waters of a municipal sewer collection system. *Water Res.* **2014**, *58*, 221–229. [CrossRef] [PubMed]
7. Nguyen, T.T.; Westerhoff, P.K. Drinking water vulnerability in less-populated communities in Texas to wastewater-derived contaminants. *NPJ Clean Water* **2019**, *2*, 1–9. [CrossRef]
8. Abbing-Karahagopian, V.; Huerta, C.; Souverein, P.C.; De Abajo, F.; Leufkens, H.G.M.; Slattery, J.; Alvarez, Y.; Miret, M.; Gil, M.; Oliva, B.; et al. Antidepressant prescribing in five European countries: Application of common definitions to assess the prevalence, clinical observations, and methodological implications. *Eur. J. Clin. Pharmacol.* **2014**, *70*, 849–857. [CrossRef] [PubMed]
9. Bachmann, C.J.; Aagaard, L.; Burcu, M.; Glaeske, G.; Kalverdijk, L.J.; Petersen, I.; Schuiling-Veninga, C.C.; Wijlaars, L.; Zito, J.M.; Hoffmman, F. Trends and patterns of antidepressantuse in children and adolescents fromfivewestern countries, 2005–2012. *Eur. Neuropsychopharmacol.* **2016**, *26*, 411–419. [CrossRef]
10. Ford, A.T.; Herrera, H. 'Prescribing' psychotropic medication to our rivers and estuaries. *BJ Psych. Bull.* **2019**, *43*, 147–150. [CrossRef]
11. Vandermeersch, G.; Lourenço, H.M.; Alvarez-Muñoz, D.; Cunha, S.; Diogène, J.; Cano-Sancho, G.; Sloth, J.J.; Kwadijk, C.; Barcelo, D.; Allegaert, W.; et al. Environmental contaminants of emerging concern in seafood—European database on contaminant levels. *Environ. Res.* **2015**, *143*, 29–45. [CrossRef]
12. Sehonova, P.; Svobodova, Z.; Dolezelova, P.; Vosmerova, P.; Faggio, C. Effects of waterborne antidepressants on non-target animals living in the aquatic environment: A review. *Sci. Total Environ.* **2018**, *631–632*, 789–794. [CrossRef] [PubMed]
13. Martin, J.M.; Saaristo, M.; Tan, H.; Bertram, M.G.; Nagarajan-Radha, V.; Dowling, D.K.; Wong, B.B.M. Field-realistic antidepressant exposure disrupts group foraging dynamics in mosquitofish. *Biol. Lett.* **2019**, *15*, 20190615. [CrossRef]

14. Loos, R.; Carvalho, R.; António, D.C.; Comero, S.; Locoro, G.; Tavazzi, S.; Paracchini, B.; Ghiani, M.; Lettieri, T.; Blaha, L.; et al. EU-wide monitoring survey on emerging polar organic contaminants in wastewater treatment plant effluents. *Water Res.* **2013**, *47*, 6475–6487. [CrossRef] [PubMed]
15. Baresel, C.; Cousins, A.P.; Ek, M.; Ejhed, H.; Allard, A.-S.; Magnér, J.; Westling, K.; Fortkamp, U.; Wahlberg, C.; Hörsing, M.; et al. Pharmaceutical residues and other emerging substances in the effluent of sewage treatment plants. Review on concentrations, quantification, behaviour, and removal options. *Swed. Environ. Res. Inst. Rep. B* **2015**, 2226.
16. UNESCO; HELCOM. *Pharmaceuticals in the Aquatic Environment of the Baltic Sea Region—A Status Report. UNESCO Emerging Pollutants in Water Series—No. 1*; Minna, P.S.Z., Ed.; UNESCO Publishing: Paris, France, 2017; ISBN 9789231002137.
17. Deng, Y.; Zhao, R. Advanced Oxidation Processes (AOPs) in Wastewater Treatment. *Curr. Pollut. Rep.* **2015**, *1*, 167–176. [CrossRef]
18. Zhang, M.; Dong, H.; Zhao, L.; Wang, D.; Meng, D. A review on Fenton process for organic wastewater treatment based on optimization perspective. *Sci. Total Environ.* **2019**, *670*, 110–121. [CrossRef] [PubMed]
19. Bokare, A.D.; Choi, W. Review of iron-free Fenton-like systems for activating H_2O_2 in advanced oxidation processes. *J. Hazard. Mater.* **2014**, *275*, 121–135. [CrossRef]
20. Demarchis, L.; Minella, M.; Nisticò, R.; Maurino, V.; Minero, C.; Vione, D. Photo-Fenton reaction in the presence of morphologically controlled hematite as iron source. *J. Photochem. Photobiol. A Chem.* **2015**, *307–308*, 99–107. [CrossRef]
21. Nosaka, Y.; Nosaka, A. Understanding Hydroxyl Radical (•OH) Generation Processes in Photocatalysis. *ACS Energy Lett.* **2016**, *1*, 356–359. [CrossRef]
22. Nosaka, Y.; Nosaka, A.Y. Generation and Detection of Reactive Oxygen Species in Photocatalysis. *Chem. Rev.* **2017**, *117*, 11302–11336. [CrossRef] [PubMed]
23. Barakat, M.A.; Kumar, R. *Photocatalytic Activity Enhancement of Titanium Dioxide Nanoparticles*; Springer International Publishing: Cham, Switzerland, 2016; ISBN 978-3-319-24269-9.
24. Hu, K.; Li, R.; Ye, C.; Wang, A.; Wei, W.; Hu, D.; Qiu, R.; Yan, K. Facile synthesis of Z-scheme composite of TiO2 nanorod/g-C3N4 nanosheet efficient for photocatalytic degradation of ciprofloxacin. *J. Clean Prod.* **2020**, *253*, 120055. [CrossRef]
25. Kumar, S.G.; Rao, K.S.R.K. Zinc oxide based photocatalysis: Tailoring surface-bulk structure and related interfacial charge carrier dynamics for better environmental applications. *RSC Adv.* **2015**, *5*, 3306–3351. [CrossRef]
26. Paganini, M.C.; Giorgini, A.; Gonçalves, N.P.F.; Gionco, C.; Bianco Prevot, A.; Calza, P. New insight into zinc oxide doped with iron and its exploitation to pollutants abatement. *Catal. Today* **2019**, *328*, 230–234. [CrossRef]
27. Gionco, C.; Paganini, M.C.; Giamello, E.; Burgess, R.; Di Valentin, C.; Pacchioni, G. Cerium-doped zirconium dioxide, a visible-light-sensitive photoactive material of third generation. *J. Phys. Chem. Lett.* **2014**, *5*, 447–451. [CrossRef]
28. Gonçalves, N.P.F.; Minella, M.; Fabbri, D.; Calza, P.; Malitesta, C.; Mazzotta, E.; Bianco, A. Humic acid coated magnetic particles as highly efficient heterogeneous photo-Fenton materials for wastewater treatments. *Chem. Eng. J.* **2020**, *390*, 124619. [CrossRef]
29. Gonçalves, N.P.F.; Minella, M.; Mailhot, G.; Brigante, M.; Bianco Prevot, A. Photo-activation of persulfate and hydrogen peroxide by humic acid coated magnetic particles for Bisphenol A degradation. *Catal. Today* **2019**, in press.
30. Cory, W.C.; Welch, A.M.; Ramirez, J.N.; Rein, L.C. Naproxen and Its Phototransformation Products: Persistence and Ecotoxicity to Toad Tadpoles (Anaxyrus terrestris), Individually and in Mixtures. *Environ. Toxicol. Chem.* **2019**, *38*, 2008–2019. [CrossRef] [PubMed]
31. Ellepola, N.; Ogas, T.; Turner, D.N.; Gurung, R.; Maldonado-Torres, S.; Tello-Aburto, R.; Patidar, P.L.; Rogelj, S.; Piyasena, M.E.; Rubasinghege, G. A toxicological study on photo-degradation products of environmental ibuprofen: Ecological and human health implications. *Ecotoxicol. Environ. Saf.* **2020**, *188*, 109892. [CrossRef] [PubMed]
32. Boxall, A.B.A.; Sinclair, C.J.; Fenner, K.; Kolpin, D.; Maund, S.J. When Synthetic Chemicals Degrade in the Environment. *Environ. Sci. Technol.* **2004**, *70*, 368–375. [CrossRef]
33. Salimi, M.; Esrafili, A.; Gholami, M.; Jonidi, J.A.; Rezaei, K.R.; Farzadkia, M.; Kermani, M.; Sobhi, H.R. Contaminants of emerging concern: A review of new approach in AOP technologies. *Environ. Monit. Assess.* **2017**, *189*, 414–436. [CrossRef] [PubMed]
34. Chong, M.N.; Jin, B.; Chow, C.W.K.; Saint, C. Recent developments in photocatalytic water treatment technology: A review. *Water Res.* **2010**, *44*, 2997–3027. [CrossRef]
35. Xu, X.; Pliego, G.; Zazo, J.A.; Liu, S.; Casas, J.A.; Rodriguez, J.J. Two-step persulfate and Fenton oxidation of naphthenic acids in water. *J. Chem. Technol. Biotechnol.* **2018**, *93*, 2262–2270. [CrossRef]
36. Gonçalves, N.P.F.; Varga, Z.; Bouchonnet, S.; Dulio, V.; Alygizakis, N.; Bello, F.D.; Medana, C.; Calza, P. Study of the photoinduced transformations of maprotiline in river water using liquid chromatography high-resolution mass spectrometry. *Sci. Total Environ.* **2020**, *755*, 143556. [CrossRef]
37. Suh, S.K.; Smith, J.B. Maprotiline Hydrochloride. *Anal. Profiles Drug Subst. Excip.* **1986**, *15*, 393–426.
38. Blum, K.M.; Norström, S.H.; Golovko, O.; Grabic, R.; Järhult, J.D.; Koba, O.; Söderström Lindström, H. Removal of 30 active pharmaceutical ingredients in surface water under long-term artificial UV irradiation. *Chemosphere* **2017**, *176*, 175–182. [CrossRef] [PubMed]
39. Nicol, E.; Xu, Y.; Varga, Z.; Kinani, S.; Bouchonnet, S.; Lavielle, M. SPIX: A new software package to reveal chemical reactions at trace amounts in very complex mixtures from high-resolution mass spectra data sets. *Rapid Commun. Mass Spectrom.* **2020**, *35*, e9015.
40. Wells, B.G.; Pharm, D.; Gelenberg, A.J. Adverse Effects, and Efficacy of the Antidepressant Maprotiline Hydrochloride. *Pharmacother. J. Hum. Pharmacol. Drug Ther.* **1981**, *1*, 121–138. [CrossRef]

41. Wu, S.; Fisher, J.; Naciff, J.; Laufersweiler, M.; Lester, C.; Daston, G.; Blackburn, K. Framework for identifying chemicals with structural features associated with the potential to act as developmental or reproductive toxicants. *Chem. Res. Toxicol.* **2013**, *26*, 1840–1861. [CrossRef]
42. Paganini, M.C.; Dalmasso, D.; Gionco, C.; Polliotto, V.; Mantilleri, L.; Calza, P. Beyond TiO2: Cerium-Doped Zinc Oxide as a New Photocatalyst for the Photodegradation of Persistent Pollutants. *ChemistrySelect* **2016**, *1*, 3377–3383. [CrossRef]
43. Hernández-Ramírez, A.; Medina-Ramírez, I. *Photocatalytic Semiconductors*; Springer International Publishing: Cham, Switzerland, 2015; ISBN 978-3-319-10998-5.
44. Huie, R.E.; Clifton, C.L.; Neta, P. Electron tranfer reactions rates and equilibria of the carbonate and sulfate radical anions. *Radiat. Phys. Chem.* **1991**, *38*, 477–481.
45. Gligorovski, S.; Strekowski, R.; Barbati, S.; Vione, D. Environmental Implications of Hydroxyl Radicals (•OH). *Chem. Rev.* **2015**, *115*, 13051–13092. [CrossRef]
46. Neta, P.; Madhavan, V.; Zemel, H.; Fessenden, R.W. Rate Constants and Mechanism of Reaction of SO4 with Aromatic Compounds. *J. Am. Chem. Soc.* **1977**, *99*, 163–164. [CrossRef]
47. Matzek, L.W.; Carter, K.E. Activated persulfate for organic chemical degradation: A review. *Chemosphere* **2016**, *151*, 178–188. [CrossRef]
48. Ross, A.B.; Neta, P. Rate Constants for Reactions of Inorganic Radicals in Aqueous Solution. *Natl. Bur. Stand. Natl. Stand. Ref. Data Ser.* **1979**, *17*, 1027–1038.
49. Yang, Z.; Su, R.; Luo, S.; Spinney, R.; Cai, M.; Xiao, R.; Wei, Z. Comparison of the reactivity of ibuprofen with sulfate and hydroxyl radicals: An experimental and theoretical study. *Sci. Total Environ.* **2017**, *590–591*, 751–760. [CrossRef]
50. Technical Overview of Ecological Risk Assessment—Analysis Phase: Ecological Effects Characterization | Pesticide Science and Assessing Pesticide Risks | US EPA. Available online: https://www.epa.gov/pesticide-science-and-assessing-pesticide-risks/technical-overview-ecological-risk-assessment-0 (accessed on 25 November 2020).
51. Minguez, L.; Farcy, E.; Ballandonne, C.; Lepailleur, A.; Serpentini, A.; Lebel, J.M.; Bureau, R.; Halm-Lemeille, M.P. Acute toxicity of 8 antidepressants: What are their modes of action? *Chemosphere* **2014**, *108*, 314–319. [CrossRef] [PubMed]
52. Yakubu, A.M. Determination of lindane and its metabolites by HPLC-UV-Vis and MALDI-TOF. *J. Clin. Toxicol.* **2012**, *71*, 310–319. [CrossRef]
53. Guler, Y.; Ford, A.T. Anti-depressants make amphipods see the light. *Aquat. Toxicol.* **2010**, *99*, 397–404. [CrossRef] [PubMed]
54. Weinberger, J.; Klaper, R. Environmental concentrations of the selective serotonin reuptake inhibitor fluoxetine impact specific behaviors involved in reproduction, feeding and predator avoidance in the fish Pimephales promelas (fathead minnow). *Aquat. Toxicol.* **2014**, *151*, 77–83. [CrossRef]
55. Benfenati, E.; Manganaro, A.; Gini, G. VEGA-QSAR: AI inside a platform for predictive toxicology. *CEUR Workshop Proc.* **2013**, *1107*, 21–28.

Article

Assessing an Integral Treatment for Landfill Leachate Reverse Osmosis Concentrate

Javier Tejera [1], Daphne Hermosilla [2,3,*], Ruben Miranda [1], Antonio Gascó [2], Víctor Alonso [4], Carlos Negro [1] and Ángeles Blanco [1,*]

[1] Department of Chemical Engineering and Materials, Chemistry Science Faculty, Ciudad Universitaria s/n, Complutense University of Madrid, 28040 Madrid, Spain; jttejo@ucm.es (J.T.); rmiranda@ucm.es (R.M.); cnegro@ucm.es (C.N.)

[2] Department of Forest and Environmental Engineering and Management, Universidad Politécnica de Madrid, E.T.S.I. Montes, Forestal y del Medio Natural, C/ José Antonio Novais 10, 28040 Madrid, Spain; antonio.gasco@upm.es

[3] Department of Agricultural and Forest Engineering, University of Valladolid, EIFAB, Campus Duques de Soria, 42005 Soria, Spain

[4] Department of Applied Physics, University of Valladolid, EIFAB, Campus Duques de Soria, 42005 Soria, Spain; victor.alonso.gomez@uva.es

* Correspondence: daphne.hermosilla@upm.es (D.H.); ablanco@ucm.es (Á.B.); Tel.: +34-91-06-71657 (D.H.); +34-91-394-4247 (Á.B)

Received: 26 October 2020; Accepted: 23 November 2020; Published: 28 November 2020

Abstract: An integral treatment process for landfill leachate reverse osmosis concentrate (LLROC) is herein designed and assessed aiming to reduce organic matter content and conductivity, as well as to increase its biodegradability. The process consists of three steps. The first one is a coagulation/flocculation treatment, which best results were obtained using a dosage of 5 g L^{-1} of ferric chloride at an initial pH = 6 (removal of the 76% chemical oxygen demand (COD), 57% specific ultraviolet absorption (SUVA), and 92% color). The second step is a photo-Fenton process, which resulted in an enhanced biodegradability (i.e., the ratio between the biochemical oxygen demand (BOD$_5$) and the COD increased from 0.06 to 0.4), and an extra 43% of the COD was removed at the best trialed reaction conditions of [H$_2$O$_2$]/COD = 1.06, pH = 4 and [H$_2$O$_2$]/[Fe]$_{mol}$ = 45. An ultra violet-A light emitting diode (UVA-LED) lamp was tested and compared to conventional high-pressure mercury vapor lamps, achieving a 16% power consumption reduction. Finally, an optimized 30 g L^{-1} lime treatment was implemented, which reduced conductivity by a 43%, and the contents of sulfate, total nitrogen, chloride, and metals by 90%. Overall, the integral treatment of LLROC achieved the removal of 99.9% color, 90% COD, 90% sulfate, 90% nitrogen, 86% Al, 77% Zn, 84% Mn, 99% Mg, and 98% Si; and significantly increased biodegradability up to BOD$_5$/COD = 0.4.

Keywords: coagulation; landfill leachate; reverse osmosis concentrate; photo-Fenton; LED; lime precipitation

---

1. Introduction

Landfill leachate is an extremely harmful wastewater for the environment because it contains a high amount of organic and inorganic pollutants. Several authors have studied different treatment alternatives to manage this highly contaminant-loaded wastewater, including biological [1], physical [2,3], and chemical processes [4,5]. Nowadays, physical processes using pressure-driven membranes (microfiltration and/or ultrafiltration followed by nanofiltration or reverse osmosis) [6–8] are widely used because of the high quality of the permeate water that is obtained. However, these technologies have the drawback of producing a leachate concentrate containing the retained pollutants [9]. This concentrate is usually characterized by low biodegradability, high specific ultraviolet absorbance (SUVA) [10,11], and high concentration of inorganic and bio-recalcitrant organic compounds. Therefore, its treatment by conventional biological processes [12] is not possible, and the concentrate is recirculated back to the landfill in most of the cases [13,14]. This reinjection increases the content of persistent organic compounds, salts, nitrogen, and phosphorous in the leachate [15], which then increases membrane fouling and progressively limits the efficiency of the treatment in the medium to the long term [16].

Consequently, the implementation of complementary treatment alternatives for landfill leachate reverse osmosis concentrate (LLROC) is necessary. Some reported treatment possibilities are: Membrane distillation [17], evaporation [18], and different advanced oxidation process (AOPs), such as electro-oxidation [12,19,20], ozonation [21,22], sulfate radical-based AOPs [23,24], and Fenton processes [25–27]. For example, Qi et al. (2015) [17] optimized a vacuum distillation membrane system for LLROC, achieving a 97% chemical oxygen demand (COD) removal at a vacuum pressure of 0.08 MPa and a temperature of 75 °C for a flux of 200 L h^{-1}. However, membrane distillation requires a large investment, and the system suffers high corrosion levels because of the associated evaporation of corrosive gases through the installation. Similarly, Yang et al. (2018) [18] achieved the complete elimination of organics by a co-bio-evaporation process, mixing LLROC with food waste in the following optimum conditions: 1:1.1 LLROC:food waste ratio, and a flux of 0.035 m^3 kg^{-1} h^{-1}. 86% of the water was evaporated, but severe corrosion troubles in the equipment were predicted.

Although AOPs have been reported to have the potential advantage of increasing the biodegradability of LLROC [28] and different types of wastewater, such as, for example, textile effluents [29], the state of the art shows that the high concentration of organic compounds makes AOPs an expensive technology in terms of the reagents dosages and power consumption. In addition, the biodegradability increase may not be enough [30–33]. For example, in the case of electro-oxidation processes using boron-doped diamond electrodes, 89.6–100% of the COD was removed, but biodegradability only increased up to BOD$_5$/COD = 0.2, by the application of 50–100 mA cm^{-2} for 6 h [12,20]. Different new materials and Fenton-like processes, including the coupling with other technologies, are being developed at laboratory scale to increase treatment efficiency, and reduce sludge production and treatment cost. For example, Wang et al. (2018) [27] studied FePC amorphous alloys to increase the catalytic efficiency, and Liang et al. (2018) [24] reported the use of Fe$_{78}$Si$_9$B$_{13}$ as metallic glass to generate hydroxyl and persulfate radicals when H$_2$O$_2$, persulfate and monopersulfate were used as oxidants.

In order to reduce the cost of AOP treatments, a coagulation/flocculation pretreatment, which may even be easily applied at industrial scale, has been proposed to reduce COD, color, and UV absorbance at 254 nm [4,34,35]. UV absorbance reduction significantly enhances ultraviolet radiation penetration in the media, and, thus, the efficiency of the photo-oxidation processes. Zhou et al. (2011) [35] reported an increase in the removal of contaminants by combining coagulation with different AOPs applied to treat an urban wastewater with a total organic carbon (TOC) of 18 mg L^{-1}. In this case, 49% of the TOC was removed by combining ferric chloride coagulation with an UVA/H$_2$O$_2$ process; whereas less than 3% of the TOC was removed when the UVA/H$_2$O$_2$ treatment was applied standalone.

The use of UV radiation (photo-Fenton) to recycle the iron catalyst for the Fenton reaction and decarboxylate ferric carboxylates [36,37], might be a good alternative to save iron addition and generate less sludge, provided it is sufficiently cheap [38]. Thus, coupling ferric chloride coagulation/flocculation pretreatment with a photo-Fenton process might be a promising alternative for the treatment of LLROC. In fact, the amount of Fe^{3+} that remains in the leachate after coagulation might be used as the catalyst of the posterior photo-Fenton process. Complementarily, this treatment may increase the biodegradability of the outflowing effluent from the oxidation process, measured in terms of the ratio of the biochemical oxygen demand to the chemical oxygen demand (BOD$_5$/COD) [39]. Therefore, it might finally be biologically treated with success. Amor et al. (2015) [39] reported an 89% COD removal of a landfill leachate by combining a coagulation pre-treatment with 2 g L^{-1} of ferric chloride at pH = 5 with a solar photo-Fenton process, when just a 63% COD removal was achieved by the standalone solar photo-Fenton treatment. However, this treatment increases the conductivity of the leachate caused by the addition of iron and chloride, which would inhibit the activity of microorganisms in a biological reactor. Therefore, in this study, we considered a final lime precipitation step aiming to reduce conductivity by removing inorganics, metal ions, more organics, and ammonia nitrogen by air stripping induced by the increase of the pH, which, as a result, will enable its final biological treatment. Some authors have reported the application of lime in a first step of the treatment, or even before a reverse osmosis process [40]; but the addition of lime produced a considerable pH increase, which is a disadvantage for the photo-Fenton process because it works better under acidic conditions.

High-pressure immersion mercury vapor lamps have been widely used as the UV irradiation source for the photo-Fenton treatment step, although these lamps have a low energy efficiency, with only 20% of the consumed energy actually contributing to the emission of photons within the UV range [41], and the greater rest consumed in infrared (IR) irradiation losses. This is why other radiation sources have gained importance lately, such as light-emitting diode (LED) technology [42] or solar radiation with the aid of compound parabolic collectors (CPC) or thin parabolic collectors [43,44]. Silva et al. (2016) [44] compared an artificial UV source of radiation with CPCs, reporting that the use of CPCs was 0.4 € m^{-3} cheaper than the applied artificial UV source in the photo-Fenton treatment of 100 m^3 day^{-1} of urban landfill leachate, but they required a very large installation area and a great investment.

Based on the state of the art, the efficiency of an integrated treatment for LLROC is herein studied. It consists of three steps (namely, coagulation/flocculation, photo-Fenton, and lime addition) with the aim of accomplishing the following objectives: (1) To maximize the removal of COD, TOC, color, and SUVA by a coagulation pre-treatment; (2) to optimize the photo-Fenton oxidation step to maximize COD removal and enhance biodegradability, and to assess a range of pre-treatment options; and (3) to achieve a sufficient reduction of the conductivity of the oxidized LLROC by lime addition to enable a potential final biological treatment. In addition, the reduction of costs by using LED technology as the source of UV radiation in the Fenton process is assessed.

2. Results and Discussion

2.1. Coagulation Pretreatment, Effect of pH and Coagulant Dosage

The pH played an important role in the coagulation process. At lower pH values, the efficiency of coagulation in terms of COD removal increased, as it has previously been reported by other authors [34,39]. In particular, it is well-known that the COD of LLROC will be reduced by decreasing the pH in the medium [45] due to the precipitation of humic acids [41,46]. Unfortunately, the high alkalinity of the sampled LLROC (44,125 mg L^{-1} $CaCO_3$, Table 1) required the addition of a relatively large quantity of sulfuric acid, namely: 16, 23, and 24.5 g L^{-1} of H_2SO_4, to reduce the pH from its initial value of 8.13 (Table 1) to 7.0, 6.0, and 5.0, respectively, before performing the coagulation treatment. The amount of sulfate in the solution consequently increased as the pH was lowered, rising from an initial value of 2.4 g L^{-1} (Table 1) to 15 g L^{-1} at pH = 7; 25 g L^{-1} at pH = 6; and 28 g L^{-1} at pH = 5, respectively.

Conductivity also increased, as expected, as the pH value of the solution was set lower to increase coagulation efficiency, increasing from an initial value of 87.3 mS cm^{-1} (Table 1) to 90.3, 94.4, and 95.0 mS cm^{-1} when the pH was adjusted to 7.0, 6.0, and 5.0, respectively. This may negatively affect the efficiency of the posterior photo-Fenton treatment that is designed to be applied [47], as well as the potential biological treatment that might be performed to the oxidized LLROC afterwards.

Coagulation with ferric chloride achieved COD removals of 43% at the natural pH value of the LLROC (8.13; Table 1); 78% at pH = 7.0; 76% at pH = 6.0; and 71% at pH = 5.0; respectively (Figure 1). In addition, whereas a 30 g L^{-1} dose of $FeCl_3$ were required to optimize coagulation at pH values at or above 7.0, only 5 g L^{-1} was required at the acidic pH values (i.e., 5.0 and 6.0) because the optimum pH for coagulation with $FeCl_3$ is about 4–5 [41], and the addition of $FeCl_3$ produces a significant reduction of pH with the subsequent precipitation of humic acids. In short, decreasing the initial pH value of LLROC to acid values not only favored a higher coagulation efficiency, but it also significantly reduced the optimum coagulant dose to one-sixth of that required at neutral and basic pH values, which results in a significantly lower treatment cost and reduces chloride content.

Table 1. Initial characteristics of the landfill leachate reverse osmosis concentrate (LLROC).

Parameter [1] (LLROC)	Value [2]	Parameter (Dissolved Fraction)	Value [2]
pH	8.13 ± 0.10	Chloride, mg L^{-1}	8968 ± 897
Conductivity, mS cm^{-1}	87.30 ± 0.90	Sulfate, mg L^{-1}	2431 ± 243
UV-254, cm^{-1}	150 ± 10	Aluminum, mg L^{-1}	4.20 ± 0.60
Color, mg Pt L^{-1}	28,100 ± 900	Iron, mg L^{-1}	2.30 ± 0.30
COD, mg O_2 L^{-1}	21,220 ± 750	Chromium, mg L^{-1}	1.00 ± 0.20
BOD_5, mg O_2 L^{-1}	1273 ± 100	Sodium, mg L^{-1}	6769 ± 677
BOD_5/COD	0.06 ± 0.01	Potassium, mg L^{-1}	3157 ± 316
TOC, mg C L^{-1}	9980 ± 150	Magnesium, mg L^{-1}	245 ± 25
TS, mg L^{-1}	51,270 ± 1620	Calcium, mg L^{-1}	19 ± 2
TSS, mg L^{-1}	360 ± 32	Silicon, mg L^{-1}	29 ± 3
TDS, mg L^{-1}	50,910 ± 1230	Zinc, mg L^{-1}	0.60 ± 0.09
Alkalinity, mg $CaCO_3$ L^{-1}	44,125 ± 1023	Nickel, mg L^{-1}	0.38 ± 0.06
TN_b, mg N L^{-1}	3000 ± 150	Strontium, mg L^{-1}	1.50 ± 0.30

[1] COD (chemical oxygen demand); BOD_5 (biochemical oxygen demand); TOC (total organic carbon); TS (total solid); TSS (total suspended solid); TDS (total dissolved solid); TN_b (total nitrogen bonded). [2] Values are m ± sd.

Aluminum sulfate (alum) achieved lower maximum COD removal efficiencies than $FeCl_3$, that is: 25% at pH = 8.13; 30% at pH = 7.0; 60% at pH = 6.0; and 43% at pH = 5.0 (Figure 1). Optimal $Al_2(SO_4)_3$ dosages were also six times higher (30 g L^{-1}) at neutral to basic pH values than at acidic initial process conditions (5 g L^{-1}), as it was the case with $FeCl_3$. Therefore, acidic initial conditions also result in a lower treatment cost to coagulate LLROC more efficiently, but the drawback in this case is the increased sulfate content.

The application of $Al_2(SO_4)_3 \cdot 18H_2O$ or $FeCl_3$ under acidic conditions achieved worse COD removal treatment efficiency results at pH = 5 than at pH = 6 (17% lower COD removal in the case of alum, and only 5% in the case of $FeCl_3$; Figure 1). This effect is explained by the fact that ferric chloride has better efficiencies in COD removal at pH values around 4 [45], and at pH = 6 and 5, with a dose of 5 g L^{-1}, the final pH values after coagulation were 4.07 and 3.30, respectively. The lower final pH at initial pH = 5 reduced the efficiency of ferric chloride coagulation. In the case of alum, the optimal pH for the coagulation process is around 5.5 [48], and the coagulation with 5 g L^{-1} ended in a final pH of 4.15 and 4.99, for initial pH values of 5 and 6, respectively. So, at an initial pH of 5 and a dose of 5 g L^{-1}, the final pH is far from the optimal for alum coagulation than at initial pH = 6.

In general, the best LLROC coagulation results were achieved at a coagulant dose of 5 g L^{-1} at pH = 6.0 for both coagulants, but ferric chloride coagulation gave significantly ($p < 0.001$ ***) better removal of COD, TOC, UV-254, SUVA, and color (Table 2). The differences in COD removal (60% and 76% with alum and ferric chloride, respectively) and in TOC removal (65% with alum and 75% with ferric chloride, respectively) are important in order to reduce the cost of reagents in a posterior photo-Fenton treatment. These differences are directly related to the final pH value reached after coagulation, that is: pH = 4.99 for alum and pH = 4.07 for ferric chloride, since precipitation of humic acids is produced under pH = 5 [45].

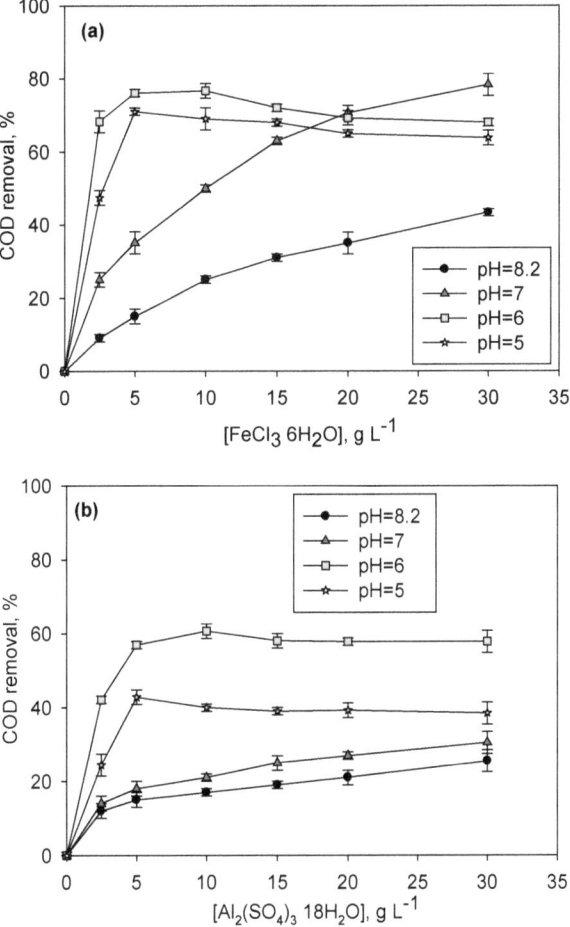

Figure 1. Coagulation of LLROC at different pH values and coagulant dosage: (**a**) Ferric chloride, and (**b**) alum (values are m ± sd).

Table 2. Characteristics of the LLROC after an optimum coagulation treatment with different coagulants ($Al_2(SO_4)_3 \cdot 18H_2O$ or $FeCl_3 \cdot 6H_2O$; pH = 6.0 and dosage of 5 g L^{-1} in both cases).

Parameters [1]	LLROC [2]	$Al_2(SO_4)_3 \cdot 18H_2O$ [2]	$FeCl_3 \cdot 6H_2O$ [2]
H_2SO_4, g L^{-1}	-	25	25
Initial coagulation pH	8.13 ± 0.10	6.00 ± 0.10	6.00 ± 0.10
Final pH	8.13 ± 0.10	4.99 ± 0.10	4.07 ± 0.10
Conductivity, mS cm^{-1}	87.30 ± 0.90	90.20 ± 0.90	92.30 ± 0.90
COD, mg O_2 L^{-1}	21,220 ± 1000	8488 ± 700 (60%) [1]	5092 ± 500 (76%)
TOC, mg C L^{-1}	9980 ± 100	3493 ± 100 (65%)	2495 ± 100 (75%)
UV-254, cm^{-1}	150 ± 10	49 ± 1 (67%)	16 ± 1 (89%)
SUVA, L mg^{-1} m^{-1}	1.50 ± 0.50	1.40 ± 0.50 (7%)	0.64 ± 0.50 (57%)
Color, mg Pt L^{-1}	28,100 ± 1000	7700 ± 400 (73%)	2160 ± 200 (92%)
BOD_5/COD	0.06 ± 0.01	0.06 ± 0.01	0.06 ± 0.01

[1] SUVA = 100·UV-254/TOC. [2] Values are mean (m) ± standard deviation (sd). Removal results are in brackets.

The efficiency of a posterior photo-Fenton treatment may be highly affected by the presence of color and 254 nm UV quenching substances, which are scavengers of this advanced oxidation reaction [10,11]. In fact, the efficiency of the photo-Fenton process is highly related to the penetration of UV radiation into the reaction media, achieving better COD removal at a lower cost in terms of energy consumed (lower reaction time thanks to a greater radiation efficiency) and H_2O_2 dosage [10,11]. In regard to these issues, ferric chloride achieved better color removal results than alum (92% vs. 73% at its best), resulting in 2160 and 7700 mg Pt L^{-1}, respectively (Table 2). In addition, UV-254 removal was also higher when ferric chloride was used as the coagulant agent (89% vs. 67% using alum), resulting in a final value of 16 cm^{-1} (Table 2). Additionally, whereas SUVA was almost not altered after coagulation with alum (only a 7% removal), it was reduced by a 57% after $FeCl_3$ coagulation.

These results are in agreement with Chen et al. (2019) [49] who achieved optimal COD removal values of 30%, 61%, 75%, and 79% after coagulating LLROC with alum, PAC, PFS, and ferrous sulfate, respectively. UV-254 removal was reported ranging from 52% to 85% following the same positive correlation with COD removal results, as it has been addressed in the present study using ferric chloride and alum. Differences among the results reported by Chen et al. (2019) [49] can be attributed to different wastewater characteristics.

As expected, there were no changes in the BOD_5/COD ratio ($\approx$0.06; Table 2) after the essayed coagulation treatments [4]. Considering the photo-Fenton process as the next treatment step in order to enhance biodegradability, we chose to investigate $FeCl_3$-coagulated LLROC. Firstly, this option gave the best coagulation results. Secondly, the residual iron should serve as the catalyst for this advanced oxidation process, thus avoiding the addition of iron from other external sources.

2.2. Photo-Fenton Treatment

The optimization of the conventional 100 W immersion mercury lamp photo-Fenton reaction was therefore just carried out for the two best essayed coagulation pre-treatments, namely those performed at pH = 6 using 5 and 10 g L^{-1} of ferric chloride. In both cases, the initial COD before the photo-Fenton treatment was 5000 ± 100 mg L^{-1}. The main differences between these cases were the initial pH of the oxidation step (pH = 4.07 and 2.75 for 5 and 10 g L^{-1} of coagulant, respectively) and the remaining dissolved iron content in the media after the coagulation pre-treatment (168 and 800 mg L^{-1} for 5 and 10 g L^{-1} of $FeCl_3 \cdot 6H_2O$, respectively). It is well known from the literature that the best conditions to perform a Fenton treatment are pH $\approx$ 2.8 and low H_2O_2/Fe concentration ratio (H_2O_2/Fe = 20), which implies adding a high iron content [50]. Nevertheless, an initial pH below 5.0 was considered possibly sufficient to obtain reasonable efficiency due to the production of organic acids through the Fenton reaction itself, which would decrease the pH to values close to the optimum [51].

The photo-Fenton treatment of previously coagulated LLROC with 5 g L^{-1} of $FeCl_3 \cdot 6H_2O$ achieved a 40% reduction of the COD under the best tested reaction conditions ($[H_2O_2]/[COD]$ = both 2.125 and 1.063; Figure 2a); whereas it resulted in 43% and 49% COD reductions at $[H_2O_2]/[COD]$ = 2.125 and 1.063, respectively, after the coagulation pre-treatment with 10 g L^{-1} of $FeCl_3 \cdot 6H_2O$ (Figure 2b). The higher efficiency achieved at higher $FeCl_3$ dose may be explained by the different initial pH value (4.07 and 2.75 for 5 and 10 g L^{-1} of $FeCl_3 \cdot 6H_2O$, respectively) and different dissolved iron concentrations present in the media at the beginning of the oxidation process (168 and 800 mg L^{-1} for 5 and 10 g L^{-1} of $FeCl_3 \cdot 6H_2O$, respectively).

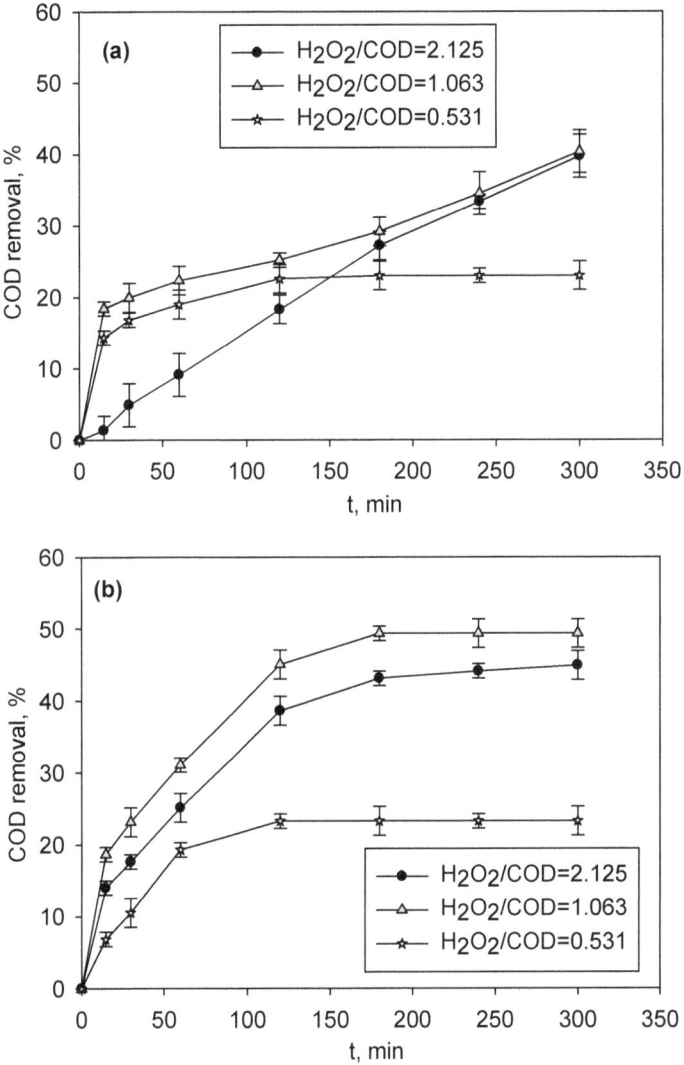

Figure 2. Chemical oxygen demand (COD) removal results of the photo-Fenton treatment of LLROC previously coagulated with ferric chloride using a 100 W UV mercury lamp. Coagulant dose: (**a**) 5 g L^{-1} and (**b**) 10 g L^{-1} (values are m ± sd).

The reported optimum pH to perform the Fenton process is about 2.8 [52], and the presence of more dissolved iron after the 10 g L^{-1} coagulation process may have sped up the process. In addition, the reaction time that was required to finish the oxidation process at an [H_2O_2]/[COD] ratio of 1.063 (i.e., zero peroxide content in the media) was 5 h for LLROC pre-coagulated with 5 g L^{-1} of FeCl$_3$·6H$_2$O, and 3 h for the 10 g L^{-1} dose. This is explained by the different [H_2O_2]/[Fe] ratio that results in different conditions. For 5 g L^{-1} of ferric chloride pre-coagulation: [H_2O_2]/[Fe] ratios of 90, 45, and 23 were achieved for [H_2O_2]/[COD] of 2.125, 1.063, and 0.531, respectively; and for 10 g L^{-1} ferric chloride pre-coagulation: [H_2O_2]/[Fe] of 19, 10, and 5 were obtained for [H_2O_2]/[COD] of 2.125, 1.063, and 0.531, respectively.

The resulting reduction of SUVA indicates that high molecular weight substances have been degraded to smaller ones [53]. SUVA values of 0.18, 0.21, and 0.33 L mg^{-1} m^{-1} were obtained for [H$_2$O$_2$]/COD = 2.125, 1.063, and 0.531, respectively, from a previous initial value of 0.64 L mg^{-1} m^{-1}, in the photo-Fenton treatment of pre-coagulated LLROC with 5 g L^{-1} of FeCl$_3$·6H$_2$O (Figure 3a). On the other hand, the photo-Fenton treatment of LLROC pre-coagulated with 10 g L^{-1} of FeCl$_3$·6H$_2$O achieved SUVA values of 0.43, 0.47, and 0.51 L mg^{-1} m^{-1} at [H$_2$O$_2$]/COD ratio values of 2.125, 1.063, and 0.531, respectively, from an initial SUVA = 0.74 (Figure 3b). After 3 h of treatment, SUVA value was constant, indicating that UV quenching substances were not being further removed.

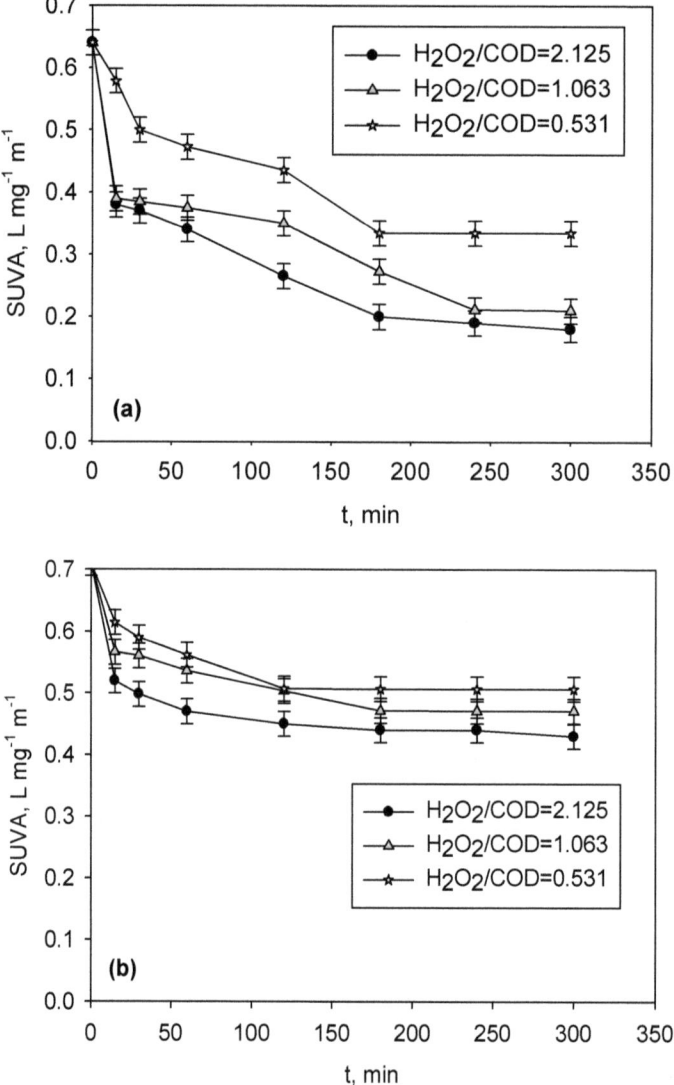

Figure 3. Specific ultraviolet absorption (SUVA) removal by the photo-Fenton treatment of pre-coagulated LLROC with: (**a**) 5 g L^{-1}, and (**b**) 10 g L^{-1} FeCl$_3$·6H$_2$O (values are m ± sd).

Biodegradability enhancement during the photo-Fenton treatment mainly occurred at the beginning of the photo-oxidation process (20 min; Figure 4a), which would enable reducing oxidation treatment time and cost in a potential combination of this treatment sequence with a final biological step. Performing a photo-Fenton treatment with an [H_2O_2]/COD ratio of 1.063 will be enough to significantly ($p < 0.001$ ***) enhance biodegradability up to $BOD_5/COD \approx 0.4$ if LLROC was previously coagulated with just 5 g L^{-1} of $FeCl_3 \cdot 6H_2O$. The higher content of iron in the media, in the case of using 10 g L^{-1} of $FeCl_3 \cdot 6H_2O$ in the coagulation step, did not significantly accelerated the oxidation process, and biodegradability was just improved up to $BOD_5/COD \approx 0.3$ (Figure 4b). This difference in the biodegradability increment can be explained by the more intensive oxidation that resulted for pre-coagulated leachate with 10 g L^{-1}, and by the fact that the final products are less biodegradable than the final products when the pre-coagulation is carried out with 5 g L^{-1}. In addition, the post-management of iron sludge will be cheaper coagulating with less amount of ferric chloride; and the added chloride to the media will be lower itself.

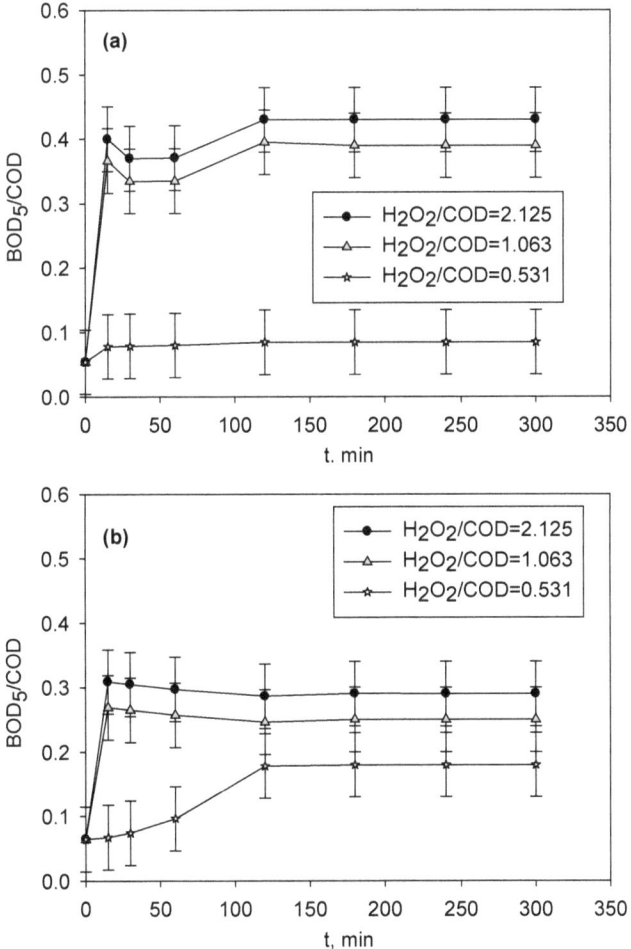

Figure 4. Biodegradability evolution along the photo-Fenton treatment of coagulated LLROC with: (a) 5 g L^{-1}, and (b) 10 g L^{-1} of $FeCl_3 \cdot 6H_2O$ (values are m ± sd).

Considering the source of radiation to perform the photo-Fenton treatment, the power consumption of immersion mercury lamps represents an important part of the cost of the treatment. For this reason, the photo-Fenton treatment using a 100 W immersion mercury lamp was compared with the use of a novel UVA-LED lamp designed to be more efficient, and, therefore, a significant reduction of the cost of this treatment is expected. As a result, COD removal and final biodegradability reached the same figures with both sources of radiation at the selected optimal conditions to perform the treatment.

The radiated energy in the UV region of the used mercury immersion lamp is reported by the manufacturer to be 15%, whereas new mercury immersion lamps account for 40% radiation emitted in the UV region; so the power consumption of the mercury lamp technology was recalculated, as included in our previous work [41], and proper technological comparison could be made. In short, the application of LED technology resulted in a significant ($p < 0.001$ ***) monetary saving of the 16.46% in terms of power consumption (as calculated from data on Table 3, which is further discussed in Section 2.4), besides lower costs of investment and disposal (environmental hazard of mercury) shall also be considered; as well as further optimization of the LED technology use might also contribute to even further reduce the final cost of treatment.

Table 3. LLROC treatment cost considering the same coagulation pre-treatment (pH = 6; coagulant dose = 5 g L^{-1} of ferric chloride) and lime precipitation post-treatment (calcium hydroxide dose = 30 g L^{-1}) but different light sources (100 W mercury-vapor immersion lamp and UVA-LED lamp) for the photo-Fenton treatment.

Treatment Costs	100 W Mercury Lamp	UVA LED Lamp
Power consumption, € m^{-3}	15.8	13.2
Hydrogen peroxide, € m^{-3}	1.9	1.9
Photo-Fenton, € m^{-3}	17.7	15.1
Pre-treatment, € m^{-3}	4.2	4.2
Post-treatment, € m^{-3}	3.9	3.9
Total, € m^{-3}	25.8	23.2
Total, € kg COD	1.4	1.2

Mean values are provided, $n = 3$; standard deviation values were lower than 1%.

2.3. Lime Treatment

Considering the results (Figure 5), 30 g L^{-1} of calcium hydroxide (lime, $Ca(OH)_2$) was selected as the optimum dosage, for which the pH reached values of 12.38 and 12.46 for the oxidized LLROC that was previously coagulated with 5 and 10 g L^{-1} of $FeCl_3 \cdot 6H_2O$, respectively. Higher doses of lime produce the dissolution of calcium, thus increasing calcium concentration in the medium, and, therefore, limiting the removal of sulfate.

Lime precipitation did not affect the biodegradability of the previously coagulated and oxidized LLROC in any tested case, although an extra 10–20% removal of COD and TOC were produced at best. All other results for this treatment step (conductivity, color, SUVA, TN_b, [Cl^-], and [SO_4^{2-}] (Figure 5)) followed the same tendency for the two coagulant doses that were used in the pre-treatment of LLROC (5 and 10 g L^{-1} of $FeCl_3 \cdot 6H_2O$) before its oxidation treatment. In both cases, SUVA was reduced down to 0.05 L mg^{-1} m^{-1}, and conductivity and color were removed by a 43% (final value ≈ 50 mS cm^{-1}) and 99.9%, respectively, from raw LLROC figures. In addition, an ammonium stripping process occurred as pH turned to basic values during lime treatment just by stirring the sample [54], achieving a 90% removal of the total nitrogen content from the wastewater in this last treatment step.

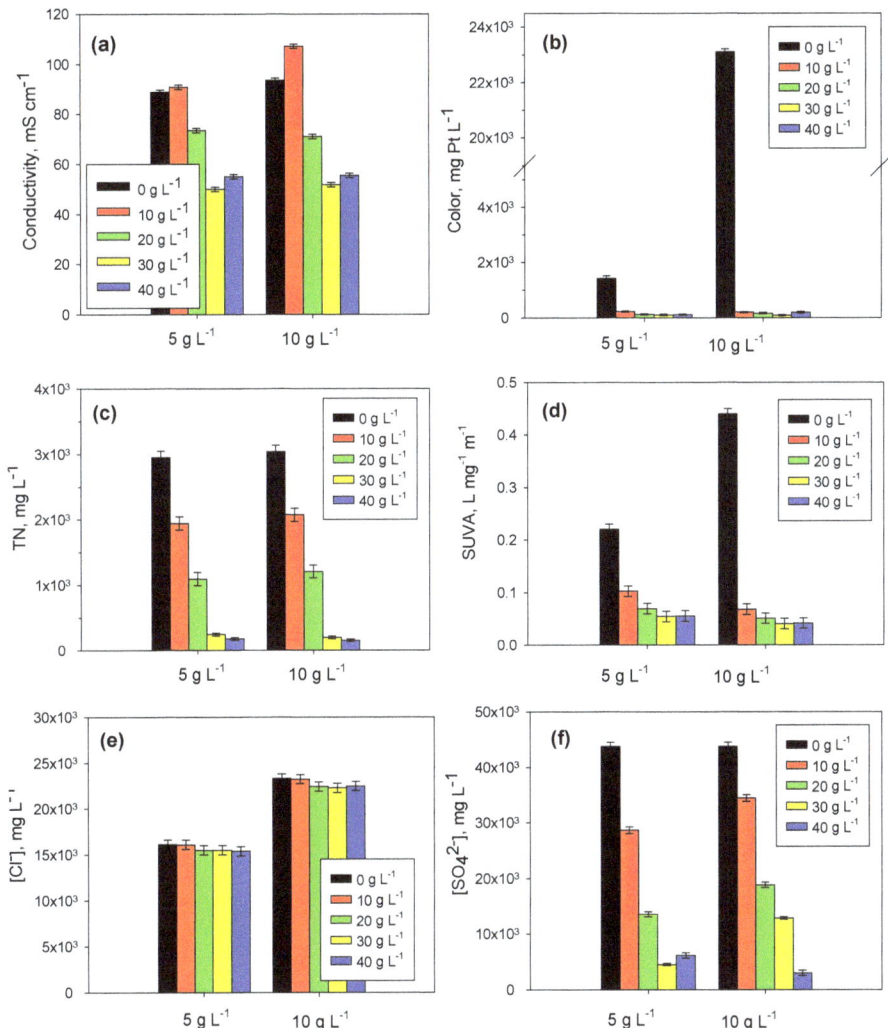

Figure 5. Lime precipitation results of previously oxidized coagulated LLROC ([H_2O_2]/COD = 1.063; coagulant doses of 5 and 10 g L^{-1} of $FeCl_3·6H_2O$): (**a**) Conductivity; (**b**) color; (**c**) TNb; (**d**) SUVA; (**e**) chloride; and (**f**) sulfate (values are m ± sd).

The reduction of conductivity by lime precipitation is in accordance with the removal of sulfate and total nitrogen contents from the oxidized post-coagulated LLROC. The reduction of sulfate content (Figure 5f) is explained by the induced precipitation of calcium sulfate. Initial sulfate levels were not so important ([SO_4^{2-}] = 2431 mg L^{-1}), but after the adjustment of pH with sulfuric acid, before the coagulation step, the reached value supposes an increment of sulfate concentration ([SO_4^{2-}] = 43,758 mg L^{-1}). Chloride is another ion present in the wastewater contributing to build up conductivity, but its content was not affected by this treatment step (Figure 5e). Metals content also contributes to produce conductivity and, in the case of iron, color as well. The increase of pH along this lime precipitation step produced the precipitation of iron in the form of $Fe(OH)_3$. As a result, the final concentration of iron was below 0.5 mg L^{-1}, whether its initial concentration was 168 or 800 mg L^{-1} after the photo-Fenton treatment (5 or 10 g L^{-1} of ferric chloride coagulation, respectively). Moreover, a considerable fraction of other metal, transition metal, or metalloid ions were precipitated, namely: 86% of Al, 77% of Zn, 84% of Mn, 99% of Mg, and 98% of Si.

2.4. Economic Assessment

Although a more economical result may be expected at industrial scale, the treatment cost of the herein assessed treatment at lab scale has been estimated in order to set reference values (Table 3). The following average values have been adopted as reference prices for each industrial grade reagent: Ferric chloride (200 € ton^{-1}, 40 wt%), alum (180 € ton^{-1}, pure), hydrogen peroxide (350 € ton^{-1}, 35 wt%), concentrated sulphuric acid (130 € ton^{-1}), and calcium hydroxide (130 € ton^{-1}). The cost of power was calculated using the average cost of electricity in Spain (0.11 € kWh^{-1}).

The application of UVA-LED technology produced an approximately 16% reduction in the cost of energy consumption in the photo-Fenton treatment step, which represents a 10% reduction in the total treatment cost. A cheaper implementation cost and the lower environmental hazard of its final disposal, as well as the potential to design an even more efficient LED lamp, may further foster the use of this technology for these environmental applications. In fact, Rodriguez-Chueca et al. (2016) [42] have similarly reported UVA-LED as a promising technology to more efficiently assist the photo-Fenton treatment. Moreover, the combination of coagulation (pH = 5.0; 2 g L^{-1} of FeCl$_3$) with UVA-LED photo-Fenton oxidation was assessed to cost 1.56 € kg COD^{-1} in a previous study [41], in which lime precipitation was not considered, denoting the importance of process optimization at lab scale and technology development in the short term.

3. Material and Methods

3.1. Landfill Leachate ROC

The LLROC sample was collected from "Las Dehesas—Valdemingómez" landfill (Madrid, Spain). The leachate of this landfill is treated by a three-stage RO system, achieving a 52% total recovery in the treatment of 200 m^3 per day. In this process, LLROC, of which the main characteristics are shown in Table 1, is partially recirculated back to the landfill leachate tank, and the rest is collected by an external waste manager, implying a very high cost. Therefore, new alternatives are necessary.

3.2. Chemicals

Aluminum sulfate decaoctahydrate (Al$_2$(SO$_4$)$_3$·18H$_2$O; 99.5% min. purity) in a 50% *w/w* water solution, and ferric chloride (FeCl$_3$·6H$_2$O; 99% min. purity) in a 69% *w/w* water solution were used as coagulants. Both chemicals were purchased from Sigma-Aldrich (Highland, CA, USA). An anionic flocculant of high molecular weight (A-120HMW) from Kemira (Helsinki, Finland) was used in a 0.025% *w/w* water solution. Lime (Ca(OH)$_2$; 95% min. purity) was purchased from Panreac Química S.L.U. (Castellar del Vallés, Spain), and used as powder. Sulfuric acid (H$_2$SO$_4$, 96–98% purity) from Sigma-Aldrich (Highland, CA, USA) was used for pH adjustment without further purification. Hydrogen peroxide (35% *w/w*; Sigma-Aldrich, Highland, CA, USA) was used as oxidant in the photo-Fenton treatment of LLROC.

3.3. Analytical Determinations and Data Processing

All analyses were performed according to the Standard Methods for the Examination of Water and Wastewater [55]. Conductivity and pH were measured using a Sension™ + MM374 pH-meter (Hach, CO, USA) equipped with pH and conductivity probes. COD, sulfate, and chloride contents were measured following the Nanocolor® test methods (Macherey-Nagel GmbH, Düren, Germany) and using an Aquamate UV-Vis spectrophotometer (Thermo Fisher Scientific, Waltham, MA, USA) to perform measurements. A Hanna LP 2000-11 turbidity-meter (Hanna Instruments, Laval, Canada) was used to determine turbidity. UV-254 absorbance was measured using a Varyan Cary 50 UV-visible spectrophotometer (Varian, Palo Alto, CA, USA) using 1-cm-pathway quartz-cuvettes (Hellma,

Müllheim, Germany). Total organic carbon (TOC) and total nitrogen (TN_b) were determined by the combustion-infrared method using a Multi N/C® 3100 TOC/TN analyzer (Analytik Jena AG, Jena, Germany) with catalytic oxidation on cerium oxide at 850 °C. H_2O_2 concentration was determined by Pobiner's titanium sulfate spectrophotometric method [56]. Dissolved iron, aluminum, and calcium were measured by atomic absorption spectrometry (3111B, 3111E) with a Varian SpectrAA 220 spectrophotometer (Varian, Palo Alto, CA, USA). 5-day biochemical oxygen demand (BOD_5) and total alkalinity were respectively determined following standard methods 5210B and 2320B [55]. Samples were filtered through 0.45 μm, and dilution was applied when necessary.

All analyses were carried out in triplicate; hence, all data provided in the text, figures (error bars), or tables are expressed in terms of the mean value (m) ± the standard deviation (sd) of measurements and results. ANOVA was performed to address the statistical significance (p-value < 0.05 *, 0.01 **, or 0.001 ***) of resulting differences when necessary.

3.4. Coagulation/Flocculation Pretreatment

Both coagulants, $Al_2(SO_4)_3 \cdot 18H_2O$ and $FeCl_3 \cdot 6H_2O$, were tested following a jar-test methodology at four different initial pH values (8.2, 7.0, 6.0, and 5.0) and several dosages ranging from 0 to 30 g L^{-1}, aiming to determine the optimum conditions for the removal of COD, color, and SUVA. Experiments were performed in a 500 mL beaker filled with 250 mL of sample in a flocculation tester. Fast mixing (150 rpm) was applied for 10 min after the coagulant was added. Then, the flocculant was added followed by 30 min of slow mixing (50 rpm). Finally, the samples were left to settle for 60 min to achieve maximum clarification.

3.5. Photo-Fenton Treatment

Two different set-ups were used under the same experimental conditions to perform the photo-Fenton treatment after the coagulation/flocculation pretreatment of the LLROC. In the first setup, a high-pressure mercury vapor immersion lamp of 100 W (Model 7825-30 from ACE Glass, Vineland, NJ, USA) that was covered by a quartz glass cooling jacket, and vertically positioned in the reactor, was used as the source of radiation, as it has previously been described [41]. A total photon flux of 6.6×10^{18} photon s^{-1} m^{-2} was measured to be emitted inside the reactor using a radiometer (UV-Elektronik, UV-VIS Radiometer RM-21, Ettlingen, Germany) [57]. 1.75 L of coagulated leachate was mechanically stirred in the reactor. The appropriate amount of H_2O_2 (35% w/v) corresponding to initial $[H_2O_2]/[COD]$ ratios of 2.125 (stoichiometric theoretical optimum [51]), 1.063 and 0.531, was then added.

In the second photo-Fenton treatment setup, the source of radiation consisted of an 8W UVA LED lamp made up of 10,365 nm LED emitters (CUN6GB1A, Seoul Viosys, Asan, Korea) uniformly disposed in series over the 9 cm diameter reactor, which was filled with 100 mL of the post-coagulated LLROC that were magnetically stirred. The lamp was located at 4.5 cm from LLROC surface. 250 mA of current intensity were applied to generate a total photon flux in the solution contained in the reactor of 1.0×10^{21} photon s^{-1} m^{-2}, as measured by potassium ferrioxalate actinometry [58,59]. Trials were carried out in triplicate, at room temperature, and without adjusting the initial pH value.

3.6. Lime Precipitation Step

Calcium hydroxide powder was added to the previously oxidized coagulated LLROC at dosages ranging from 0 to 40 g L^{-1}. The samples were mechanically stirred at 150 rpm for 16 h, and then were left to settle for 60 min. This treatment was carried out using the same equipment than in the coagulation/flocculation treatment.

4. Conclusions

The pre-treatment of coagulation/flocculation with ferric chloride achieved better COD removal results than with alum (76% vs. 60%), and the remaining dissolved iron in the LLROC after coagulation was efficiently used for its post-photo-Fenton treatment step, avoiding the need to add more iron to the solution to catalyze the process.

Photo-Fenton treatment after optimum coagulation with ferric chloride was efficient in enhancing the biodegradability of LLROC to BOD_5/COD ratios of $\approx$ 0.3–0.4, as well as in the removal of COD (45%), TOC (44%), and SUVA (46%).

The most important difference between optimal coagulant doses, with regard to a posterior photo-Fenton process, was the dissolved iron content that remained in the medium (168 and 800 mg L^{-1} of Fe for 5 and 10 g L^{-1} of $FeCl_3 \cdot 6H_2O$ coagulation, respectively) affecting the kinetics of the reaction; so that the time of irradiation was reduced from 300 min to 180 min, respectively.

The cost of managing higher amounts of residual iron in the medium after treatment, the better biodegradability improvement results (BOD_5/COD = 0.4), and the lower chloride addition to the media, indicate that, in these terms, coagulation with 5 g L^{-1} of $FeCl_3 \cdot 6H_2O$ may be preferred for the coagulation step.

The use of UVA-LED technology as the source of irradiation in the photo-Fenton treatment may reduce the cost of power consumption by up to 16% compared with using 100 W UV mercury immersion lamps, and this is significant because longer reaction time (thus longer lamp use) is required when lower iron content remains in the medium after coagulation. The substitution of a 100 W UV mercury immersion lamp by UVA-LED technology addressed an overall 10% cheaper treatment, which may be further enhanced with further optimization of this new technology.

A final step of lime precipitation (30 g L^{-1}, 24 h) lowered conductivity by 50%, sulfate and nitrogen content by 90%, and iron content by 99.9%, in the oxidized-coagulated-LLROC.

The assessed best combination of treatments (ferric chloride coagulation + UVA-LED photo-Fenton + lime addition) for LLROC achieved an overall removals of 90% COD, 43% conductivity, 86% aluminum, 77% zinc, 84% manganese, 99% magnesium, and 98% silicon; as well as an enhancement of biodegradability, increasing the BOD_5/COD ratio from 0.06 to 0.4.

Author Contributions: Conceptualization, J.T., D.H., R.M., A.G., V.A., C.N. and Á.B.; methodology, J.T., D.H., R.M., A.G., V.A., C.N. and Á.B.; formal analysis, J.T., D.H. and A.G.; investigation, J.T., D.H., R.M., A.G. and V.A.; data curation, J.T., and A.G.; writingoriginal dra-ft preparation, J.T., D.H., R.M., A.G., V.A., C.N. and Á.B.; writing-review and editing, J.T., D.H., R.M., A.G., V.A., C.N. and Á.B.; supervision, D.H., R.M., A.G., C.N. and Á.B; funding acquisition, Á.B. All authors have read and agreed to the published version of the manuscript.

Funding: This piece of research has been funded by the Spanish Ministry of Economy, Industry and Competitiveness granting project CTM2016-77948-R, and by the Community of Madrid by funding project RETOPROSOST-2 (S2018/EMT-4459).

Acknowledgments: The authors acknowledge to Las Dehesas—Valdemingómez landfill (Madrid, Spain) for kindly supplying the ROCLL used in this study.

Conflicts of Interest: The authors declare no conflict of interest.

References

1. Koc-Jurczyk, J.; Jurczyk, L. Biological treatment of landfill leachate at elevated temperature in the presence of polyurethane foam of various porosity. *Clean Soil Air Water* **2017**, *45*, 8. [CrossRef]
2. Ye, Z.L.; Hong, Y.; Pan, S.; Huang, Z.; Chen, S.; Wang, W. Full-scale treatment of landfill leachate by using the mechanical vapor recompression combined with coagulation pretreatment. *Waste Manag.* **2017**, *66*, 88–96. [CrossRef] [PubMed]
3. Teh, C.Y.; Budiman, P.M.; Shak, K.P.Y.; Wu, T.Y. Recent advancement of oagulation-flocculation and its application in wastewater treatment. *Ind. Eng. Chem. Res.* **2016**, *55*, 4363–4389. [CrossRef]
4. Torretta, V.; Ferronato, N.; Katsoyiannis, I.A.; Tolkou, A.K.; Airoldi, M. Novel and conventional technologies for landfill leachates treatment: A Review. *Sustainability* **2017**, *9*, 9. [CrossRef]
5. Oller, I.; Malato, S.; Sánchez-Pérez, J.A. Combination of advanced oxidation processes and biological treatments for wastewater decontamination—A review. *Sci. Total Environ.* **2011**, *409*, 4141–4166. [CrossRef]

6. Trebouet, D.; Schlumpf, J.P.; Jaouen, P.; Quemeneur, F. Stabilized landfill leachate treatment by combined physicochemical–nanofiltration processes. *Water Res.* **2001**, *35*, 2935–2942. [CrossRef]
7. Bohdziewicz, J.; Bodzek, M.; Górska, J. Application of pressure-driven membrane techniques to biological treatment of landfill leachate. *Process. Biochem.* **2001**, *36*, 641–646. [CrossRef]
8. Weber, B.; Holz, F. Landfill leachate treatment by reverse osmosis. In *Effective Industrial Membrane Processes: Benefits and Opportunities*; Turner, M.K., Ed.; Springer: Dordrecht, The Netherlands, 1991; pp. 143–154.
9. Van der Bruggen, B.; Lejon, L.; Vandecasteele, C. Reuse, treatment, and discharge of the concentrate of pressure-driven membrane processes. *Environ. Sci. Technol.* **2003**, *37*, 3733–3738. [CrossRef]
10. Zhao, R.; Jung, C.; Trzopek, A.; Torrens, K.; Deng, Y. Characterization of ultraviolet-quenching dissolved organic matter (DOM) in mature and young leachates before and after biological pre-treatment. *Environ. Sci. Water Res. Technol.* **2018**, *4*, 731–738. [CrossRef]
11. Iskander, S.M.; Novak, J.T.; Brazilb, B.; He, Z. Percarbonate oxidation of landfill leachates towards removal of ultraviolet quenchers. *Environ. Sci. Water Res. Technol.* **2017**, *3*, 1162–1170. [CrossRef]
12. Zhou, B.; Yu, Z.; Wei, Q.; Long, H.Y.; Xie, Y.; Wang, Y. Electrochemical oxidation of biological pretreated and membrane separated landfill leachate concentrates on boron doped diamond. *Appl. Surf. Sci.* **2016**, *377*, 406–415. [CrossRef]
13. Morello, L.; Cossu, R.; Raga, R.; Pivato, A.; Lavagnolo, M.C. Recirculation of reverse osmosis concentrate in lab-scale anaerobic and aerobic landfill simulation reactors. *Waste Manag.* **2016**, *56*, 262–270. [CrossRef] [PubMed]
14. Calabrò, P.S.; Gentili, E.; Meoni, C.; Orsi, S.; Komilis, D. Effect of the recirculation of a reverse osmosis concentrate on leachate generation: A case study in an Italian landfill. *Waste Manag.* **2018**, *76*, 643–651. [CrossRef] [PubMed]
15. Labiadh, L.; Fernandes, A.; Ciriaco, L.; Pacheco, M.J.; Gadri, A.; Ammar, S. Electrochemical treatment of concentrate from reverse osmosis of sanitary landfill leachate. *J. Environ. Manag.* **2016**, *181*, 515–521. [CrossRef] [PubMed]
16. Talalaj, I.A.; Biedka, P. Impact of concentrated leachate recirculation on effectiveness of leachate treatment by reverse osmosis. *Ecol. Eng.* **2015**, *85*, 185–192. [CrossRef]
17. Qi, X.X.; Zhang, C.J.; Zhang, Y. Treatment of landfill leachate RO concentrate by VMD. In Proceedings of the International Conference on Circuits and Systems, Madrid, Spain, 20–21 July 2015; Volume 9, pp. 13–17.
18. Yang, B.Q.; Yang, J.M.; Yang, H.; Liu, Y.M.; Li, X.K.; Wang, Q.Z.; Pan, X.J. Co-bioevaporation treatment of concentrated landfill leachate with addition of food waste. *Biochem. Eng. J.* **2018**, *130*, 76–82. [CrossRef]
19. Bagastyo, A.Y.; Radjenovic, J.; Mu, Y.; Rozendal, R.A.; Batstone, D.J.; Rabaey, K. Electrochemical oxidation of reverse osmosis concentrate on mixed metal oxide (MMO) titanium coated electrodes. *Water Res.* **2011**, *45*, 4951–4959. [CrossRef]
20. Pérez, G.; Fernández-Alba, A.R.; Urtiaga, A.M.; Ortiz, I. Electro-oxidation of reverse osmosis concentrates generated in tertiary water treatment. *Water Res.* **2010**, *44*, 2763–2772. [CrossRef]
21. He, Y.; Zhang, H.; Li, J.J.; Zhang, Y.; Lai, B.; Pan, Z. Treatment of landfill leachate reverse osmosis concentrate by catalytic ozonation with γ-Al_2O_3. *Environ. Eng. Sci.* **2018**, *35*, 501–511. [CrossRef]
22. Mojiri, A.; Ziyang, L.; Hui, W.; Gholami, A. Concentrated landfill leachate treatment by electro-ozonation. In *Advanced Oxidation Processes (AOPs) in Water and Wastewater Treatment*; Aziz, H.A., Abu Amr, S.S., Eds.; IGI Global: Hershey, PA, USA, 2019; pp. 150–170.
23. Cui, Y.H.; Xue, W.J.; Yang, S.Q.; Tu, J.L.; Guo, X.L.; Liu, Z.Q. Electrochemical/peroxydisulfate/Fe^{3+} treatment of landfill leachate nanofiltration concentrate after ultrafiltration. *Chem. Eng. J.* **2018**, *353*, 208–217. [CrossRef]
24. Liang, S.X.; Jia, Z.; Zhang, W.C.; Lia, X.F.; Wang, W.M.; Line, H.C.; Zhang, L.C. Ultrafast activation efficiency of three peroxides by $Fe_{78}Si_9B_{13}$ metallic glass under photo-enhanced catalytic oxidation: A comparative study. *Appl. Catal. B* **2018**, *221*, 108–118. [CrossRef]
25. Hermosilla, D.; Cortijo, M.; Huang, C.P. Optimizing the treatment of landfill leachate by conventional Fenton and photo-Fenton processes. *Sci. Total Environ.* **2009**, *407*, 3473–3481. [CrossRef] [PubMed]
26. Liang, S.X.; Jia, Z.; Liu, Y.J.; Zhang, W.; Wang, W.; Lu, J.; Zhang, L.C. Compelling rejuvenated catalytic performance in metallic glasses. *Adv. Mater.* **2018**, *30*, 1802764. [CrossRef] [PubMed]
27. Wang, Q.; Chen, M.; Lin, P.; Cui, Z.; Chu, C.; Shen, B. Investigation of FePC amorphous alloys with self-renewing behaviour for highly efficient decolorization of methylene blue. *J. Mater. Chem. A* **2018**, *6*, 10686. [CrossRef]

28. Martín, M.B.; López, J.C.; Oller, I.; Malato, S.; Pérez, J.S. A comparative study of different tests for biodegradability enhancement determination during AOP treatment of recalcitrant toxic aqueous solutions. *Ecotoxicol. Environ. Saf.* **2010**, *73*, 1189–1195. [CrossRef] [PubMed]
29. Paździor, K.; Bilińska, L.; Ledakowicz, S. A review of the existing and emerging technologies in the combination of AOPs and biological processes in industrial textile wastewater treatment. *Chem. Eng. J.* **2019**, *376*, 120597. [CrossRef]
30. Umar, M.; Roddick, F.; Fan, L.H. Recent advancements in the treatment of municipal wastewater reverse osmosis concentrate—An overview. *Crit. Rev. Environ. Sci. Technol.* **2015**, *45*, 193–248. [CrossRef]
31. Dialynas, E.; Mantzavinos, D.; Diamadopoulos, E. Advanced treatment of the reverse osmosis concentrate produced during reclamation of municipal wastewater. *Water Res.* **2008**, *42*, 4603–4608. [CrossRef]
32. Liu, K.; Roddick, F.A.; Fan, L. Impact of salinity and pH on the UVC/H_2O_2 treatment of reverse osmosis concentrate produced from municipal wastewater reclamation. *Water Res.* **2012**, *46*, 3229–3239. [CrossRef]
33. Zheng, H.; Pan, Y.; Xiang, X. Oxidation of acidic dye Eosin Y by the solar photo-Fenton processes. *J. Hazard. Mater.* **2007**, *141*, 457–464. [CrossRef]
34. Aziz, H.A.; Alias, S.; Adlan, M.N.; Faridah; Asaari, A.H.; Zahari, M.S. Colour removal from landfill leachate by coagulation and flocculation processes. *Bioresour. Technol.* **2007**, *98*, 218–220. [CrossRef] [PubMed]
35. Zhou, T.; Lim, T.T.; Chin, S.S.; Fane, A.G. Treatment of organics in reverse osmosis concentrate from a municipal wastewater reclamation plant: Feasibility test of advanced oxidation processes with/without pretreatment. *Chem. Eng. J.* **2011**, *166*, 932–939. [CrossRef]
36. Kavitha, V.; Palanivelu, K. The role of ferrous ion in Fenton and photo-Fenton processes for the degradation of phenol. *Chemosphere* **2004**, *55*, 1235–1243. [CrossRef] [PubMed]
37. Safarzadeh-Amiri, A.; Bolton, J.R.; Cater, S.R. Ferrioxalate-mediated photodegradation of organic pollutants in contaminated water. *Water Res.* **1997**, *31*, 787–798. [CrossRef]
38. Hermosilla, D.; Cortijo, M.; Huang, C.P. The role of iron on the degradation and mineralization of organic compounds using conventional Fenton and photo-Fenton processes. *Chem. Eng. J.* **2009**, *155*, 637–646. [CrossRef]
39. Amor, C.; de Torres-Socias, E.; Peres, J.A.; Maldonado, M.I.; Oller, I.; Malato, S.; Lucas, M.S. Mature landfill leachate treatment by coagulation/flocculation combined with Fenton and solar photo-Fenton processes. *J. Hazard. Mater.* **2015**, *286*, 261–268. [CrossRef] [PubMed]
40. Renou, S.; Givaudan, J.G.; Poulain, S.; Dirassouyan, F.; Moulin, P. Treatment process adapted to stabilized leachates: Lime precipitation–prefiltration–reverse osmosis. *J. Membr. Sci.* **2008**, *313*, 9–22. [CrossRef]
41. Tejera, J.; Miranda, R.; Hermosilla, D.; Urra, I.; Negro, C.; Blanco, A. Treatment of a mature landfill leachate: Comparison between homogeneous and heterogeneous photo-Fenton with different pretreatments. *Water* **2019**, *11*, 1849. [CrossRef]
42. Rodríguez-Chueca, J.; Amor, C.; Fernandes, J.R.; Tavares, P.B.; Lucas, M.S.; Peres, J.A. Treatment of crystallized-fruit wastewater by UV-A LED photo-Fenton and coagulation–flocculation. *Chemosphere* **2016**, *145*, 351–359. [CrossRef]
43. Gomes, A.I.; Silva, T.F.C.V.; Duarte, M.A.; Boaventura, R.A.R.; Vilar, V.J.P. Cost-effective solar collector to promote photo-Fenton reactions: A case study on the treatment of urban mature leachate. *J. Clean. Prod.* **2018**, *199*, 369–382. [CrossRef]
44. Silva, T.F.; Fonseca, A.; Saraiva, I.; Boaventura, R.A.; Vilar, V.J. Scale-up and cost analysis of a photo-Fenton system for sanitary landfill leachate treatment. *Chem. Eng. J.* **2016**, *283*, 76–88. [CrossRef]
45. Chow, C.W.K.; van Leeuwen, J.A.; Fabris, R.; Drikas, M. Optimised coagulation using aluminium sulfate for the removal of dissolved organic carbon. *Desalination* **2009**, *245*, 120–134. [CrossRef]
46. Renou, S.; Givaudan, J.G.; Poulain, S.; Dirassouyan, F.; Moulin, P. Landfill leachate treatment: Review and opportunity. *J. Hazard. Mater.* **2008**, *150*, 468–493. [CrossRef] [PubMed]
47. Bacardit, J.; Stötzner, J.; Chamarro, E.; Esplugas, S. Effect of salinity on the photo-Fenton process. *Ind. Eng. Chem. Res.* **2007**, *46*, 7615–7619. [CrossRef]
48. Qin, J.J.; Oo, M.H.; Kekre, K.A.; Knops, F.; Miller, P. Impact of coagulation pH on enhanced removal of natural organic matter in treatment of reservoir water. *Sep. Purif. Technol.* **2006**, *49*, 295–298. [CrossRef]
49. Chen, W.; Gu, Z.; Wen, P.; Li, Q. Degradation of refractory organic contaminants in membrane concentrates from landfill leachate by a combined coagulation-ozonation process. *Chemosphere* **2019**, *217*, 411–422. [CrossRef] [PubMed]

50. Biglarijoo, N.; Mirbagheri, S.A.; Ehteshami, M.; Ghaznavi, S.M. Optimization of Fenton process using response surface methodology and analytic hierarchy process for landfill leachate treatment. *Process. Saf. Environ. Prot.* **2016**, *104*, 150–160. [CrossRef]
51. Hermosilla, D.; Merayo, N.; Ordóñez, R.; Blanco, A. Optimization of conventional Fenton and ultraviolet-assisted oxidation processes for the treatment of reverse osmosis retentate from a paper mill. *Waste Manag.* **2012**, *32*, 1236–1243. [CrossRef]
52. Liu, R.; Chiu, H.M.; Shiau, C.S.; Yeh, R.Y.L.; Hung, Y.T. Degradation and sludge production of textile dyes by Fenton and photo-Fenton processes. *Dyes Pigm.* **2007**, *73*, 1–6. [CrossRef]
53. Iskander, S.M.; Zhao, R.; Pathak, A.; Gupta, A.; Pruden, A.; Novak, J.T.; He, Z. A review of landfill leachate induced ultraviolet quenching substances: Sources, characteristics, and treatment. *Water Res.* **2018**, *145*, 297–311. [CrossRef]
54. De, S.; Hazra, T.; Dutta, A. Sustainable treatment of municipal landfill leachate by combined association of air stripping, Fenton oxidation, and enhanced coagulation. *Environ. Monit. Assess.* **2019**, *191*, 49. [CrossRef] [PubMed]
55. APHA; AWWA; WPCF. *Standard Methods for the Examination of Water and Wastewater*; APHA: Washington, DC, USA, 2005.
56. Pobiner, H. Determination of hydroperoxides in hydrocarbon by conversion to hydrogen peroxide and measurement by titanium complexing. *Anal. Chem.* **1961**, *33*, 1423–1426. [CrossRef]
57. Liang, X.; Zhu, X.; Butler, E.C. Comparison of four advanced oxidation processes for the removal of naphthenic acids from model oil sands process water. *J. Hazard. Mater.* **2011**, *190*, 168–176. [CrossRef] [PubMed]
58. Hatchard, C.; Parker, C.A. A new sensitive chemical actinometer. II. Potassium ferrioxalate as a standard chemical actinometer. *Proc. R. Soc. A.* **1956**, *235*, 518–536.
59. Montalti, M.; Credi, A.; Prodi, L.; Gandolfi, M.T. *Handbook of Photochemistry*, 1st ed.; CRC Press: Boca Raton, FL, USA, 2006; p. 650.

Publisher's Note: MDPI stays neutral with regard to jurisdictional claims in published maps and institutional affiliations.

© 2020 by the authors. Licensee MDPI, Basel, Switzerland. This article is an open access article distributed under the terms and conditions of the Creative Commons Attribution (CC BY) license (http://creativecommons.org/licenses/by/4.0/).

Article

Phytochemical-Assisted Green Synthesis of Nickel Oxide Nanoparticles for Application as Electrocatalysts in Oxygen Evolution Reaction

Vidhya Selvanathan [1], M. Shahinuzzaman [2,*], Shankary Selvanathan [3], Dilip Kumar Sarkar [1], Norah Algethami [4], Hend I. Alkhammash [5], Farah Hannan Anuar [6], Zalita Zainuddin [7], Mohammod Aminuzzaman [8,9], Huda Abdullah [10] and Md. Akhtaruzzaman [1,*]

1. Solar Energy Research Institute (SERI), Universiti Kebangsaan Malaysia (UKM), Bangi 43600, Selangor, Malaysia; vidhya@ukm.edu.my (V.S.); dilipks551@gmail.com (D.K.S.)
2. School of Computer Science and Information Technology, Central University of Science and Technology, Dhaka 1216, Bangladesh
3. Department of Chemistry, Faculty of Science, University of Malaya, Kuala Lumpur 50603, Malaysia; shankaryselvanathan@gmail.com
4. Department of Physics, Faculty of Science, Taif University, Taif 21944, Saudi Arabia; n.alkthamy@tu.edu.sa
5. Department of Electrical Engineering, College of Engineering, Taif University, Taif 21944, Saudi Arabia; Khamash.h@tu.edu.sa
6. Department of Chemistry, Faculty of Science and Technology, Universiti Kebangsaan Malaysia (UKM), Bangi 43600, Selangor, Malaysia; farahhannan@ukm.edu.my
7. Department of Physics, Faculty of Science and Technology, Universiti Kebangsaan Malaysia (UKM), Bangi 43600, Selangor, Malaysia; zazai@ukm.edu.my
8. Department of Chemical Science, Faculty of Science, Universiti Tunku Abdul Rahman (UTAR), Perak Campus, Jalan Universiti, Bandar Barat, Kampar 31900, Perak, Malaysia; mohammoda@utar.edu.my
9. Centre for Photonics and Advanced Materials Research (CPAMR), Universiti Tunku Abdul Rahman (UTAR), Jalan Sungai Long, Bandar Sungai Long, Kajang 43000, Selangor, Malaysia
10. Department of Electrical, Electronic & Systems Engineering, Faculty of Engineering & Built Environment, Universiti Kebangsaan Malaysia (UKM), Bangi 43600, Selangor, Malaysia; huda.abdullah@ukm.edu.my
* Correspondence: shahinchmiu@gmail.com (M.S.); akhtar@ukm.edu.my (M.A.); Tel.: +603-89118587 (M.S. & M.A.)

Abstract: Electrocatalytic water splitting is a promising solution to resolve the global energy crisis. Tuning the morphology and particle size is a crucial aspect in designing a highly efficient nanomaterials-based electrocatalyst for water splitting. Herein, green synthesis of nickel oxide nanoparticles using phytochemicals from three different sources was employed to synthesize nickel oxide nanoparticles (NiO$_x$ NPs). Nickel (II) acetate tetrahydrate was reacted in presence of *aloe vera* leaves extract, papaya peel extract and dragon fruit peel extract, respectively, and the physicochemical properties of the biosynthesized NPs were compared to sodium hydroxide (NaOH)-mediated NiO$_x$. Based on the average particle size calculation from Scherrer's equation, using X-ray diffractograms and field-emission scanning electron microscope analysis revealed that all three biosynthesized NiO$_x$ NPs have smaller particle size than that synthesized using the base. Aloe-vera-mediated NiO$_x$ NPs exhibited the best electrocatalytic performance with an overpotential of 413 mV at 10 mA cm^{-2} and a Tafel slope of 95 mV dec^{-1}. Electrochemical surface area (ECSA) measurement and electrochemical impedance spectroscopic analysis verified that the high surface area, efficient charge-transfer kinetics and higher conductivity of aloe-vera-mediated NiO$_x$ NPs contribute to its low overpotential values.

Keywords: green synthesis; nickel oxide; nanoparticles; oxygen evolution reaction; electrocatalysts

1. Introduction

One of the greatest challenges of the technological era is the inflating global energy demand as electricity becomes a necessity in every facet of life. The non-renewable nature of fossil-fuel-based energy production has stimulated the pursuit of sustainable energy sources such as solar and wind energy. To overcome the intermittency issues of these renewable energy sources, production of hydrogen fuel from electrocatalytic water splitting is considered a strategic approach. Under a standard environment, 1.23 V of thermodynamic potential is required to drive electrochemical water splitting [1]. Nonetheless, in practical conditions, the sluggish kinetics of oxygen evolution reaction (OER) mandates high overpotential to generate the required current density. Hence, most practical electrolyzers consume more than 1.23 V to drive the desired reaction.

Electrocatalysts are usually employed to attain satisfactory water-splitting efficiency, but by far best electrocatalytic performance was demonstrated by noble-metal-based materials (e.g., RuO_2 and IrO_2). The high cost and low abundancy of noble metals are not conducive for large-scale water-splitting applications. Therefore, non-precious transition metal oxides and hydroxides were explored as alternative electrocatalysts. Nickel oxide (NiO_x) is one such promising material for electrocatalytic water splitting due to its high electron transport capacity, good chemical stability and facile method of synthesis. Particularly, in the case of NiO_x NPs, the synthetic route adapted influences the electrical and structural properties of the material.

Traditionally, metal oxide nanoparticles are synthesized using a strong base as a reducing agent, followed by reaction with a capping agent or reducing agent. Sometimes, additional organic solvents are required to dissolve these capping agents. Alternatively, a green synthesis approach can be utilized to reduce the usage of unnecessary toxic chemicals [2]. In green synthesis, primary and secondary metabolites of plants are used to aid the synthesis of metal oxide nanoparticles [3]. These metabolites form a coating layer or stabilizing layer on the surface of the metal oxide nanoparticles and prevent intensive aggregation, thus yielding well-defined morphologies and tunable particle size [4]. The choice of plant species and its respective parts play an important role in determining the morphology and particle size of the NPs as they contain distinct phytochemical profiles [5].

One of the major advantages of chemical capping agents is the tunability of the species to yield nanoparticles with different particle size and morphology. For example, in the case of synthetic-polymers-based capping agents, factors such as the nature of polymer (functional group, branching), molecular weight and concentration were found to affect its function as a stabilizer [6]. However, in the context of green synthesis, much work is dedicated to exploring the effect of using a single plant extract formulation. Since each plant species comes with its unique cocktail of phytochemicals, it is useful to compare the effect of using different plant extracts on the overall size and morphology of the nanoparticles. Thus, in this study we have explored three different plant extracts, namely, *aloe vera* leaves extract, papaya fruit peel extract and dragon fruit peel extract, as capping agents for NiO_x NPs.

Aloe vera gel contains mostly water and polysaccharides, such as pectins, cellulose, hemicellulose, glucomannan and acemannan [7,8]. In addition to that, the aloe latex is made up of hydroxyanthracenic derivatives and anthraquinone glycosides, along with emodin. Papaya peel is rich in a variety of vitamins including β-carotene, vitamin B (thiamine, riboflavin, niacin and folate), vitamin C and vitamin E [9]. On the other hand, dragon fruit peel is reported to contain wide compositions of antioxidants such as betasianin, flavonoids, phenols, terpenoids, thiamine, niacin, pyridoxine and cobalamin [10]. The effect of these different phytochemical compositions of extracts on the structural and morphological properties of the NiO_x NPs and, consequently, their electrocatalytic activities for OER were studied and reported in detail.

2. Results and Discussion
2.1. Structural Properties

Figure 1 shows the structural properties of the synthesized NiO_x nanoparticles using different extracts after calcination at 300 °C and 600 °C. As shown in the diffractograms, all the samples exhibited diffraction peaks at 2θ values of 37.2, 43.2, 62.8, 75.4 and 79.4° attributed to (101), (012), (104), (113) and (006) crystal planes of cubic NiO_x [11,12]. These peaks were in agreement with the standard JCPDS card no. 00-004-1159. Significant differences in the peak intensities were evident between samples calcined at 300 °C and 600 °C. At lower calcination temperature, the presence of organic residues from the plant extract used during NiO_x synthesis reduced the crystallinity of the nanoparticles. Upon calcination at 600 °C, high purity NiO_x nanoparticles were produced, leading to sharp and intense XRD peaks.

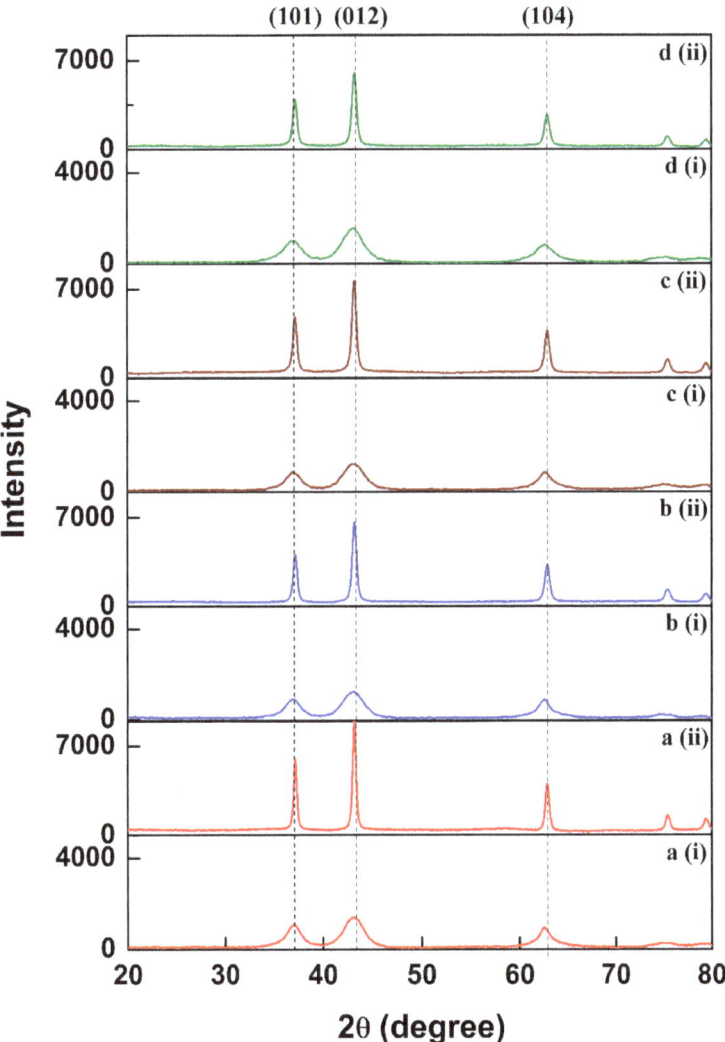

Figure 1. X-ray diffractogram of (**a**) NaOH-, (**b**) aloe-vera-, (**c**) dragon-fruit- and (**d**) papaya-extract-mediated NiO_x nanoparticles calcined at (**i**) 300 °C and (**ii**) 600 °C.

The average crystal sizes (D) were calculated based on the width of the peak due to the (200) planes by using the Scherrer's formula [13]:

$$D = \frac{0.94\lambda}{\beta Cos\theta} \quad (1)$$

From Table 1, it could be seen that the average crystallite size of the nanoparticles synthesized with only NaOH was 15.9 nm, whereas *aloe vera*, papaya and dragon fruit extracts was 11.5 nm, 8.1 nm and 10.2 nm, respectively. The reduced crystallite size of NiO_x nanoparticles synthesized in presence of plant extracts emphasizes the role of phytochemicals that are able to chelate with metal ions, thus acting as stabilizers or capping agents. Figure 2 illustrates the plausible mechanism for formation of NiO_x nanoparticles in this study. Upon dissolution, nickel acetate tetrahydrate dissociated to release Ni^{2+} ions which then combined with hydroxide ions from the base to form $Ni(OH)_2$. The continuous stirring under a heated environment enabled $Ni(OH)_2$ to thermally decompose into NiO nuclei [14]. The phytochemicals present in each plant extract served as capping ligands which formed interactions with the nanoparticles' surface, leading to steric hindrance and providing eventual stability to the nanocomposite. Hence, the presence of biomass-derived extracts controlled the particle size and minimized agglomeration of the nanoparticles.

Table 1. Particle size calculation based on Debye–Scherrer's equation.

Material	2θ	FWHM	βcosθ	Crystallite Size (nm)	Average Crystallite Size (nm)
NiO–NaOH	37.4	0.51	0.0085	16.3	15.9
	43.2	0.51	0.0083	16.8	
	62.8	0.63	0.0094	14.7	
NiO–Aloe	37.4	0.71	0.0117	11.8	11.5
	43.2	0.63	0.0101	13.6	
	62.8	1.03	0.0153	9.0	
NiO–Papaya	37.4	0.90	0.0149	9.3	10.2
	43.2	0.68	0.0109	12.6	
	62.8	1.08	0.0160	8.6	
NiO–Dragon	37.4	1.62	0.0269	5.2	8.1
	43.2	0.68	0.0109	12.6	
	62.8	1.45	0.0216	6.4	

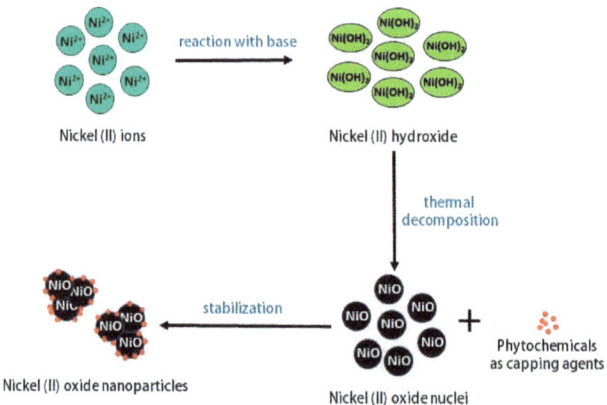

Figure 2. Plausible mechanism for green synthesis of NiOx nanoparticles.

2.2. Morphological Properties

The morphological properties of the synthesized NiO_x nanoparticles are shown in Figure 3. From the images, it is clear that the NiO_x nanoparticles were spherical-shaped, densely packed and intercalated with each other. Dragon fruit and papaya-extract-mediated NiO_x NPs depict well-defined nanostructures with fewer intercalations. The particle size distribution for the NiO_x NPs synthesized without extract ranged between

40 and 60 nm, whereas plant-extract-mediated NiO$_x$ NPs were between 20 and 30 nm in size. The trend in particle size coincided with the trend in crystallite size derived from the XRD analysis. For all the samples, the particle size revealed in morphological analysis was higher than average crystallite size estimated from XRD analysis, hence indicating that the single NiO$_x$ particle is made up of few crystallites.

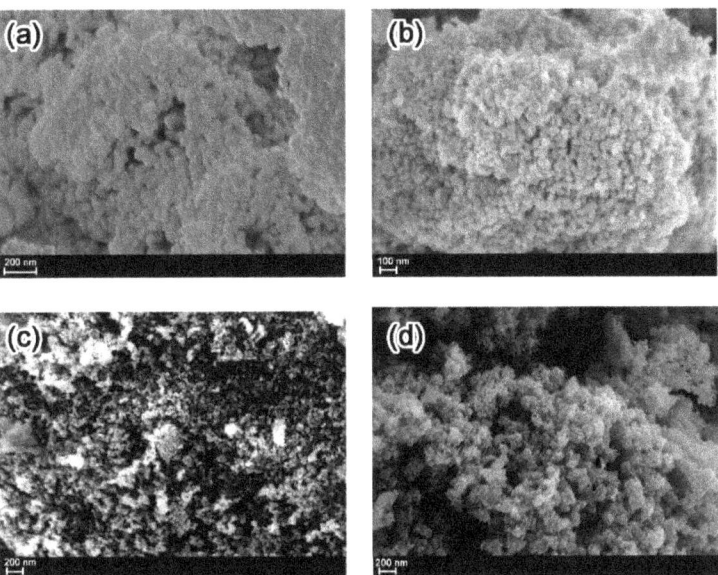

Figure 3. FESEM images of (**a**) NaOH-, (**b**) aloe-vera-, (**c**) dragon-fruit- and (**d**) papaya-extract-mediated NiO$_x$ nanoparticles.

The elemental composition of the nanoparticles was verified from EDX analysis. As shown in Figure 4, the choice of plant extract influenced the composition of nickel and oxygen in the samples. NiO$_x$ NPs synthesized using *aloe vera* were more oxygen deficient compared to papaya and dragon fruit. Aloe-vera-mediated nickel oxide also exhibited the highest amount of residual carbon, despite calcining the samples at 600 °C. It is deduced that this could be due to the more viscous nature of the *aloe vera* extract, which adheres to the surface of the nanoparticles during synthesis. The absence of residual carbon originating from the phytochemicals was evident in dragon-fruit- and papaya-extract-mediated NiO$_x$ NPs, which could explain the occurrence of fewer intercalations, as shown in the FESEM images.

2.3. Optical Properties

The absorption spectrum of the NiO$_x$ nanoparticles, as shown in Figure 5a, was recorded at room temperature within the range of 200–800 nm. There was a peak observed at around 300–350 nm for all synthesized nanopowders, which can be assigned to the exciton transitions [15]. The band gap energies were estimated by extrapolation using Tauc plots, and the energy curves are shown in Figure 5. It was seen that the band gap of NiO$_x$ nanoparticle synthesis in different samples were ranging from 3.09 to 3.22 eV and the band gap was slightly increased for plant-extract-assisted synthesized nanoparticles. Particularly, the decrease in particle size, as well as the increase in band gap energy of the as-prepared nanoparticles, signified the size quantization effects [16]. When particle size reached the nanosized scale, the overlapping of adjacent energy levels minimized and, subsequently, the width of energy band widened [17,18].

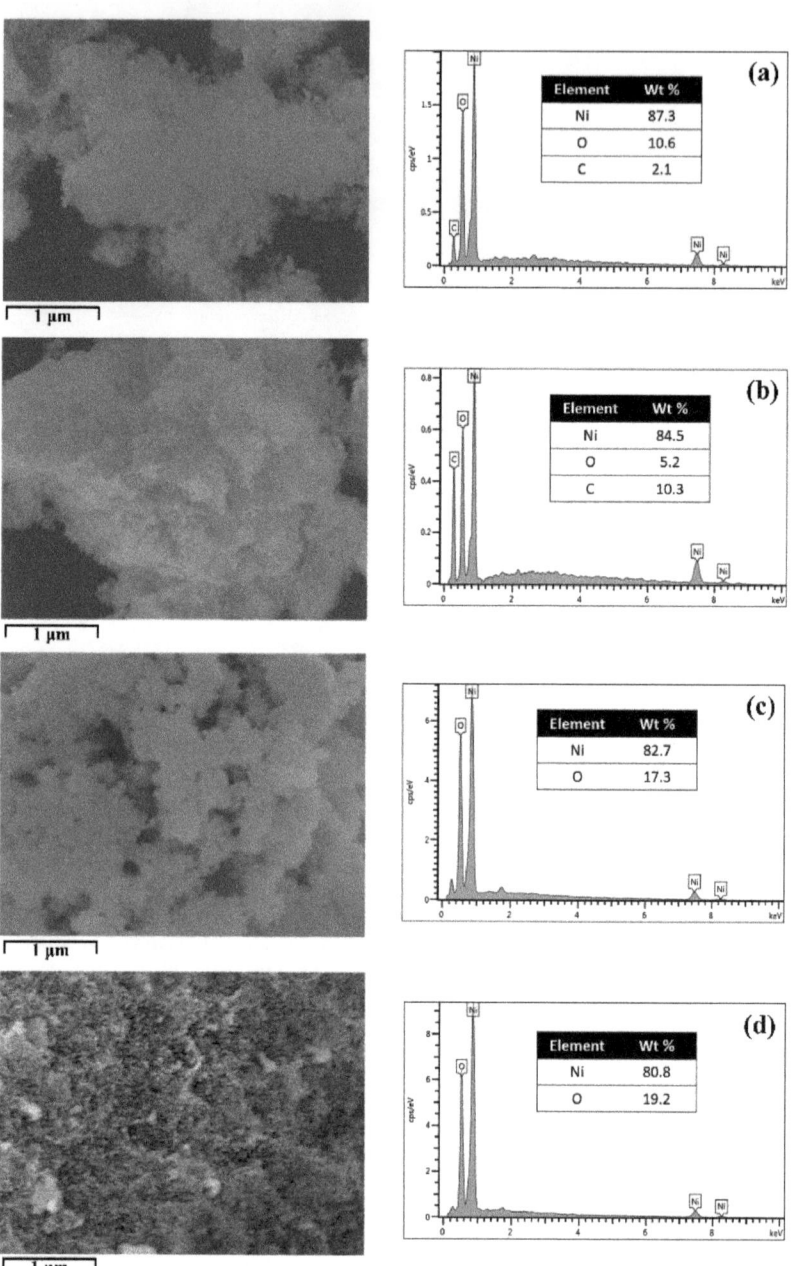

Figure 4. EDX spectrum of (**a**) NaOH-, (**b**) aloe-vera-, (**c**) dragon-fruit- and (**d**) papaya-extract-mediated NiO_x nanoparticles.

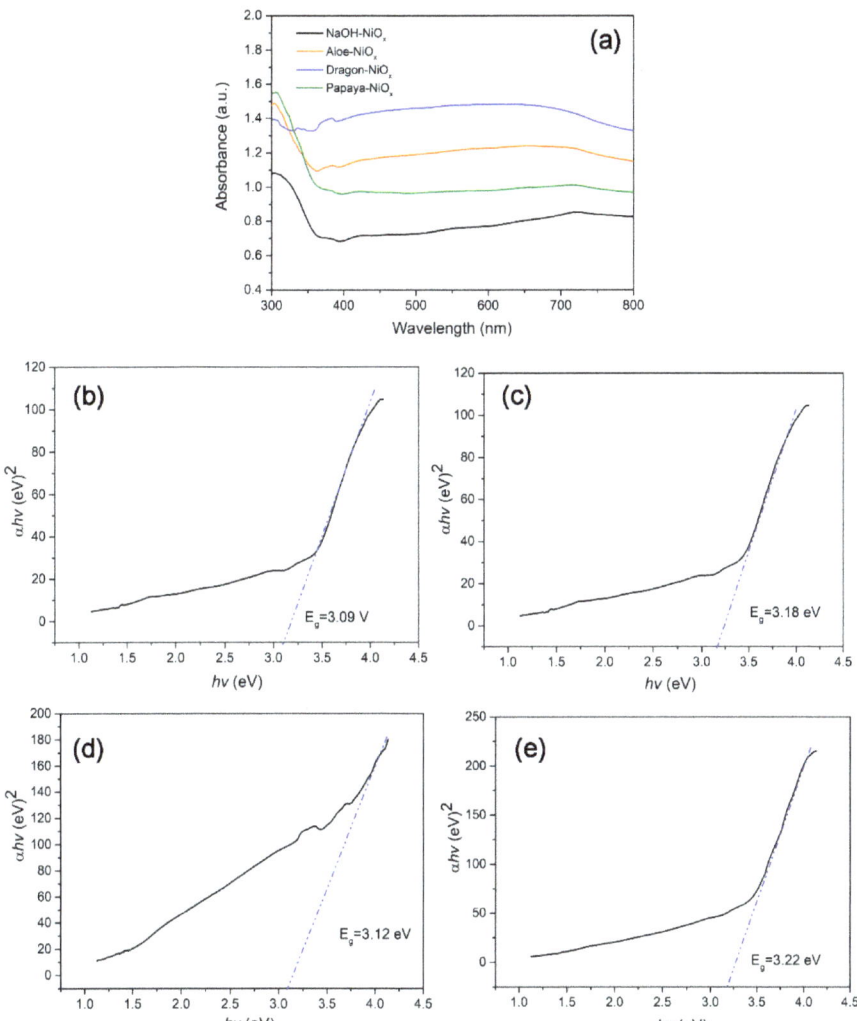

Figure 5. (**a**) UV-Vis absorption spectrum and the corresponding Tauc plot of (**b**) NaOH-, (**c**) aloe-vera-, (**d**) dragon-fruit- and (**e**) papaya-extract-mediated NiO$_x$ nanoparticles.

2.4. Electrocatalytic Properties for Oxygen Evolution Reaction (OER)

The electrocatalytic activity of the synthesized NiO$_x$ NPs towards OER was explored by performing linear scanning voltammetry (LSV) in an aqueous solution of 1.0 M KOH. Generally, the oxygen evolution reaction in alkaline medium is represented as follows [19–21]:

$$4OH^- \rightarrow O_2 + 2H_2O + 4e^- \tag{2}$$

The four-electron reaction mechanism can be described sequentially via the following equations:

$$OH^- \rightarrow OH^* + e^- \tag{3}$$

$$OH^* + OH^- \rightarrow O^* + H_2O + e^- \tag{4}$$

$$O^* + OH^- \rightarrow OOH^* + e^- \tag{5}$$

$$OOH^* + OH^- \rightarrow O_2 + 2H_2O + e^- \tag{6}$$

where OH*, O* and OOH* indicate the species adsorbed to the active site of the catalyst.

Based on the polarization curves shown in Figure 6a, to attain the current density of 10 mA cm^{-2}, aloe-vera- and papaya-extract-mediated NiO$_x$ NPs exhibited the overpotential values of 416 mV and 433 mV, respectively, which are lower than NaOH-based NiO$_x$, which records 474 mV. Interestingly, dragon-fruit-extract-mediated NiO$_x$ showed the overpotential of 501 mV, the highest overpotential compared to the previous three samples. As listed in Table 2, the onset potential, which indicates the voltage at which current begins to rise, was lowest for aloe-vera-mediated NPs and highest for dragon fruit.

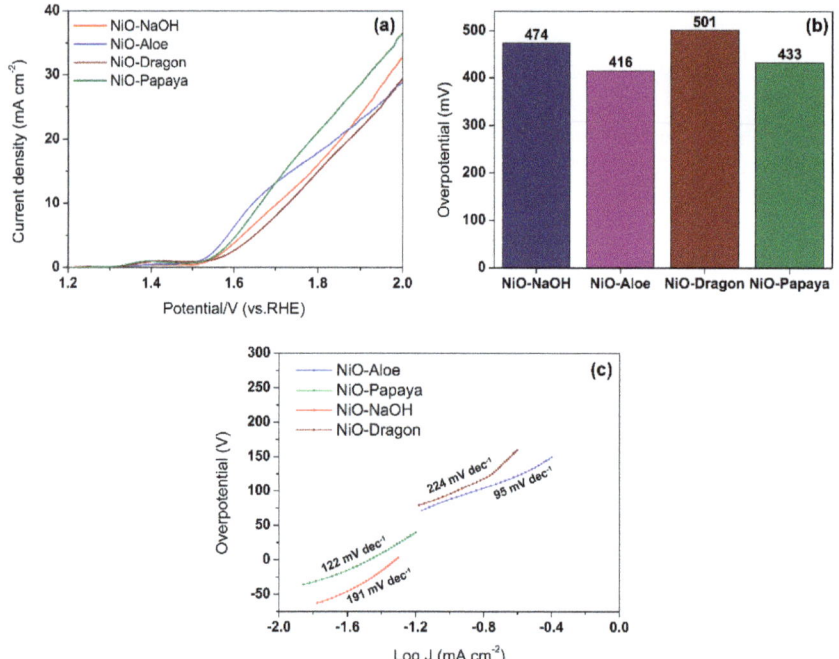

Figure 6. (**a**) Polarization curve by linear sweep voltammetry, (**b**) overpotential at 10 mA cm^{-2} and (**c**) Tafel plot.

Table 2. Electrocatalytic parameters of NiO$_x$ nanoparticles.

Material	Onset Potential (V)	Overpotential at 10 mA cm^{-2} (mV)	Tafel Slope (mV dec^{-1})
NiO–NaOH	1.57	474	191
NiO–Aloe	1.54	416	95
NiO–Papaya	1.56	433	122
NiO–Dragon	1.59	501	224

The plausible reaction mechanism of NiO$_x$ for OER is outlined as follows:

$$Ni^{2+} + 3OH^- \leftrightarrow NiOOH + H_2O + e^- \tag{7}$$

$$NiOOH + OH^- \leftrightarrow NiO(OH)_2 + e^- \tag{8}$$

$$NiO(OH)_2 + 2OH^- \leftrightarrow NiOO_2 + 2H_2O + 2e^- \tag{9}$$

$$NiOO_2 + 2OH^- \leftrightarrow NiOOH + O_2 + e^- \tag{10}$$

Throughout the electrocatalytic process, Ni^{2+} ions that can be present on surface of the NiO_x NPs will undergo oxidation to form Ni^{3+} ions which then aid to catalyze the oxidation of OH^- ions, leading to the formation of water [22].

To further comprehend the reaction kinetics of the nickel-oxide-based electrocatalysts, the Tafel plot was plotted (Figure 6c), and the linear section of Tafel slope was derived from the following equation:

$$\eta = b \log j + a \qquad (11)$$

where η is the overpotential, b is the Tafel slope and j refers to current density. NaOH-, aloe-vera-, papaya- and dragon-fruit-mediated NiO_x recorded slope values of 191, 95, 122 and 224 mV/decade, respectively. The Tafel slope values indicate the rate-limiting step which is attributed to the reaction with first electron transfer. Thus, the lower Tafel slope value of aloe-vera-mediated NiO_x NPs translates to more efficient reaction kinetics for OER. One of the proposed explanations for the superior electrocatalytic performance of aloe-vera-mediated NiO_x NPs can be related to the relatively high oxygen deficiency in the samples, as evidenced by EDX measurements. The non-stoichiometric composition most likely led to the presence of surface defects, which have been found to enhance the electrocatalytic performance of metal oxides for OER [23,24].

Another key factor that governs the performance of electrocatalyst is the surface area available as active sites for catalytic activities. In OER, electrochemically active surface area (ECSA) is used as the parameter to assess the specific catalytic activity of an electrocatalyst, and it can be derived from electrochemical double-layer capacitance (C_{dl}). To calculate C_{dl} values, the samples were subjected to cyclic voltammetry (CV) at different scan rates in the non-Faradaic segment where the voltage was scanned from 1.46 to 1.56 V vs. RHE (Figure 7). The double-layer capacitance was then derived based on the gradient of linear plot of $\Delta J = (J_a - J_c)$ (where J_a and J_c are anodic and cathodic current density) at 1.52 V vs. RHE as a function of the scan rate [25]. The ECSA was then calculated from the double-layer capacitance based on the following equation:

$$ECSA = \frac{C_{dl}}{C_s} \qquad (12)$$

where C_s is the specific capacitance which is equated to 0.040 mF cm^{-2} based on the conventional value for metal electrode in KOH solution. The roughness factor (R_f) can then be evaluated by dividing ECSA by the electrode area. Table 3 summarizes the ECSA and roughness factor for all the samples. Papaya-mediated NiO_x NPs exhibit the highest value of ECSA and R_f of 5.50 mFcm^{-2} and 1936, respectively, thus indicating that it has the highest catalytically active surface area. This could be the boosting factor that supports the low overpotential value recorded by papaya-mediated NiO_x NPs despite its high Tafel slope. The variation of active surface area of the electrocatalysts prepared using different plant extracts further emphasizes the role of phytochemicals in fine-tuning the morphology and size distribution of nanoparticles during the synthesis process.

Table 3. Electrochemical active surface area and double-layer capacitance of NiO_x nanoparticles.

Material	C_{dl} (mF cm^{-2})	ECSA (cm^2)	R_f
NiO–NaOH	4.50	112.50	1584
NiO–Aloe	4.80	120.00	1690
NiO–Papaya	5.50	137.50	1936
NiO–Dragon	2.25	56.25	792

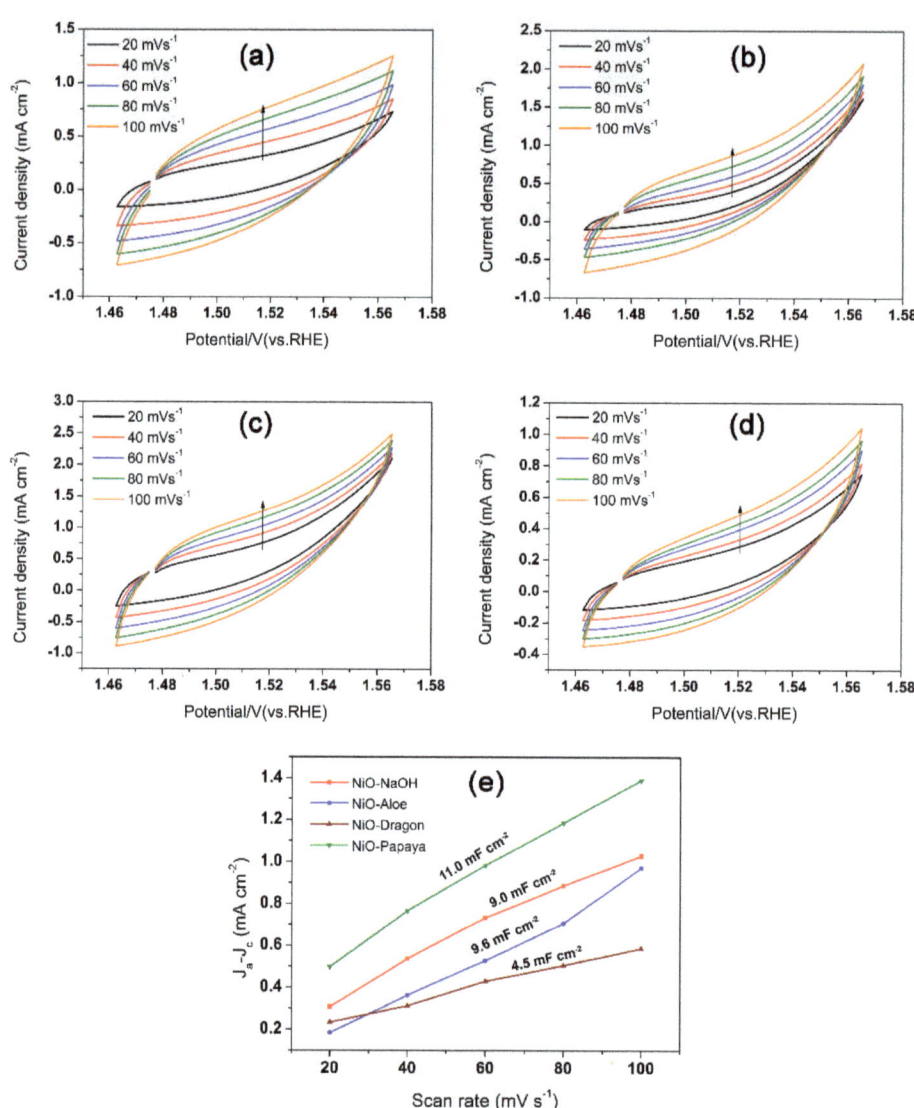

Figure 7. Cyclic voltammogram of (**a**) NaOH-, (**b**) aloe-vera-, (**c**) dragon-fruit- and (**d**) papaya-extract-mediated NiO$_x$ nanoparticles and (**e**) linear plot between current density and scan rate.

Electrochemical impedance spectroscopy (EIS) is a powerful technique that allows us to obtain insights on electrode–electrolyte interface dynamics. The EIS measurements for all the samples were performed with a frequency range from 0.1 Hz to 100 kHz, and the resulting Nyquist plots comprising of semicircles are presented in Figure 8. The starting point of the semicircle signifies the internal resistance (also known as solution resistance (R_S)) of the catalyst, whereas the diameter of the semicircle denotes charge-transfer resistance between the electrode–electrolyte interfaces (R_{CT}) [26]. In terms of equivalent circuit, the depressed semicircle is represented by R_S connected in series to the capacitor (Q_{DL}) and R_{CT} in a parallel arrangement, as shown in Figure 8. The following equation evaluates impedance values according to the described equivalent circuit [27]:

$$Z_{DSSC} = R_S + \frac{R_{CT}}{1 + (j\omega)^{n_{CT}} R_{CT} Q_{CT}} \quad (13)$$

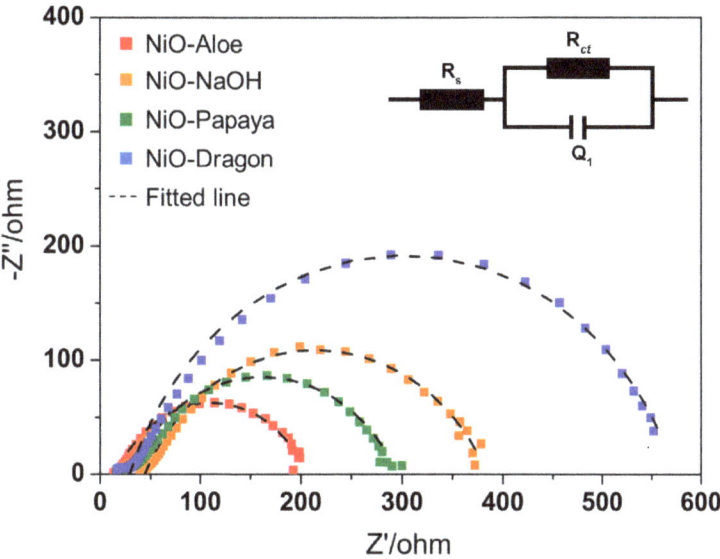

Figure 8. Nyquist plots from EIS analysis.

The equation was used to fit the Nyquist plots and the solution resistance, charge-transfer resistance and double-layer capacitance based on the equivalent circuit fitting are tabulated in Table 4. Typically, charge-transfer resistance (R_{CT}) is considered the most significant parameter as it dictates the extent of electronic transfers between the reactant and electrocatalyst surface. A lower R_{CT} value also indicates higher conductivity of the electrocatalyst material which favors electron mobility during the electrocatalytic reaction [28]. In this study, aloe-vera-mediated NiO_x NPs showed the lowest solution resistance and charge-transfer resistance. This can be attributed to better conductivity of the NiO_x NPs due to its higher oxygen deficiency compared to other green-synthesized nanoparticles. Therefore, it is evident that efficient charge-transfer kinetics and improved conductivity is one of the crucial factors for the higher electrocatalytic performance of aloe-vera-mediated NiO_x NPs compared to its dragon fruit and papaya counterparts.

Table 4. Equivalent circuit fitting parameters for the EIS analysis.

Material	R_S (Ω)	R_{CT} (Ω)	Q_{DL} (F)
NiO–NaOH	44.5	339.9	3.04×10^{-4}
NiO–Aloe	17.0	187.9	3.59×10^{-4}
NiO–Papaya	29.2	265.6	2.72×10^{-4}
NiO–Dragon	29.4	544.5	1.54×10^{-4}

To evaluate the stability of aloe-vera-mediated NiO_x NPs, chronoamperometry analysis was performed for 24 h at an overpotential of 415 mV. Within the first 2 h, it was observed that there was a gradual increase in current density. This is attributed to the activation of nickel oxide into nickel oxyhydroxide (NiOOH) during the initial part of the analysis [22]. Higher conductivity of NiOOH compared to NiO rendered the gradual increase in current density. As the chronoamperometry was extended for 24 h, the current density remained stable without any significant decay.

The stability of the final product was further verified by subjecting the NiO_x-modified glassy carbon electrode to cyclic voltammetry (CV) analysis for 500 continuous cycles, which were carried out between 1.2 and 1.8 V at a scan rate of 100 mVs^{-1}. As shown in Figure 9, after 500 CV cycles, the overpotential of aloe-vera-mediated NiO_x NPs increased from 416 mV to 428 mV with a difference of only 0.12 mV at the current density of 10 mA cm^{-2}. This observation validates the high electrocatalytic stability of the green-synthesized nickel oxide nanoparticle for OER activity.

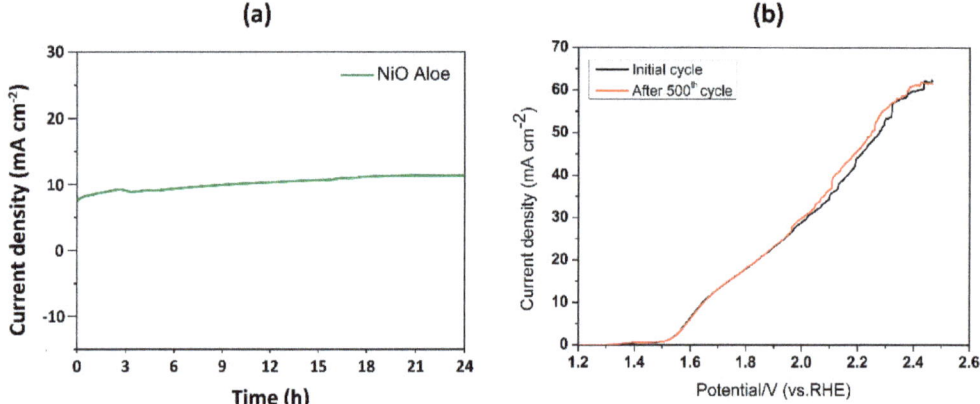

Figure 9. (a) Chronoamperometric stability tests at 1.65 V (vs the RHE) and (b) LSV curves obtained before and after 500 CV cycles of aloe-vera-mediated NiOx NPs.

3. Materials and Methods

3.1. Chemicals

Nickel acetate tetrahydrate (Ni(CH$_3$COO)$_2$.4H$_2$O), NaOH and KOH were purchased from Sigma Aldrich, Hamburg, Germany. Deionized (DI) water was used to prepare all types of solution. All the other chemicals were of analytical reagent grade and used without any further purification.

3.2. Preparation of Plant Extract

The NiO_x nanoparticles were synthesized using various plant extracts such as *aloe vera* leaf extract, dragon fruit peel extract and papaya peel extract. NiO_x NPs without extract were also synthesized. Different plant samples were collected from the local market at Hentian Kajang, Selangor, Malaysia. In order to prepare the plant extract, at first, 50 g of different plant samples were washed with DI water, cut into small pieces and were taken in three different 500 mL glass beakers and 250 mL DI water was added to all the beakers. Then the mixtures were heated at 60–70 °C for 30 min with continuous stirring. After heating, the extracts were filtered with vacuum filter and the filtrates were collected in three different 500 mL reagent bottles.

3.3. Synthesis of NiO_x Nanoparticle

The NiO_x nanoparticles were synthesized using the green synthesis technique. In this technique, at first, 30 mL (1 M) nickel acetate solutions were heated in four different 250 mL glass beakers at 60–70 °C temperature with continuous stirring. Then, 30 mL DI water was added to the first beaker (without extract) and 30 mL of three different plant extracts to the other three beakers was added in the precursor solution with constant heating

(60–70 °C temperature) and stirring. Then, 1 M NaOH solution was added, dropwise, into the reaction mixture with vigorous stirring and heating for 2 h. After completion of the reaction, the mixtures were washed with DI water several times and filtered using Whatman 1 filter paper. Finally, the synthesized NPs were dried in an oven at 100 °C overnight, calcined at 300 °C and 600 °C for 6 h, powdered using mortar-pastel and stored in the glass vial for further analysis. Figure 10 represents the schematic view of green synthesis process of NiO_x NPs.

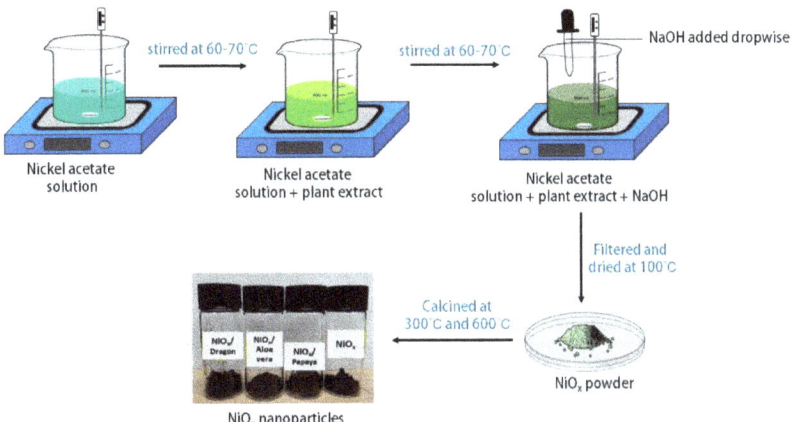

Figure 10. Reaction scheme for green synthesis process of NiOx NPs.

3.4. Characterization of Materials

The morphological and structural properties of synthesized NiO_x nanoparticles were examined using various characterization techniques such as X-ray diffraction (XRD), absorption spectra (UV-Vis) and field-emission scanning electron microscopy (FESEM), along with energy dispersive X-ray (EDX). The XRD patterns were taken in the 2θ ranging from 10° to 80° using BRUKER aXS-D8 Advance Cu-Kα diffractometer (Bruker, MA, USA). The FESEM model LEO 1450 Vp was used for investigating the surface morphology, as well as grain size and growth of the NiO_x nanoparticles. The optoelectronic properties of the synthesized nanocatalyst, such as the optical transmittance, absorbance and optical band gap, were calculated using UV-Vis spectrometer Perkin Elmer Instruments Lambda35 (PerkinElmer, Waltham, MA, USA).

3.5. Electrochemical Characterization

Electrochemical characterization was performed using Metrohm Autolab workstation (Metrohm, Herisau, Switzerland) in a three-electrode setup in which platinum electrode and silver/silver chloride were used as the counter electrode and reference electrode, respectively. The NiO_x NPs were drop casted onto glassy carbon electrode with an active area of 0.07 cm² to function as the working electrode. A 5 mg amount of the NiO_x NPs was mixed with 250 µL ethanol, 250 µL deionized water and 50 µL of Nafion (5 wt%). The dispersion was sonicated for an hour and 10 µL of the dispersion was drop casted on top of the glassy carbon electrode surface followed by drying. Linear sweep voltammetry between 0 and 1 V with scan rate of 5 mV s^{-1} was performed to evaluate the oxygen evolution reaction. The measured potential vs. Ag/AgCl was converted to a reversible hydrogen electrode (RHE) based on the following equation:

$$E_{RHE} = E_{Ag/AgCl} + 0.059\, pH + E^0_{Ag/AgCl} \tag{14}$$

where $E^0_{Ag/AgCl}$ is 0.1976 V at 25 °C and $E_{Ag/AgCl}$ is measured potential vs. Ag/AgCl. The overpotential is calculated by measuring the difference between measured potential vs. RHE at current density of 10 mA cm^{-2} and the standard value of 1.23 V.

4. Conclusions

Utilization of different plant extracts as capping agents affects the structural and morphological properties of NiO$_x$ NPs. FESEM results show that NiO$_x$ NPs synthesized without extract had a particle size between 40 and 60 nm, whereas plant-extract-mediated NiO$_x$ were between 20 and 30 nm. The reduction in particle size upon introducing plant extracts was further supported by the shifting of the band gap to a higher value based on optical characterization. EDX analysis reveal that *aloe vera*, papaya and dragon fruit extract yielded NiO$_x$ NPs with different oxygen and nitrogen stoichiometry. The deficiency of oxygen is higher in *aloe vera*-mediated NiO$_x$ NPs, which might elevate its conductivity and thus lead to lower charge-transfer resistance in EIS analysis. ECSA calculation based on cyclic voltammetry indicated that the highest electrocatalytic surface area was possessed by papaya-extract-mediated NiO$_x$ NPs, followed by *aloe vera* and dragon fruit extract. Overall, the best electrocatalytic performance is attributed to *aloe vera*-extract-mediated NiO$_x$ NPs, which only required 433 mV overpotential to reach 10 mAcm^{-2} with a Tafel slope of 95 mVdec^{-1}.

Author Contributions: Conceptualization, V.S, M.S. and S.S.; methodology, M.S., D.K.S. and M.A. (Mohammod Aminuzzaman); formal analysis, F.H.A. and Z.Z.; writing—original draft preparation, V.S. and S.S.; writing—review and editing, H.I.A. and N.A.; supervision, M.A. (Md. Akhtaruzzaman); project administration, H.A. and M.A. (Md. Akhtaruzzaman); funding acquisition, H.I.A. and N.A. All authors have read and agreed to the published version of the manuscript.

Funding: The authors are highly grateful to Universiti Kebangsaan Malaysia for supporting this study through the Modal Insan fellowship (RFA1). Authors also extend their appreciation to The University Researchers Supporting Project Number (TURSP-2020/264), Taif University, Taif, Saudi Arabia. Authors are thankful to Universiti Tunku Abdul Rahman (UTAR) providing necessary research facilities to carry out this research work successfully.

Data Availability Statement: Not applicable.

Acknowledgments: The authors are highly grateful to Universiti Kebangsaan Malaysia for supporting this study through the Modal Insan fellowship (RFA1). Authors also extend their appreciation to The University Researchers Supporting Project Number (TURSP-2020/264), Taif Uni-versity, Taif, Saudi Arabia.

Conflicts of Interest: The authors declare no conflict of interest.

References

1. You, B.; Sun, Y. Innovative strategies for electrocatalytic water splitting. *Acc. Chem. Res.* **2018**, *51*, 1571–1580. [CrossRef]
2. Pal, G.; Rai, P.; Pandey, A. Chapter 1—Green synthesis of nanoparticles: A greener approach for a cleaner future. In *Green Synthesis, Characterization and Applications of Nanoparticles*; Shukla, A.K., Iravani, S., Eds.; Elsevier: Amsterdam, The Netherlands, 2019; pp. 1–26. [CrossRef]
3. Jadoun, S.; Arif, R.; Jangid, N.K.; Meena, R.K. Green synthesis of nanoparticles using plant extracts: A review. *Environ. Chem. Lett.* **2021**, *19*, 355–374. [CrossRef]
4. Shafey, A.M.E. Green synthesis of metal and metal oxide nanoparticles from plant leaf extracts and their applications: A review. *Green Process. Synth.* **2020**, *9*, 304–339. [CrossRef]
5. Singh, J.; Dutta, T.; Kim, K.-H.; Rawat, M.; Samddar, P.; Kumar, P. 'Green' synthesis of metals and their oxide nanoparticles: Applications for environmental remediation. *J. Nanobiotechnol.* **2018**, *16*, 84. [CrossRef]
6. Madkour, M.; Bumajdad, A.; Al-Sagheer, F. To what extent do polymeric stabilizers affect nanoparticles characteristics? *Adv. Colloid Interface Sci.* **2019**, *270*, 38–53. [CrossRef]
7. Tippayawat, P.; Phromviyo, N.; Boueroy, P.; Chompoosor, A. Green synthesis of silver nanoparticles in aloe vera plant extract prepared by a hydrothermal method and their synergistic antibacterial activity. *PeerJ* **2016**, *4*, e2589. [CrossRef]
8. Quispe, C.; Villalobos, M.; Bórquez, J.; Simirgiotis, M. Chemical composition and antioxidant activity of *Aloe vera* from the Pica Oasis (Tarapacá, Chile) by UHPLC-Q/Orbitrap/MS/MS. *J. Chem.* **2018**, *2018*, 6123790. [CrossRef]

9. Phang, Y.-K.; Aminuzzaman, M.; Akhtaruzzaman, M.; Muhammad, G.; Ogawa, S.; Watanabe, A.; Tey, L.-H. Green synthesis and characterization of CuO nanoparticles derived from papaya peel extract for the photocatalytic degradation of palm oil mill effluent (POME). *Sustainability* **2021**, *13*, 796. [CrossRef]
10. Hendra, R.; Masdeatresa, L.; Abdulah, R.; Haryani, Y. Red dragon peel (*Hylocereus polyrhizus*) as antioxidant source. *AIP Conf. Proc.* **2020**, *2243*, 030007. [CrossRef]
11. Manigandan, R.; Dhanasekaran, T.; Padmanaban, A.; Giribabu, K.; Suresh, R.; Narayanan, V. Bifunctional hexagonal Ni/NiO nanostructures: Influence of the core–shell phase on magnetism, electrochemical sensing of serotonin, and catalytic reduction of 4-nitrophenol. *Nanoscale Adv.* **2019**, *1*, 1531–1540. [CrossRef]
12. Sekar, S.; Kim, D.Y.; Lee, S. Excellent oxygen evolution reaction of activated carbon-anchored NiO nanotablets prepared by green routes. *Nanomaterials* **2020**, *10*, 1382. [CrossRef]
13. Munna, F.T.; Selvanathan, V.; Sobayel, K.; Muhammad, G.; Asim, N.; Amin, N.; Sopian, K.; Akhtaruzzaman, M. Diluted chemical bath deposition of CdZnS as prospective buffer layer in CIGS solar cell. *Ceram. Int.* **2021**, *47*, 11003–11009. [CrossRef]
14. Tripathi, R.M.; Bhadwal, A.S.; Gupta, R.K.; Singh, P.; Shrivastav, A.; Shrivastav, B.R. ZnO nanoflowers: Novel biogenic synthesis and enhanced photocatalytic activity. *J. Photochem. Photobiol. B Biol.* **2014**, *141*, 288–295. [CrossRef]
15. Lamba, P.; Singh, P.; Singh, P.; Kumar, A.; Singh, P.; Bharti; Kumar, Y.; Gupta, M. Bioinspired synthesis of nickel oxide nanoparticles as electrode material for supercapacitor applications. *Ionics* **2021**, *27*, 5263–5276. [CrossRef]
16. Cao, Y.; Hu, P.; Jia, D. Solvothermal synthesis, growth mechanism, and photoluminescence property of sub-micrometer PbS anisotropic structures. *Nanoscale Res. Lett.* **2012**, *7*, 668. [CrossRef]
17. Talluri, B.; Prasad, E.; Thomas, T. Ultra-small (r < 2 nm), stable (>1 year) copper oxide quantum dots with wide band gap. *Superlattices Microstruct.* **2018**, *113*, 600–607. [CrossRef]
18. Selvanathan, V.; Aminuzzaman, M.; Tey, L.-H.; Razali, S.A.; Althubeiti, K.; Alkhammash, H.I.; Guha, S.K.; Ogawa, S.; Watanabe, A.; Shahiduzzaman, M.; et al. *Muntingia calabura* leaves mediated green synthesis of CuO nanorods: Exploiting phytochemicals for unique morphology. *Materials* **2021**, *14*, 6379. [CrossRef]
19. Liang, Q.; Brocks, G.; Bieberle-Hütter, A. Oxygen evolution reaction (OER) mechanism under alkaline and acidic conditions. *J. Phys. Energy* **2021**, *3*, 026001. [CrossRef]
20. Jin, H.; Wang, X.; Tang, C.; Vasileff, A.; Li, L.; Slattery, A.; Qiao, S.-Z. Stable and highly efficient hydrogen evolution from seawater enabled by an unsaturated nickel surface nitride. *Adv. Mater.* **2021**, *33*, 2007508. [CrossRef]
21. Wang, C.; Qi, L. Hollow nanosheet arrays assembled by ultrafine ruthenium–cobalt phosphide nanocrystals for exceptional pH-universal hydrogen evolution. *ACS Mater. Lett.* **2021**, *3*, 1695–1701. [CrossRef]
22. Rani, B.J.; Ravi, G.; Yuvakkumar, R.; Ravichandran, S.; Ameen, F.; Al-Sabri, A. Efficient, highly stable Zn-doped NiO nanocluster electrocatalysts for electrochemical water splitting applications. *J. Sol-Gel Sci. Technol.* **2019**, *89*, 500–510. [CrossRef]
23. Zhu, K.; Shi, F.; Zhu, X.; Yang, W. The roles of oxygen vacancies in electrocatalytic oxygen evolution reaction. *Nano Energy* **2020**, *73*, 104761. [CrossRef]
24. Wang, Y.; Liang, Z.; Zheng, H.; Cao, R. Recent progress on defect-rich transition metal oxides and their energy-related applications. *Chemistry* **2020**, *15*, 3717–3736. [CrossRef]
25. Wang, H.; Xie, A.; Li, X.; Wang, Q.; Zhang, W.; Zhu, Z.; Wei, J.; Chen, D.; Peng, Y.; Luo, S. Three-dimensional petal-like graphene Co3.0Cu1.0 metal organic framework for oxygen evolution reaction. *J. Alloys Comp.* **2021**, *884*, 161144. [CrossRef]
26. Santos, H.L.S.; Corradini, P.G.; Medina, M.; Dias, J.A.; Mascaro, L.H. NiMo–NiCu Inexpensive Composite with High Activity for Hydrogen Evolution Reaction. *ACS Appl. Mater. Interfaces* **2020**, *12*, 17492–17501. [CrossRef]
27. Selvanathan, V.; Ruslan, M.H.; Aminuzzaman, M.; Muhammad, G.; Amin, N.; Sopian, K.; Akhtaruzzaman, M. Resorcinol-Formaldehyde (RF) as a Novel Plasticizer for Starch-Based Solid Biopolymer Electrolyte. *Polymers* **2020**, *12*, 2170. [CrossRef]
28. Akbayrak, M.; Önal, A.M. Metal oxides supported cobalt nanoparticles: Active electrocatalysts for oxygen evolution reaction. *Electrochim. Acta* **2021**, *393*, 139053. [CrossRef]